I0065755

# Organic Chemistry: Synthesis, Structures and Mechanisms

# Organic Chemistry: Synthesis, Structures and Mechanisms

# Organic Chemistry: Synthesis, Structures and Mechanisms

Edited by
Allegra Smith

WILLFORD PRESS

www.willfordpress.com

Published by Willford Press,
118-35 Queens Blvd., Suite 400,
Forest Hills, NY 11375, USA

Copyright © 2017 Willford Press

This book contains information obtained from authentic and highly regarded sources. Copyright for all individual chapters remain with the respective authors as indicated. All chapters are published with permission under the Creative Commons Attribution License or equivalent. A wide variety of references are listed. Permission and sources are indicated; for detailed attributions, please refer to the permissions page and list of contributors. Reasonable efforts have been made to publish reliable data and information, but the authors, editors and publisher cannot assume any responsibility for the validity of all materials or the consequences of their use.

**Trademark Notice:** Registered trademark of products or corporate names are used only for explanation and identification without intent to infringe.

ISBN: 978-1-68285-374-0

**Cataloging-in-Publication Data**

Organic chemistry : synthesis, structures and mechanisms / edited by Allegra Smith.
    p. cm.
Includes bibliographical references and index.
ISBN 978-1-68285-374-0
1. Chemistry, Organic. 2. Organic compounds--Synthesis. 3. Organic reaction mechanisms.
4. Chemistry. I. Smith, Allegra.
QD251.3 .O65 2017
547--dc23

For information on all Willford Press publications
visit our website at www.willfordpress.com

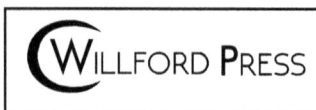

WILLFORD PRESS

Printed in the United States of America.

# Contents

# Preface

Since its emergence, organic chemistry has been rapidly expanding. Developments in organic chemistry are helping to improve productivity in industries like medicine, rubber, detergents, coatings, paints and biotechnology etc. This book outlines the process and applications of this field in detail. It unfolds the innovative aspects of organic chemistry which will crucial for the progress of this discipline in the future. This book elucidates new techniques and their applications in a multidisciplinary approach. It includes contributions of experts and scientists which will provide innovative insights to readers. This book, with its detailed analyses and data, will prove immensely beneficial to professionals and students involved in organic chemistry at various levels.

This book unites the global concepts and researches in an organized manner for a comprehensive understanding of the subject. It is a ripe text for all researchers, students, scientists or anyone else who is interested in acquiring a better knowledge of this dynamic field.

I extend my sincere thanks to the contributors for such eloquent research chapters. Finally, I thank my family for being a source of support and help.

**Editor**

# Cobalt-catalysed site-selective intra- and intermolecular dehydrogenative amination of unactivated sp$^3$ carbons

Xuesong Wu[1], Ke Yang[2], Yan Zhao[1], Hao Sun[2], Guigen Li[2,3] & Haibo Ge[1]

Cobalt-catalysed sp$^2$ C–H bond functionalization has attracted considerable attention in recent years because of the low cost of cobalt complexes and interesting modes of action in the process. In comparison, much less efforts have been devoted to the sp$^3$ carbons. Here we report the cobalt-catalysed site-selective dehydrogenative cyclization of aliphatic amides via a C–H bond functionalization process on unactivated sp$^3$ carbons with the assistance of a bidentate directing group. This method provides a straightforward synthesis of monocyclic and spiro β- or γ-lactams with good to excellent stereoselectivity and functional group tolerance. In addition, a new procedure has been developed to selectively remove the directing group, which enables the synthesis of free β- or γ-lactam compounds. Furthermore, the first cobalt-catalysed intermolecular dehydrogenative amination of unactivated sp$^3$ carbons is also realized.

[1] Department of Chemistry and Chemical Biology, Indiana University-Purdue University Indianapolis, 402 N. Blackford Street, Indianapolis, Indiana 46202, USA. [2] Institute of Chemistry and BioMedical Sciences, Nanjing University, Nanjing 210093, P.R. China. [3] Department of Chemistry and Biochemistry, Texas Tech University, Lubbock, Texas 79409-1061, USA. Correspondence and requests for materials should be addressed to G.L. (email: guigen.li@ttu.edu) or to H.G. (email: geh@iupui.edu).

Transition metal-catalysed direct functionalization of relatively unreactive C–H bonds has emerged as a major topic of research in organic chemistry[1-13]. This method does not require the use of prefunctionalized materials, and thus provides an attractive alternative to traditional cross-coupling reactions. Within this reaction class, cobalt-catalysed processes have received special interest due to the low cost and toxicity of cobalt complexes, and their interesting modes of action[14-17]. In the 1950s, Murahashi et al. demonstrated the chelation-assisted C–H functionalization process on benzaldimines and azobenzenes, and it is well accepted that the catalytic cycle is initiated by oxidative addition of a low-valent cobalt species to the aromatic C–H bonds[18-21]. Recently, azoles, benzamides and 2-phenylpyridines were also proven to be effective substrates through a similar reaction pathway[22-26]. Furthermore, Co[II] or Co[III]-catalysed direct C–H functionalization of azole, 2-phenylpyridine, indole and benzamide derivatives was also demonstrated[27-30]. In this case, the C–H bond activation process is believed to proceed through either an electrophilic aromatic substitution or concerted metalation-deprotonation pathway. Moreover, the cobalt-catalysed hydroacylation of olefins has also been demonstrated via an $sp^2$ C–H functionalization process in the absence of chelation assistance[14,31].

In comparison with the well-established cobalt-catalysed direct functionalization on $sp^2$ carbons, there are only a few examples of direct functionalization on $sp^3$ C–H bonds (Fig. 1). Cenini and co-workers reported the intermolecular amination of relatively reactive $sp^3$ carbons with moderate yields in 1999 (refs 32,33). Recently, more efficient intramolecular version of this transformation was developed on electron-deficient $sp^3$ carbons in Zhang's laboratory (Fig. 1a)[34,35]. These reactions were proposed to proceed via the outer-sphere mechanism, in which a carbon–metal bond is not involved[5,36,37]. Instead, the $sp^3$ C–H bonds were indirectly activated by an inter- or intramolecular hydrogen atom transfer of the radical intermediates. In contrast to this, Brookhart and co-workers reported intramolecular hydrogen transfer of cyclic amines in 2007 via an inner-sphere mechanism, in which a carbon–cobalt bond was formed by oxidative addition of a cobalt species to the α-$sp^3$ C–H bond (Fig. 1b)[36-40].

Inspired by the reports of transition metal-catalysed bidentate ligand-directed $sp^3$ C–H functionalization process[41,42], we have explored and demonstrated here the cobalt-catalysed site-selective direct C–H functionalization on unactivated $sp^3$ carbons with the aid of a bidentate directing group (Fig. 1c). In addition, a novel two-step procedure has been developed under oxidative conditions to remove the directing group, which enables an efficient access to β-lactam, γ-lactam or β-amino amide derivatives.

## Results

**Reaction condition optimization of intramolecular amidation.** Synthesis of lactams via transition metal-catalysed C–H functionalization is of current research interest because of the biological importance of these molecules[43,44]. In the past two years, Pd-, Cu- or Ni-catalysed process for the formation of β- or γ-lactams has been achieved[45-51]. However, all of these approaches suffer from their own limitations on the substrate scope. To provide a complementary method and demonstrate the feasibility of cobalt catalysis on unactivated $sp^3$ carbons, we carried out the study of cobalt-catalysed bidentate ligand-directed intramolecular cyclization of propanamides. Our investigation began with oxidative cyclization of 2-ethyl-2-methyl-N-(quinolin-8-yl)pentanamide (**1a**) in the presence of catalytic amount of CoCl₂ by using Ag₂CO₃ as the oxidant (Table 1). After an initial solvent screening, chlorobenzene turned out to be optimal, producing the β-lactam compound **2a** in 33% yield (entry 1). Notably, this reaction proceeded in a highly site-selective manner, favouring the C–H bond of a β-methyl group over those of β-methylene and γ-methyl groups. Next, a screening on oxidants was carried out. It was observed that the reaction could also be performed with several other oxidants with lower efficiency (entries 2–4). Further optimization of reaction conditions showed that the reaction could be significantly improved by using Co(OAc)₂ as the catalyst and sodium benzoate as the base (entry 15). Considering that PhCO₂Na could potentially compete with amide **1** for coordination to the cobalt complex, we further reduced the loading of this base. Delightfully, the chemical yield of this reaction was significantly improved (entry 17).

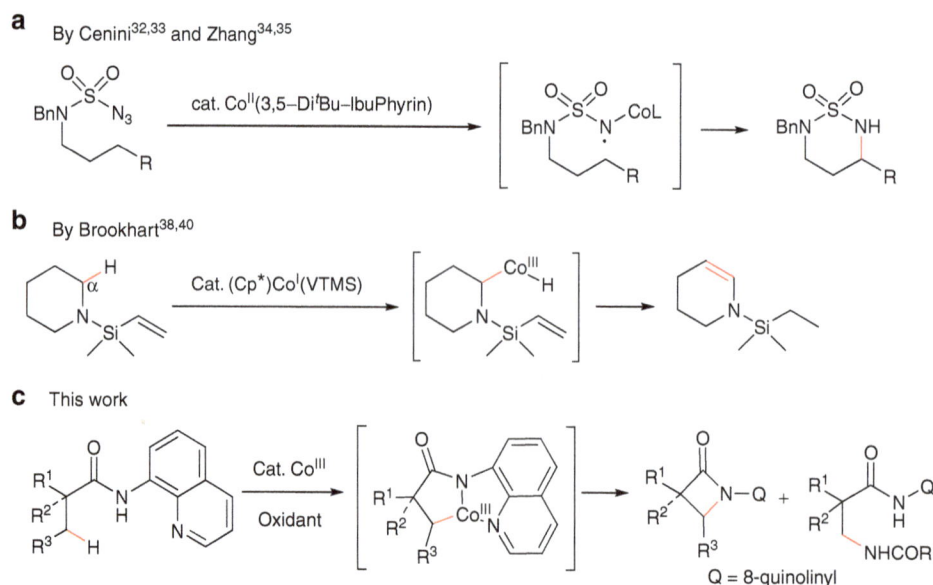

**a** By Cenini[32,33] and Zhang[34,35]

**b** By Brookhart[38,40]

**c** This work

Q = 8-quinolinyl

**Figure 1 | Cobalt-catalysed $sp^3$ C–H bond functionalization.** (**a**) Out-sphere mechanism. (**b**) Inner-sphere mechanism (via oxidative addition to a C–H bond). (**c**) Inner-sphere mechanism (via cyclometalation of an $sp^3$ carbon).

## Table 1 | Optimization of reaction conditions.

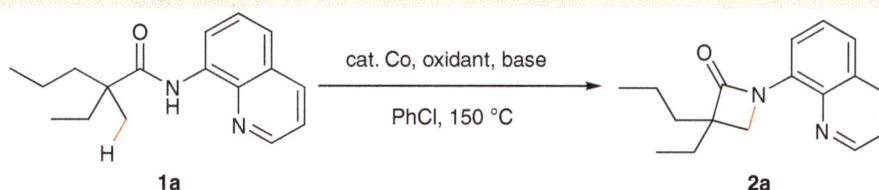

| Entry | Co source (mol%) | Oxidant (2.5 eq) | Base (eq) | Yield (%) |
|---|---|---|---|---|
| 1 | CoCl$_2$ (10) | Ag$_2$CO$_3$ | K$_2$HPO$_4$ (1) | 33 |
| 2 | CoCl$_2$ (10) | Ce(SO$_4$)$_2$ | K$_2$HPO$_4$ (1) | 7 |
| 3 | CoCl$_2$ (10) | AgOAc | K$_2$HPO$_4$ (1) | 24 |
| 4 | CoCl$_2$ (10) | Ag$_2$O | K$_2$HPO$_4$ (1) | <5 |
| 5 | CoBr$_2$ (10) | Ag$_2$CO$_3$ | K$_2$HPO$_4$ (1) | 12 |
| 6 | CoF$_2$ (10) | Ag$_2$CO$_3$ | K$_2$HPO$_4$ (1) | 22 |
| 7 | Co(acac)$_2$ (10) | Ag$_2$CO$_3$ | K$_2$HPO$_4$ (1) | 28 |
| 8 | Co(PhCO$_2$)$_2$ (10) | Ag$_2$CO$_3$ | K$_2$HPO$_4$ (1) | 38 |
| 9 | Co(OAc)$_2$ (10) | Ag$_2$CO$_3$ | K$_2$HPO$_4$ (1) | 45 |
| 10 | CoF$_3$ (10) | Ag$_2$CO$_3$ | K$_2$HPO$_4$ (1) | 27 |
| 11 | Co(acac)$_3$ (10) | Ag$_2$CO$_3$ | K$_2$HPO$_4$ (1) | 30 |
| 12 | Co(OAc)$_2$ (10) | Ag$_2$CO$_3$ | Na$_2$HPO$_4$ (1) | 56 |
| 13 | Co(OAc)$_2$ (10) | Ag$_2$CO$_3$ | Na$_2$CO$_3$ (1) | 19 |
| 14 | Co(OAc)$_2$ (10) | Ag$_2$CO$_3$ | NaOAc (1) | 42 |
| 15 | Co(OAc)$_2$ (10) | Ag$_2$CO$_3$ | PhCO$_2$Na (1) | 75 |
| 16 | Co(OAc)$_2$ (10) | Ag$_2$CO$_3$ | — | 33 |
| 17 | Co(OAc)$_2$ (10) | Ag$_2$CO$_3$ | PhCO$_2$Na (0.5) | 90 (87) |

Reaction conditions: **1a** (0.3 mmol), Co source (10 mol%), oxidant (2.5 eq), base, 0.6 ml of solvent, 150 °C, 24 h. Yields are based on **1a**, determined by $^1$H NMR using dibromomethane as the internal standard. Isolated yield is in brackets based on three runs.

**Substrate scope of intramolecular amination**. With optimized conditions in hand, the substrate scope studies were carried out (Fig. 2; also see Supplementary Figs 1 and 2 for the structures of substrates). Gratifyingly, the reaction showed great generality with 2,2-disubstituted propanamides bearing either linear or cyclic chains on α-carbons with predominant selectivity for C–H bonds of β-methyl groups (**2b**, **2d–j**). However, with α-phenyl substituted substrates, a preference of C–H bond functionalization of sp$^2$ carbons was observed, providing indolin-2-one derivative as the major products (**2k** and **2l**). Noticeably, the replacement of the quinolyl group with 5-methoxyquinolyl group had no apparent effect on the reaction (**2c**). Moreover, the removability of 5-methoxyquinolyl moiety of β- or γ-lactams has been well documented[45,48–51].

Furthermore, substrates bearing a trifluoromethyl, cyano, ethoxycarbonyl, sulfonyl or phthalimidyl group on an α-carbon showed good compatibility (**2m–q**). In addition, although α-methoxy and acetoxy-substituted amides failed to provide the desired products (**2r**), substrates with an acetoxy or benzene-carboxy group on β-carbons produced β-lactams **2s** and **2t** in good yields. It was also noticed that the reaction favours the C–H bond of the β-methyl over that of the more reactive benzyl group (**2u**). Moreover, the β-benzylic sp$^3$ C–H bonds could also be effectively functionalized (**2v-ab**).

Next, we carried out compatibility studies of α-monosubstituted propanamide derivatives (Fig. 3; also see Supplementary Figs 1 and 2 for the structures of substrates). To our delight, introduction of a relatively bulky group on the α-carbon could effectively initiate the process (**3a–k**). Furthermore, excellent diastereoselectivity was observed with β-phenyl substituted substrates (**3g–k**).

Interestingly, a great preference of functionalizing the C–H bonds of γ-benzylic carbons over those of β-methyl carbons was observed during the course of substrate scope studies, providing γ-lactams as the major products (Fig. 4; also see Supplementary Figs 1 and 2 for the structures of substrates). However, α-mono-substituted substrates failed to provide any γ-lactams (**4d–e**).

**Mechanistic investigation**. To gain some insights on the mechanism of this reaction, the deuterium-labelling experiments were carried out (Fig. 5). With the deuterium-labelled compound [D$_3$]-**1d**, an apparent deuterium-proton exchange occurred with both the substrate and product (Fig. 5a). More interestingly, the product has a lower deuterium ratio compared with the recovered starting material, which is presumably due to the different reaction rates of proton- and deuterium-containing starting materials. Furthermore, a primary kinetic isotope effect was also observed for **1d** based on the early relative rate of parallel reactions (see Supplementary Methods), indicating that the sp$^3$ C–H bond cleavage of amide **1d** is the rate-limiting step in the catalytic process (Fig. 5b).

We then carried out a series of control experiments with 2-ethyl-2-methyl-N-(quinolin-8-yl)butanamide (**1d**). As shown in Table 2, the reaction failed to provide the desired product without a cobalt catalyst under the standard or modified conditions based on Shi's study (entries 2 and 3)[52]. It was then noticed that the oxidant, Ag$_2$CO$_3$, could be replaced with Ce(SO$_4$)$_2$, albeit with a low yield (entries 4 and 5). Furthermore, no desired product **2d** was obtained with stoichiometric amounts of commercially available Co(acac)$_3$ or CoF$_3$ in the absence of Ag$_2$CO$_3$ (entries 6 and 7). On the other hand, the reaction could be performed with a catalytic amount of Co(acac)$_3$ or CoF$_3$ in the presence of Ag$_2$CO$_3$ (entries 8 and 9). These results suggest that the C–H bond activation process could be initiated by a Co$^{III}$ species[28–30], but the product is unlikely generated from reductive elimination of a Co$^{III}$ complex. It was also observed that addition of the radical inhibitor, TEMPO, had no significant effect on the reaction, indicating that a radical intermediate may not be involved in the catalytic cycle (entries 10 and 11).

Next, a series of control experiments with 1-phenethyl-N-(quinolin-8-yl)cyclohexane-1-carboxamide (**1d**) were carried out to explore the plausible reaction pathway for the formation of γ-lactams (Supplementary Table 1). It was found that neither a cobalt nor a silver species is required for this reaction (entries

Co(OAc)₂ (10 mol%), Ag₂CO₃ (2.5 eq)

PhCO₂Na (0.5 eq), PhCl, 150 °C

**2b**, 83%  **2c**, 86%  **2d**, 85%  **2e**, 88%  **2f**, 80%

**2g**, 85%  **2h**, 69%  **2i**, 80%  **2j**, 83%

**2k1**, 35% + **2k2**, 55%  **2l1**, 18% + **2l2**, 70%

**2m**, 90%  **2n**, 41%[a,b]  **2o**, 83%  **2p**, 73%

**2q**, 81%  **2r**, R = Ac or Me, 0%  **2s**, 78%  **2t**, 74%

**2u1**, 75% + **2u2**, 11%  **2v**, 65%[a]  **2w**, 59%[a]  **2x**, 67%[a]

**2y**, 64%[a]  **2z**, 51%[a]  **2aa**, 61%[a]  **2ab**, 68%[a]

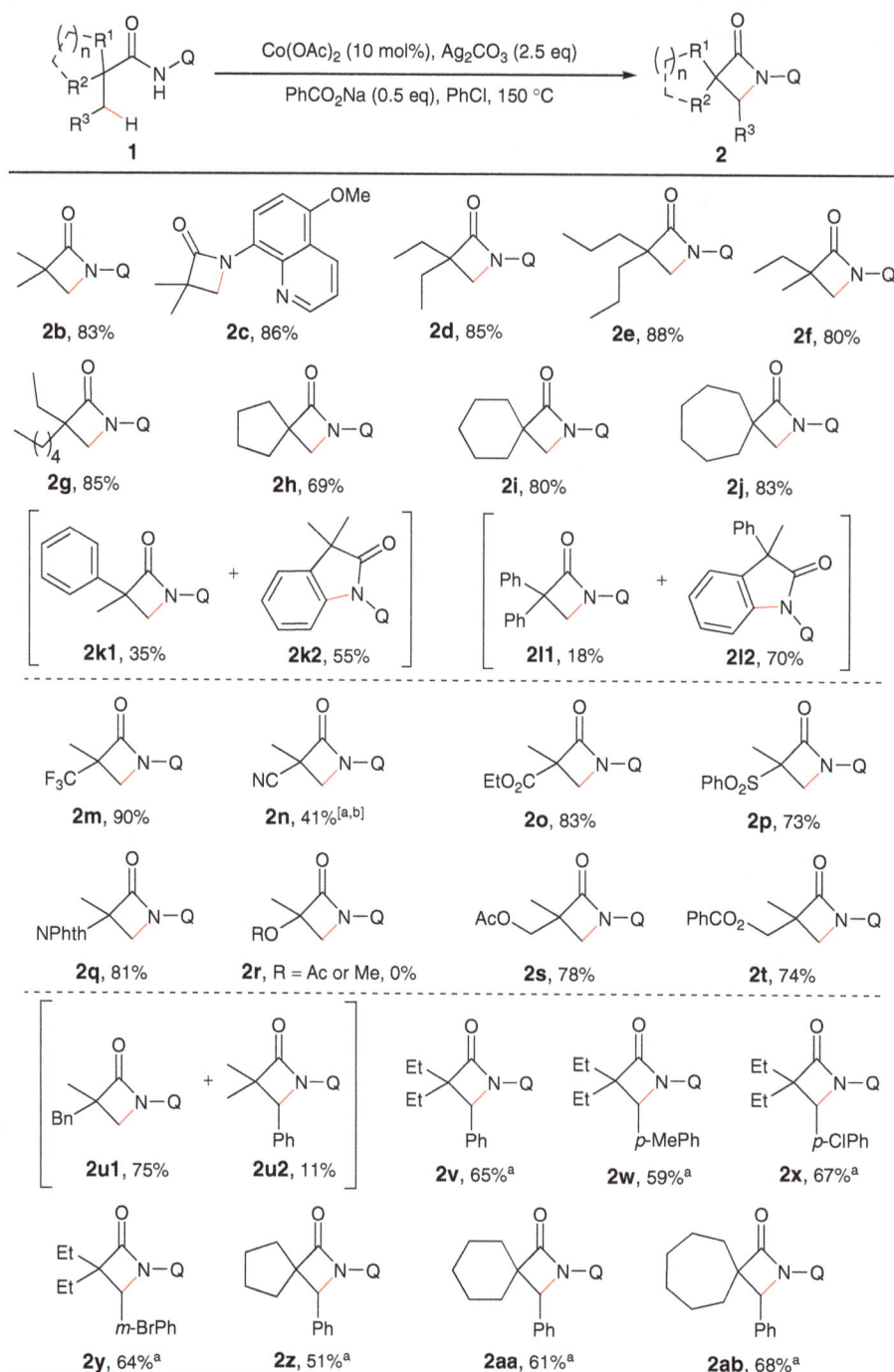

**Figure 2 | Scope of α,α-disubstituted propanamides.** Reaction conditions: **1** (0.3 mmol), Co(OAc)₂ (0.03 mmol), Ag₂CO₃ (0.75 mmol), PhCO₂Na (0.15 mmol), 0.6 ml PhCl, 150 °C, 24 h. Isolated yield based on three runs of each reaction. [a]Run for 48 h. [b]Co(OAc)₂ (0.06 mmol). Q = 8-quinolinyl.

2–5). However, the efficiency of the reaction was significantly decreased without these species. Furthermore, reaction yield was dramatically decreased by the addition of TEMPO, which indicates that an alkyl radical intermediate generated from a single electron oxidation process may be involved in the reaction (entries 6 and 7).

On the basis of the above results, a plausible catalytic cycle for the formation of β-lactams is proposed (Fig. 6)[14–17,53,54]. The Co^III complex **A** is initially generated by coordination of amide **1** to a cobalt species followed by a ligand exchange process under basic conditions. Cyclometalation of this intermediate produces the intermediate **B**, which is believed to be an irreversible step based on the kinetic isotope effect studies. In this process,

benzoate might act as a ligand to coordinate to the Co^III complex, and subsequently facilitates the C–H bond cleavage via concerted metallation-deprotonation[7,55,56]. Oxidation of intermediate **B** with Ag₂CO₃ gives rise to the Co^IV complex **C**, which produces the β-lactam compound **2** upon reductive elimination[57,58]. The newly generated Co^II species could then be re-oxidized to the Co^III species to furnish the catalytic cycle. It is noteworthy that the catalytic Co^II/Co^IV cycle could not be excluded, which involves cyclometalation of amide **1** with a Co^II species followed by oxidation to generate the intermediate **C**. It should also be mentioned that a competing side reaction, protonation of the Co^IV complex **C**, is also possible in the process, giving the Co^IV species **D**. Furthermore, although a radical-mediated process

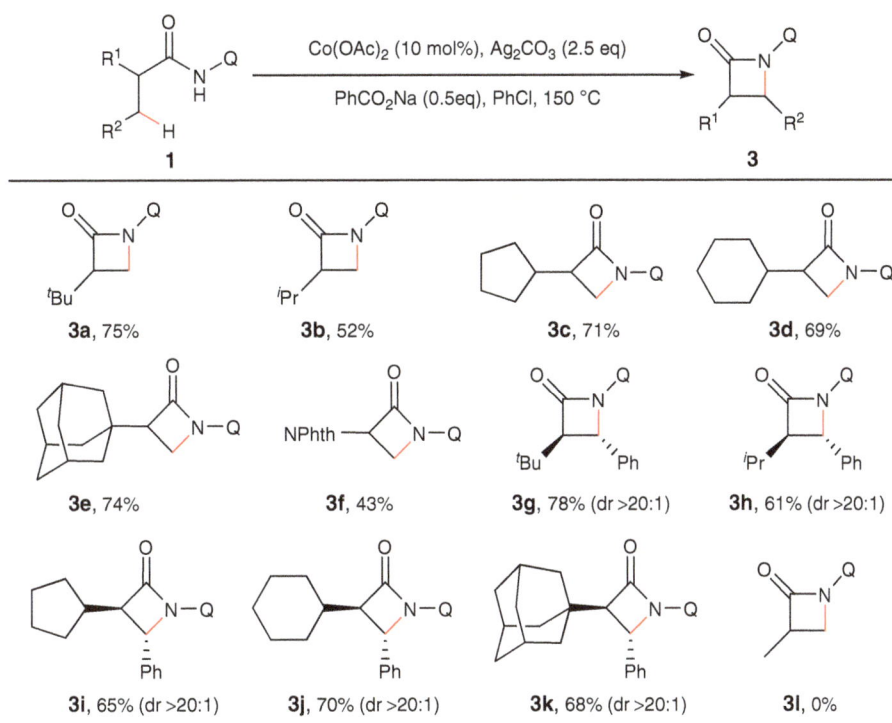

**Figure 3 | Scope of α-monosubstituted propanamides.** Reaction conditions: same as in Fig. 2. Isolated yield based on three runs of each reaction.

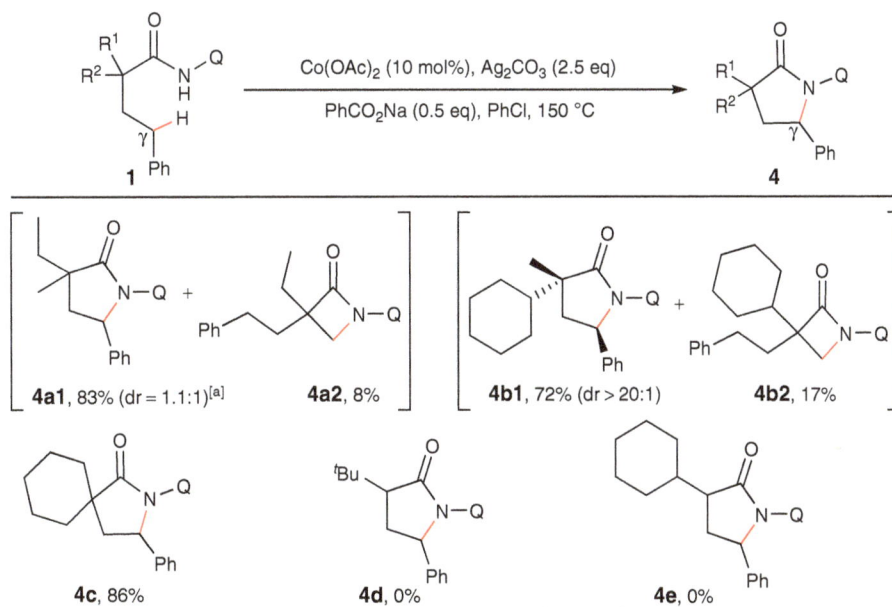

**Figure 4 | Functionalization of sp³ carbons.** Reaction conditions: same as in Fig. 2. Isolated yield based on three runs of each reaction. [a]Unisolated diastereoisomers.

could not be excluded, the observed high selectivity of the β-methyl over the β-benzylic C–H bonds suggests that the catalytic cycle is unlikely performed with a radical intermediate. However, in the case of the formation of γ-lactam derivatives, a radical or cationic species is believed to be involved in the catalytic cycle because of the predominant preference of functionalization of the γ-benzylic over the β-methyl C–H bonds.

To broaden the synthetic applications of this method, we carried out studies on the selective removal of the directing group (Fig. 7). It was found that the quinolyl group could be cleaved by the introduction of a methoxy group on the C5 position of this moiety under oxidative conditions followed by oxidative cleavage

of the newly generated 5-methoxyquinolyl moiety with ammonium cerium(IV) nitrate (CAN). Under these conditions, α-mono and α,α-di-substituted β-lactams, and α,α-di-substituted γ-lactams were all effective substrates, which enables the efficient synthesis of free β- or γ-lactam compounds.

**Intermolecular amination.** Direct intermolecular amination of sp³ carbons is an important research topic because of the importance of the products in pharmaceutical industry[5,59–61]. As one of the most efficient synthetic methods, the transition metal-catalysed ligand-directed approach has attracted

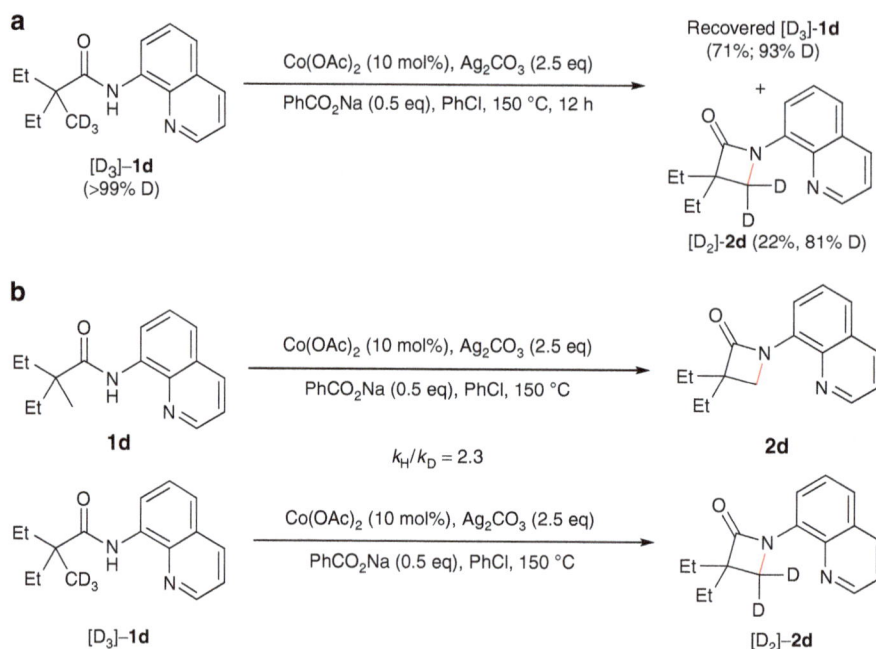

**Figure 5 | Deuterium labelling experiments.** (**a**) The deuterium–proton exchange experiment. (**b**) The kinetic isotope effect experiments.

**Table 2 | Control experiments of β-lactam formation.**

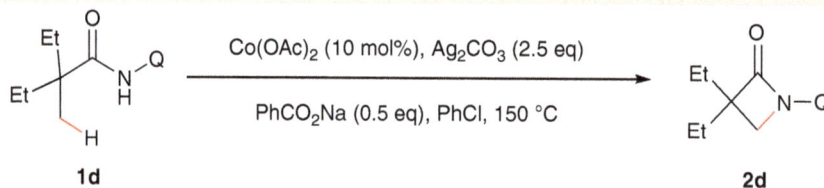

| Entry | Change from the 'standard conditions'* | Yield of 2d (%)† |
|---|---|---|
| 1 | None | 88 |
| 2 | No Co(OAc)$_2$ | 0 |
| 3 | Shi's conditions‡ instead of the standard conditions | 0 |
| 4 | Co(OAc)$_2$ (1 eq), no Ag$_2$CO$_3$ | 0 |
| 5 | Ce(SO$_4$)$_2$ (6 eq) instead of Ag$_2$CO$_3$ | 25 |
| 6 | Co(acac)$_3$ (1 eq) instead of Co(OAc)$_2$, no Ag$_2$CO$_3$ | 0 |
| 7 | CoF$_3$ (1 eq) instead of Co(OAc)$_2$, no Ag$_2$CO$_3$ | 0 |
| 8 | Co(acac)$_3$ (10%) instead of Co(OAc)$_2$ | 53 |
| 9 | CoF$_3$ (10%) instead of Co(OAc)$_2$ | 46 |
| 10 | Addition of TEMPO (1 eq) | 63 |
| 11 | Addition of TEMPO (2 eq) | 56 |

*Reaction conditions: **1d** (0.3 mmol), Co(OAc)$_2$ (0.03 mmol), Ag$_2$CO$_3$ (0.75 mmol), PhCO$_2$Na (0.15 mmol), 0.6 ml PhCl, 150 °C, 24 h.
†Yields are based on **1d**, determined by $^1$H NMR using dibromomethane as the internal standard.
‡Shi's conditions: **1d** (0.25 mmol), AgOAc (20 mol%), PhI(TFA)$_2$ (0.50 mmol), 4,4'-di-t-butyl-2,2'-bipyridine (20 mol%), K$_2$CO$_3$ (0.50 mmol), PhCl/DCE (1.5 ml/1.5 ml), 120 °C, 12 h.

considerable attention and significant progress has been achieved in recent years[62–67]. However, within this reaction category, reports of dehydrogenative aminations are rare[62,63] Encouraged by the above results, we carried out the study of a cobalt-catalysed intermolecular dehydrogenative amination of *N*-(quinolin-8-yl)propanamide derivatives (Fig. 8; also see Supplementary Figs 1 and 2 for the structures of substrates). After an extensive screening, trifluoroacetamide was proved to be an effective coupling partner (**5a**), whereas many other nitrogen sources such as acetamide, benzamide, phthalimide, sulfonamide, morpholine and aniline failed to produce the desired products (for optimization of reaction conditions, see Supplementary Table 2). Furthermore, replacement of trifluoroacetamide with

heptafluorobutanamide significantly improved the reaction (**5b**). As expected, 2,2-disubstituted propanamides bearing either linear or cyclic chains on α-carbons proceeded smoothly to give the corresponding products **5c–j** with a predominant selectivity for C–H bonds of β-methyl groups. It was also noticed that with α-phenyl-substituted substrate **1k**, C–H bond functionalization of sp$^2$ carbons was favoured, providing indolin-2-one derivative as the major product (**5k1** and **2k2**).

We then carried out a series of control experiments with 2-ethyl-2-methyl-*N*-(quinolin-8-yl)butanamide (**1d**) to gain some insights on the reaction mechanism. As shown in Supplementary Table 3, the reaction failed to provide the desired product in the absence of a cobalt or silver species (entries 2–4). Furthermore,

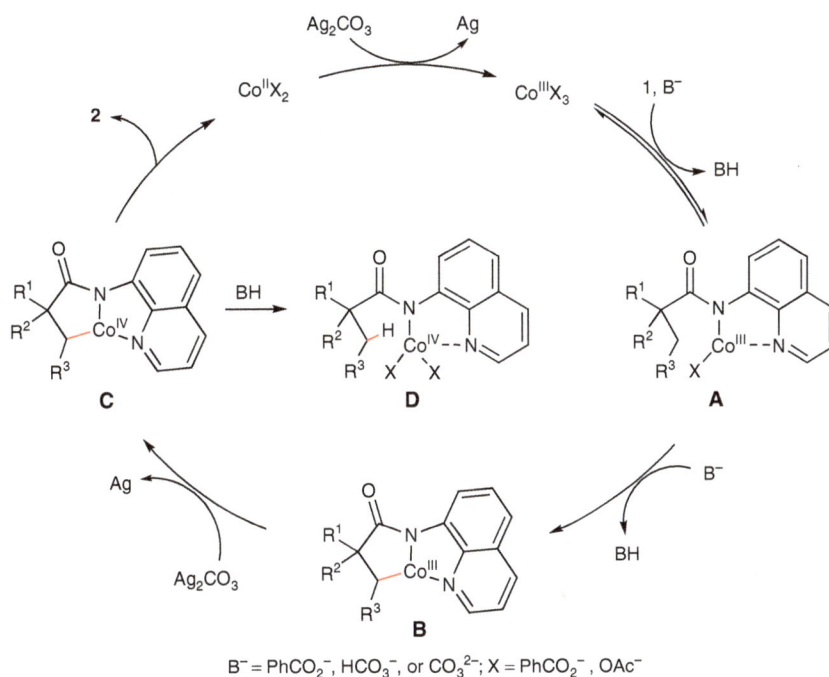

**Figure 6 | Proposed catalytic cycle of β-lactam formation.** The possible mechanism involves the Co-catalysed sp$^3$ C–H activation, oxidation and subsequent reductive elimination.

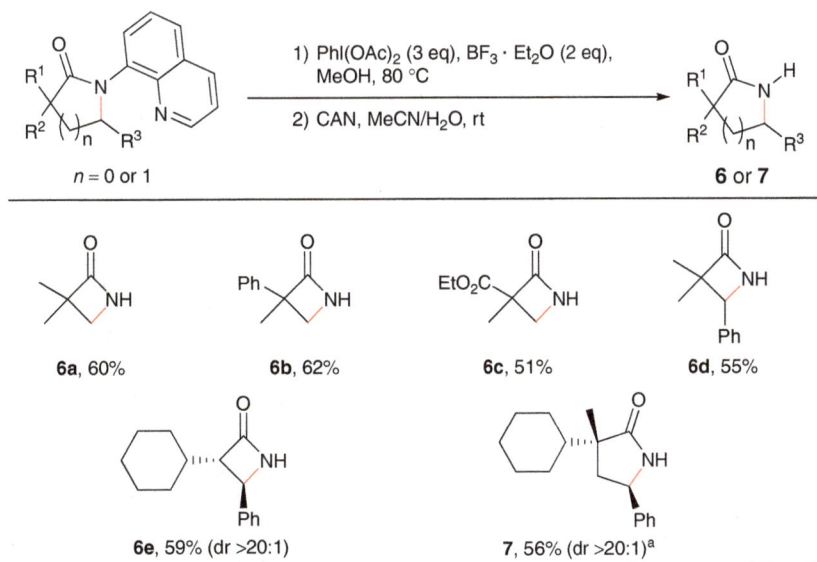

**Figure 7 | Removal of the quinolyl group.** Reaction conditions: (1) amide (0.15 mmol), BF$_3$ · Et$_2$O (0.30 mmol), PhI(OAc)$_2$ (0.45 mmol), MeOH (0.5 ml), 80 °C, 3–8 h; (2) CAN (0.45 mmol), MeCN/H$_2$O(2.0 ml/0.4 ml), room temperature, 6 h. Isolated yield based on three runs of each reaction. [a]The methoxylation step was carried out at 50 °C.

the addition of TEMPO had no apparent effect on the reaction, indicating that an alkyl radical intermediate may not be involved in this process (entries 5 and 6).

## Discussion

As described, a highly regioselective intramolecular amination of propionamide and butyramide derivatives with an 8-aminoquinolinyl group as the bidentate directing group was developed via a cobalt-catalysed sp$^3$ C–H bond functionalization process. The reaction favours the C–H bonds of β-methyl groups over those of the β-methylene and γ- or δ-methyl groups, providing the β-lactam derivatives in a highly site- and diastereo-selective manner. Interestingly, a predominant preference for the

functionalization of γ-benzylic over β-methyl C–H bonds was observed, producing γ-lactams as the major products. On the basis of these results, it is believed that two distinct reaction pathways are involved in the formation of these four- and five-membered ring products. As mentioned earlier, synthesis of lactams has been demonstrated recently via a Pd-, Cu- or Ni-catalysed C–H functionalization process. However, the Cu- or Ni-catalysed process is restricted to substrates with an α-quaternary carbon and the formation of β-lactams. On the other hand, Pd-catalysed synthesis of β-lactams either is restricted to α-unsubstituted or α,β-cyclic substrates, or suffers from the irremovability of the directing group, whereas synthesis of γ-lactams is limited to β-substituted substrates. Therefore,

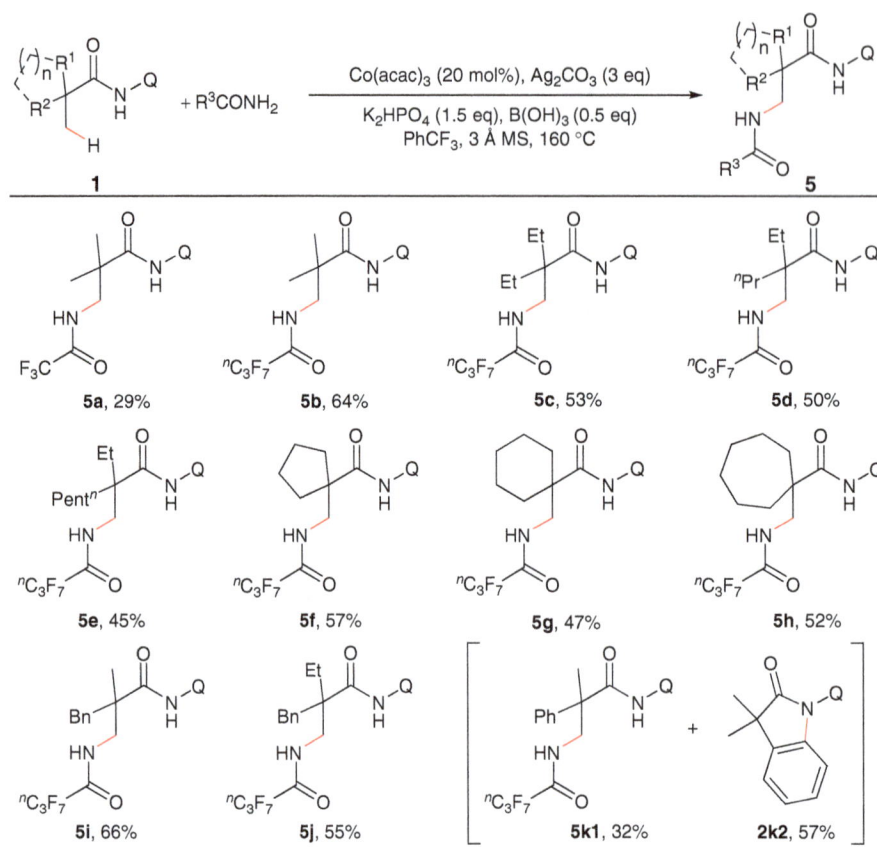

**Figure 8 | Intermolecular amination of α,α-disubstituted propanamides.** Reaction conditions: **1** (0.15 mmol), amide (0.45 mmol), Co(acac)$_3$ (0.03 mmol), Ag$_2$CO$_3$ (0.45 mmol), K$_2$HPO$_4$ (0.23 mmol), B(OH)$_3$ (0.075 mmol), 2.0 ml PhCF$_3$, 160 °C, 24 h. Isolated yield based on three runs of each reaction.

this method provides a complementary approach to access monocyclic and spiro β- or γ-lactams. Furthermore, the cobalt-catalysed ligand-directed intermolecular amination of unactivated sp$^3$ carbons was realized for the first time. The detailed mechanistic investigations of these transformations are currently undergoing in our laboratory. In the meanwhile, N-phosphonyl and phosphinyl groups will be investigated for the present intra- and intermolecular dehydrogenative amination reactions for achieving the GAP work-up[68,69].

## Methods

**General methods.** For $^1$H and $^{13}$C NMR spectra of compounds in this manuscript and details of the synthetic procedures, see Supplementary Figs 3–162 and Supplementary Methods.

**General procedure for intramolecular amination.** A 20-ml tube was charged with α,α,α-trisubstituted N-(quinolin-8-yl)acetamides (**1**, 0.30 mmol), Co(OAc)$_2$ (5.3 mg, 0.030 mmol), Ag$_2$CO$_3$ (207 mg, 0.75 mmol), PhCO$_2$Na (21.6 mg, 0.15 mmol) and 0.60 ml of PhCl. The reaction mixture was stirred rigorously open to the air at 150 °C for 24 h. Then, the mixture was cooled to room temperature, diluted with EtOAc (2 ml), filtered through a celite pad and concentrated *in vacuo*. The residue was purified by flash chromatography on silica gel (gradient eluent of 2–5% EtOAc in hexanes, v/v) to give the desired product.

**General procedure for intermolecular amination.** A 20-ml tube was charged with α,α,α-trisubstituted N-(quinolin-8-yl)acetamides (**1**, 0.15 mmol), hepta-fluorobutyramide (95.9 mg, 0.45 mmol), Co(acac)$_3$ (10.7 mg, 0.030 mmol), Ag$_2$CO$_3$ (75.1 mg, 0.45 mmol), K$_2$HPO$_4$ (39.2 mg, 0.23 mmol), B(OH)$_3$ (4.6 mg, 0.075 mmol), 3 Å MS (200 mg) and 2.0 ml of PhCF$_3$. Then the vial was sealed, and stirred rigorously at 160 °C for 24 h. The mixture was cooled to room temperature, diluted with CH$_2$Cl$_2$ (5 ml), filtered through a celite pad and concentrated *in vacuo*. The residue was purified by flash chromatography on silica gel (gradient eluent of 2–5% EtOAc in hexanes, v/v) to give the desired product.

## References

1. Daugulis, O., Do, H.-Q. & Shabashov, D. Palladium- and copper-catalyzed arylation of carbon–hydrogen bonds. *Acc. Chem. Res.* **42**, 1074–1086 (2009).
2. Chen, X., Engle, K. M., Wang, D.-H. & Yu, J.-Q. Palladium(II)-catalyzed C–H activation/C–C cross-coupling reactions: versatility and practicality. *Angew. Chem. Int. Ed.* **48**, 5094–5115 (2009).
3. Colby, D. A., Bergman, R. G. & Ellman, J. A. Rhodium-catalyzed C – C bond formation via heteroatom-directed C–H bond activation. *Chem. Rev.* **110**, 624–655 (2010).
4. Lyons, T. W. & Sanford, M. S. Palladium-catalyzed ligand-directed C–H functionalization reactions. *Chem. Rev.* **110**, 1147–1169 (2010).
5. Jazzar, R., Hitce, J., Renaudat, A., Sofack-Kreutzer, J. & Baudoin, O. Functionalization of organic molecules by transition-metal-catalyzed C(sp$^3$)–H activation. *Chem. Eur. J.* **16**, 2654–2672 (2010).
6. Yeung, C. S. & Dong, V. M. Catalytic dehydrogenative cross-coupling: forming carbon – carbon bonds by oxidizing two carbon – hydrogen bonds. *Chem. Rev.* **111**, 1215–1292 (2011).
7. Ackermann, L. Carboxylate-assisted transition-metal-catalyzed C–H bond functionalizations: mechanism and scope. *Chem. Rev.* **111**, 1315–1345 (2011).
8. Davis, H. M. L., Du Bois, J. & Yu, J.-Q. C–H functionalization in organic synthesis. *Chem. Soc. Rev.* **40**, 1855–1856 (2011).
9. Gutekunst, W. R. & Baran, P. S. C–H functionalization logic in total synthesis. *Chem. Soc. Rev.* **40**, 1976–1991 (2011).
10. Hartwig, J. F. Regioselectivity of the borylation of alkanes and arenes. *Chem. Soc. Rev.* **40**, 1992–2002 (2011).
11. White, M. C. Adding aliphatic C-H bond oxidations to synthesis. *Science* **335**, 807–809 (2012).
12. Yamaguchi, J., Yamaguchi, A. D. & Itami, K. C–H bond functionalization: emerging synthetic tools for natural products and pharmaceuticals. *Angew. Chem. Int. Ed.* **51**, 8960–9009 (2012).
13. Collins, K. & Glorius, F. A robustness screen for the rapid assessment of chemical reactions. *Nat. Chem* **5**, 597–601 (2013).
14. Hess, W., Treutwein, J. & Hilt, G. Cobalt-catalysed carbon-carbon bond-formation reactions. *Synthesis* 3537–3562 (2008).
15. Kulkarni, A. A. & Daugulis, O. Direct conversion of carbon-hydrogen into carbon-carbon bonds by first-row transition-metal catalysis. *Synthesis* 4087–4109 (2009).

16. Yoshikai, N. Chiral-auxiliary-controlled diastereoselective epoxidations. *Synlett* 1047–1051 (2011).

17. Gao, K. & Yoshikai, N. Low-valent cobalt catalysis: new opportunities for C-H functionalization. *Acc. Chem. Res.* **47**, 1208–1219 (2014).

18. Murahashi, S. Synthesis of phthalimidines from schiff bases and carbon monoxide. *J. Am. Chem. Soc.* **77**, 6403–6404 (1955).

19. Murahashi, S. & Horiie, S. The reaction of azobenzene and carbon monoxide. *J. Am. Chem. Soc.* **78**, 4816–4817 (1956).

20. Funk, J., Yennawar, H. & Sen, A. Cobalt-catalyzed carbonylation of *N*-alkylbenzaldimines to 'N'-alkylphthalimidines' (2,3-dihydro-1H-isoindol-1-ones) via tandem C-H activation and cyclocarbonylation. *Helv. Chim. Acta* **89**, 1687–1695 (2006).

21. Lee, P.-S., Fujita, T. & YoshikaI, N. Cobalt-catalyzed, room-temperature addition of aromatic imines to alkynes via directed C-H bond activation. *J. Am. Chem. Soc.* **133**, 17283–17295 (2011).

22. Gao, K., Lee, P.-S., Fujita, T. & Yoshikai, N. Cobalt-catalyzed hydroarylation of alkynes through chelation-assisted C–H bond activation. *J. Am. Chem. Soc.* **132**, 12249–12251 (2010).

23. Ding, Z.-H. & Yoshikai, N. Cobalt-catalyzed addition of azoles to alkynes. *Org. Lett.* **12**, 4180–1283 (2010).

24. Chen, Q., Ilies, L. & Nakamura, E.-I. Cobalt-catalyzed *ortho*-alkylation of secondary benzamide with alkyl chloride through directed C–H bond activation. *J. Am. Chem. Soc.* **133**, 428–429 (2011).

25. Li, B. *et al.* Direct cross-coupling of C-H bonds with Grignard reagents through cobalt catalysis. *Angew. Chem. Int. Ed.* **50**, 1109–1113 (2011).

26. Gao, K. & YoshikaI, N. Cobalt-catalyzed *ortho*-alkylation of aromatic imines with primary and secondary alkyl halides. *J. Am. Chem. Soc.* **135**, 9279–9282 (2013).

27. Yao, T., Hirano, K., Satoh, T. & Miura, M. Nickel- and cobalt-catalyzed direct alkylation of azoles with N-tosylhydrazones bearing unactivated alkyl groups. *Angew. Chem. Int. Ed.* **51**, 775–779 (2012).

28. Yoshino, T., Ikemoto, H., Matsunaga, S. & Kanai, M. A cationic high-valent Cp*Co$^{III}$ complex for the catalytic generation of nucleophilic organometallic species: directed C-H bond activation. *Angew. Chem. Int. Ed.* **52**, 2207–2211 (2013).

29. Ikemoto, H., Yoshino, T., Sakata, K., Matsunaga, S. & Kanai, M. Pyrroloindolone synthesis via a Cp*Co$^{III}$-catalyzed redox-neutral directed C-H alkenylation/annulation sequence. *J. Am. Chem. Soc.* **136**, 5424–5431 (2014).

30. Grigorjeva, L. & Daudulis, O. Cobalt-catalyzed, aminoquinoline-directed C(sp$^2$)-H bond alkenylation by alkynes. *Angew. Chem. Int. Ed.* **53**, 10209–10212 (2014).

31. Lenges, C. P., White, P. S. & Brookhart, M. Mechanistic and synthetic studies of the addition of alkyl aldehydes to vinylsilanes catalyzed by Co(I) complexes. *J. Am. Chem. Soc.* **120**, 6965–6979 (1998).

32. Cenini, S., Tollari, S., Penoni, A. & Cereda, C. Catalytic amination of unsaturated hydrocarbons: reactions of *p*-nitrophenyl azide with alkenes catalyzed by metalloporphyrins. *J. Mol. Catal* **137**, 135–146 (1999).

33. Ragaini, F. *et al.* Amination of benzylic C-H bonds by arylazides catalyzed by CoII-porphyrin complexes: a synthetic and mechanistic study. *Chem. Eur. J.* **9**, 249–259 (2003).

34. Lu, H.-J. & Zhang, X. P. Catalytic C-H functionalization by metalloporphyrins: recent developments and future directions. *Chem. Soc. Rev.* **40**, 1899–1909 (2011).

35. Lu, H.-J., Li, C.-Q., Jiang, H.-L., Lizardi, C. L. & Zhang, X. P. Chemoselective amination of propargylic C(sp$^3$)-H bonds by cobalt(II)-based metalloradical catalysis. *Angew. Chem. Int. Ed.* **53**, 7028–7032 (2014).

36. Dick, A. R. & Sanford, M. S. Transition metal catalyzed oxidative functionalization of carbon-hydrogen bonds. *Tetrahedron* **62**, 2439–2463 (2006).

37. Boutadla, Y., Davies, D. L., Macgregor, S. A. & Poblador-Bahamonde, A. I. Mechanisms of C-H bond activation: rich synergy between computation and experiment. *Dalton Trans.* 5820–5831 (2009).

38. Bolig, A. D. & Brookhart, M. Activation of sp$^3$ C-H bonds with cobalt(I): catalytic synthesis of enamines. *J. Am. Chem. Soc.* **129**, 14544–14545 (2007).

39. Hung-Low, F., Krogman, J. P., Tye, J. W. & Bradley, C. A. Development of more labile low electron count Co(I) sources: mild, catalytic functionalization of activated alkanes using a [(Cp*Co)$_2$-μ-(η$^4$:η$^4$-arene)] complex. *Chem. Commun.* **48**, 368–370 (2012).

40. Lyons, T. W. & Brookhart, M. Cobalt-catalyzed hydrosilation/hydrogen-transfer cascade reaction: a new route to silyl enol ethers. *Chem. Eur. J.* **19**, 10124–10127 (2013).

41. Corbet, M. & De Campo, F. 8-Aminoquinoline: a powerful directing group in metal-catalyzed direct functionalization of C-H bonds. *Angew. Chem. Int. Ed.* **52**, 9896–9898 (2013).

42. Rouquet, G. & Chatani, N. Catalytic functionalization of C(sp$^2$)-H and C(sp3)-H bonds by using bidentate directing groups. *Angew. Chem. Int. Ed.* **52**, 11726–11743 (2013).

43. Tahlan, K. & Jensen, S. E. J. Origins of the β-lactam rings in natural products. *Antibiotic* **66**, 401–410 (2013).

44. Ye, L.-W., Shu, C. & Gagosz, F. Recent progress towards transition metal-catalyzed synthesis of γ-lactams. *Org. Biomol. Chem.* **12**, 1833–1845 (2014).

45. He, G., Zhang, S.-Y., Nack, W. A., Li, Q. & Chen, G. Use of a readily removable auxiliary group for the synthesis of pyrrolidones by the palladium-catalyzed intramolecular amination of unactivated γ C(sp$^3$)-H bonds. *Angew. Chem. Int. Ed.* **52**, 11124–11128 (2013).

46. Zhang, Q. *et al.* Stereoselective synthesis of chiral α-amino-β-lactams through palladium(II)-catalyzed sequential monoarylation/amidation of C(sp$^3$)-H bonds. *Angew. Chem. Int. Ed.* **52**, 13588–13592 (2013).

47. McNally, A., Haffemayer, B., Collins, B. S. & Gaunt, M. J. Palladium-catalysed C-H activation of aliphatic amines to give strained nitrogen heterocycles. *Nature* **510**, 129–133 (2014).

48. Sun, W.-W. *et al.* Palladium-catalyzed unactivated C(sp$^3$)-H bond activation and intramolecular amination of carboxamides: a new approach to β-lactams. *Org. Lett.* **16**, 480–483 (2014).

49. Wang, Z., Ni, J.-Z., Kuninobu, Y. & Kanai, M. Copper-catalyzed intramolecular C(sp$^3$)-H and C(sp$^2$)-H amidation by oxidative cyclization. *Angew. Chem. Int. Ed.* **53**, 3496–3499 (2014).

50. Wu, X.-S., Zhao, Y., Zhang, G.-W. & Ge, H.-B. Copper-catalyzed site-selective intramolecular amidation of unactivated C(sp$^3$)-H bonds. *Angew. Chem. Int. Ed.* **53**, 3706–3710 (2014).

51. Wu, X.-S., Zhao, Y. & Ge, H.-B. Nickel-catalyzed site-selective amidation of unactivated C(sp$^3$)-H bonds. *Chem. Eur. J.* **20**, 9530–9533 (2014).

52. Yang, M. *et al.* Silver-catalysed direct amination of unactivated C-H bonds of functionalized molecules. *Nat. Commun* **5**, 4707–4712 (2014).

53. Lewis, R. A. *et al.* Synthesis of a cobalt(IV) ketimide with a squashed tetrahedral geometry. *Chem. Commun.* **49**, 2888–2890 (2013).

54. Nguyen, A. I., Hadt, R. G., Solomon, E. I. & Tilley, T. D. Efficient C-H bond activations via O$_2$ cleavage by a dianionic cobalt(II) complex. *Chem. Sci.* **5**, 2874–2878 (2014).

55. Théveau, L. *et al.* Mechanism selection for regiocontrol in base-assisted, palladium-catalysed direct C-H coupling with halides: first approach for oxazole- and thiazole-4-carboxylates. *Chem. Eur. J.* **17**, 14450–14463 (2011).

56. Aihara, Y. & Chatani, N. Nickel-catalyzed direct arylation of C(sp$^3$)-H bonds in aliphatic amides via bidentate-chelation assistance. *J. Am. Chem. Soc.* **136**, 898–901 (2014).

57. Brunschwig, B. S., Chou, M.-H., Creutz, C., Ghosh, P. & Sutin, N. Mechanisms of water oxidation to oxygen: cobalt(IV) as an intermediate in the aquocobalt(II)-catalyzed reaction. *J. Am. Chem. Soc.* **105**, 4832–4833 (1983).

58. Anson, F. C., Collins, T. J., Coots, R. J., Gipson, S. L. & Richmond, T. G. Synthesis and characterization of stable cobalt(IV) coordination complexes: molecular structure of *trans*-[η$^4$-1,2-bis(3,5-dichloro-2-hydroxybenzamido)-4,5-dichlorobenzene]bis(4-tert-butylpyridine)cobalt(IV). *J. Am. Chem. Soc.* **106**, 5037–5038 (1984).

59. Collet, F., Dodd, R. H. & Dauban, P. Catalytic C-H amination: recent progress and future directions. *Chem. Commun.* 5061–5074 (2009).

60. Ramirez, T. A., Zhao, B. & Shi, Y. Recent advances in transition metal-catalyzed sp$^3$ C-H amination adjacent to double bonds and carbonyl groups. *Chem. Soc. Rev.* **41**, 931–942 (2012).

61. Jeffrey, J. L. & Sarpong, R. Intramolecular C(sp$^3$)-H amination. *Chem. Sci* **4**, 4092–4106 (2013).

62. Thu, H. Y., Yu, W. Y. & Che, C. M. Intermolecular amidation of unactivated sp$^2$ and sp$^2$ C-H bonds via palladium-catalyzed cascade C-H activation/nitrene insertion. *J. Am. Chem. Soc.* **128**, 9048–9049 (2006).

63. Pan, J., Su, M. & Buchwald, S. L. Palladium(0)-catalyzed intermolecular amination of unactivated C(sp$^3$)-H bonds. *Angew. Chem. Int. Ed.* **50**, 8647–8651 (2011).

64. Iglesias, Á., Álvarez, R., de Lera, Á. R. & Muñiz, K. Palladium-catalyzed intermolecular C(sp$^3$)-H amidation. *Angew. Chem. Int. Ed.* **51**, 2225–2228 (2012).

65. Kang, T., Kim, Y., Lee, D., Wang, Z. & Chang, S. Iridium-catalyzed intermolecular amidation of sp$^3$ C-H bonds: late-stage functionalization of an unactivated methyl group. *J. Am. Chem. Soc.* **136**, 4141–4144 (2014).

66. Wang, N. *et al.* Rhodium(III)-catalyzed intermolecular amidation with azides via C(sp$^3$)-H functionalization. *J. Org. Chem.* **79**, 5379–5385 (2014).

67. Kang, T., Kim, H., Kim, J. G. & Chang, S. Facile one step-synthesis and characterization of high aspect ratio core-shell copper-polyaniline nanowires. *Chem. Commun.* **50**, 1207–1212 (2014).

68. Sun, H., Zhang, H.-W., Han, J.-L., Pan, Y. & Li, G. Asymmetric C-C Bond Formation between Chiral N-Phosphonyl Imines and Ni(II)-Complex of Glycine Schiff Base Provides a GAP Synthesis of α,β-*syn*-diamino acid derivatives. *Eur. J. Org. Chem.* **22**, 4744–4747 (2013).

69. Pindi, S., Wu, J.-B. & Li, G. Design, synthesis and applications of new chiral N-2-phenyl-2-propyl sulfinylimines for Group-Assisted Purification (GAP) asymmetric synthesis. *J. Org. Chem.* **78**, 4006–4012 (2013).

## Acknowledgements

We gratefully acknowledge Indiana University-Purdue University Indianapolis and NSF CHE-1350541 for financial support. We are also grateful for financial support from the Robert A. Welch Foundation (D-1361, USA), NSFC (No. 21332005), and the Jiangsu Innovation Programs (China)

## Author contributions

X.W. and K.Y. performed the experiments and analysed the data. All authors conceived and designed the experiments, contributed to discussions and wrote the manuscript.

## Additional information

**Competing financial interests:** The authors declare no competing financial interests.

**2**

# Non-covalent synthesis of supermicelles with complex architectures using spatially confined hydrogen-bonding interactions

Xiaoyu Li[1,*,†], Yang Gao[1,*], Charlotte E. Boott[1], Mitchell A. Winnik[2] & Ian Manners[1]

Nature uses orthogonal interactions over different length scales to construct structures with hierarchical levels of order and provides an important source of inspiration for the creation of synthetic functional materials. Here, we report the programmed assembly of monodisperse cylindrical block comicelle building blocks with crystalline cores to create supermicelles using spatially confined hydrogen-bonding interactions. We also demonstrate that it is possible to further program the self-assembly of these synthetic building blocks into structures of increased complexity by combining hydrogen-bonding interactions with segment solvophobicity. The overall approach offers an efficient, non-covalent synthesis method for the solution-phase fabrication of a range of complex and potentially functional supermicelle architectures in which the crystallization, hydrogen-bonding and solvophobic interactions are combined in an orthogonal manner.

[1] School of Chemistry, University of Bristol, Bristol BS8 1TS, UK. [2] Department of Chemistry, University of Toronto, Toronto, Ontario M5S 3H6, Canada. * These authors contributed equally to this work. † Present address: Department of Polymer Materials, School of Material Science and Technology, Beijing Institute of Technology, Beijing 100081, China. Correspondence and requests for materials should be addressed to I.M. (email: ian.manners@bristol.ac.uk).

The intricate hierarchical structures achieved by nature[1] using orthogonal interactions on different length scales have inspired the pursuit of artificial materials with complexity and functionality using analogous synthetic concepts[2,3]. Recent breakthroughs in colloidal particle synthesis have enabled their use as building blocks for higher-level assembly[4]. In particular, block copolymers provide a diverse array of soft-matter nanoparticles of varied shape and size[5-10], and intricately designed structures such as Janus[11-15] and patchy particles[16-19]. Several reports have described the assembly of these nanoparticles through electrostatic[20] or solvophobic interactions[11-19] into discrete multicomponent composite supermicelles[11,12] or extended superstructures[13-19].

Hydrogen-bonding (H-bonding) interactions, which play pivotal roles in the spontaneous organization of building blocks into hierarchical structures in natural systems, have been extensively used to assemble artificial structural units, as illustrated by work on small molecules[21-25], DNA origami[26-28], polymer/small molecule pairs[29-31], polymer/polymer pairs[32] and polymer/inorganic particles[33]. However, the creation of nanoscale architectures with similar levels of precision to nature by low-cost solution-phase processing remains a key challenge.

Here, we describe the use of cylindrical block comicelles with crystalline cores as building blocks to create complex hierarchical supermicelle structures via H-bonding interactions. Specifically, we demonstrate that fine control of the length and position of H-bonding donor and acceptor segments, made possible by the living crystallization-driven block copolymer self-assembly method, allows the interactions to be confined and the subsequent hierarchical organization of the resulting cylindrical block comicelles to be precisely directed.

## Results

### The basic polymer components and building blocks.

In Fig. 1a, we show the five different polymers that form the block copolymers studied. All of the block copolymers possessed a crystallizable, core-forming poly(ferrocenyldimethylsilane), PFS block together with a corona-forming segment that was either an H-bonding donor block (hydroxyl-functionalized poly(methylvinylsiloxane), PMVSOH, $H_D$), an H-bonding acceptor block (poly(2-vinylpyridine), P2VP, $H_A$), a non-interactive block (poly(*tert*-butyl acrylate), PtBA, N) or a crosslinkable block (poly(methylvinylsiloxane), PMVS, X). For convenience, as the crystalline PFS core was a common feature, all of the micelles are depicted in an abbreviated form that reflects their coronal chemistry (for example, triblock comicelle M(PFS$_{20}$-*b*-PtBA$_{280}$)-*b*-M(PFS$_{32}$-*b*-P2VP$_{448}$)-*b*-M(PFS$_{20}$-*b*-PtBA$_{280}$) is described as having segments of N, $H_A$ and N and is written as N-$H_A$-N). Cylindrical block comicelle building blocks such as N-$H_A$-N were prepared via living crystallization-driven self-assembly (CDSA)[33-38]. In this process (see Fig. 1b (i)), dissolved block copolymer unimers with a crystallizable core-forming block and different corona-forming blocks can be added sequentially to short cylindrical micelles (seeds), which were prepared by sonication. This allowed the formation of monodisperse cylindrical micelles with segmented structures and lengths controlled from 50 nm to 5 μm (detailed characteristics of all seeds and triblock comicelles used in this study are shown in Supplementary Table 1).

### Initial studies of hierarchical assembly via H-bonding interactions.

The H-bonding interactions between hydroxyl and pyridyl groups in polymers have been exploited extensively[29-32,39], and our initial studies focused on exploring their potential use in supermicelle formation. We found that the addition of

$H_D$ homopolymer to block copolymer seeds with an $H_A$ corona in isopropanol (*i*-PrOH, a selective solvent for both PMVSOH and P2VP blocks) led to the formation of an insoluble precipitate (Supplementary Fig. 2k). On the basis of this observation, we investigated the more controlled interactions between the $H_D$ homopolymer and N-$H_A$-N triblock comicelles, in which the $H_A$ polymer chains were confined to the corona of the central segment of the cylindrical building block (Fig. 1b). From transmission electron microscopy (TEM) analysis, addition of $H_D$ homopolymer to the *i*-PrOH solution of the triblock comicelles at a mole ratio of hydroxyl/pyridyl groups of 5:1 led to assembly of the comicelles through parallel aggregation of the central $H_A$ coronal segments forming 'fish-spine' supermicelles (Fig. 1b (ii),c; Supplementary Fig. 2g). The $H_D$ homopolymer chains appeared to function as a 'glue' to enable the triblock comicelles to assemble via H-bonding interactions. Although a distribution of very similar structures was formed in each case, substantial control could be imposed by adjusting various parameters. For example, use of either a very low or very high mole ratio of hydroxyl/pyridyl groups led to a significant reduction of the aggregation number (or arm distribution) in each supermicelle (Supplementary Fig. 2a–e). Furthermore, we were also able to tune the steric hindrance arising from the N segment by using several N segment-forming block polymers (PFS-*b*-PtBA) with different PtBA coronal block chain lengths. As shown in Supplementary Fig. 2f–j, with increasing PtBA block length and thus coronal spatial extent, the aggregation number also decreased.

Next, we replaced $H_D$ homopolymer with block copolymer seeds bearing an $H_D$ corona (Fig. 1b (iii) and (iv)) and studied how this changed the assembly behaviour. As an initial experiment, we mixed seeds with an $H_D$ corona ($L_n = 37$ nm) and seeds with a corona of $H_A$ ($L_n = 45$ nm) in *i*-PrOH. Seconds after mixing, a precipitate was observed in the solution (Supplementary Fig. 2l,m), which indicated the formation of large aggregates via uncontrolled H-bonding interactions. We then explored the analogous assembly process for the $H_D$ seeds with N-$H_A$-N triblock comicelles. At a ratio of hydroxyl/pyridyl groups of 20:1, TEM analysis revealed the formation of composite block comicelles with the $H_D$ seeds attached to the surface of the central $H_A$ segments. The presence of a small number of free $H_D$ seeds was also detected (Fig. 1b (iii),d). Interestingly, when the ratio was decreased to 1:2, instead of forming composite block comicelles, large micelle bundles were obtained, in which each $H_D$ seed was bound to several N-$H_A$-N triblock comicelles via H-bonding interactions (Fig. 1b (iv),e; Supplementary Fig. 2n). This behaviour contrasts to that observed in the case involving the addition of $H_D$ homopolymer to N-$H_A$-N triblock comicelles, which showed no aggregation at the same ratio (Supplementary Fig. 2a). We attribute the more extensive assembly to the larger size of $H_D$ seeds compared with the homopolymer chains.

### Hierarchical assembly of triblock comicelles via H-bonding.

The studies described above suggested that the H-bonding interactions between hydroxyl groups on $H_D$ chains and the pyridyl groups on $H_A$ chains were sufficiently strong to build hierarchical structures with a higher level of complexity. We therefore investigated how confined H-bonding interactions would direct the assembly process for triblock comicelles. First, we explored a 'condensation polymerization' of $H_D$-N-$H_D$ and $H_A$-N-$H_A$ triblock comicelles where the interacting $H_D$ and $H_A$ segments were both placed at the termini and were therefore only shielded by N segments on one side. To distinguish the two kinds of triblock comicelles in the resulting supermicelles, different N central segment lengths were used (680 and 240 nm

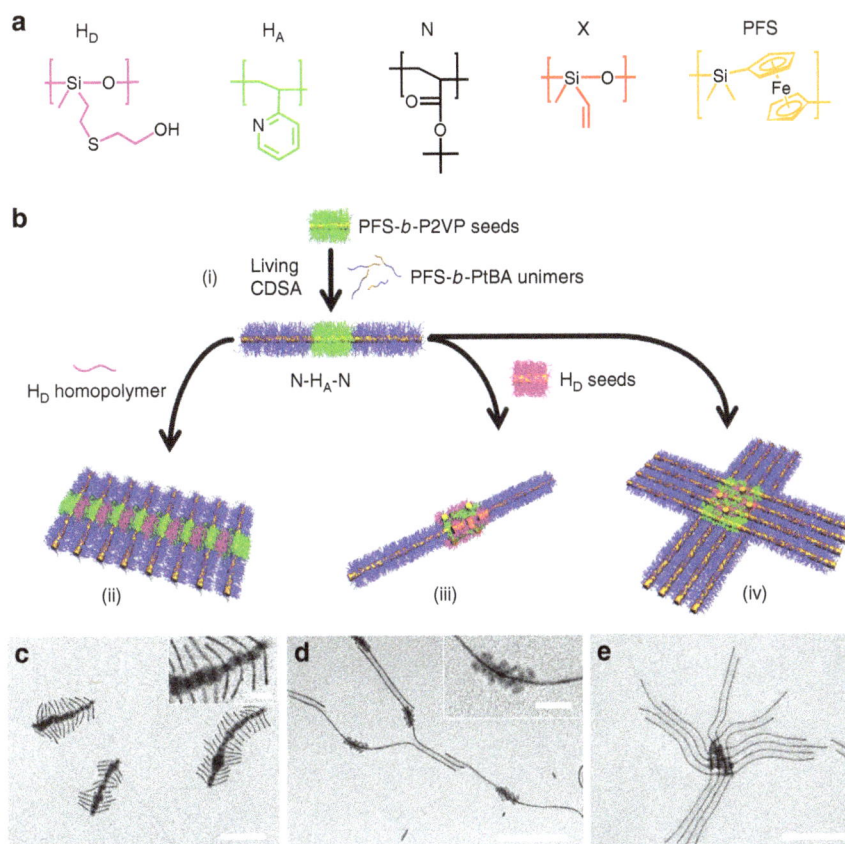

**Figure 1 | Initial studies of hierarchical assembly via H-bonding interactions.** (a) Chemical structures of PMVSOH ($H_D$), P2VP ($H_A$), PtBA (N), PMVS (X) and PFS. (b) Schematic of the assembly processes. TEM images after solvent evaporation of supermicelles formed by (c) mixing $H_D$ homopolymer (PMVSOH$_{105}$) with N-$H_A$-N (540 nm) triblock comicelles (hydroxyl/pyridyl group ratio = 5:1); and $H_D$ seeds (M(PFS$_{20}$-b-PMVSOH$_{120}$), 37 nm) with N-$H_A$-N (1.4 μm) triblock comicelles (hydroxyl/pyridyl group ratio: (d) 20:1 and (e) 1:2) in i-PrOH. Insets in **c** and **d** are high magnification images of the $H_A$ segments after the adsorption of $H_D$ homopolymer and seeds, respectively. Scale bar, 500 or 100 nm (inset). Segments: $H_A$ = M(PFS$_{32}$-b-P2VP$_{448}$) and N = M(PFS$_{20}$-b-PtBA$_{280}$).

for $H_D$-N-$H_D$ and $H_A$-N-$H_A$, respectively. See Supplementary Fig. 1 and Supplementary Table 1). Figure 2a,b show TEM images of the 'ABA' and 'BAB' supermicelles using a hydroxyl/pyridyl group ratio of 3:1. However, the intermicellar H-bonding interactions were not fully controlled as 'branched' supermicelles were also detected (Supplementary Fig. 3a,b).

To achieve more effective confinement for the H-bonding interactions, the $H_A$ segments were placed at the centre of a triblock comicelle. Subsequent mixing of $H_D$-N-$H_D$ and N-$H_A$-N triblock comicelles (hydroxyl/pyridyl group ratio of 3:1) in i-PrOH led to 'T'-shaped supermicelles (Fig. 2c). The supermicelles could be extended in length to form dimer, trimer, tetramer and even 'polymeric' supermicelles (Fig. 2d–f; Supplementary Fig. 3c), but no branched structures were observed. When the N-$H_A$-N triblock comicelles were in slight excess, predominantly 'T'-shaped supermicelles were produced (Supplementary Fig. 3d).

When the $H_D$ and $H_A$ segments were both placed in the centre of the triblock comicelles and the resulting N-$H_D$-N and N-$H_A$-N cylindrical comicelles mixed in i-PrOH (hydroxyl/pyridyl group ratio of 2:1) 'cross' supermicelles were formed (Fig. 2g). Unsymmetrical examples were prepared using N-$H_D$-N and N-$H_A$-N triblock comicelles with different lengths with high efficiency (yield = 80%; Fig. 2h,i). This indicated that the balance between the attractive H-bonding between the $H_D$ and $H_A$ central segments and the repulsive, steric interactions of the coronas of the terminal N segments led only to one-to-one complexes. However, when this balance was disturbed, no 'cross'

supermicelles were formed. For example, compared with the samples shown in Fig. 2g–i, when the P2VP coronal block for the central $H_A$ segment was reduced in length (from 448 to 250 repeat units) the weakened attraction between $H_D$ and $H_A$ segments was insufficient to generate 'cross' supermicelles (Supplementary Fig. 3e). In contrast, multiple-to-one complex 'cross' supermicelles were formed when an increased $H_A$ segment length was used (109 nm, larger than the value of $L_n = 45$ nm in the cases of Fig. 2g–i) in the N-$H_A$-N triblock comicelles (Supplementary Fig. 3f).

**Programmed stepwise hierarchical assembly.** We also explored the use of a combination of H-bonding and solvophobic interactions to build even more complex, higher-level micelle architectures. First, $H_A$-X-$H_A$ triblock comicelles with crosslinkable central segments were prepared by living CDSA. After switching to a more polar solvent via dialysis (from i-PrOH/n-hexane = 4:1 (v/v), to MeOH), these triblock comicelles formed 'cross' supermicelles, due to the aggregation of X segments via solvophobic interactions (Fig. 3a). The yield of 'cross' supermicelles was ca. 90% (Supplementary Fig. 4a). The X segments were then subject to covalent crosslinking[40] using UV irradiation to create $^{XL}$X regions to make the 'cross' supermicelles permanent (Fig. 3b; Supplementary Fig. 4b). To build the next level of structure, N segments were grown from the four termini of these supermicelles by living CDSA (Fig. 3c; Supplementary Fig. 4c). To impart further robustness to the supermicelles and to hinder

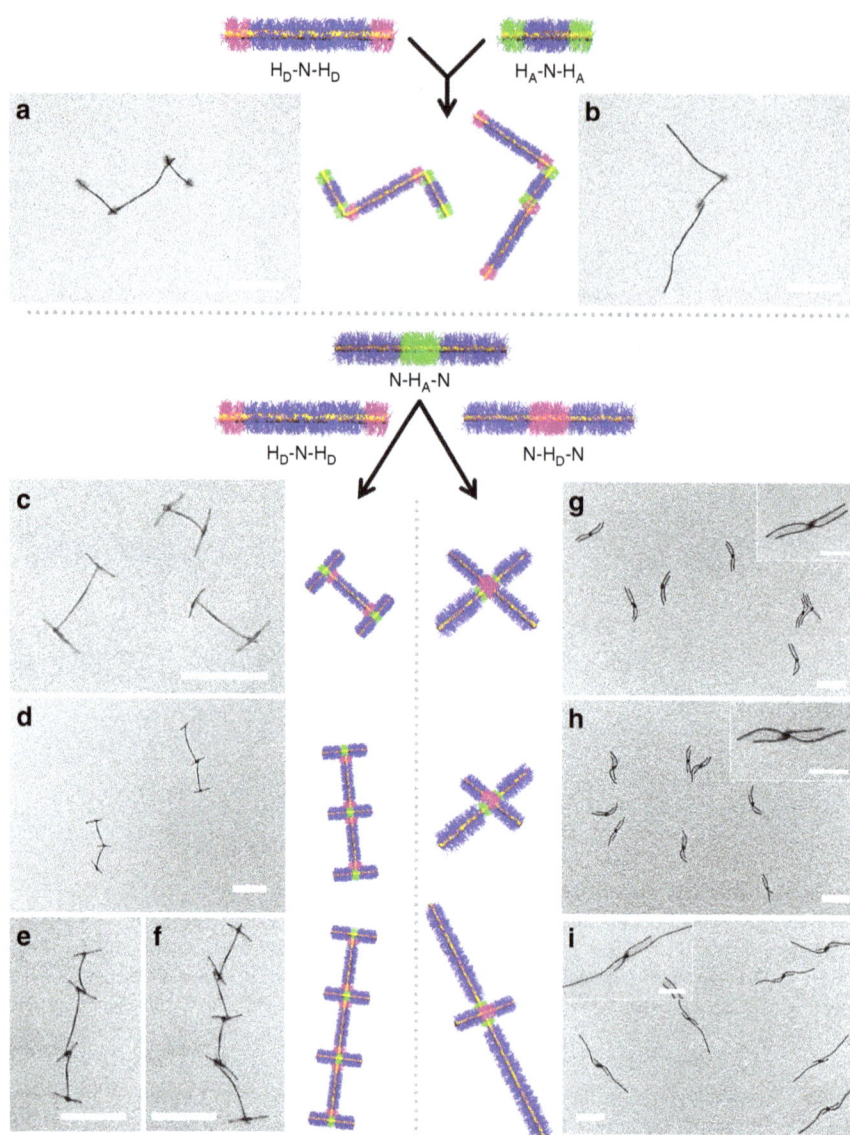

**Figure 2 | Hierarchical assembly of triblock comicelles via H-bonding.** TEM images (after solvent evaporation) and corresponding schematic representations of (**a**) 'ABA' and (**b**) 'BAB' supermicelles from $H_D$-N-$H_D$ (910 nm) and $H_A$-N-$H_A$ (380 nm) triblock comicelles (hydroxyl/pyridyl groups of 3:1) in $i$-PrOH. (**c**-**f**) TEM images (after solvent evaporation) and corresponding schematic representations of **c**, 'I'-shaped structures; **d**, dimers; **e**, trimers; **f**, tetramers from $H_D$-N-$H_D$ (480 nm) and N-$H_A$-N (320 nm) triblock comicelles in $i$-PrOH (hydroxyl/pyridyl groups of 2:1). (**g**-**i**) TEM images (after solvent evaporation) of unsymmetrical micelles from N-$H_D$-N and N-$H_A$-N triblock comicelles in $i$-PrOH with total comicelle lengths of **g**, 540 and 540 nm; **h**, 330 and 540 nm; **i**, 330 nm and 1.3 µm, respectively. Scale bars are 500 nm and the scale bars in the insets are 200 nm. Segment $H_D$ = M(PFS$_{20}$-$b$-PMVSOH$_{120}$), $H_A$ = M(PFS$_{32}$-$b$-P2VP$_{448}$) and N = M(PFS$_{20}$-$b$-PtBA$_{280}$).

the acceptor ability of the P2VP chains during subsequent transformations, the $H_A$ coronal segments from the initial $H_A$-X-$H_A$ triblock comicelle were crosslinked using Pt nanoparticles[13] to yield $^{XL}H_A$ segments (Fig. 3d; Supplementary Fig. 4d). As a result of the crosslinking, the electron density of $^{XL}H_A$ and $^{XL}X$ segments increased and they appeared darker than the other regions by TEM. A further $H_A$ segment was then created at each of the cross supermicelle termini by living CDSA (Fig. 3e), and seeds with an $H_D$ corona were then added to bind to the uncrosslinked $H_A$ segments via H-bonding (Fig. 3f). The resulting composite 'cross' supermicelles were uniform in size and architecture, demonstrating the precision of this controlled hierarchical assembly method (see low resolution TEM images in Fig. 3g,h). To further illustrate the level of control, we grew $H_A$ segments with a longer P2VP corona-forming block at the termini of 'cross' supermicelles and we were then able to bind more $H_D$ seeds and to form larger bundles on the arms of

composite 'cross' supermicelles (Supplementary Fig. 5a–c). Moreover, the bound $H_D$ seeds were found to be still active to living CDSA[33] and addition of the PFS-$b$-PtBA unimer led to the formation of N segments (Fig. 4a). Interestingly, the subsequent growth of the N segments led to their mutual repulsion and resulted in a perpendicular alignment of the N-$H_D$-N comicelles relative to the cross supermicelle arms. This afforded hierarchical 'windmill-like' supermicelles with high uniformity in size, architecture and yield (Fig. 4a,b; Supplementary Fig. 4e).

Direct characterization of the supermicelles in solution was achieved using laser scanning confocal microscopy (LSCM) analysis of 'windmill'-shaped supermicelle analogues with N segments labelled with red or green fluorescent BODIPY dyes. 'Cross' micelles with red-dye-labelled N segments in the arms were prepared first (Supplementary Fig. 5d–g). Finally, green-dye-labelled N segments were grown from the $H_D$ seeds by living CDSA (Fig. 4c). A key issue for supramolecular, H-bonded

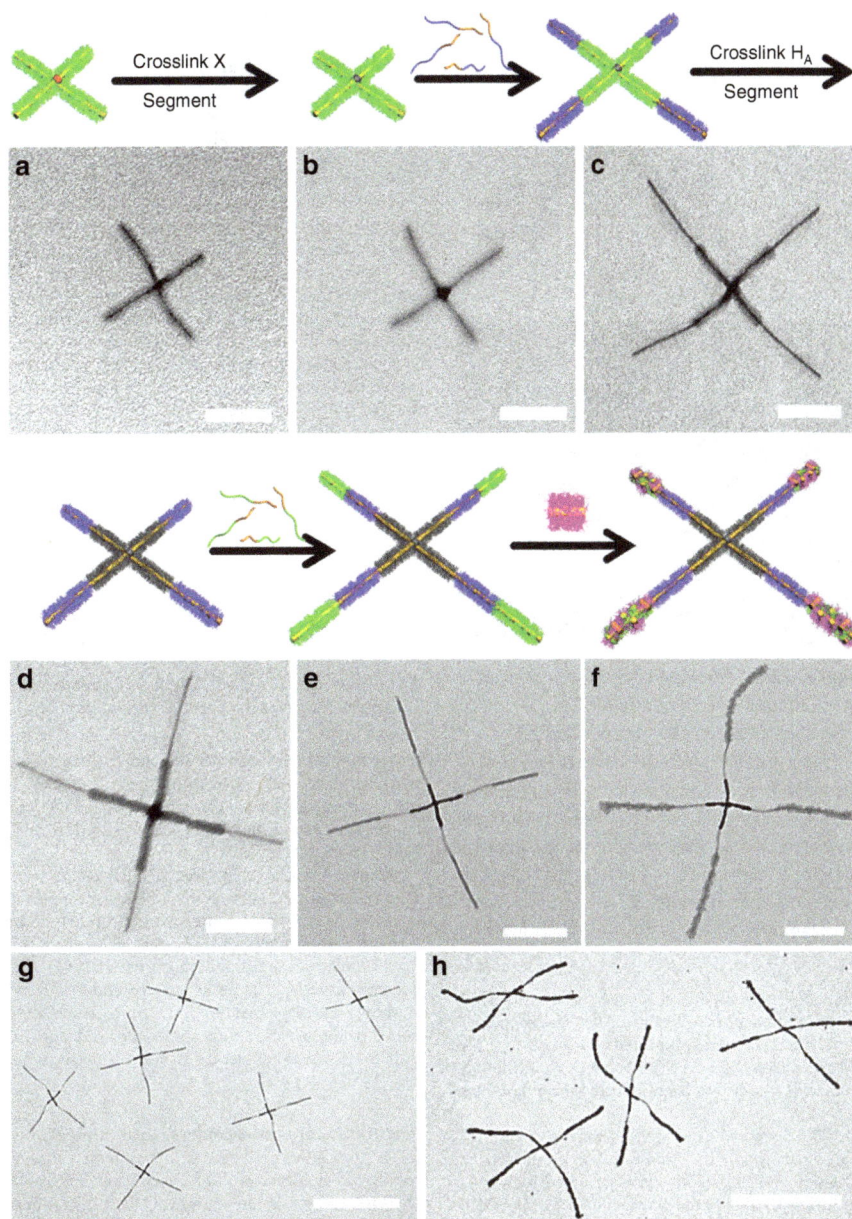

**Figure 3 | Programmed stepwise hierarchical assembly.** TEM images (after solvent evaporation) and corresponding schematic representations of (**a**) 'cross' supermicelles from the $H_A$-X-$H_A$ (420 nm) triblock comicelles; and the 'cross' supermicelles (**b**) with crosslinked $^{XL}$X segments; (**c**) after the growth of N segments; (**d**) with crosslinked $^{XL}H_A$ segments; (**e**) after the growth of another $H_A$ segments; and (**f**) after the further addition of $H_D$ seed micelles in $i$-PrOH. Scale bars are 200 nm in images **a–d**, 500 nm in images (**e,f**) and 2 μm in lower magnification images (**g,h**). Segment X = M(PFS$_{36}$-$b$-PMVS$_{324}$), $H_A$ = M(PFS$_{25}$-$b$-P2VP$_{250}$), N = M(PFS$_{20}$-$b$-PtBA$_{280}$) and $H_D$ = M(PFS$_{20}$-$b$-PMVSOH$_{120}$).

assemblies is their potential for dynamic behaviour. To explore the possible lability of the perpendicularly arranged N-$H_D$-N comicelles near the termini of the 'cross' supermicelle arms, we mixed supermicelles in which the corresponding N segments possessed green dyes with analogues containing only red dyes in $i$-PrOH. No significant exchange between the two kinds of 'windmill' supermicelles was detected by LSCM even after 10 days at 22 °C, which indicated that they are static under these conditions and exist as kinetically stable, non-equilibrium structures (Fig. 4d; Supplementary Fig. 6g–j; for control experiments see Supplementary Fig. 6a–f).

## Discussion

In summary, these results demonstrate the programmed formation of complex supermicelles by the use of confined regions of hydrogen bonding in cylindrical micelle precursors in orthogonal combination with crystallization and solvophobic interactions. This non-covalent synthesis approach allows the high-yield formation of hierarchical architectures that are persistent in solution, kinetically stable, easy to functionalise and sufficiently robust to allow for further processing. Additional stabilization can also be introduced through covalent crosslinking methods. The novel architectures accessible may be useful as models of biological structures (for example, chromosomes in the case of unsymmetrical cross-micelles) or as tectons for the creation of yet more complex hierarchical materials. As the living CDSA method used to create the micelle-building blocks has been shown to be applicable to a variety of crystallizable block copolymers and related π-stacking molecular amphiphiles, including those based on electroactive[41–45], bioactive[46,47] and bioinert[48] characteristics, potential future applications for the resulting assemblies in sensing, nano-electronics, catalysis and biomedicine can be envisaged.

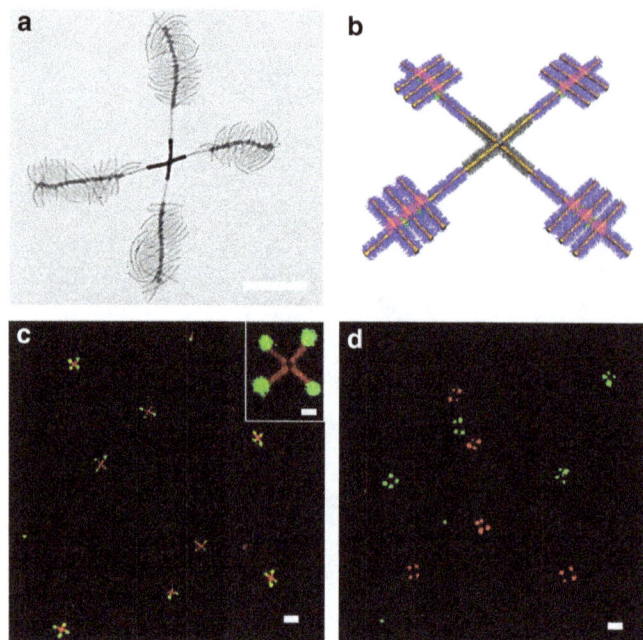

**Figure 4 | 'Windmill'-like supermicelles.** (**a**) TEM image (after solvent evaporation) of a 'windmill'-like supermicelle formed in *i*-PrOH. (**b**) Schematic illustration of the supermicelles. (**c**) LSCM image of fluorescent dye-labelled large 'supermicelles' with the inner N segments labelled with a red dye and the outer N segments labelled with a green dye in *i*-PrOH. Inset is a zoom-in LSCM image of a 'windmill'-like supermicelle. (**d**) LSCM images of the mixture of 'windmill'-shaped supermicelles labelled with either red or green dyes (after 10 days at 22 °C) in *i*-PrOH. Scale bars are 500 nm in (**a**), 5 μm in (**c,d**) and 1 μm in the inset.

## Methods

**Polymer synthesis.** The diblock copolymers used in this study were synthesized via anionic polymerization in an inert atmosphere glovebox. The detailed polymerization procedures for these polymers have been reported elsewhere[49-51]. The PFS-*b*-PMVSOH polymer was prepared from PFS-*b*-PMVS polymer via a thiol-ene click reaction (with 2-mercaptoethanol), which has been reported recently by our group[52].

Fluorescent dye-labelled PFS-*b*-PtBA was obtained by post-polymerization modification. The PFS-*b*-PtBA precursor with a hydroxyl terminal group was synthesized via anionic polymerization and the anions were quenched with ethylene oxide and subsequently with 4-*tert*-butylphenol (yield 99% by $^1$H NMR). Green dye (BODIPY FL, 3 mg, $8 \times 10^{-3}$ mmol of carboxyl groups)[38] was mixed with 80 mg of the PFS-*b*-PtBA with a hydroxyl terminal group ($2 \times 10^{-3}$ mmol of hydroxyl groups), 1 mg of dicyclohexylcarbodiimide ($8 \times 10^{-3}$ mmol) and 0.8 mg of 4-dimethylaminopyridine ($1 \times 10^{-3}$ mmol) in dry tetrahydrofuran (THF) (2 ml)[38]. The solution was stirred at room temperature for 3 days before the polymer was precipitated into a mixed solvent of MeOH and water (8:2, v/v) for three times and further by dialysis against THF. The red dye-containing polymer was prepared the same way except BODIPY 630/650 was used[38].

**Polymer characterization.** Gel permeation chromatography (GPC) measurements were carried out on a Viscotek VE 2001 triple-detector gel permeation chromatograph equipped with an automatic sampler, a pump, an injector, an inline degasser and a column oven (30 °C). The elution columns consist of styrene/divinyl benzene gels with pore sizes of 500 and 100,000 Å. Detection was conducted by means of a VE 3580 refractometer, a four-capillary differential viscometer, and 90° and low angle (7°) laser light ($\lambda_0 = 670$ nm) scattering detectors, VE 3210 and VE 270. THF (Fisher) was used as the eluent, with a flow rate of 1.0 ml min$^{-1}$. Samples were dissolved in the eluent (2 mg ml$^{-1}$) and filtered with a Ministart SRP 15 filter (polytetrafluoroethylene membrane of 0.45 μm pore size) before analysis. The calibration was conducted using a PolyCALTM polystyrene standard from Viscotek. To determine the molar mass of the block copolymers, aliquots of the first block were taken and the molar mass of the first block was determined by gel permeation chromatography. The polymerization degrees of the two blocks were then determined by combining the molecular weight $M_n$ of the first block with the block ratio of the diblock copolymer, which was obtained by integration of the $^1$H NMR spectrum. Polydispersities of the PFS homopolymers and block copolymers were in the range 1.08–1.16 by GPC.

**Transmission electron microscopy.** Copper grids from Agar Scientific, mesh 400, were coated with a carbon film. Carbon coating was done using an Agar TEM Turbo Carbon Coater in which carbon was sputtered onto mica sheets before deposition on the grids via flotation on water. The samples for electron microscopy were prepared by drop-casting one drop ($\sim 10$ μl) of the micelle colloidal solution onto a carbon-coated copper grid. Bright-field TEM micrographs were obtained on a JEOL 1200-EX II microscope operating at 120 kV and equipped with an SIS MegaView III digital camera. Images were analysed using the ImageJ software package developed at the US National Institutes of Health. As the micelles reported here are trapped kinetically because of crystallization of the core block, the morphologies observed from dried samples by TEM are anticipated to match closely those observed in solution, as previously shown for 1D systems[36]. For the statistical length analysis, $\sim 300$–400 objects were measured to determine the contour length. Histograms of the length distribution were constructed. From these data, values of the s.d. of the length distribution 'sigma' were determined. The number average micelle length ($L_n$) and weight average micelle length ($L_w$) were calculated using the equations (1) and (2) below from measurements of the contour lengths ($L_i$) of individual micelles, where $N_i$ is the number of micelles of length $L_i$, and $n$ is the number of micelles examined in each sample.

$$L_n = \frac{\sum_{i=1}^{n} N_i L_i}{\sum_{i=1}^{n} N_i} \tag{1}$$

$$L_w = \frac{\sum_{i=1}^{n} N_i L_i^2}{\sum_{i=1}^{n} N_i L_i} \tag{2}$$

The distribution of micelle lengths is characterized by both $L_w/L_n$ and the ratio $\sigma/L_n$, where $\sigma$ is the s.d. of the length distribution.

**Experimental evidence for H-bonding interactions.** PMVSOH$_{192}$ and P2VP$_{20}$ homopolymers were dried from their *i*-PrOH solution and dried in vacuum oven before characterization. The mixture of PMVSOH and P2VP (mole ratio of hydroxyl/pyridyl group = 1:1) was prepared by mixing the two solutions and dried before characterization.

Fourier Transform Infrared (FTIR) spectra were recorded with an attenuated total reflection accessory. As can be seen from Supplementary Fig. 2o, a new peak appeared at 3,264 cm$^{-1}$ after mixing the two homopolymers suggesting the formation of H-bonds[53,54]. The peak at 3,390 cm$^{-1}$ originated from the residual *i*-PrOH molecules and the homopolymers (O-H stretching of H-bonded and free hydroxyl groups). The peaks at 1,565 and 1,580 cm$^{-1}$ (from P2VP homopolymer, as shown in Supplementary Fig. 2p), corresponding to the stretching of double bonds in the pyridine ring, shifted to 1,631 and 1,658 cm$^{-1}$ after mixing with PMVSOH homopolymers, a typical response to hydrogen bonding of pyridine ring[54,55].

**Preparation of seed cylinders.** Seed cylinders were prepared by ultrasonicating (50 W sonication processor equipped with a titanium sonotrode) the block copolymer cylinders at low temperatures. PFS-*b*-PMVS seed cylinders were obtained via sonication at $-78$ °C for 1 h in *n*-hexane, and sonication at 0 °C for 1 h in *i*-PrOH was used to obtain other seed cylinders (M(PFS-*b*-P2VP) and M(PFS-*b*-PMVSOH)). The original cylinders were obtained by directly dispersing the diblock copolymers into the corresponding solvents and heating at 70 °C for 1 h before the solution was allowed to cool to room temperature. The lengths of seed cylinders could be increased by adding a solution of fresh unimers (the polymer dissolved in THF) into the seed micelle solution.

**Preparation of triblock comicelles.** The triblock comicelles were prepared via the seeded growth method. For example, to prepare triblock comicelle M(PFS$_{20}$-*b*-PtBA$_{280}$)-*b*-M(PFS$_{32}$-*b*-P2VP$_{448}$)-*b*-M(PFS$_{20}$-*b*-PtBA$_{280}$), 0.1 ml of M(PFS$_{32}$-*b*-P2VP$_{448}$) seed cylinders *i*-PrOH solution (0.5 mg ml$^{-1}$) was diluted with 0.4 ml of *i*-PrOH, and then 5 μl of a THF solution of PFS$_{20}$-*b*-PtBA$_{280}$ unimers was added. The PFS$_{20}$-*b*-PtBA$_{280}$ unimers then grew epitaxially from the active termini of the M(PFS$_{32}$-*b*-P2VP$_{448}$) seed cylinders to form triblock comicelles.

**Preparation of supermicelles via H-bonding interactions.** Two kinds of cylinders, one bearing M(PFS-*b*-P2VP) segments and the other bearing M(PFS-*b*-PMVSOH) segments were mixed with each other at the desired feed ratio in *i*-PrOH. The solution was gently shaken for 20 s and kept still overnight before analysis by TEM.

The resulting supermicelles (such as those shown in Fig. 1c, the large micelle bundles in Fig. 1e and the linear 'oligomers' in Fig. 2c–f) were reproducibly formed as a distribution of very closely related structures (for example, see low resolution TEM images in Supplementary Figs 2c,n and 3). For example, in the case of the micelle bundles, these were relatively uniform in size but the aggregation number (that is, arm distribution) generally varied between 5 and 10.

The yields of the simple symmetrical 'cross' micelles and unsymmetrical 'cross' micelles (Fig. 2) by TEM were *ca.* 80 and 90%, respectively.

**Preparation of complex 'cross' and 'windmill' supermicelles.** The first level 'cross' supermicelles prepared from $H_A$-N-$H_A$ triblock comicelle (M(PFS$_{25}$-$b$-P2VP$_{250}$)-$b$-M(PFS$_{36}$-$b$-PMVS$_{324}$)-$b$-M(PFS$_{25}$-$b$-P2VP$_{250}$)), which were initially prepared in a mixed solvent consisting of $n$-hexane/$i$-PrOH = 1:4 (v/v) by the seeded growth method. The solvent was slowly switched to MeOH by changing the solvent stepwise (100% of $i$-PrOH, $i$-PrOH/MeOH = 8:2, 6:4, 4:6, 2:8 and eventually 100% of MeOH). This slow solvent switching protocol was designed to gradually induce the aggregation of insoluble block and thus the formation of 'cross' supermicelles. By counting from TEM images, $ca.$ 80% of the objects observed from TEM are 'cross' supermicelles. Most of ($ca.$ 90%) the $H_A$-N-$H_A$ triblock comicelles formed 'cross' supermicelles, and the others either remained as individual cylindrical comicelles in the solution or formed larger aggregates consisting of multiple triblock comicelles.

The MeOH solution of the 'cross' supermicelles (1.0 ml) was degassed by $N_2$ bubbling for 30 min before being transferred into a glovebox, where 10 µl of 2,2-dimethoxy-2-phenylacetophenone in MeOH (10 mg/ml) was added. The solution was subsequently subjected to UV irradiation for 30 min with occasional agitation. The collapsed and densely packed PMVS chains were crosslinked on UV irradiation.

After crosslinking of the PMVS block, the sample solution was dialyzed against $i$-PrOH to remove MeOH. After crosslinking, the 'cross' supermicelles were stable in $i$-PrOH (Supplementary Fig. 4b). M(PFS$_{20}$-$b$-PtBA$_{280}$) segments were grown via seeded growth by adding THF solution of PFS$_{20}$-$b$-PtBA$_{280}$ unimers into the solution. The M(PFS$_{25}$-$b$-P2VP$_{250}$) segments were further crosslinked by Karstedt's catalyst. Typically, a mass ratio of 1:2 (pyridyl to Pt) was used. Subsequently, another M(PFS$_{25}$-$b$-P2VP$_{250}$) segment was grown onto the open termini of the M(PFS$_{20}$-$b$-PtBA$_{280}$) segments via seeded growth, after which M(PFS$_{20}$-$b$-PMVSOH$_{120}$) seeds were added into the solution with a ratio of 25:1 (hydroxyl/pyridyl). The M(PFS$_{20}$-$b$-PMVSOH$_{120}$) seeds adsorbed onto the uncrosslinked M(PFS$_{25}$-$b$-P2VP$_{250}$) segments due to H-bonding interactions. Into the solution fresh PFS$_{20}$-$b$-PtBA$_{280}$ was added again, which grew from the open termini, and finally 'windmill'-shaped supermicelles were obtained.

As a result of the efficiency of living CDSA, the 'cross' supermicelles were successfully and quantitatively converted into 'windmill' supermicelles based on TEM images. Thus, about 90% of the supermicelles were 'windmill' supermicelles. However, to avoid having the $H_D$ seeds interact with multiple composite 'cross' supermicelles, a slight excess of $H_D$ seeds was used (Fig. 3f,h). The $H_D$ seeds that did not become attached to the composite 'cross' supermicelles were also able to initiate the growth of N segments and form N-$H_D$-N triblock comicelles once PFS-$b$-PtBA unimers were added into the solution (Supplementary Fig. 4e, marked with red circles). Thus, by counting from TEM images, $ca.$ 75% of the objects observed from TEM were 'windmill' supermicelles. By counting the individual N-$H_D$-N triblock comicelles and those present on the 'windmill' supermicelles from TEM images, we estimate that $ca.$ 95% of the N-$H_D$-N triblock comicelles present were incorporated into the 'windmill' supermicelles.

**Preparation of fluorescent 'windmill' supermicelles.** The fluorescent 'windmill' supermicelles were prepared in the same way as the normal ones, except fluorescent dye-containing PFS$_{20}$-$b$-PtBA$_{280}$ unimers were used. Specifically, after the crosslinking of the $H_A$ segments, as shown in Fig. 3d, dye-terminated PFS$_{20}$-$b$-PtBA$_{280}$-red unimers were added to extend the PFS-$b$-PtBA arms. Subsequently, PFS$_{20}$-b-P2VP$_{250}$ unimers were added to further extend the arms, after which M(PFS$_{20}$-$b$-PMVSOH$_{120}$) seeds were added into the solution with a ratio of 25:1 (hydroxyl/pyridyl). The M(PFS$_{20}$-$b$-PMVSOH$_{120}$) seeds adsorbed onto the uncrosslinked M(PFS$_{25}$-$b$-P2VP$_{250}$) segments. Finally, dye-terminated PFS$_{20}$-$b$-PtBA$_{280}$ unimers were added into the solution to generate the final 'windmill' supermicelles.

**General techniques.** Photolysis experiments were carried out with Pyrex-glass filtered emission from a water-cooled, 125-W medium-pressure mercury lamp (Photochemical Reactors Ltd.), emitting predominantly 365 nm. The emission lines of the mercury lamp were 577–579, 546, 436, 408–405, 366–365, 334, 313, 302, 297, 289, 280, 270, 265 and 254 nm. An ethylene glycol/deionized water bath in conjunction with a thermostat was used to maintain constant temperatures during photoirradiation. All the photoreactions were carried out at 20 °C.

All the $^1$H NMR characterization was carried out on a Varian 400 MHz instrument, with chemical shifts referenced to tetramethylsilane in $d$-chloroform (CDCl$_3$). FTIR spectra were recorded with Cary 630 FTIR spectrometer fitted with an attenuated total reflection accessory (Agilent Technologies).

**Confocal fluorescence microscopy.** Confocal imaging was performed using a Leica SP5 system attached to a Leica DMI6000 inverted epifluorescence microscope with a × 63 (NA 1.4) oil immersion objective lens. Fluorophores from PFS-$b$-PtBA-red and PFS-$b$-PtBA-green were excited using a HeNe laser operating at 594 nm and an argon laser operating at 488 nm, respectively. Confocal images were obtained using digital detectors with observation windows of 605–700 nm for the red dye and 500–570 nm for the green dye. The resulting outputs were obtained as digital false-colour images, which are colour coded using the chromaticity of each micelle block under UV irradiation at 365 nm, as observed by fluorescence

spectroscopy. For imaging of multicomponent supermicelles, the output power of each laser was varied until the fluorescence of all blocks could be observed at approximately equal brightness.

## References

1. Fratzl, P. & Weinkamer, R. Nature's hierarchical materials. *Prog. Mater. Sci.* **52**, 1263–1334 (2007).
2. Whitesides, G. M. & Grzybowski, B. Self-assembly at all scales. *Science* **295**, 2418–2421 (2002).
3. Mann, S. Self-assembly and transformation of hybrid nano-objects and nanostructures under equilibrium and non-equilibrium conditions. *Nat. Mater.* **8**, 781–792 (2009).
4. Glotzer, S. C. & Solomon, M. J. Anisotropy of building blocks and their assembly into complex structures. *Nat. Mater.* **6**, 557–562 (2007).
5. Zhang, L. F. & Eisenberg, A. Multiple morphologies of crew-cut aggregates of polystyrene-$b$-poly(acrylic acid) block-copolymers. *Science* **268**, 1728–1731 (1995).
6. Zhang, L. F., Yu, K. & Eisenberg, A. Ion-induced morphological changes in "crew-cut" aggregates of amphiphilic block copolymers. *Science* **272**, 1777–1779 (1996).
7. Pochan, D. J. et al. Toroidal triblock copolymer assemblies. *Science* **306**, 94–97 (2004).
8. Cui, H. G., Chen, Z. Y., Zhong, S., Wooley, K. L. & Pochan, D. J. Block copolymer assembly via kinetic control. *Science* **317**, 647–650 (2007).
9. Li, Z. B., Kesselman, E., Talmon, Y., Hillmyer, M. A. & Lodge, T. P. Multicompartment micelles from ABC miktoarm stars in water. *Science* **306**, 98–101 (2004).
10. Dupont, J., Liu, G. J., Niihara, K., Kimoto, R. & Jinnai, H. Self-assembled ABC triblock copolymer double and triple helices. *Angew. Chem. Int. Ed.* **48**, 6144–6147 (2009).
11. Erhardt, R. et al. Amphiphilic Janus micelles with polystyrene and poly(methacrylic acid) hemispheres. *J. Am. Chem. Soc.* **125**, 3260–3267 (2003).
12. Gröschel, A. H. et al. Facile, solution-based synthesis of soft, nanoscale Janus particles with tunable Janus balance. *J. Am. Chem. Soc.* **134**, 13850–13860 (2012).
13. Cheng, L., Zhang, G. Z., Zhu, L., Chen, D. Y. & Jiang, M. Nanoscale tubular and sheetlike superstructures from hierarchical self-assembly of polymeric Janus particles. *Angew. Chem. Int. Ed.* **47**, 10171–10174 (2008).
14. Nie, L., Liu, S. Y., Shen, W. M., Chen, D. Y. & Jiang, M. One-pot synthesis of amphiphilic polymeric Janus particles and their self-assembly into supermicelles with a narrow size distribution. *Angew. Chem. Int. Ed.* **46**, 6321–6324 (2007).
15. Walther, A. et al. Self-assembly of Janus cylinders into hierarchical superstructures. *J. Am. Chem. Soc.* **131**, 4720–4728 (2009).
16. Fang, B. et al. Undulated multicompartment cylinders by the controlled and directed stacking of polymer micelles with a compartmentalized corona. *Angew. Chem. Int. Ed.* **48**, 2877–2880 (2009).
17. Gröschel, A. H. et al. Precise hierarchical self-assembly of multicompartment micelles. *Nat. Commun.* **3**, 710 (2012).
18. Gröschel, A. H. et al. Guided hierarchical co-assembly of soft patchy nanoparticles. *Nature* **503**, 247–251 (2013).
19. Qiu, H. B., Hudson, Z. M., Winnik, M. A. & Manners, I. Multidimensional hierarchical self-assembly of amphiphilic cylindrical block comicelles. *Science* **347**, 1329–1332 (2015).
20. Kostiainen, M. A., Kasyutich, O., Cornelissen, J. J. L. M. & Nolte, R. J. M. Self-assembly and optically triggered disassembly of hierarchical dendron-virus complexes. *Nat. Chem.* **2**, 394–399 (2010).
21. Hirschberg, J. H. K. K. et al. Helical self-assembled polymers from cooperative stacking of hydrogen-bonded pairs. *Nature* **407**, 167–170 (2000).
22. Prins, L. J., De Jong, F., Timmerman, P. & Reinhoudt, D. N. An enantiomerically pure hydrogen-bonded assembly. *Nature* **408**, 181–184 (2000).
23. Sijbesma, R. P. et al. Reversible polymers formed from self-complementary monomers using quadruple hydrogen bonding. *Science* **278**, 1601–1604 (1997).
24. Hartgerink, J. D., Beniash, E. & Stupp, S. I. Self-assembly and mineralization of peptide-amphiphile nanofibers. *Science* **294**, 1684–1688 (2001).
25. Albertazzi, L. et al. Probing exchange pathways in one-dimensional aggregates with super-resolution microscopy. *Science* **344**, 491–495 (2014).
26. Douglas, S. M. et al. Self-assembly of DNA into nanoscale three-dimensional shapes. *Nature* **459**, 414–418 (2009).
27. Lo, P. K. et al. Loading and selective release of cargo in DNA nanotubes with longitudinal variation. *Nat. Chem.* **2**, 319–328 (2010).
28. Goodman, R. P. et al. Rapid chiral assembly of rigid DNA building blocks for molecular nanofabrication. *Science* **310**, 1661–1665 (2005).
29. Valkama, S. et al. Self-assembled polymeric solid films with temperature-induced large and reversible photonic-bandgap switching. *Nat. Mater.* **3**, 872–876 (2004).

30. Ruokolainen, J. *et al.* Switching supramolecular polymeric materials with multiple length scales. *Science* **280**, 557–560 (1998).

31. Zhao, Y. *et al.* Small-molecule-directed nanoparticle assembly towards stimuli-responsive nanocomposites. *Nat. Mater.* **8**, 979–985 (2009).

32. Tang, C. B., Lennon, E. M., Fredrickson, G. H., Kramer, E. J. & Hawker, C. J. Evolution of block copolymer lithography to highly ordered square arrays. *Science* **322**, 429–432 (2008).

33. Jia, L. *et al.* A design strategy for the hierarchical fabrication of colloidal hybrid mesostructures. *Nat. Commun.* **5**, 3882 (2014).

34. Hudson, Z. M., Lunn, D. J., Winnik, M. A. & Manners, I. Colour-tunable fluorescent multiblock micelles. *Nat. Commun.* **5**, 3372 (2014).

35. Rupar, P. A., Chabanne, L., Winnik, M. A. & Manners, I. Non-centrosymmetric cylindrical micelles by unidirectional growth. *Science* **337**, 559–562 (2012).

36. Wang, X. S. *et al.* Cylindrical block copolymer micelles and co-micelles of controlled length and architecture. *Science* **317**, 644–647 (2007).

37. Gilroy, J. B. *et al.* Monodisperse cylindrical micelles by crystallization-driven living self-assembly. *Nat. Chem.* **2**, 566–570 (2010).

38. Hudson, Z. M. *et al.* Tailored hierarchical micelle architectures using living crystallization-driven self-assembly in two dimensions. *Nat. Chem.* **6**, 893–898 (2014).

39. Freitas, L. D., Jacobi, M. M., Gonçalves, G. & Stadler, R. Microphase separation induced by hydrogen bonding in a poly(1,4-butadiene)-*block*-poly(1,4-isoprene) diblock copolymers—an example of supramolecular organization via tandem interactions. *Macromolecules* **31**, 3379–3382 (1998).

40. O'Reilly, R. K., Hawker, C. J. & Wooley, K. L. Cross-linked block copolymer micelles: functional nanostructures of great potential and versatility. *Chem. Soc. Rev.* **35**, 1068–1083 (2006).

41. Qian, J. *et al.* Uniform, high aspect ratio fiber-like micelles and block co-micelles with a crystalline pi-conjugated polythiophene core by self-seeding. *J. Am. Chem. Soc.* **136**, 4121–4124 (2014).

42. Zhang, W. *et al.* Supramolecular linear heterojunction composed of graphite-like semiconducting nanotubular segments. *Science* **334**, 340–343 (2011).

43. Ogi, S., Sugiyasu, K., Manna, S., Samitsu, S. & Takeuchi, M. Living supramolecular polymerization realized through a biomimetic approach. *Nat. Chem.* **6**, 188–195 (2014).

44. Bu, L., Dawson, T. J. & Hayward, R. C. Tailoring ultrasound-induced growth of perylene diimide nanowire crystals from solution by modification with poly(3-hexyl thiophene). *ACS Nano* **9**, 1878–1885 (2015).

45. Görl, D., Zhang, X., Stepanenko, V. & Würthner, F. Supramolecular block copolymers by kinetically controlled co-self-assembly of planar and core-twisted perylene bisimides. *Nat. Commun.* **6**, 7009 (2015).

46. Pal, A. *et al.* Controlling the structure and length of self-synthesizing supramolecular polymers through nucleated growth and disassembly. *Angew. Chem. Int. Ed.* **54**, 7852 (2015).

47. Sun, L. *et al.* Structural reorganization of cylindrical nanoparticles triggered by polylactide stereocomplexation. *Nat. Commun.* **5**, 6746 (2014).

48. Schmelz, J., Schedl, A. E., Steinlein, C., Manners, I. & Schmalz, H. Length control and block-type architectures in worm-like micelles with polyethylene cores. *J. Am. Chem. Soc.* **134**, 14217–14225 (2012).

49. Massey, J. A. *et al.* Self-assembly of organometallic block copolymers: the role of crystallinity of the core-forming polyferrocene block in the micellar morphologies formed by poly(ferrocenylsilane-*b*-dimethylsiloxane) in n-alkane solvents. *J. Am. Chem. Soc.* **122**, 11577–11584 (2000).

50. Wang, H., Winnik, M. A. & Manners, I. Synthesis and self-assembly of poly(ferrocenyldimethylsilane-*b*-2-vinylpyridine) diblock copolymers. *Macromolecules* **40**, 3784–3789 (2007).

51. McGrath, N. *et al.* Synthesis and crystallization-driven solution self-assembly of polyferrocenylsilane diblock copolymers with polymethacrylate corona-forming blocks. *Polym. Chem.* **5**, 1923–1929 (2014).

52. Lunn, D. J. *et al.* Controlled thiol-ene functionalization of polyferrocenylsilane-*block*-polyvinylsiloxane copolymers. *Macromol. Chem. Phys.* **214**, 2813–2820 (2013).

53. Lee, L.-T., Woo, E. M., Hou, S. S. & Förster, S. Miscibility with positive deviation in $T_g$-composition relationship in blends of poly(2-vinyl pyridine)-*block*-poly(ethylene oxide) and poly(*p*-vinyl phenol). *Polymer* **47**, 8350–8359 (2006).

54. Cesteros, L. C., Isasi, J. R. & Katime, I. Hydrogen-bonding in poly(4-vinylpyridine)/poly(vinyl acetate-*co*-vinyl alcohol) blends. An infrared study. *Macromolecules* **26**, 7256–7262 (1993).

55. Cesteros, L. C., Isasi, J. R. & Katime, I. Study of the miscibility of poly(vinyl pyridines) with poly(vinyl acetate), poly(vinyl alcohol) and their copolymers. *J. Polym. Sci. Pol. Phys.* **32**, 223–230 (1994).

## Acknowledgements

X.L. is grateful to the European Union (EU) for a Marie Curie Postdoctoral Fellowship. C.E.B. thanks the Bristol Chemical Synthesis Centre for Doctoral Training funded by the Engineering and Physical Sciences Research Council (EPSRC) for a PhD studentship. M.A.W. thanks the Natural Sciences and Engineering Research Council of Canada for financial support. I.M. thanks the EU for a European Research Council Advanced Investigator Grant. We also thank the Wolfson Bioimaging Facility at the University of Bristol for the use of the confocal microscopy facilities.

## Author contributions

X.L., Y.G. and I.M. conceived the project and wrote the manuscript with input from M.A.W. X.L., Y.G. and C.E.B. carried out the experiments. I.M. supervised the project with input from M.A.W.

## Additional information

**Competing financial interests:** The authors declare no competing financial interests.

# 3

# Design and synthesis of digitally encoded polymers that can be decoded and erased

Raj Kumar Roy[1], Anna Meszynska[1], Chloé Laure[1], Laurence Charles[2], Claire Verchin[1] & Jean-François Lutz[1]

Biopolymers such as DNA store information in their chains using controlled sequences of monomers. Here we describe a non-natural information-containing macromolecule that can store and retrieve digital information. Monodisperse sequence-encoded poly(alkoxyamine amide)s were synthesized using an iterative strategy employing two chemoselective steps: the reaction of a primary amine with an acid anhydride and the radical coupling of a carbon-centred radical with a nitroxide. A binary code was implemented in the polymer chains using three monomers: one nitroxide spacer and two interchangeable anhydrides defined as 0-bit and 1-bit. This methodology allows encryption of any desired sequence in the chains. Moreover, the formed sequences are easy to decode using tandem mass spectrometry. Indeed, these polymers follow predictable fragmentation pathways that can be easily deciphered. Moreover, poly(alkoxyamine amide)s are thermolabile. Thus, the digital information encrypted in the chains can be erased by heating the polymers in the solid state or in solution.

[1] Precision Macromolecular Chemistry, Institut Charles Sadron, UPR22-CNRS, BP84047, 23 rue du Loess, 67034 Strasbourg Cedex 2, France. [2] Aix-Marseille Université - CNRS, UMR 7273, Institute of Radical Chemistry, 13397 Marseille Cedex 20, France. Correspondence and requests for materials should be addressed to J.-F.L. (email: jflutz@unistra.fr).

Synthetic polymers are today widely used in a large range of commodity and specialty products. Still, in most known applications, polymers are chosen for their unique physicochemical, structural and mechanical properties. However, in principle, synthetic polymers could exhibit more sophisticated molecular properties comparable to those of biopolymers. DNA, for example, stores information in its chains using a molecular code based on four different monomer units[1]. It was recently shown that artificial DNA oligomers can be used to store and retrieve a large amount of information[2,3]. However, DNA is not the only polymer allowing molecular encoding. In theory, any copolymer containing ordered co-monomer sequences could be used to store molecular information[4–7]. Indeed, a string of information can be implemented in a polymer using two co-monomers intentionally defined as 0-bit and 1-bit. However, reliable approaches for writing and reading molecular information in synthetic polymers are still missing[8,9]. Here we report a facile and rapid iterative method for preparing sequence-encoded polymers. This strategy relies on the use of two successive chemoselective coupling steps, that is, the reaction of a primary amine with a symmetric acid anhydride and the radical coupling of a carbon-centred radical with a nitroxide. This approach allows synthesis of monodisperse sequence-encoded poly(alkoxyamine amide)s. These polymers contain labile alkoxyamine linkages and are therefore easy to sequence using tandem mass spectrometry. A universal methodology for reading their encoded sequences is proposed in this work. Moreover, these polymers are thermolabile and therefore dynamic at temperatures above 60 °C. Thus, the molecular information encrypted in the chains can be erased by heating the polymers in the solid state or in solution.

## Results

**Chemoselective iterative synthesis of oligo(alkoxamine amide)s.** In order to store molecular information, synthetic polymers have to be monodisperse and exhibit controlled sequences of co-monomers[4]. Various strategies for preparing sequence-defined polymers have been reported[8]. For instance, monodisperse sequence-defined oligomers can be synthesized using template strategies[10,11] or molecular machines[12]. However, the most-widespread approach relies on the stepwise coupling of monomer units on a solid support[13]. This approach is efficient but suffers from limitations: each coupling step should be nearly quantitative and defect-free; protecting groups are often required to control the reactivity of the monomers; and time-consuming cycles involving coupling, washing and deprotection steps are needed for each monomer attachment. While the former aspect is mandatory in iterative chemistry, the latter two can be avoided. Protecting groups are necessary for the synthesis of artificial biopolymers such as peptides and oligonucleotides but are not strictly required for the synthesis of non-natural oligomers, which can be achieved using chemoselective strategies[14–18]. Moreover, the duration of coupling cycles can be reduced by selecting fast reactions. For instance, the chemistry used in the present work is shown in Fig. 1a. It involves two different types of building blocks, each of them containing two kinds of reactive groups. The first building block comprises an acyclic symmetric acid anhydride and alkyl bromides, whereas the other contains a nitroxyl radical and a primary amine. Our concept is based on the fact that two chemoselective reactions can occur using these co-monomers. The first step is the reaction of a symmetric anhydride with a primary amine to form an N-substituted amide[19]. The second step is the coupling of a carbon-centred radical, obtained by copper activation of an alkyl bromide, with a nitroxide to afford an alkoxyamine. Such a radical–radical coupling reaction

has been previously used for spin-trapping[20] and alkoxyamine synthesis[21]. The alternating iterative use of the two coupling steps allows protecting-group-free synthesis of monodisperse oligomers on an amine- or bromine-functionalized solid support. An important advantage of this strategy is the fact that both coupling steps proceed in high yields within short time frames at room temperature. For instance, it is known that the radical–radical coupling of carbon-centred radicals with 2,2,6,6-tetramethylpiperidinyloxy (TEMPO) derivatives occur within a few minutes without noticeable by-product formation[22]. The reaction of an acid anhydride with a primary amine is also generally fast, but the obtained yields depend on experimental conditions. Supplementary Fig. 1 shows the characterization of model reactions, in which cyclohexylamine was reacted with the anhydride of either 2-bromo-isobutyric acid (**a-1**) or 2-bromopropionic acid (**a-0**). In the presence of $K_2CO_3$ and of an anhydride excess[23], quantitative yields were obtained in a few minutes. Moreover, no imide by-products could be detected using mass spectrometry analysis.

These preliminary considerations indicated that the two coupling steps shown in Fig. 1a are suitable for iterative solid-phase synthesis. In order to validate this experimentally, the alternating iterative oligomerization of amino-TEMPO (T) and **a-1** was first studied on a commercial glycine-loaded Wang resin (**S1**). Successive coupling steps were performed repeatedly on the resin. After reaching a given number of steps, the formed oligomers were cleaved from the solid support and analysed using mass spectrometry, size-exclusion chromatography (SEC) and NMR. Electrospray high-resolution mass spectrometry (ESI-HRMS) indicated that the formed oligomers are monodisperse (Supplementary Table 1 and Fig. 2d). This was also confirmed using the SEC analysis. Figure 2a shows the SEC chromatograms recorded for oligomers isolated after three, five, seven and nine iterative steps. In all cases, a sharp monomodal peak with a narrow molecular weight distribution ($1.005 < M_w/M_n < 1.01$) was observed. It should be noted that the apparent polydispersity index of monodisperse polymers is never equal to 1 in SEC because of axial dispersion and peak broadening[24,25]. Moreover, the molecular structure of the oligo(alkoxyamine amide)s was confirmed with NMR analysis (Supplementary Fig. 2 and Supplementary Fig. 3). It is interesting to note that, despite the fact that **1** generates a tertiary radical, the presence of unsaturated species due to H-transfer side reaction could not be detected in $^{13}C$ NMR and was only marginally observed in ESI-HRMS[26]. These observations are in agreement with previous radical–radical coupling studies involving tertiary radicals, which indicate that this side reaction is prominent at elevated temperatures but disfavoured at milder temperatures[21,27]. On the whole, these results confirm that cycles involving successive amine–anhydride and radical–radical coupling steps are adequate for iterative synthesis. Still, it should be mentioned that, after performing a certain number of iterative steps (that is, from step 6 and beyond), incomplete yields (that is, ~95%) were often observed for the radical coupling reaction when very short reaction times were used. However, this problem was easily solved by performing this reaction under microwave irradiation.

Besides conventional solid-phase chemistry, the oligo(alkoxyamine amide)s can also be easily prepared in other conditions. The iterative oligomerization of amino-TEMPO and **a-1** was also tested on polystyrene-based soluble supports. It is known that soluble polymer supports allow better reagent accessibility and sometimes higher yields than resins[28]. A glycine-loaded Wang soluble polystyrene support (**S2**) was tested for the synthesis of the oligo(alkoxyamine amide)s. It was found that this support allows a reliable synthesis of monodisperse alternating oligomers (Supplementary Table 1). As shown in Fig. 2b and Supplementary

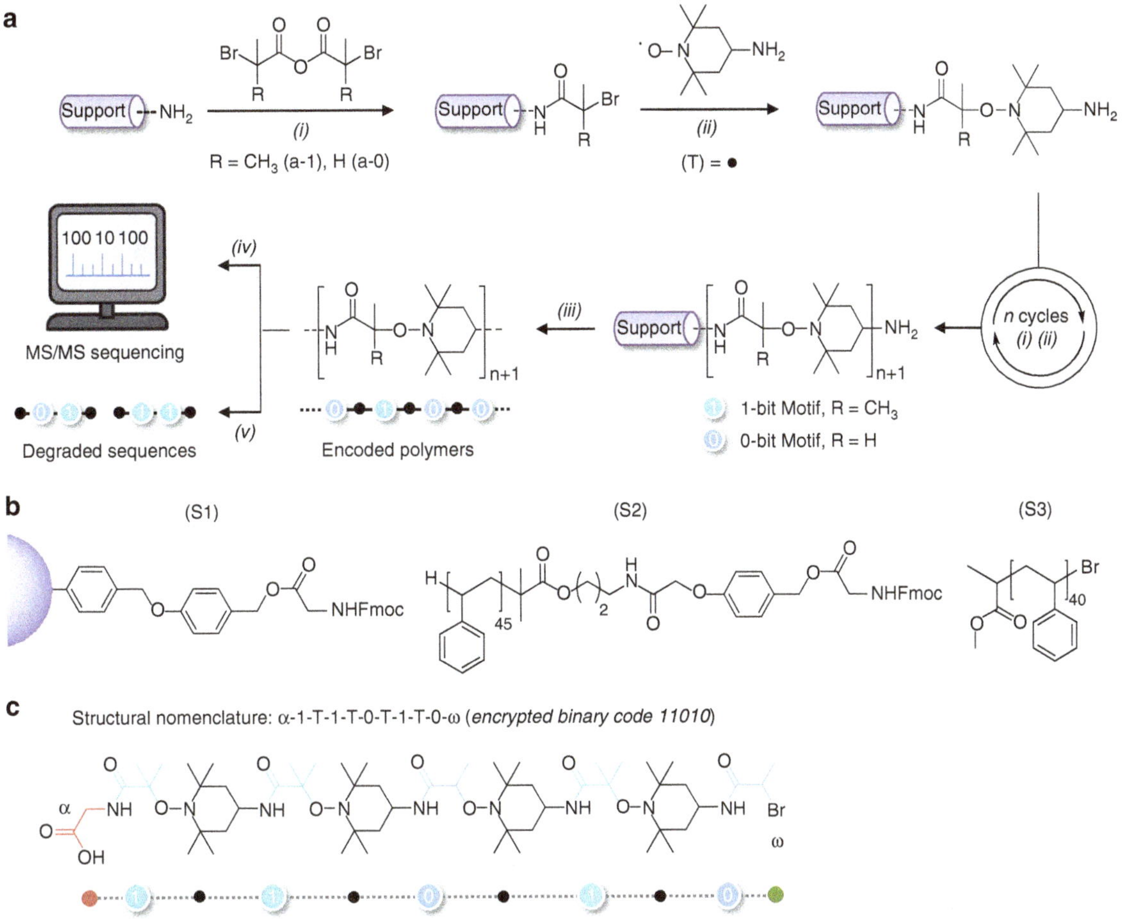

**Figure 1 | General concept for the synthesis of sequence-encodable polymers. (a)** Strategy studied for the synthesis of oligo(alkoxyamine amide)s. This approach is based on two chemoselective reactions. The first *(i)* involves a primary amine and a symmetric bromo-functionalized anhydride, whereas the second *(ii)* is the radical coupling of a carbon-centred radical (obtained *in situ* by copper activation of an alkyl bromide) with an amino-functionalized nitroxide. Two symmetric anhydrides containing motifs defined as 0-bit and 1-bit can be used in an interchangeable manner in step *(i)* to create a binary code on the polymer chains. Experimental conditions: *(i)* THF, DIPEA or $K_2CO_3$, *(ii)* CuBr, $Me_6$TREN, DMSO, *(iii)* TFA, $CH_2Cl_2$, *(iv)* ESI-MS/MS sequencing and *(v)* heating in the solid state or in solution. **(b)** Molecular structures of the solid-phase or soluble polymer supports used in this work. **(c)** Terminology and nomenclature used in the present work for the synthesis and sequencing of poly(alkoxyamine amide)s. The displayed example shows an oligomer containing the binary sequence 11010.

Fig. 4, oligomers of comparable quality could be obtained on **S1** and **S2**. The iterative chemistry was also investigated on a non-cleavable support (**S3**) prepared by atom transfer radical polymerization (ATRP)[29]. Owing to the mechanism of ATRP, **S3** is terminated by an alkyl bromide that can be used as initiator for the oligo(alkoxyamine amide) synthesis. Figure 2c shows a series of chromatograms obtained for a multistep synthesis on **S3**. It should be specified that these chromatograms were all recorded after performing the amine–anhydride step. Indeed, after the radical coupling step, the polymers possess a primary amine ω-chain end that interacts with SEC columns. A clean multistep growth can be observed in that figure. Each iteration resulted in an apparent molecular weight increase of ~300–400 g mol$^{-1}$ that roughly corresponds to the theoretical molecular weight that should be gained after performing one amine–anhydride step and one radical–radical coupling step (Supplementary Table 2). Comparable molecular weight increments were also observed on **S2** (Supplementary Fig. 4a). It is also important to note that the shape of the chromatograms does not change with the iterative growth (that is, the molecular weight distribution of the support-bound oligomer is similar to the one of the bare support). However, after 10 iteration steps on **S3**, a shoulder appears in the chromatograms at high elution volume. This peak

is not due to the iterative process but corresponds most probably to the dead chains formed during the ATRP process[30]. Indeed, these chains do not contain any terminal halogen and therefore do no initiate multistep growth. Besides SEC, NMR monitoring confirmed the iterative attachment of T and **1** to **S3**. These results account for a smooth formation of oligo(alkoxyamine amide)s on the soluble support.

**Synthesis and MS/MS sequencing of binary-coded polymers.** The chemistry shown in Fig. 1a was extended to the synthesis of sequence-encoded polymers. In order to encode a binary message in the polymer chains, three monomers were used: amino-TEMPO, **a-0** and **a-1**. The interchangeable use of **a-0** and **a-1** in the amine–anhydride step allows sequence-encoding. It should be noted that the nitroxide units could also be used to implement a code in the chains (that is, using interchangeable nitroxides in the radical coupling step). However, an anhydride-based language was chosen for the present work. At first, a series of eight model pentamers containing all possible binary triads was synthesized on **S1** (Fig. 3a). These oligomers were synthesized in five steps using three encoding building blocks and two nitroxide spacers. ESI-HRMS and SEC indicated the formation of monodisperse oligomers in all cases. It should be specified that, under similar

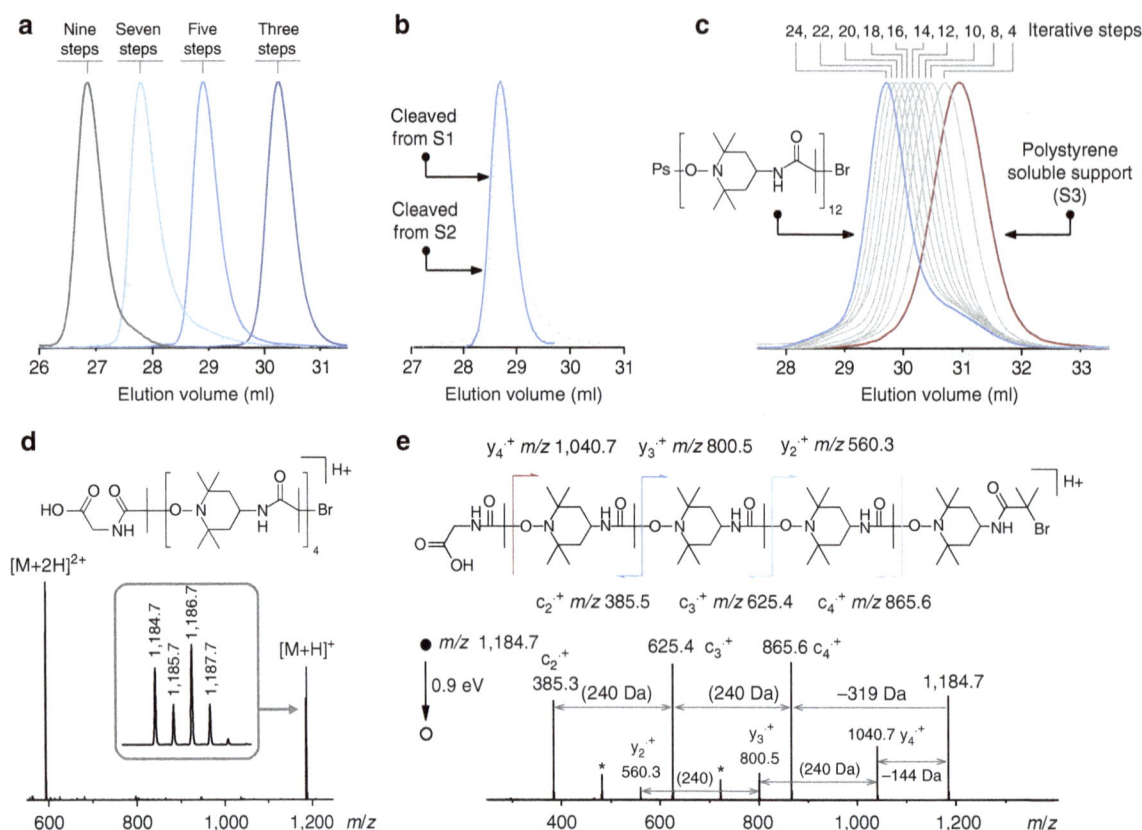

**Figure 2 | Characterization of oligo(alkoxyamine amide)s prepared using amino-TEMPO and 2-bromo-isobutyric anhydride.** (**a**) SEC chromatograms recorded in THF for oligomers cleaved from support **S1** after three, five, seven and nine iterative steps. These chromatograms correspond to Entries, 2, 11, 13 and 16 in Supplementary Table 1. (**b**) SEC chromatograms recorded in THF for oligomers obtained after five iterative steps on the solid support **S1** (dotted purple line) or on the soluble support **S2** (full dark-blue line). These data correspond to Entries 2-3 in Supplementary Table 1. (**c**) SEC chromatograms recorded in THF for the iterative of oligo(alkoxyamine amide)s on the soluble support **S3**. The numbers listed above the chromatograms denote the number of iterative steps at which the analysis was performed. It should be noted that Fig. 2a,b was recorded on a SEC set-up for oligomer analysis, whereas Fig. 2c was recorded on a SEC set-up for high molecular weight analysis. Therefore, the displayed elution volumes are not comparable in all cases. (**d**) Positive mode ESI-MS spectrum recorded for an oligomer obtained after nine iterative steps on solid support **S1** (Supplementary Table 1, Entry 16). The inset shows the isotopic pattern of the $[M+H]^+$ ion. (**e**) The ESI-MS/MS spectrum of an oligomer obtained after nine iterative steps on solid support **S1** (Supplementary Table 1, Entry 16). This MS/MS spectrum was obtained after collisional activation of the $m/z$ 1,184.7 precursor ion containing the $^{79}$Br isotope. Stars indicate secondary fragments such as $c_4^+ - 144$ and $c_3^+ - 144$.

experimental conditions, the moiety **0**-Br leads generally to lower yields than **1**-Br in the radical coupling step[21]. However, quantitative yields were obtained using an optimized amount of catalyst. Very interestingly, the sequence-encoded copolymers can be easily decoded. Various sequencing tools have been developed for proteomics and genomics during the past decades[9]. In particular, tremendous advances have been achieved in the field of biopolymers owing to tandem mass spectrometry (MS/MS) allowing their reliable, fast and automated sequencing based on universal dissociation rules[31]. Since MS/MS dissociation patterns strongly depend on the chemistry of the polymer backbone, no universal rules can however be envisaged for non-natural species. Instead, one has to refer to the specific fragmentation behaviour established for the targeted polymer family to decipher structural information from MS/MS spectra[32]. Using MS/MS, it was found here that oligo(alkoxyamine amide)s are remarkably easy to sequence. Owing to the lability of alkoxyamine linkages, there is no alternative dissociation reaction efficiently competing with the low-energy homolytic cleavage of those bonds between the TEMPO co-monomer and any **0** or **1** unit. Moreover, the nature of the coding units has no influence on this main fragmentation pathway. As a result, oligo(alkoxyamine amide)s dissociate into easy-to-read fragments (Supplementary

Methods). In order to understand the basic principles of their sequencing, isomers containing different binary triads were first studied using MS/MS (Fig. 3b,c and Supplementary Figure 5). Depending on the analysed sequence, four fragments corresponding to the release of motifs α-**0** ( − 130 Da), α-**1** ( − 144 Da), ω-**0** ( − 305 Da) or ω-**1** ( − 319 Da) as radicals were easily identified in MS/MS spectra. For longer sequences, fragment series reveal the primary structure of the polymers (Fig. 2e and Supplementary Fig. 5f): intervals of 226 or 240 Da between α- or ω-containing fragments indicate the presence in the sequence of the motif T-**0** or T-**1**, respectively. The same fragmentation scenario was found in all analysed samples and can therefore be utilized for facile poly(alkoxyamine amide) sequencing. In order to illustrate the universality of this approach, a blind sequencing experiment was performed. A sequence-coded polymer (Supplementary Table 1, Entry 19) was synthesized and given for MS/MS analysis, but the sequence was not communicated to the MS experimenter. Nevertheless, the monomer-encoded 'secret' message was easily deciphered.

**Thermal degradation of the coded polymers.** Furthermore, the formed polymers can be easily thermally modified. Indeed,

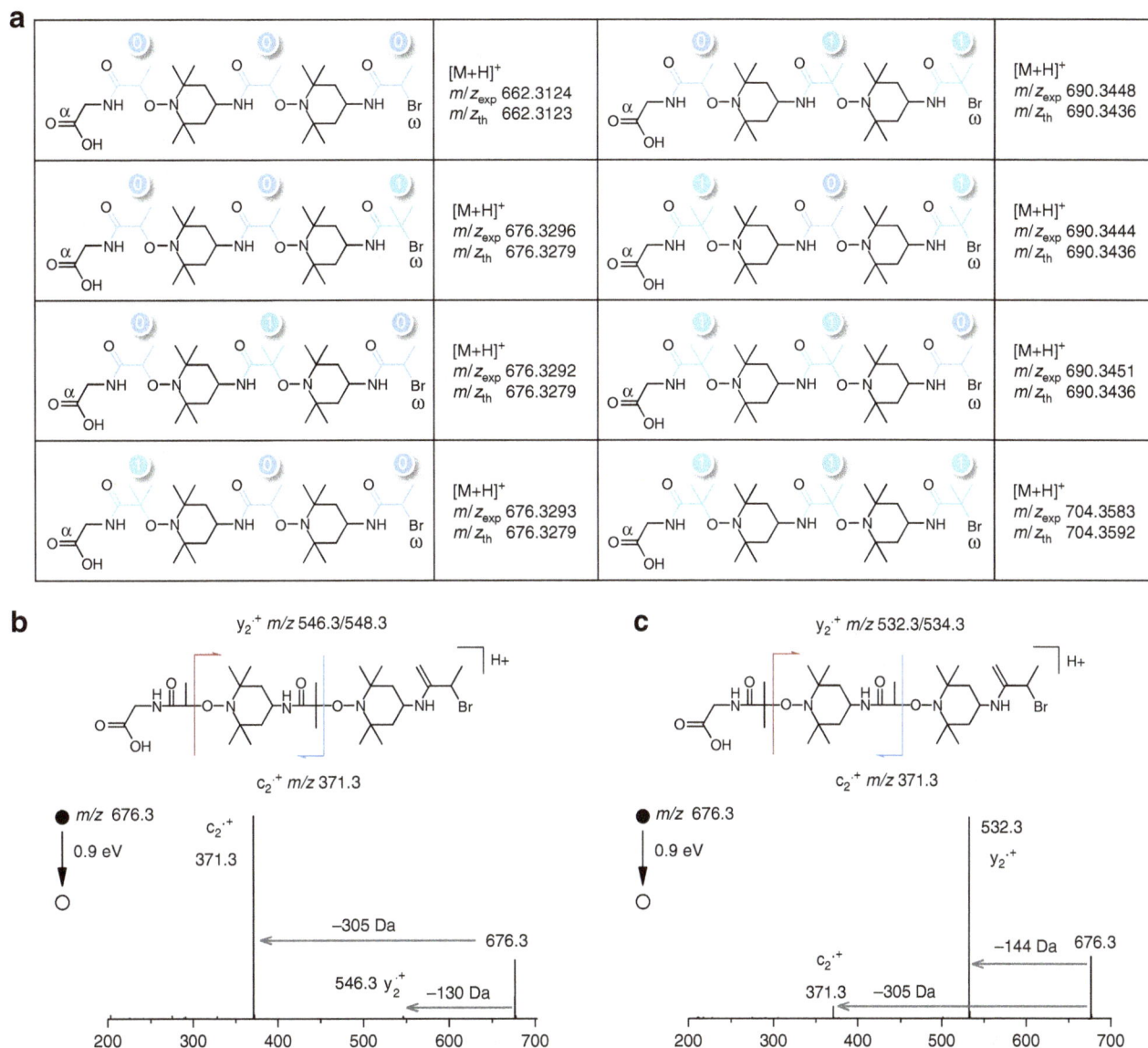

**Figure 3 | Synthesis and sequencing of sequence-encoded oligo(alkoxyamine amide)s.** (**a**) ESI-HRMS characterization of sequence-encoded oligomers containing different binary triads synthesized on solid support **S1**. (**b**) ESI-MS/MS spectrum of a pentamer encoded with the binary triad 010 (Supplementary Table 1, Entry 6). (**c**) ESI-MS/MS spectrum of a pentamer encoded with the binary triad 100 (Supplementary Table 1, Entry 7). Spectra **b** and **c** were both derived from a precursor peak at $m/z$ 676.3 corresponding to the $^{79}$Br isotope of the corresponding oligomer.

polyalkoxyamines are thermolabile polymers that usually exhibit a dynamic behaviour when heated[33,34]. This is due to the homolysis of the C–ON bond that occurs at elevated temperatures[35,36]. For instance, C-TEMPO bonds dissociate, in general, above 60–70 °C (ref. 35). Interestingly, it was observed that sequence-encoded oligo(alkoxyamine amide) do reorganize in the solid state at 90 °C (Supplementary Fig. 6a). The thermal reorganization of these polymers was also studied in tetrahydrofuran (THF) solution at different temperatures. Above 75 °C, the polymers degrade into polydisperse samples. The speed of degradation can be modulated by temperature and proceed within a few hours above 120 °C (Supplementary Fig. 6b). SEC analysis indicated the formation of multimodal species that are formed by random carbon–carbon radical coupling and other side reactions. However, this degradation process can be more finely controlled. For instance, the probability of occurrence of carbon–carbon radical coupling can be reduced when a spin trap such as TEMPO is used in excess (Fig. 4). Under such conditions, a stepwise degradation process

can be obtained (Fig. 4c). SEC analysis (Fig. 4a) indicated the formation of a controlled degradation cascade. Positive mode ESI-HRMS analysis of the final degradation products indicated the presence of the unsaturated degradation residues **6** and **7** and their corresponding hydroxylamines. Compound **8** was detected using negative mode ESI-HRMS. Overall, temperature can be used as a simple trigger to erase the sequence information stored in the polymers. It should be however noted that these materials are stable at room temperature. No sign of degradation was observed after several months of storage under standard laboratory conditions.

## Discussion

Binary-encoded poly(alkoxyamine amide)s have been identified as an interesting new class of synthetic polymers. These macromolecules permit to store a coded message that can be easily deciphered with MS/MS sequencing. In principle, a wide variety of other synthetic polymers could be used to store

**a**

**b**

| Degradation product | $M_{th}$ |
|---|---|
| 1 | 799.47 |
| 2 | 623.43 |
| 3 | 559.29 |
| 4 | 479.36 |
| 5 | 383.24 |
| 6 | 319.10 |
| 7 | 239.18 |
| 8 | 143.06 |

**c**

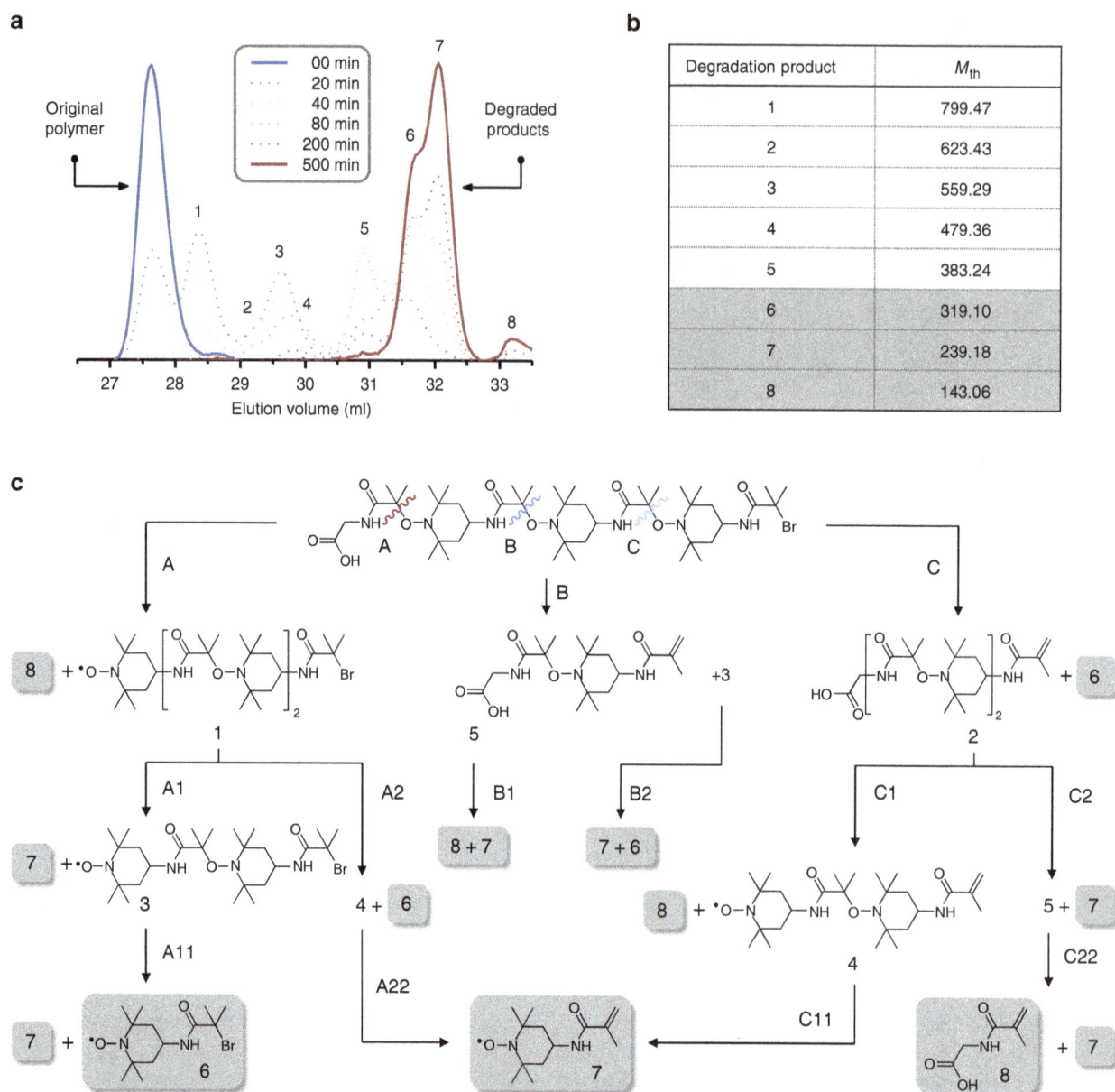

**Figure 4 | Controlled thermal degradation of a monodisperse heptamer performed in anisole solution at 120 °C and in the presence of a large excess of TEMPO.** (**a**) Kinetic monitoring of the degradation process followed by SEC in THF. Dotted blue chromatograms show the early instant of the degradation process, whereas the dotted red chromatograms show later stages. (**b**) Theoretical molecular weight values ($M_{th}$) of the degradation products that may be formed during the degradation process. The corresponding molecular formulae are shown in **c**. (**c**) Proposed degradation pathway. The structures highlighted in grey are final species. It should be noted that the hydroxylamines of 1, 3, 4, 6 and 7 may also be formed.

information[4]. However, sequence-coded poly(alkoxyamine amide)s exhibit important advantages that make them appealing for information-related applications. First of all, these polymers are easy to synthesize. As shown in the Results section, monodisperse sequence-defined poly(alkoxyamine amide)s can be prepared on different types of solid or soluble supports. Moreover, the protecting-group-free iterative approach introduced in this article is straightforward and permits to synthesize these polymers relatively rapidly. It should be also specified that the chain lengths reported here do not represent the upper limit of the approach. Longer polymers can be certainly prepared, in particular using automated robotic equipment. Still, it should be reminded that the efficiency of the coupling steps may decrease with increasing chain length and therefore that modified protocols (for example, using repeated steps or capping steps) might be needed to prepare longer polymers. Another

important advantage of the methodology is the possibility to use different monomer codes. In this first proof-of-concept, a binary code based on anhydrides **a-0** and **a-1** was chosen. However, other monomer-based alphabets can be envisaged. For instance, chain encryption could be performed using other anhydrides or using a nitroxide-based code. The latter option is interesting since different nitroxides may lead to polymers containing C–ON bonds with different homolysis rates. Such a strategy would probably permit to tune the sequencing and erasing behaviour of the polymers. It should be also remarked that the concept is not restricted to binary codes, but could be extended to more complex monomer-based codes (for example, ternary or higher).

Sequence-encoded poly(alkoxyamine amide)s are also remarkably easy to sequence because of the presence of alkoxyamine 'weak links' in their chains. These cleavable bonds enable an excellent 'readability' in MS/MS sequencing. Our results show that synthetic polymers may be as easy—or even easier—to sequence than biopolymers. Indeed, the molecular structure of synthetic polymers can be optimized for a sequencing technology, whereas in biopolymer sequencing the read-out device has to be adapted to a molecular structure that is imposed by biology. In the present case, for example, the incorporation of 'weak links' in the polymer simplify its MS/MS sequencing because no alternative dissociation reaction competes with the low-energy homolytic cleavage of the C–ON bond between a T-motif and any **0/1** motif. Moreover, the nature of the coding **0/1** units has no influence on the mechanism of this main fragmentation pathway. As a consequence, MS/MS data analysis is straightforward and can be performed in a few minutes. The analysis time can probably be further reduced with the help of a computer programme for analysing fragmentation data. Finally, it should be added that, while HRMS was used in this study for unambiguous assignment of product ions in order to establish their formation mechanism, the same sequencing results can be obtained using mass spectrometers operating with low-resolution mass analysers, such as triple quadrupole instruments that have become routine instruments in most laboratories.

The sequence-encoded polymers can also be thermally erased and degradation can be performed in the solid state or in solution. The proof-of-principle reported in this paper was provided by TEMPO-based polymers. However, as mentioned above, the thermal properties of poly(alkoxyamine amide)s (for example, storage half-life at room temperature and degradation rate at higher temperature) can be probably adjusted using other types of nitroxides. Moreover, it is imaginable to increase the storage half-life of these polymers by blending them in inorganic matrices[37,38]. These findings open up new application areas for synthetic polymers, for instance, in the fields of data storage, information processing and molecular identification. It is important to specify that these potential applications do not necessarily require very long polymer chains. For instance, a realistic, short-term, application for these coded polymers is the development of molecular barcodes for product identification. Binary-coded oligomers could be used for tagging high-value products in order to distinguish them from counterfeits. Altogether, the results shown in this article indicate that synthetic information-containing macromolecules may have a high relevance for manmade technologies.

## Methods

**Materials.** 4-Amino-2,2,6,6-tetramethylpiperidine-1-oxyl (4-amino-TEMPO, Tokyo Chemical Industry, 97%), 2,2,6,6-tetramethylpiperidine-1-oxyl (TEMPO, Alfa Aesar, 98%), 2-bromo-isobutyric acid (Alfa Aesar, 98%), 2-bromopropionic acid (Aldrich, 99%), tris(2-dimethylaminoethyl)amine (Alfa Aesar, >99%), methyl-2-bromopropionate (Aldrich, 98%), cyclohexamine (Aldrich, 99%), $N,N,N',N'',N''$-pentamethyldiethylenetriamine (PMDETA, Aldrich, 99%), tri-fluoroacetic acid (Sigma-Aldrich, 99%), $N,N$-dicyclohexylcarbodiimide (Alfa Aesar, 99%), piperidine (Sigma-Aldrich, 99%), potassium carbonate (Prolabo, 99%), $N$-ethyldiisopropylamine (DIPEA, Alfa Aesar, 99%), dichloromethane (DCM, Carlo Erba, 99.9%), THF (Aldrich, 99%, stabilized with Butylated Hydroxytoluene), dimethyl sulfoxide (DMSO, Aldrich, >99.6%) were used as received. Copper-(I)-bromide (Sigma-Aldrich, 98%) was washed with glacial acetic acid in order to remove any soluble oxidized species, filtered, washed with ethanol and dried. 2-Bromoisobutyryl anhydride (**a-1**) was synthesized following a reported procedure[39]. The synthesis of 2-bromopropionic anhydride (**a-0**) is described in the Supplementary Methods. Fmoc-Gly-Wang resin (**S1**; 0.40–1.00 mmol g$^{-1}$ loading) was purchased from Novabiochem/Merck. The cleavable Fmoc-Gly-Wang polystyrene-soluble support (**S2**) was prepared by ATRP and chain end modification as previously described ($M_n = 5,700$ g mol$^{-1}$, $M_w/M_n = 1.10$)[40,41]. The bromine-functional polystyrene-soluble support (**S3**) was prepared by ATRP in the presence of methyl-2-bromopropionate, copper-(I)-bromide and PMDETA ($M_n = 4,000$ g mol$^{-1}$, $M_w/M_n = 1.11$).

**Measurements and analysis.** Two different SEC set-ups were used in this work. The first one was equipped with four PLGel Mixed C columns (5 μm, 30 cm, Ø = 7.5 mm), a Wyatt Viscostar-II viscometer, a Wyatt TREOS light-scattering detector, a Shimadzu SPD-M20A diode array UV detector and a Wyatt Optilab

T-rEX refractometer. This set-up was used for polymer characterization (1,000–3,000,000 g mol$^{-1}$). The other set-up was equipped with a Shimadzu RiD-10 A refractometer, a Shimadzu SPD-10Avp UV detector and four monoporosity PLGel columns (5 μm, 30 cm, Ø = 7.5 mm): 50, 100, 500 and 1,000 Å. This set-up was used for oligomer and short polymer characterization (100–20,000 g mol$^{-1}$). In both set-ups, the mobile phase was THF with a flow rate of 1 ml min$^{-1}$. Toluene was used as the internal reference. The calibration was based on linear PS standards from Polymer Laboratories. $^1$H NMR and $^{13}$C NMR spectra were recorded in CDCl$_3$ on either a Bruker Avance 400 MHz or on a Bruker Avance 600 MHz spectrometers equipped with Ultrashield magnets. The 2D HSQC was realized using a Bruker 600 MHz spectrometer at 25 °C. The 2D HMBC was realized using a Bruker 400 MHz spectrometer at 25 °C. High-resolution ESI-HRMS and MS/MS experiments were performed using a QStar Elite mass spectrometer (Applied Biosystems SCIEX, Concord, ON, Canada) equipped with an ESI source operated in the positive mode. The capillary voltage was set at +5,500 V and the cone voltage at +75 V. In this hybrid instrument, ions were measured using an orthogonal acceleration time-of-flight mass analyser. In the MS mode, accurate mass measurements were performed using reference ions from a poly(propylene glycol) or a poly(ethylene glycol) internal standard. In the MS/MS mode, a quadrupole was used for selection of precursor ions to be further submitted to collision-induced dissociation in a collision cell. The precursor ion was used as the reference for accurate measurements of product ions in MS/MS spectra. In this instrument, air was used as the nebulizing gas (10 p.s.i.), while nitrogen was used as the curtain gas (20 p.s.i.) as well as collision gas. Instrument control, data acquisition and data processing were achieved using the Analyst software (QS 2.0) provided by Applied Biosystems. Oligomer solutions were prepared in MeOH supplemented with ammonium acetate (3 mM) and introduced in the ionization source with a syringe pump (flow rate: 5 μl min$^{-1}$). Detailed information about the MS/MS sequencing methodology can be found in the Supplementary Methods.

**Synthesis of sequence-encoded oligo(alkoxyamine amide)s.** The following examples describe the successive coupling of **1** (or **0**) and amino-TEMPO to the Fmoc-Gly-Wang resin **S1** and can be understood as a general procedure for the synthesis of oligo(alkoxyamine amide)s on a solid support. Steps B and C are interchangeable and should be alternated with step D. These steps can be repeated a certain number of times in order to reach a desired oligomer length. General procedures for synthesizing similar polymers on soluble supports **S2** and **S3** can be found in the Supplementary Methods. (A) Activation of **S1**: 0.1 g (0.079 mmol, 1 Eq.) of Fmoc-Gly-Wang resin **S1** (loading 0.79 mmol g$^{-1}$) was used as a solid support and placed in a fritted plastic funnel. Before the iterative steps, the resin beads were swollen by gentle shaking in DCM for 0.5 h. Next, the Fmoc was removed by treatment with piperidine/DCM (2 ml/2 ml) for 10 min. The deprotection step was repeated a second time to ensure complete removal of the Fmoc-protecting groups from the resin beads. A Kaiser test made on few resin beads confirmed the presence of deprotected primary amine groups. (B) Attachment of the **1** motif to the resin: A mixture of **a-1** (0.1248 g, 5 Eq.) and K$_2$CO$_3$ (0.2 g, 18 Eq.) was added to a fritted funnel for solid-phase synthesis containing amino-functionalized resin beads. To the reaction mixture, 4 ml of THF was added and was shaken for 50 min on a mechanical shaker. After completion, the solution was drained out from the fritted funnel. The beads were washed with MeOH-H$_2$O (1:1) to remove residual K$_2$CO$_3$ and afterwards with THF to remove the excess anhydride. (C) Attachment of the **0**-motif to the resin: A solution of **a-0** (0.1248 g, 5.3 Eq.) in THF (4 ml) was added along with DIPEA (0.25 ml, 18 Eq.) to a fritted funnel for solid-phase synthesis containing amino-functionalized resin beads. The mixture was shaken for 50 min on a mechanical shaker. After reaction, the solution was drained out from the fritted funnel. The beads were washed several times with THF. The use of DIPEA as a base is recommended for this step because it was observed that K$_2$CO$_3$ forms an inhomogeneous gel with **a-0** in THF. (D) Attachment of amino-TEMPO to the resin: Amino-TEMPO (0.08 g, 6 Eq.) and Me$_6$TREN (0.078 ml, 3.3 Eq.) were dissolved in 4 ml of anhydrous DMSO and were placed into a fritted funnel containing bromide-functionalized resin beads. The funnel was sealed with a rubber septum and the reaction mixture was purged with argon for ~10 min. Then, CuBr (0.034 g, 3 Eq.) was rapidly added. The mixture was shaken for 25 min under inert atmosphere. After reaction, the solution was drained out from the fritted funnel. The beads were washed several times with THF. It is important to mention that the efficiency of this reaction is somehow reduced after performing several iterations on the resin. Thus, from the sixth step and beyond, a microwave synthesizer (CEM Liberty 1TM, Saclay, France) was used for increasing the yields of radical coupling. Typically, the reactions were irradiated at 150 W of microwave power at 40 °C for ~90 min. (E) Cleavage of the oligomers from the resin: Cleavage of the poly(alkoxyamine amide)s from the resin was performed in TFA/DCM solution (1/1) for 2 h. After reaction, the solution was filtered, concentrated and precipitated in cold diethylether. The precipitate was filtered and redissolved in THF, and small amounts of insoluble resin fragments were separated by filtration. The filtrate was concentrated and the oligomers were isolated from cold diethylether precipitation. Trimers cannot be precipitated and were isolated by removing solvent and TFA. Example of final yield is (heptamer) 39 mg from 100 mg of **S1**; yield = 52%.

**Thermal degradation of the oligomers.** For solid state degradation tests, the oligo(alkoxyamine amide) was dissolved in a small amount of THF. The solution was poured on a glass slide and the THF was evaporated. The formed thin film was placed in a heating oven at 90 °C for several hours. For degradation tests in solution, the oligo(alkoxyamine amide) was dissolved in anisole and heated for 24 h under inert atmosphere at different temperatures, that is 50, 75, 100, 120 and 125 °C. For solution degradation tests in the presence of TEMPO, a poly(alkoxyamine amide) (Entry 14 in Supplementary Table 1, 20 mg, 1 Eq.) and TEMPO (39.6 mg, 12 Eq.) were dissolved in anisole and heated at 120 °C for several hours under argon atmosphere. In all experiments, the polymer degradation was monitored with SEC.

## References

1. Ecker, J. R. et al. Genomics: ENCODE explained. Nature 489, 52–55 (2012).
2. Church, G. M., Gao, Y. & Kosuri, S. Next-generation digital information storage in DNA. Science 337, 1628–1628 (2012).
3. Goldman, N. et al. Towards practical, high-capacity, low-maintenance information storage in synthesized DNA. Nature 494, 77–80 (2013).
4. Colquhoun, H. & Lutz, J.-F. Information-containing macromolecules. Nat. Chem. 6, 455–456 (2014).
5. Andrieux, D. & Gaspard, P. Nonequilibrium generation of information in copolymerization processes. Proc. Natl Acad. Sci. USA 105, 9516–9521 (2008).
6. Zhu, Z., Cardin, C. J., Gan, Y. & Colquhoun, H. M. Sequence-selective assembly of tweezer molecules on linear templates enables frameshift-reading of sequence information. Nat. Chem. 2, 653–660 (2010).
7. Trinh, T. T., Oswald, L., Chan-Seng, D. & Lutz, J.-F. Synthesis of molecularly encoded oligomers using a chemoselective 'AB + CD' iterative approach. Macromol. Rapid Commun. 35, 141–145 (2014).
8. Lutz, J.-F., Ouchi, M., Liu, D. R. & Sawamoto, M. Sequence-controlled polymers. Science 341, 1238149 (2013).
9. Mutlu, H. & Lutz, J.-F. Reading polymers: sequencing of natural and synthetic macromolecules. Angew. Chem. Int. Ed. 53, 13010–13019 (2014).
10. McKee, M. L. et al. Multistep DNA-templated reactions for the synthesis of functional sequence controlled oligomers. Angew. Chem. Int. Ed. 49, 7948–7951 (2010).
11. Niu, J., Hili, R. & Liu, D. R. Enzyme-free translation of DNA into sequence-defined synthetic polymers structurally unrelated to nucleic acids. Nat. Chem. 5, 282–292 (2013).
12. Lewandowski, B. et al. Sequence-specific peptide synthesis by an artificial small-molecule machine. Science 339, 189–193 (2013).
13. Merrifield, R. B. Solid Phase Synthesis (Nobel Lecture). Angew. Chem. Int. Ed. 24, 799–810 (1985).
14. Zuckermann, R. N., Kerr, J. M., Kent, S. B. H. & Moos, W. H. Efficient method for the preparation of peptoids [oligo(N-substituted glycines)] by submonomer solid-phase synthesis. J. Am. Chem. Soc. 114, 10646–10647 (1992).
15. Pfeifer, S., Zarafshani, Z., Badi, N. & Lutz, J.-F. Liquid-phase synthesis of block copolymers containing sequence-ordered segments. J. Am. Chem. Soc. 131, 9195–9197 (2009).
16. Espeel, P. et al. Multifunctionalized sequence-defined oligomers from a single building block. Angew. Chem. Int. Ed. 52, 13261–13264 (2013).
17. Solleder, S. C. & Meier, M. A. R. Sequence control in polymer chemistry through the passerini three-component reaction. Angew. Chem. Int. Ed. 53, 711–714 (2014).
18. Porel, M. & Alabi, C. A. Sequence-defined polymers via orthogonal allyl acrylamide building blocks. J. Am. Chem. Soc. 136, 13162–13165 (2014).
19. Naik, S., Bhattacharjya, G., Talukdar, B. & Patel, B. K. Chemoselective acylation of amines in aqueous media. Eur. J. Org. Chem. 2004, 1254–1260 (2004).
20. Rizzardo, E. & Solomon, D. A new method for investigating the mechanism of initiation of radical polymerization. Polym. Bull. 1, 529–534 (1979).
21. Matyjaszewski, K., Woodworth, B. E., Zhang, X., Gaynor, S. G. & Metzner, Z. Simple and Efficient synthesis of various alkoxyamines for stable free radical polymerization. Macromolecules 31, 5955–5957 (1998).
22. Kulis, J., Bell, C. A., Micallef, A. S., Jia, Z. & Monteiro, M. J. Rapid, selective, and reversible nitroxide radical coupling (nrc) reactions at ambient temperature. Macromolecules 42, 8218–8227 (2009).
23. Yang, H., Goyal, N., Ella-Menye, J. R., Williams, K. & Wang, G. Synthesis of chiral five-, six-, and seven-membered heterocycles from (S)-3-Hydroxy-γ-butyrolactone. Synthesis (Mass) 2012, 561–568 (2012).
24. Marais, L., Gallot, Z. & Benoît, H. A new method for correcting axial dispersion in GPC. J. Appl. Polym. Sci. 21, 1955–1964 (1977).
25. Striegel, A. M., Yau, W. W., Kirkland, J. J. & Bly, D. B. in Modern Size-Exclusion Liquid Chromatography Ch. 3 49–91 (John Wiley & Sons, 2009).
26. Gryn'ova, G., Lin, C. Y. & Coote, M. L. Which side-reactions compromise nitroxide mediated polymerization? Polym. Chem. 4, 3744–3754 (2013).
27. Lin, Y., Huang, B., Fu, Q., Wang, G. & Huang, J. Investigation of nitroxide radical coupling reaction in wide temperature range and different catalyst system. J. Polym. Sci. 48, 2991–2999 (2010).
28. Bayer, E. & Mutter, M. Liquid phase synthesis of peptides. Nature 237, 512–513 (1972).
29. Patten, T. E., Xia, J., Abernathy, T. & Matyjaszewski, K. Polymers with very low polydispersities from atom transfer radical polymerization. Science 272, 866–868 (1996).
30. Lutz, J.-F. & Matyjaszewski, K. Nuclear magnetic resonance monitoring of chain-end functionality in the atom transfer radical polymerization of styrene. J. Polym. Sci. 43, 897–910 (2005).
31. Biemann, K. Laying the groundwork for proteomics: Mass spectrometry from 1958 to 1988. Int. J. Mass Spectrom. 259, 1–7 (2007).
32. Wesdemiotis, C. et al. Fragmentation pathways of polymer ions. Mass Spectrom. Rev. 30, 523–559 (2011).
33. Maeda, T., Otsuka, H. & Takahara, A. Dynamic covalent polymers: Reorganizable polymers with dynamic covalent bonds. Prog. Polym. Sci. 34, 581–604 (2009).
34. Lehn, J.-M. Dynamers: dynamic molecular and supramolecular polymers. Prog. Polym. Sci. 30, 814–831 (2005).
35. Marque, S., Le Mercier, C., Tordo, P. & Fischer, H. Factors influencing the C − O − bond homolysis of trialkylhydroxylamines. Macromolecules 33, 4403–4410 (2000).
36. Bertin, D., Gigmes, D., Marque, S. R. A. & Tordo, P. Polar, steric, and stabilization effects in alkoxyamines C − ON bond homolysis: a multiparameter analysis. Macromolecules 38, 2638–2650 (2005).
37. Kaplan, M. DNA has a 521-year half-life. Nat. News doi:10.1038/nature.2012.11555 (2012).
38. Grass, R. N., Heckel, R., Puddu, M., Paunescu, D. & Stark, W. J. Robust chemical preservation of digital information on DNA in silica with error-correcting codes. Angew. Chem. Int. Ed. 54, 2552–2555 (2015).
39. Östmark, E., Harrisson, S., Wooley, K. L. & Malmström, E. E. Comb polymers prepared by ATRP from hydroxypropyl cellulose. Biomacromolecules 8, 1138–1148 (2007).
40. Meszynska, A., Badi, N., Borner, H. G. & Lutz, J.-F. 'Inverse' synthesis of polymer bioconjugates using soluble supports. Chem. Commun. 48, 3887–3889 (2012).
41. Oswald, L., Trinh, T. T., Chan-Seng, D. & Lutz, J.-F. Debromination of ATRP-made Wang soluble polymer supports. Polymer. (Guildf). doi:10.1016/j.polymer.2015.1002.1057.

## Acknowledgements

This work by supported by the European Research Council (project SEQUENCES—ERC grant agreement no. 258593), the Cluster of Excellence Chemistry of Complex Systems (LabEx CSC) and the CNRS. The post-doc position of R.K.R. and the PhD of A.M. were supported by the ERC. The PhD position of C.L. is supported by the LabEX CSC. L.C. acknowledges support from Spectropole, the Analytical Facility of Aix-Marseille University, by allowing a special access to the instruments purchased with European Funding (FEDER OBJ2142-3341). We also thank Sylvain Marque for in-depth discussions regarding thermal homolysis of C-ON bonds and Mélanie Legros for fruitful discussions on SEC peak broadening. Laurence Oswald is also acknowledged for the synthesis of S2 and Catherine Foussat for the SEC measurements.

## Author contributions

J.-F.L. conceived the idea, designed the experiments and wrote the paper. A.M., C.L., C.V. and R.K.R. conducted the experiments. L.C. performed the MS characterization and developed the sequencing methodology. All authors analysed the data.

## Additional information

Competing financial interests: A.M. and J.-F.L. are named inventors on a patent application related to this work.

# 4

# A palladium-catalysed multicomponent coupling approach to conjugated poly(1,3-dipoles) and polyheterocycles

David C. Leitch[1,*], Laure V. Kayser[1,*], Zhi-Yong Han[1,*], Ali R. Siamaki[1], Evan N. Keyzer[1], Ashley Gefen[1] & Bruce A. Arndtsen[1]

Conjugated polymers have emerged over the past several decades as key components for a range of applications, including semiconductors, molecular wires, sensors, light switchable transistors and OLEDs. Nevertheless, the construction of many such polymers, especially highly substituted variants, typically involves a multistep synthesis. This can limit the ability to both access and tune polymer structures for desired properties. Here we show an alternative approach to synthesize conjugated materials: a metal-catalysed multicomponent polymerization. This reaction assembles multiple monomer units into a new polymer containing reactive 1,3-dipoles, which can be modified using cycloaddition reactions. In addition to the synthetic ease of this approach, its modularity allows easy adaptation to incorporate a range of desired substituents, all via one-pot reactions.

[1] Department of Chemistry, McGill University, 801 Sherbrooke Street West, Montreal, Quebec, Canada H3A 0K8. * These authors contributed equally to this work. Correspondence and requests for materials should be addressed to B.A.A. (email: bruce.arndtsen@mcgill.ca).

A central goal in polymer synthesis is to directly convert simple chemical building blocks into useful materials. While a wide variety of interesting and potentially important structurally complex polymers have been discovered through recent research efforts (for example, biopolymers, advanced polymer networks, responsive materials and so on), their synthesis via traditional methods can be sufficiently involved to limit their accessibility, especially with the efficiency often demanded in polymer synthesis. One area where structure complexity has proven particularly powerful is in the field of π-conjugated polymers. The development of poly(heterocycles) (polypyrroles[1], polythiophenes[2,3] and others[4–6]) and their copolymers has sparked a renaissance in how scientists consider constructing a host of organic electronics, such as semiconductors, photovoltaic devices, or sensors[7–10]. A useful feature of conjugated polymers is their tunability. The modulation of substituents, conjugated heteroatoms or alternating backbone units can allow the construction of conjugated polymers with tailored electronic and other physical features. A number of powerful approaches have been developed to access conjugated polymers, including the now commonplace use of cross-coupling methodologies[11,12]. While very effective, these often achieve complexity from the monomers themselves, which can in some instances require a multistep synthesis, followed by halogenation and metallation, and can make accessing varied polymer structures an iterative process. The latter have made the development of alternative methods to construct conjugated polymers an area of growing relevance[13–15].

In principle, an attractive synthesis of complex conjugated polymers would be to consider their structure as arising directly from available monomers. A challenge is in how to accomplish this in an efficient fashion. One possibility is offered by multicomponent coupling reactions. Multicomponent reactions have been heavily exploited in organic synthesis to increase molecular complexity without the need for multistep synthetic sequences, with high efficiency and minimal waste[16,17]. When coupled with transition metal catalysis, these can provide methods to both activate and selectively couple several simple substrates directly into complex products[18–20]. Although metal catalysis is an established tool for activating typically unreactive components towards efficient polymerization (for example, polyolefin synthesis, ring-opening polymerization and so on), the use of metal catalysis to control the coupling of multiple different monomers into new, well-defined and more complex polymer structures is much less explored. Block terpolymer synthesis is well established (I; Fig. 1a), and a number of intriguing examples have recently emerged in the metal-catalysed assembly of alternating multicomponent polymers II (refs 21–27), including the synthesis of conjugated[22,23] and high-molecular-weight materials[27]. However, a method to directly convert multiple simple monomers into an entirely new polymer structure such as III has to our knowledge not been reported. Considering the variety of monomers relevant for polymerization (for example, diamines, diacids, alkenes, alkynes and so on), this method could provide an efficient route to access a diverse variety of structurally complex polymers, yet do so through combinations of available substrates, with high efficiency, and with facile access to structural diversity.

A polymerization as in Fig. 1a requires a multicomponent reaction that is selective, high yielding, yet relies upon available monomers. A potential transformation that fulfils these features is the palladium-catalysed synthesis of heterocycles shown in Fig. 1b. We have recently reported a palladium-catalysed route to generate 1,3-dipoles 4 (Münchnones) from imines, acid chlorides and carbon monoxide (CO)[28,29]. This reaction proceeds in high efficiency and is equally important with substrates that are all easily accessible for a polymerization: imines, acid chlorides and CO. 1,3-Dipoles, including di-Münchnones, have been used in condensation chemistry to synthesize cross-conjugated materials in low molecular weights (Fig. 1c; refs 30,31). A more attractive approach suggested by this palladium-catalysed synthesis would be to couple simple diimines, diacid chlorides and CO into a novel type of mesoionic, 1,3-dipole-containing conjugated polymer 4. Münchnones are known to undergo a variety of 1,3-dipolar cycloaddition reactions with unsaturated substrates (A≡B) to generate nitrogen-containing heterocycles[32]. As such, this platform would allow the assembly of various structurally distinct polymers from multiple combinations of available substrates.

We describe here our efforts towards the development of such a metal-catalysed multicomponent polymerization reaction. This has allowed the assembly π-conjugated poly(heterocycles) from multiple combinations of simple monomers: diimines, diacid chlorides, CO, alkenes and/or alkynes. In addition to demonstrating the feasibility of this strategy, this reaction allows access to a new class of conjugated polymer in the form of a mesoionic poly(1,3-dipole). The polymers can be easily modified by cycloaddition, providing access to families of conjugated materials in one-pot, metal-catalysed reactions.

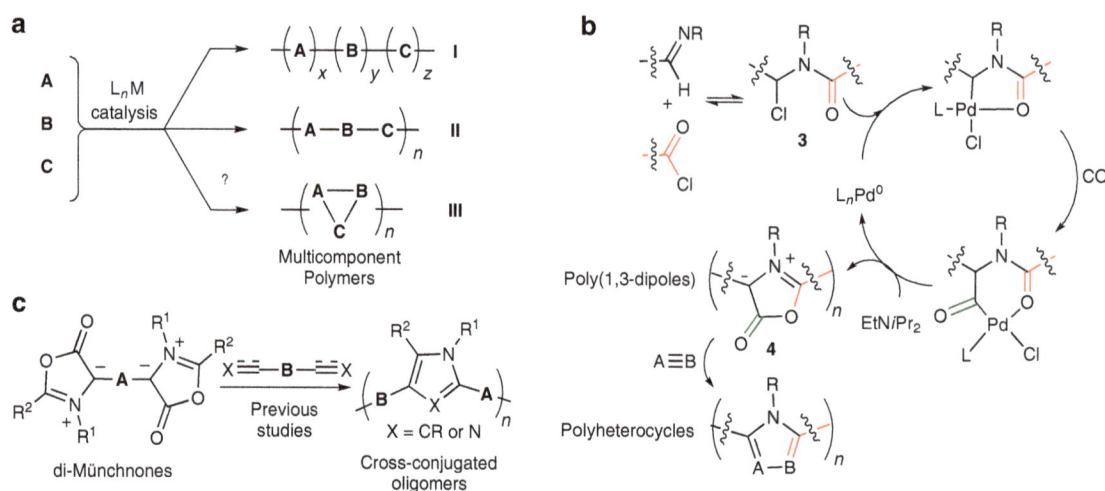

**Figure 1 | Multicomponent approaches to complex polymer synthesis. (a)** Metal-catalysed multicomponent coupling approaches to polymers. **(b)** Palladium-catalysed Münchnone formation. **(c)** Previous work involving di-Münchnones in oligomer synthesis.

## Results

**Multicomponent synthesis of conjugated polypyrroles.** As model monomers for the polymerization, we examined the palladium-catalysed coupling of terephthaloyl chloride **1a**, diimine **2a** based on dialkylfluorene and CO (Table 1). The efficiency of this polymerization was determined by reacting **4a** with the commercial alkyne dimethylacetylene dicarboxylate to form the polypyrrole **5a**. Using simple Pd(II) sources, with or without the addition of phosphines ligands, leads to no polymerization (entries 1–4), presumably owing to their slow reduction under the mild reaction conditions. Conversely, the common Pd(0) catalyst Pd$_2$dba$_3$•CHCl$_3$ (dba = dibenzylidene acetone) results in the rapid conversion of starting materials to form a poorly soluble, amorphous product (entry 7). Gel permeation chromotography (GPC) analysis shows that the tetrahydrofuran (THF) soluble fraction contains a polymer ($M_n = 7.8$ kDa) but with a broad polydispersity (PDI = 3.4). The addition of PPh$_3$ or PCy$_3$ in an attempt to attenuate the reactivity of the palladium catalyst leads instead to complete inhibition of the reaction (entries 5, 8 and 9) while P(o-tol)$_3$ restores activity but yields a similarly poorly soluble product with a broad PDI (entry 10).

The low solubility and broad PDI in this product may be the alkene-based dba on the palladium precatalyst, which could react via cycloaddition with **4a** and lead to crosslinking. To circumvent this reaction, Pd[P(o-tol)$_3$]$_2$ was used as a commercially available source of Pd(0) with only weakly associated, and unreactive, ligands. This catalyst leads to the near-complete consumption of the reagents under mild conditions (50 °C, 4 atm CO), and the generation of a polymer (**5a**) that is soluble in common solvents (entry 11). $^1$H and $^{13}$C nuclear magnetic resonance (NMR) and infrared (IR) analysis show the conversion of the monomers into the polypyrrole **5a**, which has identical spectral features to independently prepared model compounds (Supplementary Figs 5 and 16). GPC analysis shows **5a** to have a well-defined monomodal molecular weight distribution (PDI = 1.6). The catalytic activity of Pd[P(o-tol)$_3$]$_2$ can be enhanced by the addition of CuPF$_6$, which presumably acts as a phosphine scavenger to generate a mono-ligated palladium catalyst (entry 12). Alternatively, simply using high CO pressure leads to a rapid and near-quantitative polymerization, forming polypyrrole **5a** as the only observable product (entry 13). Molecular weights as high as 22.7 kDa can be obtained by increasing the concentration and reaction time (entry 14).

**Synthesis of poly(1,3-dipoles).** The reaction in Table 1 provides a new approach to synthesize pyrrole-based conjugated polymers from combinations of substrates that are either available or monomers themselves in other polymerizations (terephthaloyl chloride, CO, dialdehydes and alkynes). This platform can also be used to access new classes of conjugated materials. For example,

**Table 1 | A multicomponent synthesis of conjugated polypyrroles\*.**

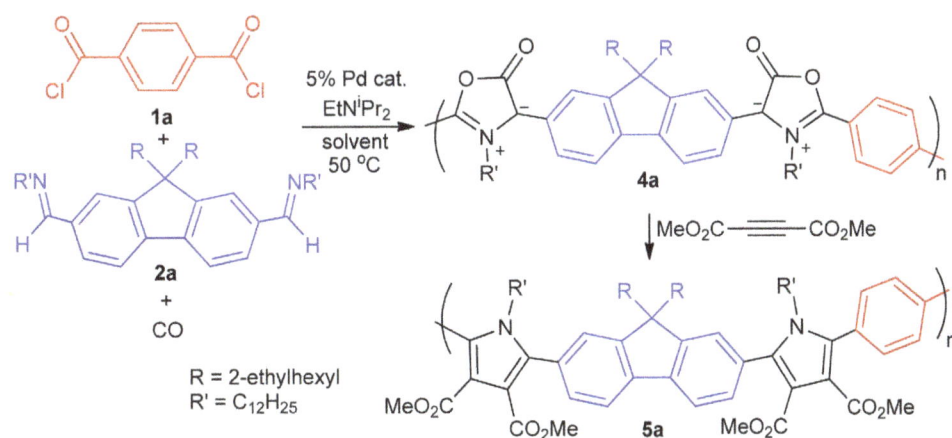

| Entry | Catalyst | Yield$^\dagger$ | $M_n$ (kDa) | $M_w$ (kDa) | PDI |
|---|---|---|---|---|---|
| 1 | PdCl$_2$(PhCN)$_2$ | NR | — | — | — |
| 2 | PdCl$_2$(PPh$_3$)$_2$ | NR | — | — | — |
| 3 | PdCl$_2$(PhCN)$_2$ /P(o-tol)$_3$ | NR | — | — | — |
| 4 | Pd(OAc)$_2$ | NR | — | — | — |
| 5 | Pd(PPh$_3$)$_4$ | NR | — | — | — |
| 6 | Pd[P(t-Bu)$_3$]$_2$ | 50 | 2.0 | 3.4 | 1.4 |
| 7 | Pd$_2$dba$_3$•CHCl$_3$ | 70 | 7.8$^\ddagger$ | 26.5 | 3.4 |
| 8 | Pd$_2$dba$_3$/PPh$_3$ | NR | — | — | — |
| 9 | Pd$_2$dba$_3$/PCy$_3$ | NR | — | — | — |
| 10 | Pd$_2$dba$_3$/P(o-tol)$_3$ | 75 | 7.4$^\ddagger$ | 28.1 | 3.8 |
| 11 | Pd[P(o-tol)$_3$]$_2$ | 60 | 5.5 | 8.8 | 1.6 |
| 12 | Pd[P(o-tol)$_3$]$_2$ $^\S$ | 69 | 7.0 | 10.5 | 1.5 |
| 13 | Pd[P(o-tol)$_3$]$_2$$^\parallel$ | 85 | 12.7 | 25.2 | 1.9 |
| 14 | Pd[P(o-tol)$_3$]$_2$$^\P$ | 73 | 22.7 | 56.6 | 2.5 |

\***1a** (0.10 mmol), **2a** (0.10 mmol), EtN$^i$Pr$_2$ (0.40 mmol), 4 atm CO, 5% Pd/pyrrole, THF, 50 °C, 24 h; then PhCOCl (0.10 mmol), 4 ml CH$_2$Cl$_2$, alkyne (0.40 mmol), room temperature, 18 h.
$^\dagger$**5a**, isolated yield, NR = no reaction.
$^\ddagger$THF extract.
$^\S$5% CuPF$_6$, 30 h.
$^\parallel$20 bar CO, THF/MeCN: 1.9/0.6 ml, 45 °C, 30 h; isolated by Soxhlet extraction.
$^\P$20 bar CO, THF/MeCN: 1.5/0.5 ml, 45 °C, 64 h

the first step in the transformation in Table 1 also generates the 1,3-dipole-containing polymer **4a**. 1,3-Dipoles such as Münchnones are typically considered reactive intermediates and used *in situ* in synthesis. However, while performing the palladium-catalysed coupling in the absence of alkyne, we noted the precipitation of a dark solid **4a** that can be easily isolated by washing with acetonitrile. This polymer is surprisingly stable (<5% mass loss at up to 180 °C under nitrogen), and can be stored at low temperature in the absence of air and moisture, although it does hydrolyse in the presence of water. In order to fully characterize the moderately soluble **4a** by NMR analysis, it was prepared in moderate molecular weight with an imine end-capping agent (**4a′**, $M_n = 6.7$ kDa; Supplementary Methods). Spectral analysis show all the signals for a Münchnone, including characteristic carbonyl resonances in the infrared (1,710 cm$^{-1}$) spectra, in the $^{13}$C NMR ($\delta$ 160.6) and others, all of which correlate with the model di-Münchnone **4a″** prepared from diimine **2a**, toluoyl chloride and CO. Therefore, **4a** represents an unusual new class of mesoionic polymer, poly(1,3-dipoles).

Donor/acceptor-conjugated polymers have become an important thrust in the recent design of polymer-based photovoltaic materials[7,8,33], although these do not typically incorporate formal charges into the backbone. Likely as a result of this charge separation, poly-Münchnone **4a** is a dark purple solid, characteristic of low-bandgap materials. Ultraviolet/visible analysis shows intense (molar absorptivity up to $5.0 \times 10^4$ l mol$^{-1}$ cm$^{-1}$) and broad absorptions in most of the visible region (Fig. 2a). These are significantly red-shifted relative to the model bis-Münchnone, and indicate that **4a** is highly conjugated. While the precise nature of this extended conjugation is still under investigation, the persistent charge-separated character of the mesoionic moieties creates a donor/acceptor motif on a single heterocycle, and may also provide a novel route to planarization as a mechanism to partially eliminate charge (for example, Fig. 2b). Estimation of the optical bandgap by absorbance onset gives values of 1.74 eV for **4a′** (1.59 eV for the high-molecular-weight **4a**), and cyclic voltammetry shows a reversible reduction and an electrochemical bandgap of 1.84 eV. As such, these are a new class of low-bandgap-conjugated materials.

In addition to their unusual electronic properties, the synthesis of **4** from diimines and diacid chlorides makes it straightforward to attenuate their structure and form a range of mesoionic polymers. Examples of the structural diversity available are shown in Fig. 3. Notably, each of the bis(acid chloride) monomers are either commercially available or easily prepared from the diacid compounds. For example, 2,5-thiophene dicarbonyl dichloride (precursor to **4c**) can be synthesized in one step from adipic acid, a commodity chemical used in the production of nylon and other polyamides, whereas 2,5-furan dicarboxylic acid is derived from carbohydrates and identified by the US Department of Energy as one of the top 10 bio-based renewable chemicals[34]. The diimines used can be similarly altered to incorporate carbazole (**4e**), a common unit in conjugated polymer production[8]. In all cases, the catalytic coupling is clean and molecular weights are limited only by the solubility of the poly(1,3-dipole).

The structural manipulations translate into the properties of these polymers (Fig. 4a,b). For example, the disubstituted carbazole-containing polymer **4e** displays visible absorbances that are blue-shifted ($\lambda_{max}$ of 539 nm) relative to **4a** ($\lambda_{max}$ of 570 nm; Fig. 4a). Alternatively, the thiophene-containing material **4b** is significantly red-shifted relative to these other polymers (absorbance onset ~768 nm), corresponding to an optical bandgap of 1.58 eV. The latter is comparable to materials currently of interest as light harvesting materials in bulk heterojunction solar cells[7,8]. As such, this provides a method to both construct low-bandgap-conjugated polymers and manipulate or tune their electronic properties by choice of the constituent components.

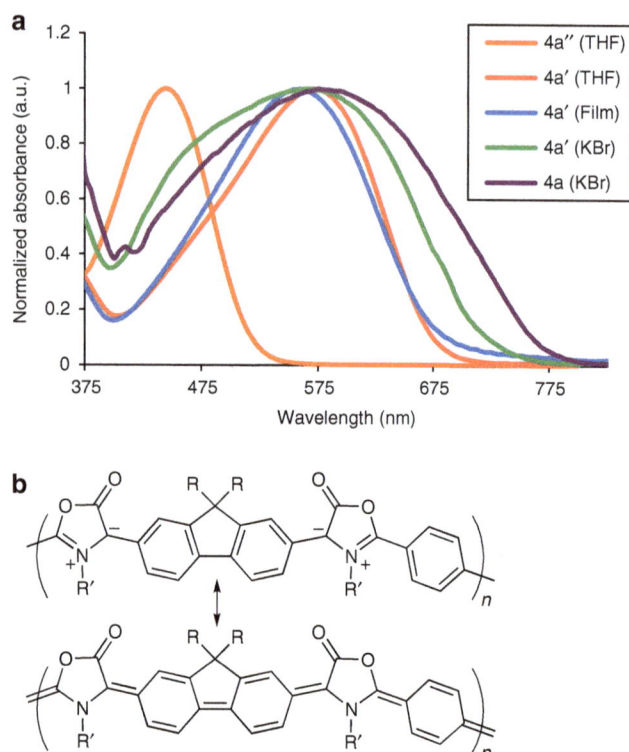

**Figure 2 | Ultraviolet/visible spectra and conjugation in poly-Münchnones.** (**a**) Ultraviolet/visible spectra of **4a** (22.7 kDa), **4a′** (6.7 kDa) and model dimer **4a″** (*n* = 1). (**b**) Potential resonance structures of **4a** leading to planarization.

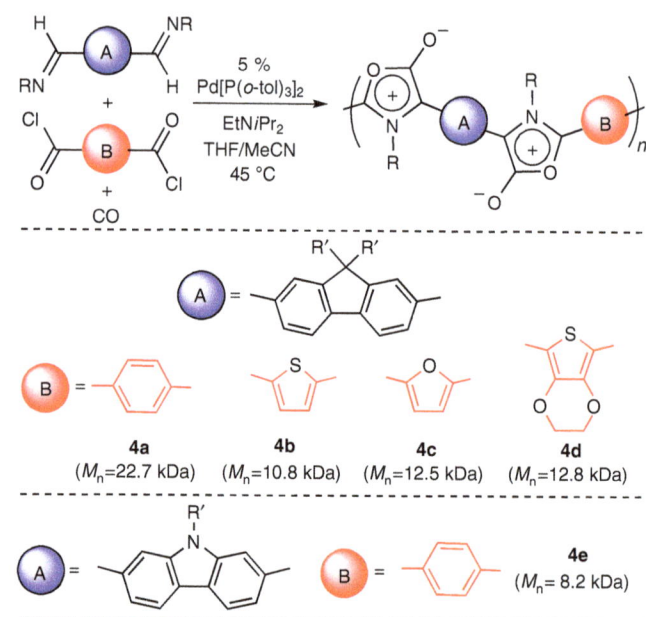

**Figure 3 | Diversity of Münchnone-containing polymers.** Owing to the moisture sensitivity of **4**, molecular weight determined by conversion to polypyrroles **5**, as in Table 1.

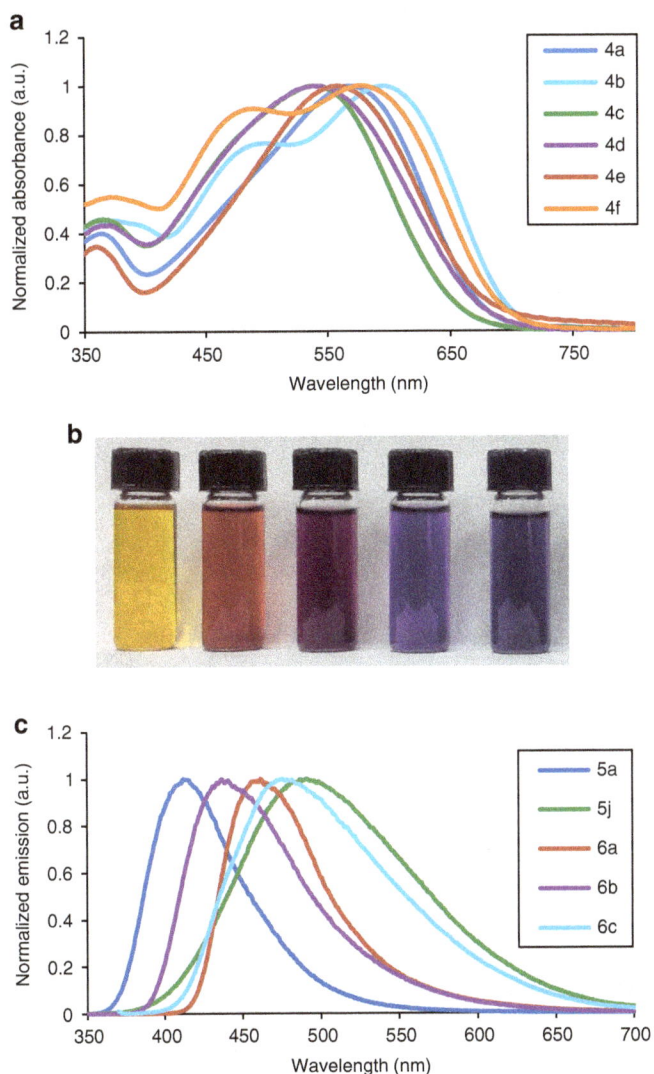

**Figure 4 | Properties of polymers.** Polymers were prepared with imine end-capping at 6–8 kDa to ensure full solubility for analyses (Supplementary Methods). (**a**) Ultraviolet/visible absorbance spectra of poly-Münchnones **4a–f** in THF. (**b**) Polymers **5a**, **4f'**, **4e'**, **4a'** and **4c'** in THF (from left to right). (**c**) Fluorescence spectra of select polymers **5–6** in THF.

**Multicomponent synthesis of poly(heterocycles).** These poly-Münchnones **4** also offer access to another deeper level of molecular complexity via their backbone reactivity. This can provide, to our knowledge, a unique route to convert one conjugated organic polymer into other backbone-conjugated polymers[35–37]. For example, the addition of phenyl methyl-propiolate to **4a** results in the transformation of the purple, low-bandgap poly-Münchnone into a moderate bandgap, blue-emitting polypyrrole **5b** (Fig. 5). [1]H NMR and infrared analysis suggest the complete disappearance of 1,3-dipole unit in this reaction, with no observable byproducts. In order to quantify the efficiency of cycloaddition, [13]C-labelled polymer **4a** was generated from isotopically enriched terephthaloyl chloride 4-$C_6H_4$([13]COCl)$_2$. NMR ([13]C) analysis shows the quantitative reaction with alkyne to form polypyrrole (>95%; see Supplementary Methods for details), and is consistent with the high cycloaddition reactivity of the 1,3-dipoles. A range of pyrrole-based polymers can be formed by this reaction. For example, changing the alkyne used can provide access to a wide variety of pyrrole-based polymers. Representative examples of

these include the diester-containing polymer **5a** or the diketone-substituted **5c**. In addition to alkynes, electron-deficient alkenes are suitable dipolarophiles to form pyrroles. This can allow the facile formation of fused ring pyrrole-based polymer **6a** (from cyclic alkene cycloaddition), the mono-substituted **6b** (from chlorocyanoethylene) or the unsubstituted polypyrrole **6c** with high yield. By using different diimines and diacid chlorides, various other polypyrroles can be obtained (for example, **5d–5 h**). As such, this provides a platform to readily incorporate desired functionalities onto conjugated polymers, all of which emanate from a single polymer **4**. The reactivity of Münchnones is also not limited to pyrroles. The addition of N-tosyl imine forms imidazole-based-conjugated polymer **7**, and backbone conjugation in these polymers can be easily quenched by the addition of alcohols, leading to the generation of polyamides **8**. Poly-Münchnones **4** can therefore be considered highly reactive and versatile conjugated materials, where a single polymer **4** can be transformed into entire families of new conjugated materials. In the case of phenyl methylpropiolate, this polymerization can be performed in a single step (Fig. 6), thereby allowing the orthogonal, four-component synthesis of a conjugated polymer.

This structural diversity can provide a further platform to modulate properties. For example, between polymers **4** and **8**, optical absorbance and polymer bandgap can be tuned across the visible spectrum, and materials can be formed that fluoresce anywhere from blue to green to not at all (Fig. 4c and Table 2; Supplementary Tables 1 and 2 for the full list of properties). Similarly, electrochemical studies show that this cycloaddition (or lack thereof) can be used to tune HOMO energies by over 1 eV. While the primary thrust of these studies was not in product design, several of the polymers reported here display notable properties. Polypyrroles and their copolymers have attracted significant interest as electronic materials[1,38]. Three of the pyrrole-based polymers (**5a**, **6a** and **6b**) have good photo-luminescent quantum efficiency (>35%) with emission maxima in the range of blue light (413–459 nm). Alternatively, the pyrrole-based imide-functionalized unit has been identified as a promising electron acceptor unit in conjugated polymers, and can be readily generated by this approach (**6a**; ref. 39), while alternating pyrrole-thiophene materials such as **6d** represent new variants of materials found to be of use in field-effect transistors[40]. The range of properties observed is a direct result of their structural diversity, which varies from the backbone-conjugated heterocycles, the spacer units, to the substituents. This level of structural attenuation would require an individual synthesis for each new monomer via typical methods. In this case, each polymer is generated in one pot, from a small pool of monomers, and with minimal waste (often only HCl and $CO_2$).

**Synthesis of polymers from vanillin.** Finally, we have examined the potential of using other renewable materials as precursors to conjugated polymers. Lignin is a major component of lignocellulosic biomass and the world's largest renewable source of aromatic compounds, thereby making it a potentially attractive, bio-based feedstock for π-conjugated polymers[41]. Lignin depolymerization yields a variety of aromatic building blocks, including the dialdehyde **9** (a dimer of vanillin). As this multicomponent polymerization uses aldehydes and carboxylic acids as monomer feedstocks, **9** can be incorporated in this palladium-catalysed polymerization to generate the conjugated poly-Münchnone **4f** (Fig. 7). Ultraviolet/visible and electrochemical studies show that **4f** is a moderately low-bandgap polymer (1.9 eV). As above, the dipole in **4f** can undergo cycloaddition reactions to generate the polypyrroles **5i** and **5j**, each of which are blue-emitting materials. These polymers are

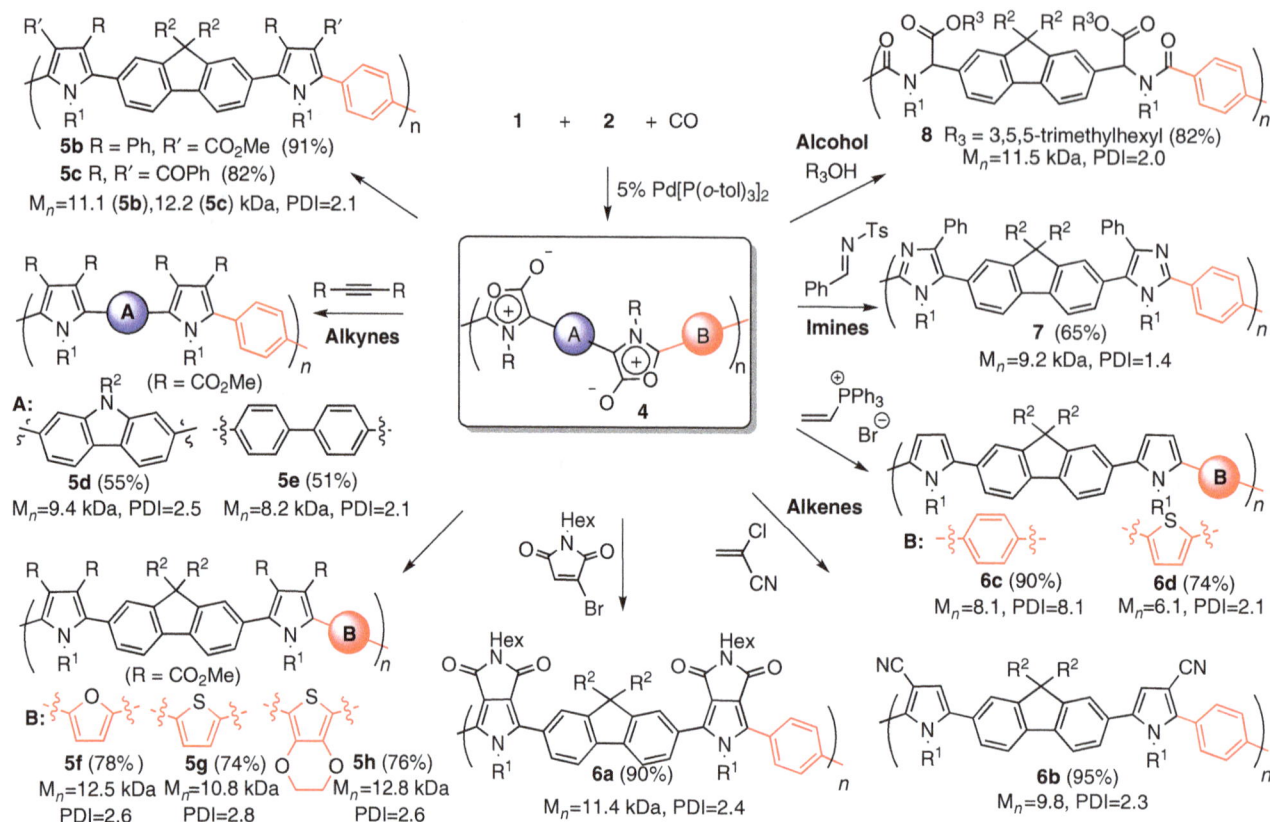

**Figure 5 | Transformation of poly-Münchnones into families of conjugated polymers.** Derivatization performed on poly-Münchnone samples of 8–12 kDa to ensure full solubility of **4a–e**, and allow the quantification of the cycloaddition.

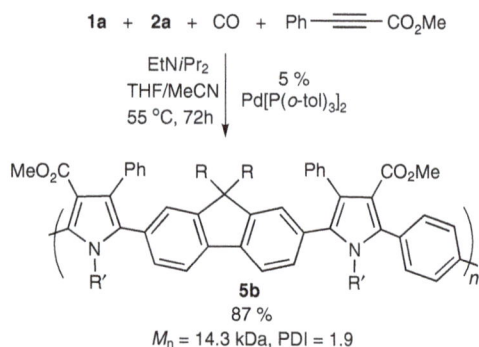

**Figure 6 | One-pot, four-component polymerization.** See Supplementary Methods for experimental details.

**Table 2 | Properties of polymers 5–8.**

| Compound | $\lambda_{max}$ (nm) | $\lambda_{em}$ (nm) | $\varphi_{PL}$ | $E_g^{opt}$ (eV) | $E_g^{opt}$ film (eV) |
|---|---|---|---|---|---|
| **5a** | 321 | 413 | 0.39 | 3.18 | 3.16 |
| **5b** | 335 | 492 | 0.27 | 2.91 | 2.87 |
| **5c** | 329 | 496 | 0.04 | 2.99 | 2.98 |
| **5d** | 325 | 474 | 0.03 | 2.92 | 2.79 |
| **5e** | 301 | 413 | 0.12 | 3.15 | 3.08 |
| **5f** | 320 | 459 | 0.14 | 3.08 | 2.97 |
| **5g** | 320 | 467 | 0.11 | 3.07 | 3.11 |
| **5h** | 321 | 467 | 0.10 | 3.05 | 2.97 |
| **5i** | 312 | 417 | 0.17 | 3.19 | 2.93 |
| **5j** | 329 | 473 | 0.02 | 2.88 | 2.92 |
| **6a** | 364 | 459 | 0.35 | 2.86 | 2.80 |
| **6b** | 345 | 431 | 0.47 | 3.01 | 2.96 |
| **6c** | 366 | 501 | 0.08 | 2.60 | 2.60 |
| **6d** | 368 | 497 | 0.06 | 2.63 | 2.47 |
| **7** | 330 | 467 | 0.12 | 2.98 | 2.80 |
| **8** | 280 | — | — | 3.45 | 3.42 |

Select physical properties of **5–8** (see Supplementary Tables 1 and 2 for further details, electrochemical studies, HOMO/LUMO energies and for properties of polymers **4a–f**).

hybrid materials derived from the following four simple substrates: vanillin, terephthaloyl chloride, a primary amine and CO, and represent, as far as we are aware, the first use of lignin in cross-conjugated polymer formation. Considering the versatility of this reaction, it should prove relevant for the controlled assembly of a range of renewable-based conjugated polymers.

## Discussion

In summary, we have described a new type of metal-catalysed multicomponent polymerization, which provides a method to convert combinations of monomers, such as diimines, diacid chlorides, CO, alkynes, alkenes and alcohols, into structurally well-defined conjugated polymers. In addition, new backbone-conjugated mesoionic polymers (**4**) can be prepared via

this coupling, which can undergo efficient post-polymerization cycloaddition to generate families of conjugated materials. In light of the variety of dipolarophiles possible for cycloaddition, as well as diimines and bis(acid chloride)s available, this can be used to construct arrays of conjugated materials, yet without the typical need to preassemble each new conjugated unit, and with the efficiency often desired in polymer synthesis. This multicomponent catalytic polymerization approach could prove equally applicable for the controlled assembly of a range of new materials

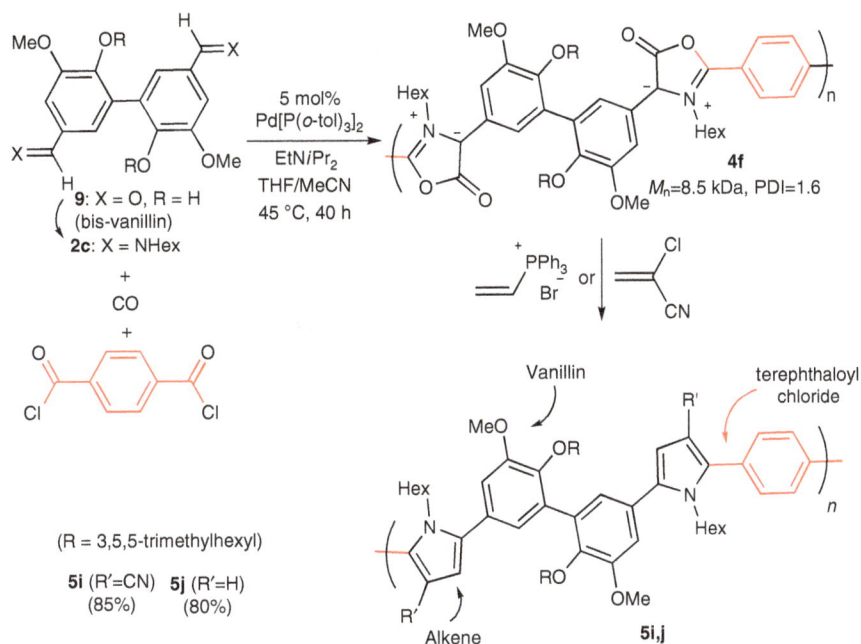

**Figure 7 | Conjugated polymers from vanillin.** Multicomponent synthesis of polymer **4f** and its derivatization to **5i** and **5j**.

from established monomers and/or other inexpensive building blocks. Experiments directed towards the latter are currently underway.

## Methods

**General procedure for the synthesis of poly-Münchnones and polyheterocycles.** To diacid chloride **1** (0.1 mmol) and diimine **2** (0.100 mmol) in THF (0.6 ml) in a 5-ml vial under $N_2$ was added N,N-diisopropylethylamine (51.7 mg, 70.0 μl, 0.400 mmol), and Pd[P(o-tol)$_3$]$_2$ (7.2 mg, 0.010 mmol) in 1.3 ml THF (1.3 ml)/MeCN (0.6 ml). The vial was placed in a 40-ml Parr steel autoclave, charged with CO (20 bar) and heated to 45 °C. (For the imine endcapped poly-Münchnone, p-tolyl(H)C = NC$_{16}$H$_{37}$ was added; see Supplementary Methods for details.) The CO was evacuated, the vessel was brought into a glovebox and the appropriate dipolarophile in 1.5 ml THF was added. The reaction was stirred at room temperature or 50 °C for 16 h, 0.2 ml water was added and the polymer product was extracted with o-dichlorobenzene using a Soxhlet extractor. The solvent was removed *in vacuo*, and the residue was dissolved with a minimum amount of hot chloroform (~1 ml) and added dropwise into methanol (~20 ml) to precipitate the polymer. The suspension was centrifuged and the methanol layer was decanted. The polymer was washed with methanol (3 × 2 ml) before drying *in vacuo* at 50 °C. See Supplementary Methods for further information. For NMR analysis, ultraviolet/visible, fluorescence and cyclic voltammetry of the molecules in this article, see Supplementary Figs 1–7.

## References

1. Audebert, P. in *Electropolymerization: Concept, Materials and Applications.* (eds Cosnier, S. & Karyakin, A.) 77–91 (Wiley-VCH, 2010).
2. Fichou, D. (ed.) *Handbook of Oligo- and Polythiophenes* (Wiley-VCH, 1999).
3. Osaka, I. & McCullough, R. D. in *Design and Synthesis of Conjugated Polymers.* (eds Leclerc, M. & Morin, J.-F.) 91–145 (Wiley-VCH, 2010).
4. Zhou, H., Yang, L. & You, W. rational design of high performance conjugated polymers for organic solar cells. *Macromolecules* **45**, 607–632 (2012).
5. Coughlin, J. E., Henson, Z. B., Welch, G. C. & Bazan, G. C. Design and synthesis of molecular donors for solution-processed high-efficiency organic solar cells. *Acc. Chem. Res.* **47**, 257–270 (2014).
6. Morin, J.-F., Leclerc, M., Adès, D. & Siove, A. Polycarbazoles: 25 years of progress. *Macromol. Rapid Commun.* **26**, 761–778 (2005).
7. Facchetti, A. π-Conjugated polymers for organic electronics and photovoltaic cell applications. *Chem. Mater.* **23**, 733–758 (2010).
8. Heeger, A. J. 25th Anniversary Article: bulk heterojunction solar cells: understanding the mechanism of operation. *Adv. Mater.* **26**, 10–28 (2014).
9. Grimsdale, A. C., Leok Chan, K., Martin, R. E., Jokisz, P. G. & Holmes, A. B. Synthesis of light-emitting conjugated polymers for applications in electroluminescent devices. *Chem. Rev.* **109**, 897–1091 (2009).

10. Ates, M. A review study of (bio)sensor systems based on conducting polymers. *Mat. Sci. Eng. C* **33**, 1853–1859 (2013).
11. Babudri, F., Farinola, G. M. & Naso, F. Synthesis of conjugated oligomers and polymers: the organometallic way. *J. Mater. Chem.* **14**, 11–34 (2004).
12. Okamoto, K. & Luscombe, C. K. Controlled polymerizations for the synthesis of semiconducting conjugated polymers. *Polym. Chem.* **2**, 2424–2434 (2011).
13. Burke, D. J. & Lipomi, D. J. Green chemistry for organic solar cells. *Energy Environ. Sci.* **6**, 2053–2066 (2013).
14. Okamoto, K., Zhang, J., Housekeeper, J. B., Marder, S. R. & Luscombe, C. K. C–H arylation reaction: atom efficient and greener syntheses of π-conjugated small molecules and macromolecules for organic electronic materials. *Macromolecules* **46**, 8059–8078 (2013).
15. Mercier, L. G. & Leclerc, M. Direct (hetero)arylation: a new tool for polymer chemists. *Acc. Chem. Res.* **46**, 1597–1605 (2013).
16. Jieping Zhu, H. B. (ed.) *Multicomponent Reactions* (Wiley-VCH, 2005).
17. Dömling, A. & Ugi, I. Multicomponent reactions with isocyanides. *Angew. Chem. Int. Ed.* **39**, 3168–3210 (2000).
18. D'Souza, D. M. & Muller, T. J. J. Multi-component syntheses of heterocycles by transition-metal catalysis. *Chem. Soc. Rev.* **36**, 1095–1108 (2007).
19. Balme, G., Bossharth, E. & Monteiro, N. Pd-assisted multicomponent synthesis of heterocycles. *Eur. J. Org. Chem* **2003**, 4101–4111 (2003).
20. Montgomery, J. Nickel-catalyzed reductive cyclizations and couplings. *Angew. Chem. Int. Ed.* **43**, 3890–3908 (2004).
21. Kakuchi, R. Multicomponent reactions in polymer synthesis. *Angew. Chem. Int. Ed.* **53**, 46–48 (2014).
22. Choi, C.-K., Tomita, I. & Endo, T. Synthesis of novel π-conjugated polymer having an enyne unit by palladium-catalyzed three-component coupling polymerization and subsequent retro-diels – alder reaction. *Macromolecules* **33**, 1487–1488 (2000).
23. Nakagawa, K. & Tomita, I. Synthesis of poly(p-phenylene-vinylene) with chiral higher-order structure from simple monomers by three-component coupling polymerization. *Macromolecules* **40**, 9212–9216 (2007).
24. Takenoya, K. & Yokozawa, T. Simultaneous construction of polymer backbone and side chains by three-component polycondensation. synthesis of polyethers with cyano side chains from dialdehydes, alkylene bis(trimethylsilyl) ethers, and cyanotrimethylsilane. *Macromolecules* **31**, 2906–2910 (1998).
25. Chan, C. Y. K. *et al.* Construction of functional macromolecules with well-defined structures by indium-catalyzed three-component polycoupling of alkynes, aldehydes, and amines. *Macromolecules* **46**, 3246–3256 (2013).
26. Lee, I.-H., Kim, H. & Choi, T.-L. Cu-catalyzed multicomponent polymerization to synthesize a library of poly(n-sulfonylamidines). *J. Am. Chem. Soc.* **135**, 3760–3763 (2013).
27. Kreye, O., Kugele, D., Faust, L. & Meier, M. A. Divergent dendrimer synthesis via the Passerini three-component reaction and olefin cross-metathesis. *Macromol. Rapid. Commun.* **35**, 317–322 (2014).

28. Dhawan, R., Dghaym, R. D. & Arndtsen, B. A. The development of a catalytic synthesis of münchnones: a simple four-component coupling approach to α-amino acid derivatives. *J. Am. Chem. Soc.* **125,** 1474–1475 (2003).

29. Lu, Y. & Arndtsen, B. A. Palladium catalyzed synthesis of münchnones from α-amidoethers: a mild route to pyrroles. *Angew. Chem. Int. Ed.* **47,** 5430–5433 (2008).

30. Siamaki, A. R., Sakalauskas, M. & Arndtsen, B. A. A palladium-catalyzed multicomponent coupling approach to pi-conjugated oligomers: assembling imidazole-based materials from imines and acyl chlorides. *Angew. Chem. Int. Ed.* **50,** 6552–6556 (2011).

31. Manecke, G. & Klawitter, J. Zur synthese von makromolekülen durch 1,3-dipolare cycloaddition aus mesoionischen oxazolonen, 3. *Die Makromol. Chem.* **175,** 3383–3399 (1974).

32. Gingrich, H. L. & Baum, J. S. in *Chemistry of Heterocyclic Compounds: Oxazoles.* (ed. Turchi, I. J.) **45,** 731–961 (John Wiley & Sons, Inc., 2008).

33. Biniek, L., Schroeder, B. C., Nielsen, C. B. & McCulloch, I. Recent advances in high mobility donor-acceptor semiconducting polymers. *J. Mater. Chem.* **22,** 14803–14813 (2012).

34. Bozell, J. J. & Petersen, G. R. Technology development for the production of biobased products from biorefinery carbohydrates-the US Department of Energy's "Top 10" revisited. *Green Chem.* **12,** 539–554 (2010).

35. Lucht, B. L., Buretea, M. A. & Tilley, T. D. A zirconocene-coupling route to substituted poly(p-phenylenedienylene)s: band gap tuning via conformational control. *J. Am. Chem. Soc.* **120,** 4354–4365 (1998).

36. He, G. *et al.* The marriage of metallacycle transfer chemistry with suzuki-miyaura cross-coupling to give main group element-containing conjugated polymers. *J. Am. Chem. Soc.* **135,** 5360–5363 (2013).

37. For non-conjugated examples to post polymerization derivatizationLeibfarth, F. A. *et al.* A facile route to ketene-functionalized polymers for general materials applications. *Nat. Chem.* **2,** 207–212 (2010).

38. Wang, L.-X., Li, X.-G. & Yang, Y.-L. Preparation, properties and applications of polypyrroles. *React. Funct. Polym.* **47,** 125–139 (2001).

39. Tamilavan, V. *et al.* Synthesis and photovoltaic properties of donor–acceptor polymers incorporating a structurally-novel pyrrole-based imide-functionalized electron acceptor moiety. *Polymer* **54,** 6125–6132 (2013).

40. Nishinaga, T. *et al.* Synthesis and structural, electronic, optical and FET properties of thiophene-pyrrole mixed hexamers end-capped with phenyl and pentafluorophenyl groups. *J. Mater. Chem.* **21,** 14959–14966 (2011).

41. Zakzeski, J., Bruijnincx, P. C. A., Jongerius, A. L. & Weckhuysen, B. M. The catalytic valorization of lignin for the production of renewable chemicals. *Chem. Rev.* **110,** 3552–3599 (2010).

## Acknowledgements

We thank NSERC, CFI and the FQRNT supported Centre for Green Chemistry and Catalysis, and the NSERC Biomaterials and Chemicals Research Network (Lignoworks) for funding this work. D.C.L. and E.N.K. thank NSERC for a PDF and a USRA, respectively. We thank Professor Perepichka for access to his cyclic voltametry, ultra-violet and fluorescence instruments. We thank CSACS for access to the GPC. We thank Julia Schneider for her support in the CV measurements.

## Author contributions

B.A.A. conceived and directed the project. D.C.L., L.V.K. and Z.-Y.H. carried out the experiments and characterized the polymers. D.C.L., L.V.K. and Z.-Y.H. contributed equally to this work. A.R.S., A.G. and E.N.K. performed preliminary experiments.

## Additional information

**Competing financial interests:** The authors declare no competing financial interests.

# 5

# Rhodium-catalysed C($sp^2$)–C($sp^2$) bond formation via C–H/C–F activation

Panpan Tian[1], Chao Feng[1,2] & Teck-Peng Loh[1,2]

Fluoroalkenes represent a class of privileged structural motifs, which found widespread use in medicinal chemistry. However, the synthetic access to fluoroalkenes was much underdeveloped with previous reported methods suffering from either low step economy or harsh reaction conditions. Here we present a Rh[III]-catalysed tandem C–H/C–F activation for the synthesis of (hetero)arylated monofluoroalkenes. The use of readily available *gem*-difluoroalkenes as electrophiles provides a highly efficient and operationally simple method for the introduction of α-fluoroalkenyl motifs onto (hetero)arenes under oxidant-free conditions. Furthermore, the employment of alcoholic solvent and the *in-situ* generated hydrogen fluoride are found to be beneficial in this transformation, indicating the possibility of the involvement of hydrogen bond activation mode with regards to the C–F bond cleavage step.

[1] Hefei National Laboratory for Physical Sciences at the Microscale and Department of Chemistry, University of Science and Technology of China, Hefei 230026, China. [2] Division of Chemistry and Biological Chemistry, Nanyang Technological University, 50 Nanyang Avenue, Singapore 639798, Singapore. Correspondence and requests for materials should be addressed to C.F. (email: fengchao@ntu.edu.sg) or to T.-P.L. (email: teckpeng@ntu.edu.sg).

**F**luorine, 'small atom with a big ego', due to its intrinsic properties such as small size and high electronegativity in comparison with other halogen atoms, has played a key role in all fields of science[1,2]. More specifically, the incorporation of fluorine or fluorine-containing structural motifs into organic molecule brings about substantial improvement in its bioactivity and provides unique chemical and physical properties, thus enabling the widespread use of this strategy in the field of medicinal chemistry[3–5]. In this context, fluoroalkenes represent a class of very important molecules owing to their biological properties and also their synthetic potential in synthetic organic chemistry[6,7]. Furthermore, as favourable peptide bond mimetics, both electrostatically and geometrically, as well as their resistant nature to enzymatic degradation, the fluoroalkene structural motifs have been attracting increasing interest in medicinal chemistry and drug-discovery research[8–12]. Albeit their great importance, compared with the development of analogous fluorination and trifluoromethylation methodologies[13–18], the synthetic access to alkenyl fluorides remains largely underdeveloped, with most of the reported protocols suffering from the need of substrate pre-activation or using non-readily available starting materials, low regio- or stereo-selectivity and poor functional group tolerance due to the employment of sensitive reagents[19–21]. By taking advantage of the Pd/Cu-catalysed C–H activation strategy, Hoarau and colleagues[22,23] reported elegant works on the fluoroalkenylation of heteroarenes either through C–H/C–Br or C–H/CO$_2$H couplings (Fig. 1d). Notwithstanding the advance attained, the development of a new synthetic method, which streamlines the access to fluoroalkene motifs using readily available building blocks while avoiding substrate pre-activation steps, would still be of meaningful importance in both the synthetic organic chemistry and pharmaceuticals development.

With explosive advancements achieved in the past decade, the directing group assisted C–H bond activation has emerged as a powerful and competent tool, which not only result in fundamental changes in the retrosynthesis but more importantly represents the state-of-the-art in the organic synthesis and shows

the direction of development beyond what traditional synthetic methodologies could bring about[24,25]. As a subclass, the Rh(III)-catalysed C–H functionalization is booming rapidly in the recent years[26–30]. Although a diverse range of synthetically useful transformations have already been attained in this context, the application of Rh(III)-catalysed C–H activation in the fluorine chemistry is unprecedented and more importantly the employment of C($sp^2$)-X as electrophilic coupling partner is totally unrecognized in high valent rhodium catalysis[31,32]. Therefore, the development of a Rh(III)-catalysed C–H functionalization protocol that enables the easy access of biologically relevant fluoroalkenes is highly desirable and the feasibility of such protocol was based on the following considerations: (i) Rh(III)-catalysed alkenylation has evolved to be a competent and reliable method for the introduction of olefin segments, although stoichiometric amount of external oxidant was always required to fulfill the redox demand (Fig. 1a)[33,34]; (ii) notwithstanding its high dissociation energy, the C–F bond could be activated by transition metal or through the formation of hydrogen bond, provided that suitable hydrogen bond donor is present (Fig. 1b)[35–40]; (iii) *gem*-difluoroalkenes represent a class of appealing synthetic intermediates with the C–C double bond being highly polarized because of the electronegativity of fluorine and also the repulsion effect stemming from its unpaired electrons[41,42]. Furthermore, it is well-known that hetero-nucleophiles could undergo facile nucleophilic addition or substitution reactions under basic conditions (Fig. 1c)[43–46]. We reasoned that the putative carbocation character of *gem*-difluoroalkenes is of critical importance in ensuring the regioselectivity of carborhodationic step and the employment of hydrogen-bond donor would result in the activation of C–F bond. Assuming the viability of our proposal, there comes with an affiliated bonus, as no external oxidant was required because of the redox neutral property of this transformation[32]. With our ongoing interest in rhodium and fluorine chemistry[47–49], herein we would like to present the chelation-assisted C–H activation strategy for the direct incorporation of the α-fluoroalkenyl unit using *gem*-difluoroalkene as the fluoroalkene donor (Fig. 1e)[50].

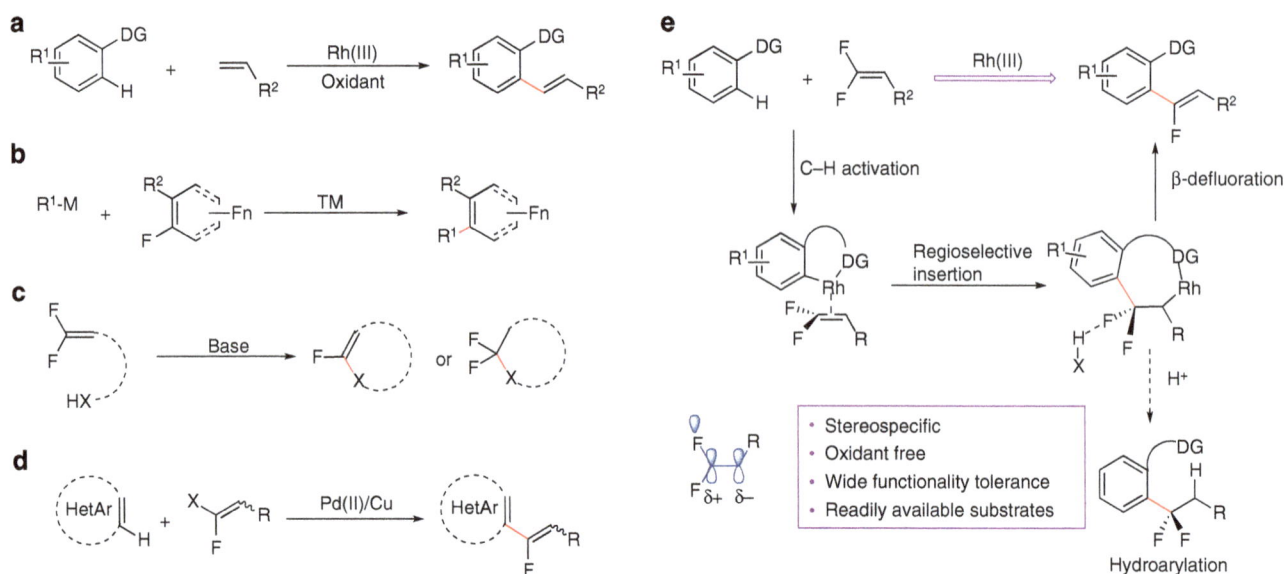

**Figure 1 | Proposed method of rhodium-catalysed α-fluoroalkenylation.** (**a**) Oxidative alkenylation through Rh(III)-catalysed C–H activation. (**b**) Transition-metal-catalysed C–C bond formation through C–F activation. M, metal; TM, transition metal. (**c**) Base-promoted inter- or intramolecular nucleophilic addition or substitution of *gem*-difluoroalkenes with heteronucleophiles. X, hetero atom. (**d**) Pd/Cu-catalysed C–H fluoroalkenylation of heteroarenes. X, Br or CO$_2$H. (**e**) In this report, oxidant-free Rh(III)-catalysed α-fluoroalkenylation of (hetero)arenes. The hydrogen bonding interaction is believed to promote the cleavage of C–F bond, which, in turn, renders this reaction redox neutral. DG, directing group.

## Results

**Reaction condition optimization for *3aa*.** To test our hypothesis, the cross-coupling between 1-(2, 2-difluorovinyl)-4-methoxybenzene (**2a**) and 1-(pyrimidin-2-yl)-1*H*-indole (**1a**) was selected as the model reaction. After examination of a considerable variety of reaction parameters, we were rather pleased to find that the anticipated fluoroalkenylation product could be obtained in 89% yield when using [RhCp*(CH₃CN)₃](SbF₆)₂ as the catalyst and methanol as solvent, and at 80 °C for 16 h (see Supplementary Tables 1–3 for details of reaction optimization). In accordance with our hypothesis mentioned above, this reaction occurred in an excellent regio- and stereo-selective manner due to the intrinsic property of *gem*-difluoro substituents. Considering the fact that hydrogen fluoride (HF) itself is a strong hydrogen-bond donor, which could provide a resultant stabilization energy in the cleavage of C($sp^3$)–F bond process, as well as to evaluate whether such an interaction was present and in turn facilitated our reaction process, control experiments with external HF sequesters were conducted[37]. As expected, the addition of exogenous bases such as NaHCO₃, Na₂CO₃ and 2,6-*di-tert*-butyl-4-methylpyridine for the neutralization of the HF generated in the reaction was proved to be detrimental to this transformation. Taking this phenomenon together with the superiority of alcoholic solvents into consideration, it is reasonable to assume that the activation of C–F bond through

hydrogen-bond formation is involved and proved to be crucial for the execution of this protocol. Furthermore, the hydroarylation product arising from the protonation of carborhodation intermediate was not observed throughout the whole reaction, implying the kinetic favourability of the β-defluorination step[51].

**Substrate scope.** Having obtained the optimized reaction conditions, the issues with respect to functional group tolerance and scope of *gem*-difluoroalkene was thus addressed and the results were summarized in Table 1. In general, with respect to aryl substituted *gem*-difluoroalkenes, both electron-donating and electron-withdrawing functional groups, regardless of the substitution patterns, on the arene moiety were all well tolerated, affording the desired products in high to excellent yield. When electron-rich substituents such as Me or OMe are present in the *ortho*, *meta* or *para* positions, the reaction proceeded well, leading to the formation of the desired products with the yields ranging from 69% to 91% (**3ab**–**3ae**). However, when **2f** was used as a substrate, the reaction afforded product **3af** in 46% yield, which may be attributed to the competitive chelation effect of nitrogen substituent in this case. Substrates with synthetically useful functional groups such as NO₂, CF₃, Ac, CO₂Me and CN (**2g**–**2k**) participated in this reaction efficiently, to deliver the

---

**Table 1 | Scope of gem-difluoroalkene for the C–H/C–F activation reaction.**

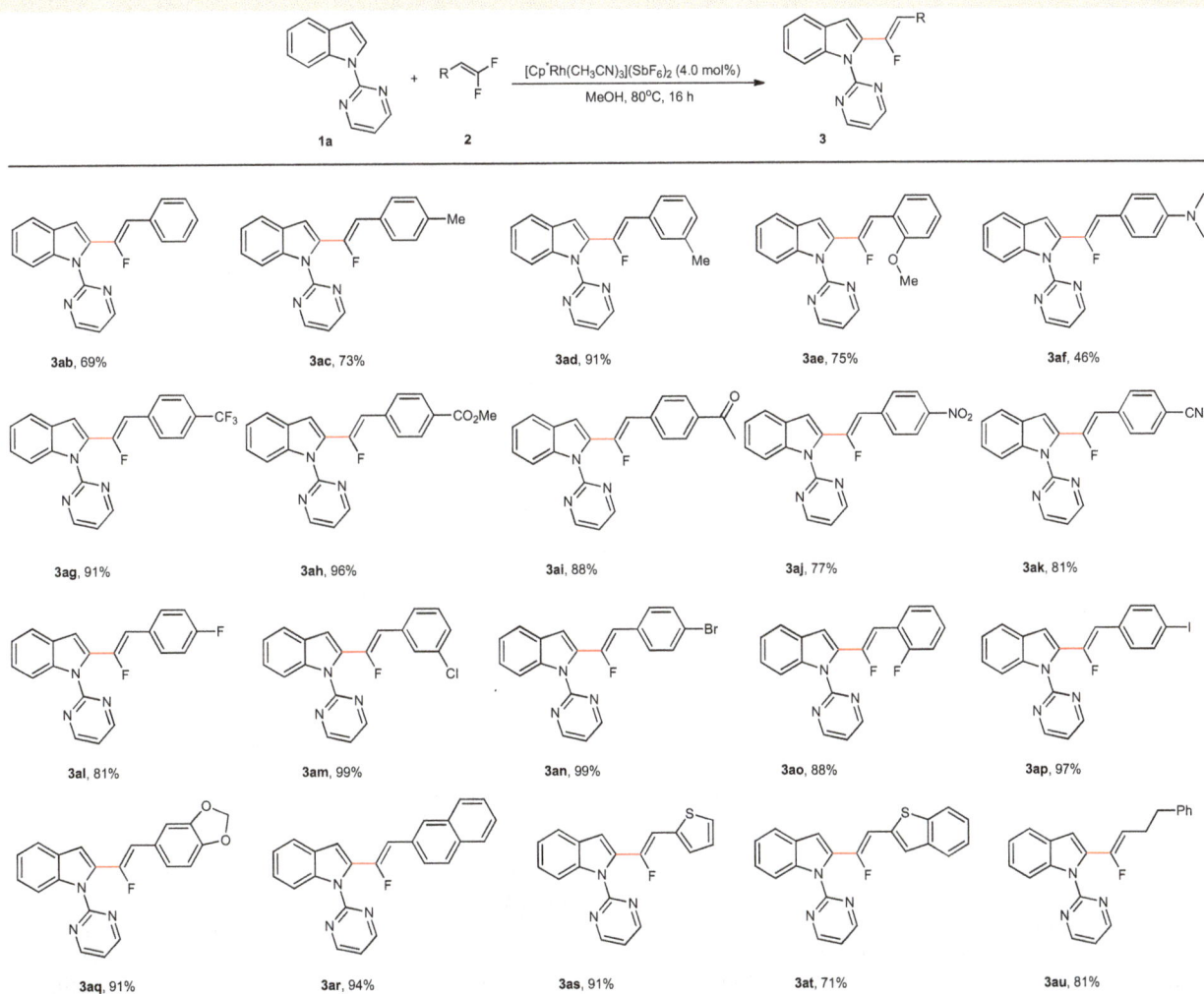

| 3ab, 69% | 3ac, 73% | 3ad, 91% | 3ae, 75% | 3af, 46% |
| 3ag, 91% | 3ah, 96% | 3ai, 88% | 3aj, 77% | 3ak, 81% |
| 3al, 81% | 3am, 99% | 3an, 99% | 3ao, 88% | 3ap, 97% |
| 3aq, 91% | 3ar, 94% | 3as, 91% | 3at, 71% | 3au, 81% |

Experiments were performed with **1** (0.1 mmol), **2a** (0.15 mmol) and [Cp*Rh(MeCN)₃](SbF₆)₂ (0.004 mmol) in MeOH (0.5 ml) for 16 h at 80 °C.

**Table 2 | Scope of indole and 2-arylpyridine derivatives in the C–H/C–F activation reaction.**

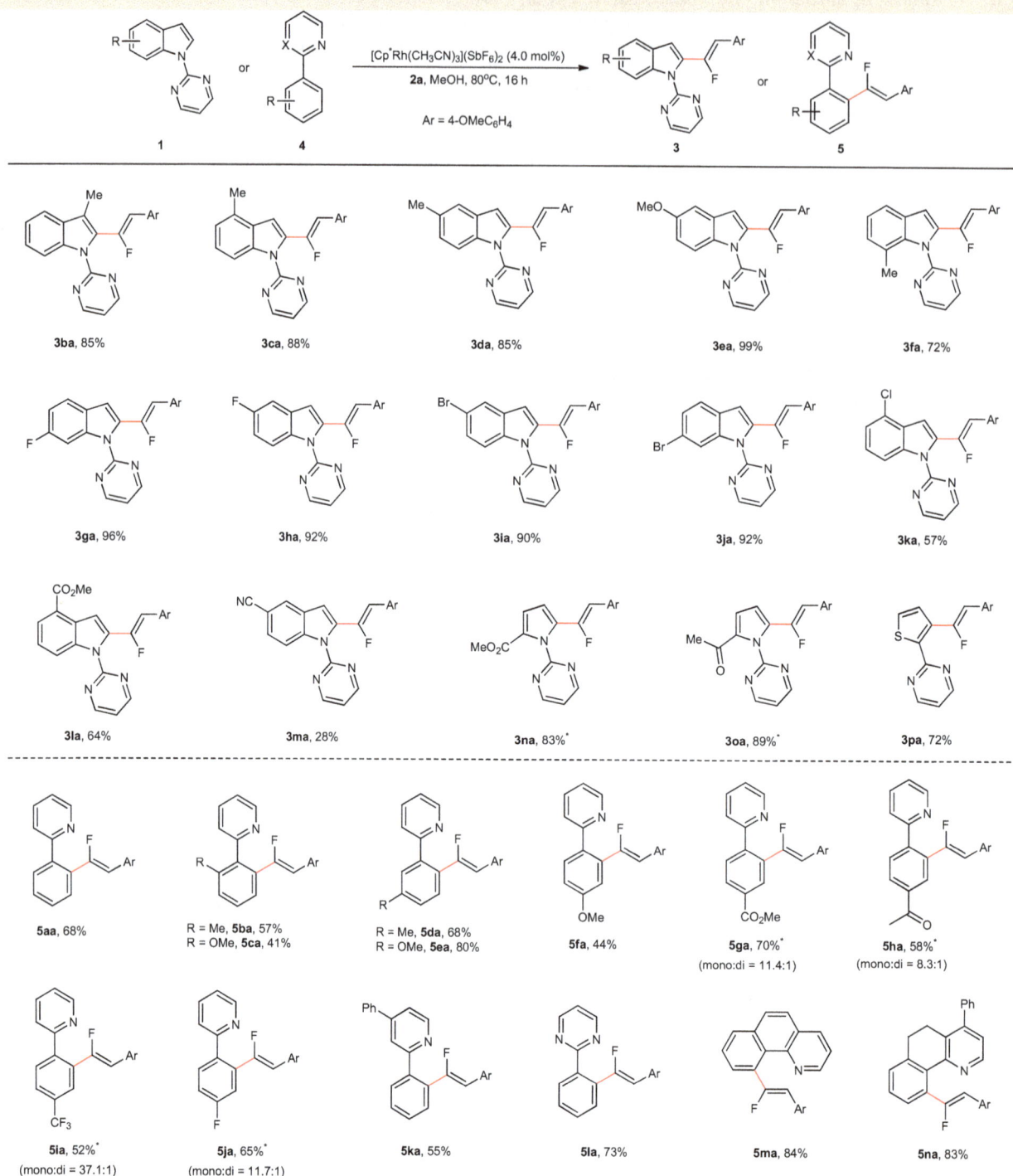

Experiments were performed with **1** (0.1 mmol), **2a** (0.15 mmol) and [Cp*Rh(MeCN)₃](SbF₆)₂ (0.004 mmol) in MeOH (0.5 ml) for 16 h at 80 °C. *Use trifluoroethanol as reaction solvent instead of MeOH.

fluoroalkenylation products in high to excellent yields. Furthermore, we were pleased to find that halogen substituents were also well tolerated in this reaction, which offers the opportunity for further synthetic elaborations (**2l–2p**). The benzo[*d*][1,3]dioxole and naphthalyl derived substrates **2q**, **2r** were also amenable to this reaction conditions to deliver the desired products in 91% and 94% yield, respectively. Notably, heterocycle based *gem*-difluoroalkenes, such as thiophene (**2s**) and benzo[*b*]thiophene (**2t**), also proved to be viable substrates, leading to valuable products in high yields. Pleasingly, this protocol also

accommodates the alkyl derived *gem*-difluoroakenes, and when **2u** was employed as substrate the desired product **2au** was obtained in 81% yield.

The reaction generality of indole derivatives was subsequently investigated and the results were summarized in Table 2. It was found that electron-releasing groups, regardless of their positions on the substrates, were proved to be beneficial, leading to the desired products in excellent yields (**3ba–3fa**). Again, halogen atoms were nicely tolerated (**3ga–3ka**). Substituents with strong electron-withdrawing ability have somewhat deleterious effect on

## Table 3 | Scope of benzamide in the C–H/C–F activation reaction.

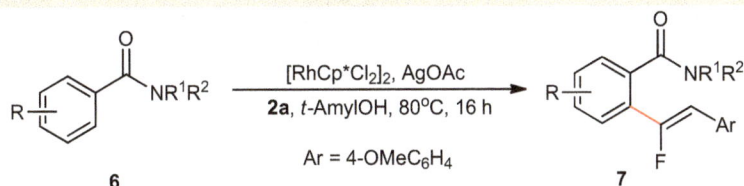

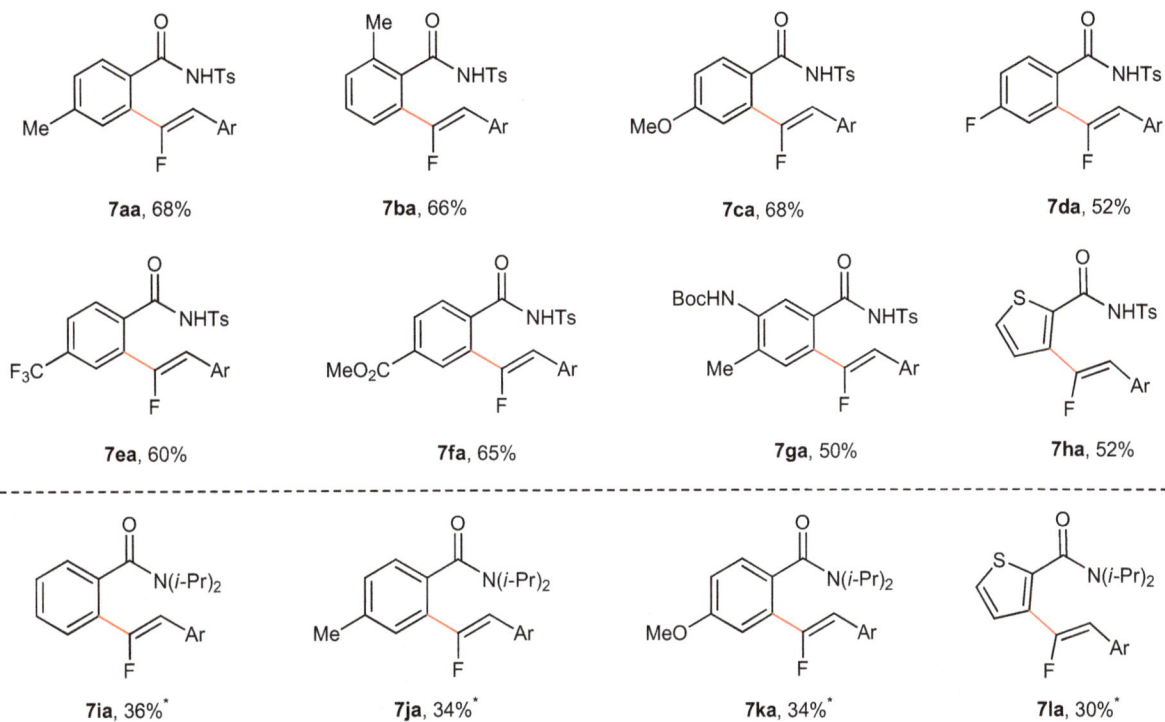

Experiments were performed with **6** (0.1 mmol), **2a** (0.15 mmol), [RhCp*Cl₂]₂ (0.0025 mmol) and AgOAc (0.02 mmol) in *t*-AmylOH (0.5 ml) for 16 h at 80 °C. *Use [Cp*Rh(MeCN)₃](SbF₆)₂ (0.004 mmol) as catalyst, 1,2-dichloroethane as reaction solvent.

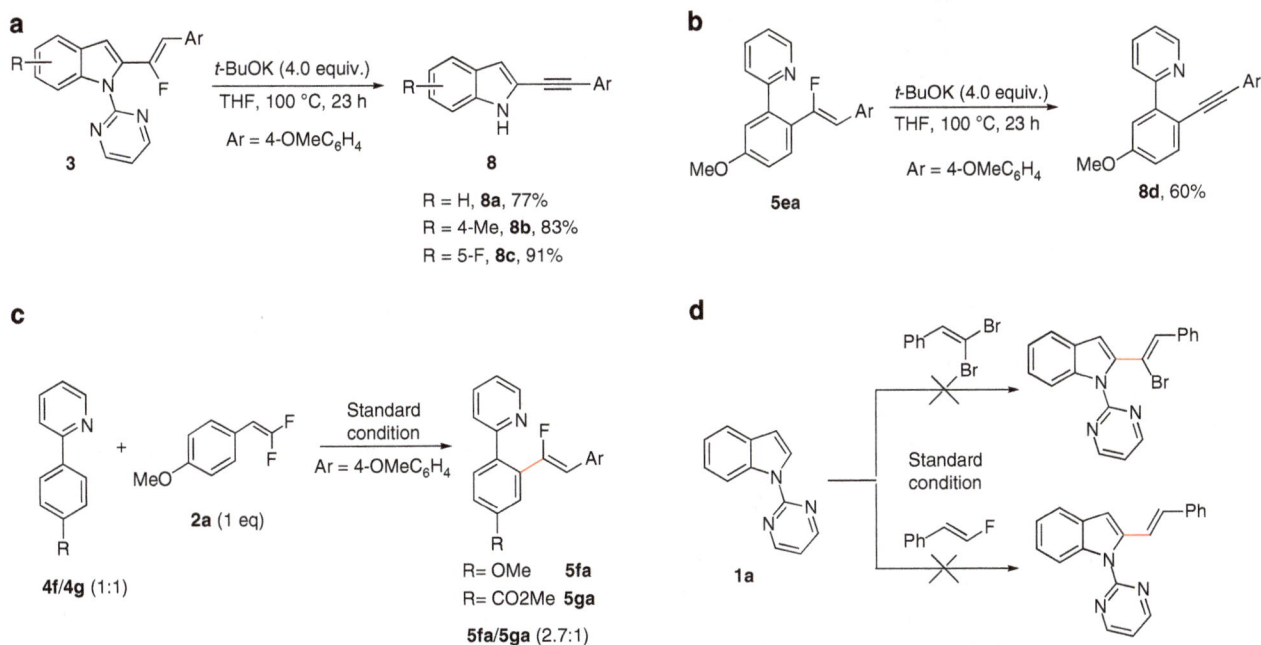

**Figure 2 | Control experiments and synthetic application.** (**a**) Base-promoted dehydrofluoration for the synthesis of 2-alkynyl indole derivatives. (**b**) Base-promoted dehydrofluoration for the synthesis of pyridine derivative. (**c**) Competitive reaction between **4f** and **4g**. (**d**) Control reactions between **1a** and *gem*-dibromoalkene and *mono*-fluoroalkene.

this reaction and the impact became more severe with regard to 5-CN-based indole substrate **1m**, which afforded the coupling product **3ma** in low yield. Furthermore, it was realized that this protocol was not restricted to indole-derived heterocycles as demonstrated in the cases of **1n** to **1p**. It needs to be noted that the switch to trifluoroethanol as solvent was proved to be critical for achieving high reactivity when pyrrole substrates **1n** and **1o** were employed[32]. The superiority of $CF_3CH_2OH$ as the reaction solvent in these cases was attributed to its ability of acting as a better hydrogen-bond donor than MeOH. When 2-(thiophen-2-yl) pyrimidine **1p** was used, the reaction also worked well to afford the desired product (**3pa**) in 72% yield. To further challenge the reaction scope of this reaction, 2-arylpyridine derivatives were subsequently examined. We were pleased to find that when simple 2-phenyl pyridine was tested, the desired product **5aa** was obtained in synthetically useful yield. For substrates containing *ortho*-substituents, such as **4b–4c**, the reaction efficiency was slightly decreased because of the steric hindrance involved, whereas in the cases of **4d** and **4e**, which have *meta*-substituents, the reaction proceeded selectively on the sterically more accessible site to afford the desired products in good yields. In accordance with the solvent effect encountered in the cases of **3na** and **3oa** formations, the use of trifluoroethanol as solvent was shown to be beneficial with respect to the reaction efficiency when using substrates containing electron-withdrawing groups, although at the expense of high selectivity for *mono*-fluoroalkenylations in some cases (**5ga–5ja**). Furthermore, substrates that possess substituents on the pyridine ring also engaged well in this reaction to furnish the desired products in moderate to good yields (**5ka** and **5na**). Finally, 2-phenylpyrimidine and benzo[*h*] quinoline were all viable substrates and participated in this reaction nicely to produce the desired products in high yields (**5la–5ma**).

To further extend the reaction scope and to gain more insight into the limitation of this reaction, substrates that do not contain strong chelation-directing groups, such as pyridine and pyrimidine, were examined and the results were presented in Table 3. Although at the present stage ketone, imine and anilide proved to be not suitable substrates, benzamide derivatives were found to be effective coupling partners, especially the Ts-imide analogues, which provide the desired fluoroalkenylation products in moderate yields, with slight modification of the reaction condition. Specifically, Ts-imide substrates with either electron-donating or electron-withdrawing groups reacted smoothly under modified reaction condition to deliver the desired products (**7aa–7ga**). The *ortho*-substitute was nicely tolerated without any deleterious effect on the reaction efficacy being observed compared with the *para*-analogue (**7aa** versus **7ba**). When substrates that contain other potential chelation groups, such as ester and carbamate were employed, the reaction selectively occurred on the *ortho*-position of Ts-imide directing group (**7fa** and **7ga**). It should be pointed out that thiophene-derived substrate also worked efficiently to produce the product **7ha** in synthetically useful yield. In contrast to the effectiveness of using Ts-imide as directing group, when the reaction was carried out with *N,N*-diisopropylbenzamides as substrates, the reaction tend to afford the desired products in relatively low yields, which was due to the low conversion of starting materials (**7ia–7la**).

**Synthetic elaboration and mechanistic investigation**. To further showcase the synthetic applicability of these fluoroalkenylation products, base-promoted dehydrofluoration protocol for the synthesis of alkynes were attempted. After the examination of a variety of reaction conditions, we were delighted to find that the alkyne product **8a** could be isolated in 77% yield when **3aa** was treated with 4 equivalent of *t*-BuOK in a solution of

tetrahydrofuran (THF) at 100 °C for 23 h. With this strategy, 2-alkynyl indoles **8b** and **8c** were readily obtained in excellent yield (Fig. 2a). Furthermore, the *t*-BuOK-promoted dehydrofluoration was also proved to be effective for the synthesis of 2-(2-(alkynyl)phenyl)pyridine derivatives and when **5ea** was employed, the desired product **8d** was generated in 60% yield (Fig. 2b). To gain insight into the electronic bias of the C–H activation step, competition experiments between **4f/4g** was examined. When an equimolar mixture of **4f**, **4g** and **2a** were subjected to the optimized reaction condition, the desired products **5fa** and **5ga** were formed in the ratio of 2.7:1, thus indicating that electron-rich substrate was more favoured in the C–H activation step in this reaction (Fig. 2c). To determine the role of *gem*-difluoro substituents in this reaction, substrates with *gem*-dibromo and *mono*-fluoro substituents were subjected to the optimized reaction condition; however, neither led to the formation of the desired products. These experiments clearly demonstrated that the *gem*-difluoro substituents was indispensable for the activation of the C–C double bond, which is in full agreement with our hypothesis (Fig. 2d).

## Discussion

A highly effective rhodium(III)-catalysed α-fluoroalkenylation of (hetero)arenes through C–H/C–F bond activations was developed. This reaction proceeds smoothly and stereoselectively under base- and oxidant-free reaction conditions to deliver a diverse range of synthetically useful and pharmaceutically relevant *cis*-alkenyl fluorides in good yields. Furthermore, the involvement of activation of the C–F bond through hydrogen-bond formation is crucial for the success of this transformation. Last but not the least, the success of incorporation of α-fluoroalkenyl unit onto (hetero)arenes through C–H/C–F activation manifold was in principle amenable to be extended to other processes wherein the key intermediates arylrhodiums were not formed by C–H activation, provided that each elemental reaction step was compatible and balanced, thereby offering a more generalized and versatile protocol for the synthesis of functionalized fluoroalkene derivatives.

## Methods

**Materials.** For 1H, 13C NMR spectra of compounds in this manuscript, see Supplementary Figs 1–66. For optimization of reaction conditions, see Supplementary Tables 1–3. For details of the synthetic procedures, see Supplementary Methods.

**Syntheses of products 3 and 5.** An oven-dried 10-ml Schlenk tube with a magnetic stirring bar was charged with **1** or **4** (0.1 mmol), [RhCp*(CH3CN)3] (SbF6)2 (3.3 mg, 0.004 mmol), **2** (0.15 mmol) in sequence, followed by adding anhydrous MeOH (0.5 ml) through syringe. The reaction tube was sealed with Teflon-coated screw cap and the reaction solution was stirred at 80 °C for 16 h. After cooling the reaction mixture to room temperature and removing the solvent *in vacuo*, the resulting residue was purified by silica gel column chromatography to afford the desired product **3** or **5**.

## References

1. Kirsch, P. *Modern Fluoroorganic Chemistry: Synthesis, Reactivity, Applications* (Wiley, 2004).
2. Hagmann, W. K. The many roles for fluorine in medicinal chemistry. *J. Med. Chem.* **51**, 4359–4369 (2008).
3. Liang, T., Neumann, C. N. & Ritter, T. Introduction of fluorine and fluorine-containing functional groups. *Angew. Chem. Int. Ed.* **52**, 8214–8264 (2013).
4. Muller, K., Faeh, C. & Diederich, F. Fluorine in pharmaceuticals: looking beyond intuition. *Science* **317**, 1881–1886 (2007).
5. O'Hagan, D. & Deng, H. Enzymatic fluorination and biotechnological developments of the fluorinase. *Chem. Rev.* **115**, 634–649 (2014).
6. Landelle, G. *et al.* Synthetic approaches to monofluoroalkenes. *Chem. Soc. Rev.* **40**, 2867–2908 (2011).

7. Jakobsche, C. E., Peris, G. & Miller, S. J. Functional analysis of an aspartate-based epoxidation catalyst with amide-to-alkene peptidomimetic catalyst analogues. *Angew. Chem. Int. Ed.* **47**, 6707–6711 (2008).

8. Niida, A. & Fujii, N. Unequivocal synthesis of (Z)-alkene and (E)-fluoroalkene dipeptide isosteres to probe structural requirements of the peptide transporter PEPT1. *Org. Lett.* **8**, 613–616 (2006).

9. Van der Veken, P. & Augustyns, K. Fluoro-olefins as peptidomimetic inhibitors of dipeptidyl peptidases. *J. Med. Chem.* **48**, 1768–1780 (2005).

10. Edmondson, S. D. *et al.* Fluoroolefins as amide bond mimics in dipeptidyl peptidase IV inhibitors. *Bioorg. Med. Chem. Lett.* **18**, 2409–2413 (2008).

11. Couve-Bonnaire, S., Cahard, D. & Pannecoucke, X. Chiral dipeptide mimics possessing a fluoroolefin moiety: a relevant tool for conformational and medicinal studies. *Org. Biomol. Chem.* **5**, 1151–1157 (2007).

12. Osada, S. *et al.* Fluoroalkene modification of mercaptoacetamide-based histone deacetylase inhibitors. *Bioorg. Med. Chem.* **18**, 605–611 (2010).

13. Furuya, T., Kamlet, A. S. & Ritter, T. Catalysis for fluorination and trifluoromethylation. *Nature* **473**, 470–477 (2011).

14. Watson, D. A. & Buchwald, S. L. Formation of ArF from LPdAr(F): catalytic conversion of aryl triflates to aryl fluorides. *Science* **325**, 1661–1664 (2009).

15. Wang, X., Truesdale, L. & Yu, J.-Q. Pd(II)-catalyzed ortho-trifluoromethylation of arenes using TFA as a promoter. *J. Am. Chem. Soc.* **132**, 3648–3649 (2010).

16. Ye, Y., Schimler, S. D., Hanley, P. S. & Sanford, M. S. Cu(OTf)₂-mediated fluorination of aryltrifluoroborates with potassium fluoride. *J. Am. Chem. Soc.* **135**, 16292–16295 (2013).

17. Cho, E. J. & Buchwald, S. L. The palladium-catalyzed trifluoromethylation of aryl chlorides. *Science* **328**, 1679–1681 (2010).

18. Fier, P. S. & Hartwig, J. F. Copper-mediated fluorination of aryl iodides. *J. Am. Chem. Soc.* **134**, 10795–10798 (2012).

19. Yang, M.-H., Matikonda, S. S. & Altman, R. A. Preparation of fluoroalkenes via the shapiro reaction: direct access to fluorinated peptidomimetics. *Org. Lett.* **15**, 3894–3897 (2013).

20. Okoromoba, O. E., Han, J., Hammond, G. B. & Xu, B. Designer HF-based fluorination reagent: highly regioselective synthesis of fluoroalkenes and *gem*-difluoromethylene compounds from alkynes. *J. Am. Chem. Soc.* **136**, 14381–14384 (2014).

21. Ghosh, A. K. & Zajc, B. High-yield synthesis of fluorinated benzothiazolyl sulfones: general synthons for fluoro-Julia olefinations. *Org. Lett.* **8**, 1553–1556 (2006).

22. Schneider, C. & Hoarau, C. Palladium- and copper-catalyzed stereocontrolled direct C-H fluoroalkenylation of heteroarenes using *gem*-bromofluoroalkenes. *Angew. Chem. Int. Ed.* **52**, 3246–3249 (2013).

23. Rousee, K., Schneider, C., Couve-Bonnaire, S., Pannecoucke, X. & Hoarau, C. Pd- and Cu-catalyzed stereo- and regiocontrolled decarboxylative/C-H fluoroalkenylation of heteroarenes. *Chem. Eur. J.* **20**, 15000–15004 (2014).

24. Colby, D. A., Bergman, R. G. & Ellman, J. A. Rhodium-catalyzed C-C bond formation via heteroatom-directed C-H bond activation. *Chem. Rev.* **110**, 624–655 (2010).

25. Wencel-Delord, J. & Glorius, F. C-H bond activation enables the rapid construction and late-stage diversification of functional molecules. *Nat. Chem.* **5**, 369–375 (2013).

26. Patureau, F. W., Wancel-Delord, J. & Glorius, F. Cp*Rh-catalyzed C-H activations: versatile dehydrogenative cross-coupling of C$_{sp2}$ C-H positions with olefins, alkynes, and arenes. *Aldrichim. Acta* **45**, 31–41 (2012).

27. Song, G., Wang, F. & Li, X. C-C, C-O and C-N bond formation via rhodium(III)-catalyzed oxidative C-H activation. *Chem. Soc. Rev.* **41**, 3651–3678 (2012).

28. Hyster, T. K. & Rovis, T. Rhodium-catalyzed oxidative cycloaddition of benzamides and alkynes via C – H/N – H activation. *J. Am. Chem. Soc.* **132**, 10565–10569 (2010).

29. Li, Y. & Shi, Z.-J. Rhodium-catalyzed direct addition of aryl C-H bonds to N-sulfonyl aldimines. *Angew. Chem. Int. Ed.* **50**, 2115–2119 (2011).

30. Stuart, D. R., Bertrand-Laperle, M., Burgess, K. M. N. & Fagnou, K. Indole synthesis via rhodium catalyzed oxidative coupling of acetanilides and internal alkynes. *J. Am. Chem. Soc.* **130**, 16474–16475 (2009).

31. Sakoda, K., Mihara, J. & Ichikawa, J. Heck-type 5-endo-trig cyclization promoted by vinylic fluorines: synthesis of 5-fluoro-3H-pyrroles. *Chem. Commun.* **41**, 4684–4686 (2005).

32. Yokota, M., Fujita, D. & Ichikawa, J. Activation of 1,1-difluoro-1-alkenes with a transition-metal complex: palladium(II)-catalyzed Friedel-Crafts-type cyclization of 4,4-(difluorohomoallyl)arenes. *Org. Lett.* **9**, 4639–4642 (2007).

33. Ueura, K., Satoh, T. & Miura, M. An efficient waste-free oxidative coupling via regioselective C – H bond cleavage: Rh/Cu-catalyzed reaction of benzoic acids with alkynes and acrylates under air. *Org. Lett.* **9**, 1407–1409 (2007).

34. Patureau, F. W. & Glorius, F. Rh catalyzed olefination and vinylation of unactivated acetanilides. *J. Am. Chem. Soc.* **132**, 9982–9983 (2010).

35. Amii, H. & Uneyama, K. C-F bond activation in organic synthesis. *Chem. Rev.* **109**, 2119–2183 (2009).

36. Ohashi, M. & Ogoshi, S. Palladium-catalyzed coupling reactions of tetrafluoroethylene with arylzinc compounds. *J. Am. Chem. Soc.* **133**, 3256–3259 (2011).

37. Champagne, P. A., Benhassine, Y., Desroches, J. & Paquin, J.-F. Friedel-Crafts reaction of benzyl fluorides: selective activation of C-F bonds as enabled by hydrogen bonding. *Angew. Chem. Int. Ed.* **53**, 13835–13839 (2014).

38. Zhu, F. & Wang, Z.-X. Nickel-catalyzed cross-coupling of aryl fluorides and organozoic reagents. *J. Org. Chem.* **79**, 4285–4292 (2014).

39. Tobisu, M., Xu, T., Shimasaki, T. & Chatani, N. Nickel-catalyzed Suzuki-Miyaura reaction of aryl fluorides. *J. Am. Chem. Soc.* **133**, 19505–19511 (2011).

40. Nakamura, Y., Yoshikai, N., Ilies, L. & Nakamura, E. Nickel-catalyzed monosubstitution of polyfluoroarenes with orgaozinc reagents using alkoxydiphosphine ligand. *Org. Lett.* **14**, 3316–3319 (2012).

41. Ichikawa, J., Yokota, M., Kudo, T. & Umezaki, S. Efficient helicene synthesis: Friedel-Crafts-type cyclization of 1,1-difluoro-1-alkenes. *Angew. Chem. Int. Ed.* **47**, 4870–4873 (2008).

42. Gao, B., Zhao, Y. & Hu, J. AgF-mediated fluorinative cross-coupling of two olefins: facile access to α-CF₃ Alkenes and β-CF₃ Ketones. *Angew. Chem. Int. Ed.* **54**, 638–642 (2015).

43. Yang, E., Reese, M. R. & Humphrey, J. M. Synthesis of α,α-difluoroethyl aryl and heteroaryl ethers. *Org. Lett.* **14**, 3944–3947 (2012).

44. Ichikawa, J., Nadano, R. & Ito, N. 5-endo Heck-type cyclization of 2 (trifluoromethyl)allyl ketone oximes: synthesis of 4-difluoromethylene-substituted 1-pyrrolines. *Chem. Commun.* **42**, 4425–4427 (2006).

45. Xiong, Y., Zhang, X., Huang, T. & Cao Song. Synthesis of N-(α-fluorovinyl) zaoles by the reaction of difluoroalkenes with azoles. *J. Org. Chem.* **79**, 6395–6402 (2014).

46. Ichikawa, J., Wada, Y., Fujiwara, M. & Sakoda, K. The nucleophilic 5-endo-trig cyclization of 1,1-difluoro-1-alkenes: ring-fluorinated hetero- and carbocycle synthesis and remarkable effect of the vinylic fluorines on the disfavored process. *Synthesis* **2002**, 1917–1936 (2002).

47. Feng, C. & Loh, T.-P. Rhodium-catalyzed C-H alkynylation of arenes at room temperature. *Angew. Chem. Int. Ed.* **53**, 2722–2726 (2014).

48. Feng, C. & Loh, T.-P. Directing group assisted copper-catalyzed olefinic trifluoromethylation of electron-deficient alkenes. *Angew. Chem. Int. Ed.* **52**, 12414–12417 (2013).

49. Feng, C. & Loh, T. -P. Copper-catalyzed olefinic trifluoromethylation of enamides at room temperature. *Chem. Sci.* **3**, 3458–3461 (2012).

50. Li, H. & Xi, Z. Intramolecular C-F and C-H bond cleavage promoted by butadienyl heavy Grignard reagents. *Nat. Commun.* **5**, 4508–4517 (2014).

51. Zhu, J., Ni, C., Gao, B. & Hu, J. Copper(0)-mediated fluoroalkylation of iodobenzene with 2-bromo-1,1,2,2-tetrafluoroethyl compounds: investigation on the influence of R substituent on the reactivity of RCF₂Cu species. *J. Fluorine Chem.* **171**, 139–147 (2015).

## Acknowledgements

This study is dedicated to Professor Chunxu Lv on the occasion of his 72th birthday. We gratefully acknowledge the funding support of the State Key Program of National Natural Science Foundation of China (21432009). We also thank the Nanyang Technological University for the funding of this research. We are grateful to Mr. Koh Peng Fei Jackson for his careful proofreading of the final manuscript.

## Author contributions

C.F. and T.P.L. directed the project. P.P.T. and C.F. conducted the experiments. C.F. wrote the manuscript with the assistance of P.P.T. C.F. conceived and designed the project. All authors participated in data analyses and discussions.

## Additional information

**Competing financial interests:** The authors declare no competing financial interests.

# 6

# Radical cascade reaction of alkynes with N-fluoroarylsulfonimides and alcohols

Guangfan Zheng[1], Yan Li[1], Jingjie Han[1], Tao Xiong[1] & Qian Zhang[1]

Cascade reactions initiated by radical addition to alkynes are synthetically very attractive because they enable access to highly complex molecular skeletons in only few synthetic steps under usually mild conditions. Here we report a general radical cascade reaction of alkynes, N-fluoroarylsulfonimides and alcohols, enabling the efficient synthesis of important α-amino-α-aryl ketones from readily available starting materials via a single operation. During this process, the highly regioselective nitrogen-centred radical addition to internal and terminal alkynes generating vinyl radicals and the next explicit migration of aryl group from the nitrogen source lead the following efficient desulfonylation, oxygenation, and semi-pinacol rearrangement. In addition, the semi-pinacol rearrangement precursors, α-alkyloxyl-α,α-diaryl imines, could also be efficiently obtained under milder conditions. This methodology might open a new entry for designing intermolecular radical cascade reaction of alkynes.

[1] Department of Chemistry, Northeast Normal University, Changchun 130024, China. Correspondence and requests for materials should be addressed to Q.Z. (email: zhangq651@nenu.edu.cn).

The regioselective construction of C–N bond under mild conditions remains an attractive research field due to the ubiquitous presence of amines in both naturally occurring and synthetic compounds, which manifest high levels of biological activity[1,2]. Alkyne functionalization, the addition of functional groups across a triple bond, exemplifies a class of reactions with significant synthetic potential. Accordingly, direct amination reaction of simple alkynes involving general intermolecular C–N bond construction step, such as hydroamiantion[3–6], diamination[7,8], aminooxygenation[9–12], aminohalogenation[13,14] and aminoacylation[15] have been successfully developed, during which nucleophilic amination was usually involved with a few strategies employing electrophilic nitrogen sources (Fig. 1a). Cascade reactions initiated by radical addition to alkynes are synthetically very attractive because they allow access to highly complex molecular skeletons in only few synthetic steps under usually mild conditions, enabling them to exhibit high functional group compatiblity[16]. Although intermolecular radical addition to alkynes generating reactive vinyl radicals to perform intramolecular cascade reactions have been well established, their intermolecular multi-component equivalents remain a formidable challenge (vide infra). It is thus not surprising in that light that even simple addition reactions of nitrogen-centred radicals to alkynes are very rare[17,18]. In fact, compared with the well-established nucleophilic and electrophilic amination reaction, the construction of C–N bonds based on nitrogen-centred radicals have not received sufficient attention. The highly reactive vinyl radical generated by the addition of nitrogen-centred radical to alkynes offers a unique platform for radical-based processes mechanistically distinct from ionic pathways. We anticipate that this novel bond-forming strategy could be harnessed for facile construction of otherwise challenging nitrogen containing molecular architectures with traditional methodologies.

Challenges for the development of general cascade reactions initiated by nitrogen-centred radical addition to alkynes mainly resulted from two reasons: (1) the usually harsh conditions for the generation of nitrogen-centred radicals and their leading

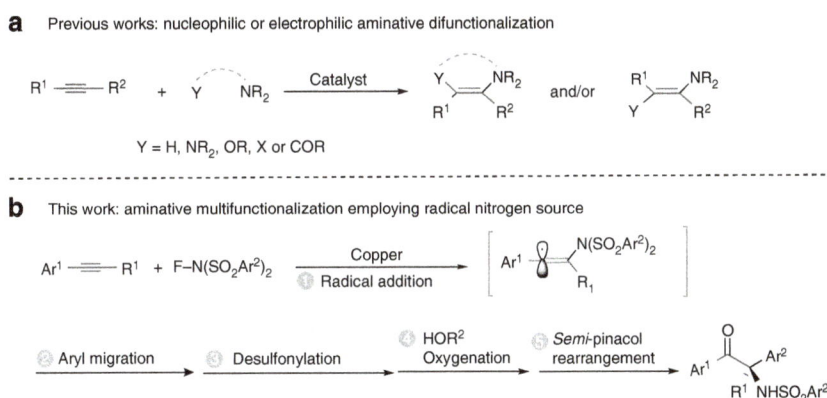

**Figure 1 | Aminative functionalization of alkynes.** (**a**) Nuclephilic or electrophilic aminative difunctionalization. (**b**) Radical cascade aminative multifunctionalization.

## Table 1 | Optimization of the reaction conditions.

| Entry | Catalyst | Additive | Solvent | Temperature (°C) | Time (h) | Yield (%) |
|---|---|---|---|---|---|---|
| 1 | $Cu(OTf)_2$ | None | $CH_3CN$ | 90 | 5 | 58 |
| 2 | CuCl | None | $CH_3CN$ | 90 | 5 | 40 |
| 3 | $Fe(OTf)_2$ | None | $CH_3CN$ | 90 | 7 | 50 |
| 4 | $Zn(OTf)_2$ | None | $CH_3CN$ | 90 | 20 | 47 |
| 5 | $Sc(OTf)_3$ | None | $CH_3CN$ | 90 | 20 | 44 |
| 6 | $Cu(OTf)_2$ | None | $CH_3CN$ | 70 | 5 | 54 |
| 7 | $Cu(OTf)_2$ | None | $CH_3CN$ | 50 | 24 | Trace |
| 8 | $Cu(OTf)_2$ | None | DCM | 70 | 24 | 34 |
| 9 | $Cu(OTf)_2$ | None | EtOAc | 70 | 24 | NR* |
| 10 | $Cu(OTf)_2$ | None | THF | 70 | 24 | —† |
| 11 | $Cu(OTf)_2$ | $C_6H_5COOH$ | $CH_3CN$ | 70 | 4 | 60 |
| 12 | $Cu(OTf)_2$ | $CH_3COOH$ | $CH_3CN$ | 70 | 3 | 61 |
| 13 | $Cu(OTf)_2$ | $CF_3SO_3H$ | $CH_3CN$ | 70 | 4 | 40 |
| 14 | $Cu(OTf)_2$ | $CF_3COOH$ | $CH_3CN$ | 70 | 5 | 78 |
| 15 | None | None | $CH_3CN$ | 70 | 24 | NR* |
| 16 | None | None | $CH_3CN$ | 130 | 10 | 32 |

Reaction conditions: **1a** (0.5 mmol), NFSI (1.5 equiv., 0.75 mmol), $CH_3OH$ (3 equiv., 1.5 mmol), catalysts (10 mol %), additives (1 equiv., 0.5 mmol), anhydrous solvents (2 ml), $N_2$ atmosphere. Isolated yield.
*NR, no reaction.
†$HN(SO_2Ph)_2$ was identified.

**Table 2 | Aminative multifunctionalization of terminal alkynes with NFSI.**

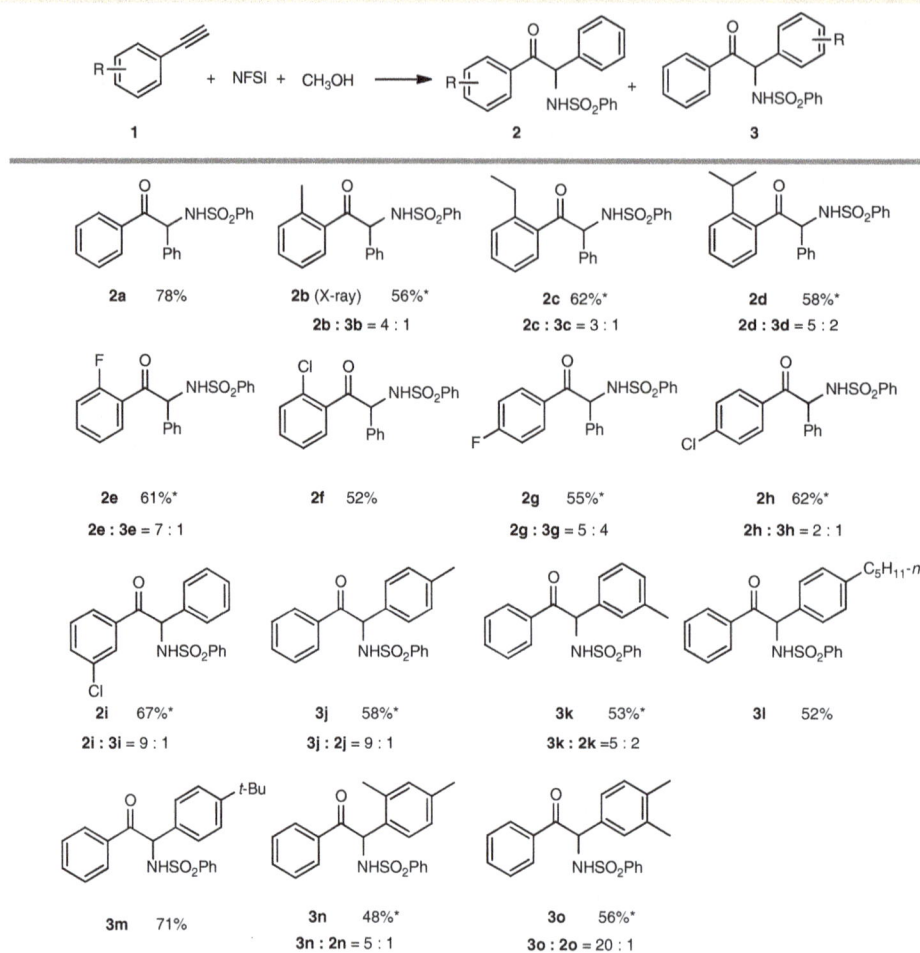

$$ 1 + NFSI + CH_3OH \longrightarrow 2 + 3 $$

| Compound | Yield | Ratio |
|---|---|---|
| 2a | 78% | |
| 2b (X-ray) | 56%* | 2b : 3b = 4 : 1 |
| 2c | 62%* | 2c : 3c = 3 : 1 |
| 2d | 58%* | 2d : 3d = 5 : 2 |
| 2e | 61%* | 2e : 3e = 7 : 1 |
| 2f | 52% | |
| 2g | 55%* | 2g : 3g = 5 : 4 |
| 2h | 62%* | 2h : 3h = 2 : 1 |
| 2i | 67%* | 2i : 3i = 9 : 1 |
| 3j | 58%* | 3j : 2j = 9 : 1 |
| 3k | 53%* | 3k : 2k = 5 : 2 |
| 3l | 52% | |
| 3m | 71% | |
| 3n | 48%* | 3n : 2n = 5 : 1 |
| 3o | 56%* | 3o : 2o = 20 : 1 |

NFSI, N-fluorobenzenesulfonimide.

Reaction condition: 1 (0.5 mmol), NFSI (1.5 equiv., 0.75 mmol), CH₃OH (3 equiv., 1.5 mmol), Cu(OTf)₂ (10 mol %) and TFA (1.0 equiv., 0.5 mmol) in CH₃CN (2 ml) at 70 °C under N₂ atmosphere for 5 h. Isolated yield.
*Mixture of two isomers. The ratio was determined by ¹H NMR analysis.

propensity for hydrogen abstraction or engaging in other degradation pathway; (2) the lack of a general intramolecular trapping manner to transfer the highly reactive incipient vinyl radical for further intermolecular cascade design. Recently, we developed copper-catalysed aminocyanation, diamination and aminoflurination reaction of alkenes via the efficient generation of nitrogen-centred radical from N-fluorobenzenesulfonimide (NFSI) under mild conditions[19,20]. As part of our continuing interest in employing NFSI as efficient amination nitrogen source[21–24], in this article, a novel aminative multifunctionalization cascade reaction of alkyne with N-fluoroarylsulfonimide (as both nitrogen and aryl source) and alcohol (as oxygen source) was developed. Utilizing this simple transformation, α-amino-α-aryl ketones could be efficiently synthesized from both terminal and internal alkynes (Fig. 1b). α-Amino-α-aryl ketones, such as N-methylwelwitindolinone C isothiocyanate[25], ketamine[26] and prasugrel[27] belong to an important class of biologically active natural products and pharmaceuticals. They are also useful precursors for the synthesis of heterocycles[28–31] and 1,2-amino alcohols[32,33]. Recently, starting from readily available substrates, interesting methods for α-amino-α-aryl ketones such as cross-aza benzoin reaction of aldehydes with aryl imines[34–38] and acyloin-type cross-coupling of aryl imines with nitriles[39] were developed. Although significant progress has been made in the formation of

C–N[40–43] and C–C(aryl)[44–48] bonds at α-position of the carbonyl group, it is a great challenge to simultaneously form C–N and C–C(aryl) bonds especially for the construction of quaternary carbon centres.

Herein, we report a cascade reaction that offers highly efficient construction of α-amino-α-aryl ketones starting from readily available alkynes, N-fluoroarylsulfonimides and alcohols via a highly efficient sequential regioselective nitrogen-centred radical addition to alkyne/aryl migration/desulfonylation/oxygenation/semi-pinacol rearrangement process (Fig. 1b). In addition, the semi-pinacol rearrangement precursors, α-alkyloxyl-α,α-diaryl imines, could also be efficiently obtained under milder conditions.

## Results

**Optimization for the sythesis of α-amino-α-aryl ketones.** On the basis of previous reports developed by us[19,20] and others[49–51], we sought to use NFSI as both nitrogen source and aryl source to investigate aminative multifunctionalization of alkynes. Our investigation commenced with the reaction of phenylacetylene (1a, 0.5 mmol) with NFSI (0.75 mmol, 1.5 equiv.) in the presence of Cu(OTf)₂ (10 mol %) at 90 °C in commercially available CH₃CN (2 ml) under N₂ atmosphere, α-amino-α-aryl ketone 2a was obtained in 30% yield after 8 h. However, when dry CH₃CN

**Table 3 | Aminative multifunctionalization of terminal alkynes with NFR1-4.**

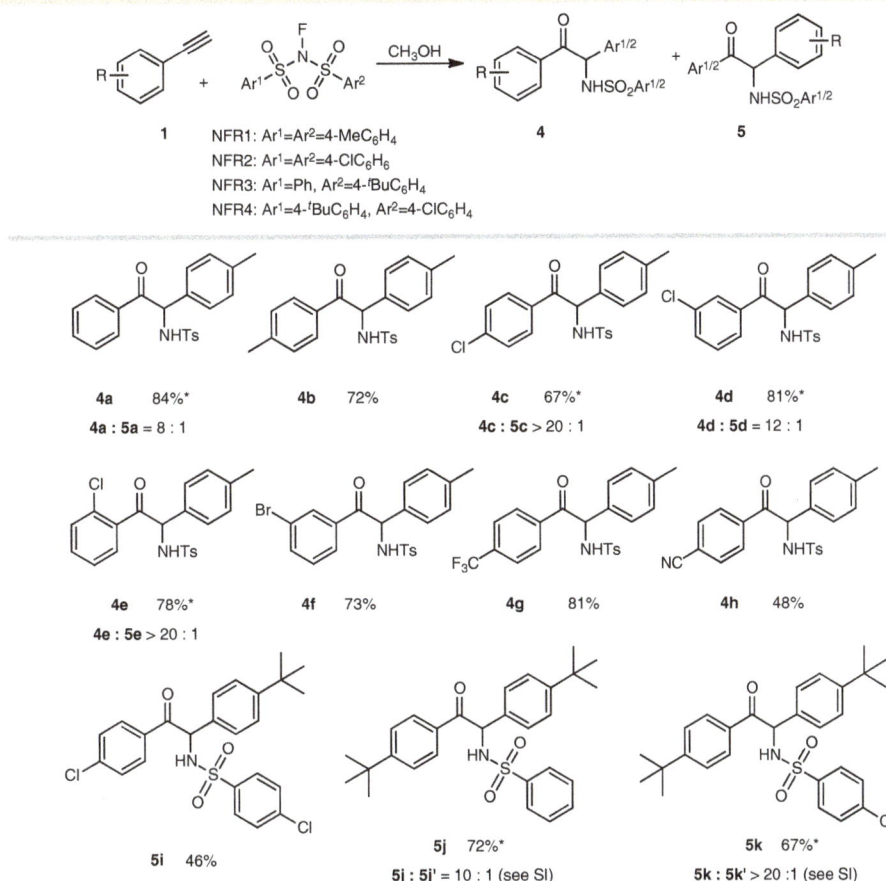

NFR1: $Ar^1=Ar^2=4\text{-MeC}_6H_4$
NFR2: $Ar^1=Ar^2=4\text{-ClC}_6H_6$
NFR3: $Ar^1=Ph$, $Ar^2=4\text{-}^tBuC_6H_4$
NFR4: $Ar^1=4\text{-}^tBuC_6H_4$, $Ar^2=4\text{-ClC}_6H_4$

4a      84%*
4a : 5a = 8 : 1

4b      72%

4c      67%*
4c : 5c > 20 : 1

4d      81%*
4d : 5d = 12 : 1

4e      78%*
4e : 5e > 20 : 1

4f      73%

4g      81%

4h      48%

5i      46%

5j      72%*
5j : 5j' = 10 : 1 (see SI)

5k      67%*
5k : 5k' > 20 :1 (see SI)

Reactions condition: **1** (0.2 mmol), NFR1-4 (1.5 equiv., 0.3 mmol), $CH_3OH$ (3 equiv., 0.6 mmol), $Cu(OTf)_2$ (10 mol %) and TFA (1.0 equiv., 0.2 mmol) in $CH_3CN$ (2 ml) at 90 °C under $N_2$ atmosphere for 8 h. Isolated yields.
*Mixture of two isomers. The ratio was determined by $^1H$ NMR analysis.

was used, no reaction occurred. Therefore, water (1.5 mmol, 3 equiv.) was added to the reaction system, α-amino-α-aryl ketone **2a** was obtained in 37% yield. In this reaction, C–N, C–C(aryl) and C=O bonds were simultaneously introduced to alkyne **1a**. Delightfully, the readily available $CH_3OH$ was viable and furnished **2a** in 58% yield (Table 1, entry 1). With pyridine-N-oxide or $CH_3COOH$ as oxygen source, no **2a** was obtained. So the reaction of **1a** with NFSI and $CH_3OH$ was used as the model to optimize the reaction conditions. As shown in Table 1, other catalysts, such as CuCl, $Fe(OTf)_2$, $Zn(OTf)_2$ and $Sc(OTf)_3$ could catalyse the reaction, but no improved result was obtained (Table 1, entry 2–5). A decrease in the temperature from 90 to 70 °C afforded **2a** in 54% yield (Table 1, entry 6). Further lowering the temperature to 50 °C resulted in sluggish reaction and only a trace amount of **2a** was observed (Table 1, entry 7). Screening of solvents (Table 1, entries 8–10) identified $CH_3CN$ as the solvent of choice. Finally, a satisfactory yield of 78% was achieved when $CF_3COOH$ was employed as additive (Table 1, entries 11–14). Considering the number of steps involved in this process, this overall yield indicates of high efficiency of this radical involved cascade. Interestingly, the reaction could also proceed at 130 °C without catalyst to provide **2a** in 32% yield (Table 1, entry 16).

**Scope of terminal alkyne and N-fluoroarylsulfonimide substrates.** With the optimized conditions at hand (Table 1, entry 14), the scope of aminative multifunctionalization of terminal alkynes was

investigated. The tested phenylethethylene derivatives **1** smoothly reacted with NFSI and $CH_3OH$ to afford the corresponding α-amino-α-aryl ketones **2** or **3** in 48–78% yields (Table 2). For α-amino-α-aryl ketones **2**, phenyl group from NFSI connected to the terminal carbon of alkynes and for α-amino-α-aryl ketones **3**, aryl group from alkynes **1** connected to the terminal carbon of alkynes. For alkynes **1b**–**1f** which bear *ortho*-substitutions, **2b**–**2f** became the major products with the less sterically hindered phenyl group selectively migrated, forming C – C (aryl) bonds. Alkynes **1g**–**1o** with electron donating or withdrawing groups afforded major products **2g**–**2i** and **3j**–**3o** in which the comparatively electron-rich aromatic ring migrated to form C – C (aryl) bonds. In addition, NFR1 (*N*-fluoro-4-methyl-*N*-tosylbenzenesulfonamide) was used instead of NFSI to further explore the scope of this alkyne aminative multifunctionalization. As expected, the reaction of **1** with NFR1 and $CH_3OH$ proceeded smoothly and provided **4a**–**h** and **5a**–**h** (with **4a**–**h** as major products) in 48–84% yield (Table 3). Similar electronic and steric effects as using NFSI were observed. Trifluoromethyl and cyano groups were compatible and provided the corresponding α-amino-α-aryl ketones **4g** (81%) and **4h** (48%). However, substrates with strong electron-donating groups on the aromatic ring, such as 1-ethynyl-4-methoxybenzene and 1-ethynyl-3-methoxybenzene, were not effective. In addition, reactions between 1-(*tert*-butyl)-4-ethynylbenzene and some other NFSI derivatives were also explored to extend scope and investigate electronic effect of aryl part of NFSI derivatives. For 4-chloro-

**Table 4 | Aminative multifunctionalization of internal alkynes with NFSI.**

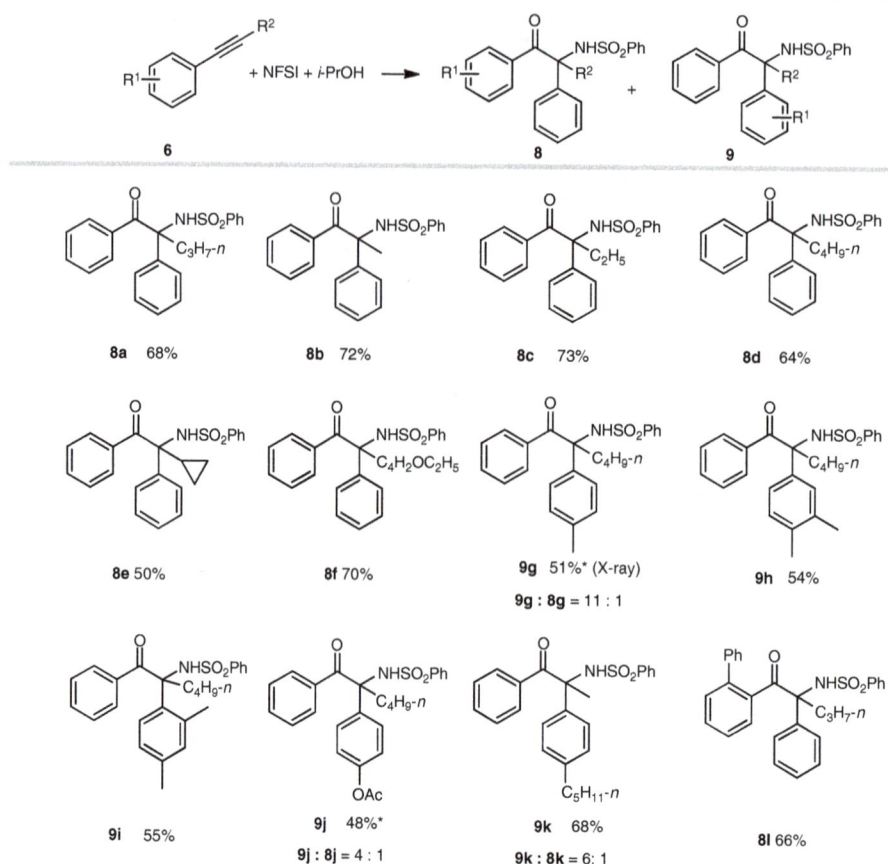

**8a** 68%  **8b** 72%  **8c** 73%  **8d** 64%

**8e** 50%  **8f** 70%  **9g** 51%* (X-ray)  **9h** 54%
 9g : 8g = 11 : 1

**9i** 55%  **9j** 48%*  **9k** 68%  **8l** 66%
 9j : 8j = 4 : 1  9k : 8k = 6 : 1

NFSI, N-fluorobenzenesulfonimide.
Reactions condition: **6** (0.5 mmol), NFSI (1.5 equiv., 0.75 mmol), i-PrOH (1.5 equiv., 0.75 mmol), CuCN (5 mol %) and ZnCl₂ (2 mol %) in CH₂Cl₂ (2 ml) at 70 °C under N₂ atmosphere for 12 h. Isolated yields.
*Mixture of two isomers. The ratio was determined by ¹H NMR analysis.

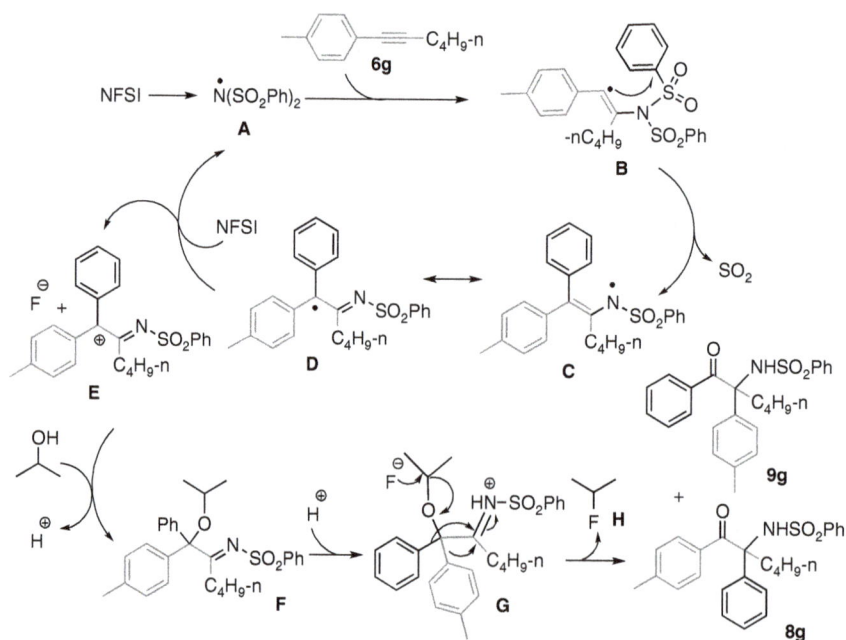

**Figure 2 | Proposed mechanism.** Sequential regioselective nitrogen-centred radical addition to alkyne/aryl migration/desulfonylation/oxygenation/semi-pinacol rearrangement were involved.

**Table 5 | Syntheses of α-alkyloxyl-α,α-diaryl imine.**

Ph—≡—R¹  +  NFSI + R²OH  →(Conditions)→  (product 7)

6 → 7

| Entry | R¹ | R²OH | Temperature (°C) | Product | Yield (%) |
|---|---|---|---|---|---|
| 1 | n-C₃H₇ | i-PrOH | 0 | 7a | 71 |
| 2 | Me | i-PrOH | 10 | 7b | 68 |
| 3 | n-C₄H₉ | i-PrOH | 0 | 7c | 63 |
| 4 | n-C₃H₇ | CH₃OH | 10 | 7d | 52 |
| 5 | n-C₃H₇ | EtOH | 10 | 7e | 62 |
| 6 | n-C₃H₇ | Butan-1-ol | 10 | 7f | 58 |
| 7 | n-C₃H₇ | Butan-2-ol | 0 | 7g | 65 |
| 8 | n-C₃H₇ | Cyclohexanol | 10 | 7h | 70 |
| 9 | n-C₃H₇ | Prop-2-yn-1-ol | 25 | 7i | 54 |
| 10 | n-C₃H₇ | (E)-but-2-en-1-ol | 25 | 7j | 46 |
| 11* | Ph | CH₃OH | 90 | 7k | 66 |
| 12* | 4-NO₂C₆H₄ | CH₃OH | 90 | 7l | 45 |
| 13* | 4-acetyl C₆H₄ | CH₃OH | 90 | 7m:7m′ = 1.5:1 | 41† |
| 14* | 4-ᵗBuC₆H₄ | CH₃OH | 90 | 7n:7n′ = 1:2 | 54† |

Reactions condition: **6** (0.5 mmol), NFSI (2 equiv., 1.0 mmol), R²OH (3 equiv., 1.5 mmol), Cu(acac)₂ (5 mol %) in CH₃CN (2 ml) under N₂ atmosphere for 48 h. Isolated yields.
*Reactions condition: **6** (0.5 mmol), NFSI (2 equiv., 1.0 mmol), CH₃OH (3 equiv., 1.5 mmol), CuCN (5 mol %) in CH₂Cl₂ (2 ml) at 90 °C under N₂ atmosphere for 48 h. Isolated yields.
†Mixture of two isomers. The ratio was determined by ¹H NMR analysis.

N-((4-chlorophenyl)sulfonyl)-N-fluorobenzenesulfonamide (NFR2), **5i** was obtained as a single isomer in yield of 46%. For N-fluoroarylsulfonimides 4-tert-butyl-N-fluoro-N-(phenylsulfonyl) benzenesulfonamide (NFR3) and 4-tert-butyl-N-(4-chloro phenylsulfonyl)-N-fluorobenzenesulfonamide (NFR4), the corresponding α-amino-α-aryl ketones were obtained in 72 and 67% yield, respectively. These results showed that the transformation was more efficient for electron-rich aromatic rings than electron-poor aromatic rings in NFSI derivatives. The halogen atom on the aromatic ring was tolerated in this process (**2e–i, 4c–f, 5i, 5k**), offering an opportunity for further elaboration.

**Scope of internal alkyne substrates.** To examine the generality of this alkyne aminative multifunctionalization, internal alkynes were subsequently examined. In the presence of 5 mol% CuCN, the reaction of pent-1-yn-1-ylbenzene (**6a**, 0.5 mmol), NFSI (1.5 equiv., 0.75 mmol) and i-PrOH (1.5 equiv., 0.75 mmol) in dichloromethane (DCM, 2 ml) was carried out at 70 °C under nitrogen atmosphere for 12 h. The expected α-amino-α-aryl ketone **8a** with a quaternary carbon at α-position was afforded in 68% yield (Table 4). As shown in Table 4, an array of α-amino-α-aryl ketones **8** and/or **9** were obtained in yields ranging from 48 to 73%. Similarly, preferential migration of electron-rich aromatic substituent (aryl on the alkyne versus Ph from NFSI) was observed in the product distribution. It should be noted that this aminative multifunctionalization of internal alkynes directly lead to the skeleton of α-tertiary amine derivatives, which are widespread in various natural products and bioactive compounds[52–55]. Quaternary carbon centres with a nitrogen substituent have been successfully constructed through molecular rearrangement[56]. However, special structures of substrates were necessary. Therefore, the directly aminative multifunctionalization of alkynes could provide a new and facile way for α-tertiary amines. Recently, Murakami and co-workers[57] reported an interesting Cu- and Rh-catalyzed aminative multifunctionalization of terminal alkynes to form α-amino-α-allyl ketones via α-imino metal carbene intermediate, during which C–C(allyl) bond formed through Claisen-type rearrangement. In their study, for internal alkynes, N-sulfonyl-1,2,3-triazoles needed to be pre-prepared.

**Mechanism investigation.** Radical scavengers were employed to probe the mechanism of the aminative multifunctionalization of alkynes. Formation of **2a** was completely inhibited when 1 equivalent of 2,6-di-tert-butyl-4-methylphenol or 2,2,6,6-tetramethyl-1-piperidinyloxy was added to the reaction. For the reaction with 2,6-di-tert-butyl-4-methylphenol as radical scavenger, 26% benzylic C–H amination product was obtained. These results suggested a possible radical mechanism. In combination with our previous finding in amination[19–24], we proposed a possible mechanism as depicted in Fig. 2. Initially, the in situ-generated nitrogen-centred radical **A** added to the triple bond of alkyne (for example **6g**) regioselectively, providing a highly reactive vinyl radical **B**. Subsequently, sequential intramolecular 1,4-aryl migration via 5-ipso cyclization and desulfonylation would produce amidyl radical **C**[58–61]. This imidyl radical exists at an equilibrium with its resonance structure α-imino carbon radical **D** which could be stabilized by two aromatic rings and a C = N double bond. Then, the single-electron oxidation of intermediate **D** by NFSI generated a carbocation intermediate **E** and a nitrogen-centred radical **A** to continue the next catalytic cycle. The reaction between intermediate **E** and ROH provided α-alkyloxyl imine **F**. Finally, the protonation and semi-pinacol rearrangement of intermediate **F** furnished aminative multifunctionalization to provide isomer **8g** and **9g**. The ratio of **8g** to **9g** depended on electronic and steric effects of the corresponding aromatic substituents. As depicted in Tables 2–4, electron-rich and the sterically less-hindered aromatic rings are more prone to migrate, which is in consistency with the requirements of semi-pinacol rearrangement. It is noted that during this transformation, trapping of the incipient vinyl radical by aromatic ring from nitrogen source was a key step leading to intermolecular cascade process, which might provide a new entry to design radical addition initiated multi-component cascade reaction of alkynes.

Since the above-mentioned mechanism invoked a semi-pinacol rearrangement from a relatively stable species α-alkyloxyl-α,α-diaryl imine **F** to the final product, we questioned if this species could be obtained separately with modification of reaction parameters. Recently, semi-pinacol rearrangement of α-hydroxy imines had been successfully applied in natural product as well as catalytic asymmetric syntheses[62–64]. To our delight, the reaction of pent-1-yn-1-ylbenzene (**6a**, 0.5 mmol), NFSI (1.0 mmol, 2.0 equiv) and propan-2-ol (1.5 mmol, 3.0 equiv.) in the presence of Cu(acac)$_2$ (5 mol %) at 0 °C in dry CH$_3$CN (2 mL) under N$_2$ atmosphere was performed for 48 h, α-alkyloxy-α,α-diaryl imine **7a** was obtained in 71% yield. Besides the reaction temperature, catalyst played an important role in obtaining this product because no desired **7a** was obtained without copper. As shown in Table 5, various alcohols could react with NFSI and alkynes to obtain the corresponding α-alkyloxyl-α,α-diaryl imine **7a–7j** in 46–71% yields. It should be noted that diaryl-substituted alkynes are also effective. Starting from 1,2-diphenylethyne (**6k**), the corresponding α-alkyloxyl-α,α-diaryl imine was obtained in 66% under relatively higher temperature (Table 5, entry 11). Interestingly, for substrate **6l**, nitrogen-centred radical highly regioselectively added to the alkyne carbon connected to the aromatic ring with strong electron-withdrawing NO$_2$ group. From substrates **6m** and **6n**, regioisomer mixtures (**7m:7m′** = 1.5:1, **7n:7n′** = 1:2) were obtained.

Identification of intermediate **F** (Fig. 1) provided strong proof to the proposed mechanism. Therefore, further experiments for more mechanistic information were also carried out. The final α-amino-α-aryl ketone **8a** could be obtained in 87% yield when heating **7a** (0.3 mmol) at 70 °C in the presence of 0.3 mmol trifluoroacetic acid (TFA) in 2 ml DCM for 4 h, lending further support for **7** as the key intermediate in the novel aminative multifunctionalization of alkynes. However, under the same conditions but adding CuCN (5 mol %) instead of TFA, no reaction occured. Instead, when ZnCl$_2$ (10 mol%) was added, **8a** was isolated in 60% yield, which showed that ZnCl$_2$ additive in Table 4 played an important role for the transformation from intermediate **F** (Fig. 2) to final α-amino-α-aryl ketones. When the semi-pinacol rearrangement of **7a** was performed in the presence of HF acid (1 equiv., 40 wt % in water) instead of TFA, **8a** could be obtained in 68% yield, along with identified side-product 2-fluoropropane (**H**, Fig. 2). Starting from **7l**, the next semi-pinacol rearrangement process was not effective, which elucidated why transformation from diaryl-substituted alkynes to the desired α-amino-α-aryl ketones could not be realized in this study. Starting from **6a**, when t-BuOH was employed instead of i-PrOH under otherwise same conditions described in Table 5, entry 9, a α-fluoro-α,α-diaryl imine could be obtained in 36% yield, which demonstrated the possible presence of intermediate **E** (Fig. 2).

In conclusion, an unprecedented cascade radical aminative multifunctionalization reaction of various aryl terminal and internal alkynes with N-fluoroarylsulfonimides and simple alcohols is developed. This methodology provides a new facile and straightforward way for both α-amino-α-aryl ketones and α-alkyloxyl-α,α-diaryl imines, especially for the construction of quaternary α-amino ketones, which might open a new entry for designing multi-component radical cascade reactions of alkynes. Further studies for the application of this transformation are ongoing in our laboratory.

## Methods

**General methods.** For $^1$H, $^{19}$F and $^{13}$C NMR spectra of compounds in this manuscript, see Supplementary Figs 1–121. For details of the synthetic procedures, tables including detail experimental, see Supplementary Information.

**Preparation of 2a.** To a solution of the NFSI (0.75 mmol, 236.5 mg) in CH$_3$CN (2.0 ml) was added the CH$_3$OH (1.5 mmol, 61 μl), TFA (0.5 mmol, 37 μl), 1-Phenylethyne (**1a**, 0.5 mmol, 54 μl) and Cu(OTf)$_2$ (0.05 mmol, 18.1 mg) in screw-cap test tube under N$_2$ atmosphere. The test tube was then sealed off with a screw-cap and the reaction mixture was stirred at 70 °C for 5.0 h. After the reaction finished, the reaction mixture was cooled to room temperature and quenched by water. The mixture was extracted with CH$_2$Cl$_2$ (3 × 5.0 ml), the combined organic phases were dried over anhydrous Na$_2$SO$_4$ and the solvent was evaporated under vacuum. The residue was purified by column chromatography (petroleum ether/ ethyl acetate (10:1 v/v)) to give the corresponding product **2a** (136.9 mg, 78%).

**Preparation of 8a.** To a solution of the NFSI (0.75 mmol, 236.5 mg) in CH$_2$Cl$_2$ (2.0 ml) was added the isopropanol (0.75 mmol, 57 μl), but-1-yn-1-ylbenzene (**6a**, 0.5 mmol, 80 μl), ZnCl$_2$ (0.01 mmol, 1.4 mg) and CuCN (0.025 mmol, 2.2 mg) in screw-cap test tube under N$_2$ atmosphere. The test tube was then sealed off with a screw-cap and the reaction was stirred at 70 °C for 12.0 h. After the reaction finished, the reaction mixture was cooled to room temperature and quenched by water. The mixture was extracted with CH$_2$Cl$_2$ (3 × 5.0 ml), the combined organic phases were dried over anhydrous Na$_2$SO$_4$ and the solvent was evaporated under vacuum. The residue was purified by column chromatography (petroleum ether/ ethyl acetate 10:1 (v/v)) to give the corresponding product **8a** (133.7 mg, 68%).

**Preparation of 7a.** To a solution of the NFSI (1.0 mmol, 314.3 mg) in CH$_3$CN (2.0 ml) was added the isopropanol (1.5 mmol, 114 μl), but-1-yn-1-ylbenzene (**6a**, 0.5 mmol, 80 μl) and Cu(acac)$_2$ (0.025 mmol, 6.5 mg) in screw-cap test tube under N$_2$ atmosphere. The test tube was then sealed off with a screw-cap and the reaction was stirred at 0 °C for 48.0 h. After the reaction finished, the reaction mixture was quenched by water. The mixture was extracted with CH$_2$Cl$_2$ (3 × 5.0 ml), the combined organic phases were dried over anhydrous Na$_2$SO$_4$ and the solvent was evaporated under vacuum. The residue was purified by column chromatography (petroleum ether/diethyl ether (25:1 v/v)) to give the corresponding product **7a** (154.5 mg, 71%).

**Preparation of 7k.** To a solution of NFSI (1.0 mmol, 315.3 mg) in CH$_2$Cl$_2$ (2.0 ml) was added methanol (1.5 mmol, 61 μl), 1,2-diphenylethyne (**6k**, 0.5 mmol, 89 mg) and CuCN (0.025 mmol, 2.2 mg) in screw-cap test tube under N$_2$ atmosphere. The test tube was then sealed off with a screw-cap and the reaction was stirred for the 48.0 h at 90 °C. After the reaction finished, the reaction mixture was cooled to room temperature and quenched by water. The mixture was extracted with CH$_2$Cl$_2$ (3 × 5.0 ml), the combined organic phases were dried over anhydrous Na$_2$SO$_4$ and the solvent was evaporated under vacuum. The residue was purified by column chromatography (petroleum ether/ethyl acetate 20:1 (v/v)) to give the corresponding product **7k** (145.6 mg, 68%).

## References

1. Hili, R. & Yudin, A. K. Making carbon-nitrogen bonds in biological and chemical synthesis. *Nat. Chem. Biol.* **2,** 284–287 (2006).
2. Henkel, T., Brunne, R. M., Müller, H. & Reichel, F. Statistical investigation into the structural complementarity of natural products and synthetic compounds. *Angew. Chem. Int. Ed.* **38,** 643–647 (1999).
3. Severin, R. & Doye, S. The catalytic hydroamination of alkynes. *Chem. Soc. Rev.* **36,** 1407–1420 (2007).
4. Pohlki, F. & Doye, S. The catalytic hydroamination of alkynes. *Chem. Soc. Rev.* **32,** 104–114 (2003).
5. Alonso, F., Beletskaya, I. P. & Yus, M. Transition-metal-catalyzed addition of heteroatom-hydrogen bonds to alkynes. *Chem. Rev.* **104,** 3079–3160 (2004).
6. Müller, T. E. & Beller, M. Metal-initiated amination of alkenes and alkynes. *Chem. Rev.* **98,** 675–703 (1998).
7. Li, J. & Neuville, L. Copper-catalyzed oxidative diamination of terminal alkynes by amidines: synthesis of 1,2,4- trisubstituted imidazoles. *Org. Lett.* **15,** 1752–1755 (2013).
8. Zeng, J., Tan, Y. J., Leow, M. L. & Liu, X.-W. Copper(II)/iron(III) Co-catalyzed intermolecular diamination of alkynes: facile synthesis of imidazopyridines. *Org. Lett.* **14,** 4386–4389 (2012).
9. He, W., Li, C. & Zhang, L. An efficient [2 + 2 + 1] synthesis of 2,5-disubstituted oxazoles via gold-catalyzed intermolecular alkyne oxidation. *J. Am. Chem. Soc.* **133,** 8482–8485 (2011).
10. Hirano, K., Satoh, T. & Miura, M. Copper-catalyzed annulative amination of ortho-alkynylphenols with hydroxylamines: synthesis of 3-aminobenzofurans by umpolung amination strategy. *Org. Lett.* **13,** 2395–2397 (2011).
11. Luo, Y., Ji, K., Li, Y. & Zhang, L. Tempering the reactivities of postulated α-Oxo gold carbenes using bidentate ligands: implication of tricoordinated gold intermediates and the development of an expedient bimolecular assembly of 2,4-disubstituted oxazoles. *J. Am. Chem. Soc.* **134,** 17412–17415 (2012).
12. Mukherjee, A. M. *et al.* Gold-catalyzed 1,2-difunctionalizations of aminoalkynes using only N- and O- containing oxidants. *J. Am. Chem. Soc.* **133,** 15372–15375 (2011).
13. Karur, S. *et al.* A catalytic reaction of alkynes via multiple-site functionalization. *J. Am. Chem. Soc.* **125,** 13340–13341 (2003).
14. Li, M., Yuan, H., Zhao, B., Liang, F. & Zhang, J. Alkyne aminohalogenation enabled by DBU-activated N-haloimides: direct synthesis of halogenated enamines. *Chem. Commun.* **50,** 2360–2363 (2014).
15. Yu, D., Sum, Y. N., Ean, A. C. C., Chin, M. P. & Zhang, Y. Acetylide ion (C$_2^{2-}$) as a synthon to link electrophiles and nucleophiles: a simple method for enaminone synthesis. *Angew. Chem. Int. Ed.* **52,** 5125–5128 (2013).
16. Wille, U. Radical cascades initiated by intermolecular radical addition to alkynes and related triple bond systems. *Chem. Rev.* **113,** 813–853 (2013).

17. Wille, U., Heuger, G. & Jargstorff, C. N-Centered radicals in self-terminating radical cyclizations: experimental and computational studies. *J. Org. Chem.* **73**, 1413–1421 (2008).

18. Wille, U., Krüger, O., Kirsch, A. & Lüning, U. Imidyl radicals, radical addition of N-bromophthalimide to linear and cyclic alkynes. *Eur. J. Org. Chem.* **1999**, 3185–3189 (1999).

19. Zhang, H. et al. Copper-catalyzed intermolecular aminocyanation and diamination of alkenes. *Angew. Chem. Int. Ed.* **52**, 2529–2533 (2013).

20. Zhang, H., Song, Y., Zhao, J., Zhang, J. & Zhang, Q. Regioselective radical aminofluorination of styrenes. *Angew. Chem. Int. Ed.* **53**, 11079–11083 (2014).

21. Xiong, T., Li, Y., Mao, L., Zhang, Q. & Zhang, Q. Palladium-catalyzed allylic C – H amination of alkenes with N-fluorodibenzenesulfonimide: water plays an important role. *Chem. Commun.* **48**, 2246–2248 (2012).

22. Ni, Z. et al. Highly regioselective copper-catalyzed benzylic C – H amination by N-fluorobenzenesulfonimide. *Angew. Chem. Int. Ed.* **51**, 1244–1247 (2012).

23. Sun, K., Li, Y., Xiong, T., Zhang, J. & Zhang, Q. Palladium-catalyzed C-H aminations of anilides with N-fluorobenzenesulfonimide. *J. Am. Chem. Soc.* **133**, 1694–1697 (2011).

24. Xiong, T., Li, Y., Lv, Y. & Zhang, Q. Remote amide-directed palladium-catalyzed benzylic C–H amination with N-fluorobenzenesulfonimide. *Chem. Commun.* **46**, 6831–6833 (2010).

25. Stratmann, K. et al. Welwitindolinones, Unusual alkaloids from the blue-green algae hapalosiphon welwitschii and westiella intricata. relationship to fischerindoles and hapalindoles. *J. Am. Chem. Soc.* **116**, 9935–9942 (1994).

26. Hijazi, Y. & Boulieu, R. Contribution of CYP3A4, CYP3A4, CYP2B6, and CYP2C9 isoforms to N-demetaylation of ketamine in human liver microsomes. *Drug Metab. Dispos.* **30**, 853–858 (2002).

27. Varenhorst, C. et al. Genetic variation of CYP2C19 affects both pharmacokinetic and pharmacodynamic responses to clopidogrel but not prasugrel in aspirin-treated patients with coronary artery disease. *Eur. Heart J.* **30**, 1744–1752 (2009).

28. Sorrell, T. N. & Allen, W. E. A regiospecific synthesis of 1,4-disubstituted imidazoles. *J. Org. Chem.* **59**, 1589–1590 (1994).

29. Langer, P. & Bodtke, A. Sequential cyclizations of 2-isothiocyanatobenzonitrile and 2-isocyanatobenzonitrile with α-aminoketones. *Tetrahedron Lett.* **44**, 5965–5967 (2003).

30. Frantz, D. E. et al. Synthesis of substituted imidazoles via organocatalysis. *Org. Lett.* **6**, 843–846 (2004).

31. Bunescu, A., Piemontesi, C., Wang, Q. & Zhu, J. Heteroannulation of arynes with N-aryl α-aminoketones for the synthesis of unsymmetrical N-aryl-2,3-disubstituted indoles: an aryne twist of Bischler–Möhlau indole synthesis. *Chem. Commun.* **49**, 10284–10286 (2013).

32. Ager, D. J., Prakash, I. D. & Schaad, R. 1,2-Amino alcohols and their heterocyclic derivatives as chiral auxiliaries in asymmetric synthesis. *Chem. Rev.* **96**, 835–876 (1996).

33. Reetz, M. T. New approaches to the use of amino acids as chiral building blocks in organic synthesis [new synthetic methods (85)]. *Angew. Chem. Int. Ed.* **30**, 1531–1546 (1991).

34. Li, G.-Q., Dai, L.-X. & You, S.-L. Thiazolium-derived N-heterocyclic carbene-catalyzed cross-coupling of aldehydes with unactivated imines. *Chem. Commun.* **8**, 852–854 (2007).

35. Murry, J. A. et al. Synthesis of α-amido ketones via organic catalysis: thiazolium-catalyzed cross-coupling of aldehydes with acylimines. *J. Am. Chem. Soc.* **123**, 9696–9697 (2001).

36. Enders, D., Henseler, A. & Lowins, S. N-Heterocyclic carbene catalyzed nucleophilic acylation of trifluoromethyl ketimines. *Synthesis* **24**, 4125–4128 (2009).

37. Mennen, S. M., Gipson, J. D., Kim, Y. R. & Miller, S. J. Thiazolylalanine-derived catalysts for enantioselective intermolecular aldehyde-imine cross-couplings. *J. Am. Chem. Soc.* **127**, 1654–1655 (2005).

38. DiRocco, D. A. & Rovis, T. Catalytic asymmetric cross-aza-benzoin reactions of aliphatic aldehydes with N-Boc-protected imines. *Angew. Chem. Int. Ed.* **51**, 5904–5906 (2012).

39. Feurer, M., Frey, G., Luu, H.-T., Kratzert, D. & Streuff, J. The cross-selective titanium(III)-catalysed acyloin reaction. *Chem. Commun.* **50**, 5370–5372 (2014).

40. Smith, A. M. R. & Hii, K. K. Transition metal catalyzed enantioselective α-heterofunctionalization of carbonyl compounds. *Chem. Rev.* **111**, 1637–1656 (2011).

41. Selig, P. The electrophilic α-amination of α-alkyl-β-ketoesters with in situ generated nitrosoformates. *Angew. Chem. Int. Ed.* **52**, 7080–7082 (2013).

42. Lv, Y. et al. nBu₄NI-catalyzed oxidative imidation of ketones with imides: synthesis of α-amino ketones. *Chem. Commun.* **50**, 2367–2369 (2014).

43. Evans, R. W., Zbieg, J. R., Zhu, S., Li, W. & MacMillan, D. W. C. Simple catalytic mechanism for the direct coupling of α-carbonyls with functionalized amines: a one-step synthesis of plavix. *J. Am. Chem. Soc.* **135**, 16074–16077 (2013).

44. Johansson, C. C. C. & Colacot, T. J. Metal-catalyzed α-arylation of carbonyl and related molecules: novel trends in C – C bond formation by C – H bond functionalization. *Angew. Chem. Int. Ed.* **49**, 676–707 (2010).

45. Bellina, F. & Rossi, R. Transition metal-catalyzed direct arylation of substrates with activated sp³-hybridized C-H bonds and some of their synthetic equivalents with aryl halides and pseudohalides. *Chem. Rev.* **110**, 1082–1146 (2010).

46. Danoun, G., Tlili, A., Monnier, F. & Taillefer, M. Direct copper-catalyzed α-arylation of benzyl phenyl ketones with aryl iodides: route towards tamoxifen. *Angew. Chem. Int. Ed.* **51**, 12815–12819 (2012).

47. Ge, S. & Hartwig, J. F. Nickel-catalyzed asymmetric α-arylation and heteroarylation of ketones with chloroarenes: effect of halide on selectivity, oxidation state, and room-temperature reactions. *J. Am. Chem. Soc.* **133**, 16330–16333 (2011).

48. Hesp, K. D., Lundgren, R. J. & Stradiotto, M. Palladium-catalyzed mono-α-arylation of acetone with aryl halides and tosylates. *J. Am. Chem. Soc.* **133**, 5194–5197 (2011).

49. Kaneko, K., Yoshino, T., Matsunaga, S. & Kanai, M. Sultam synthesis via Cu-catalyzed intermolecular carboamination of alkenes with N-Fluorobenzenesulfonimide. *Org. Lett.* **15**, 2502–2505 (2013).

50. Boursalian, G. B., Ngai, M. Y., Hojczyk, K. N. & Ritter, T. Pd-catalyzed aryl C – H imidation with arene as the limiting reagent. *J. Am. Chem. Soc.* **135**, 13278–13281 (2013).

51. Zhang, B. & Studer, A. Copper-catalyzed intermolecular aminoazidation of alkenes. *Org. Lett.* **16**, 1790–1793 (2014).

52. Du, Y. et al. Asymmetric reductive Mannich reaction to ketimines catalyzed by a Cu(I) complex. *J. Am. Chem. Soc.* **130**, 16146–16147 (2008).

53. Shintani, R., Takeda, M., Tsuji, T. & Hayashi, T. Rhodium-catalyzed asymmetric arylation of N-tosyl ketimines. *J. Am. Chem. Soc.* **132**, 13168–13169 (2010).

54. Shintani, R., Takeda, M., Soh, Y.-T., Ito, T. & Hayashi, T. Rhodium-catalyzed asymmetric addition of potassium organotrifluoroborates to N-sulfonyl Ketimines. *Org. Lett.* **13**, 2977–2979 (2011).

55. Nishimura, T., Noishiki, A., Tsui, G. C. & Hayashi, T. Asymmetric synthesis of (triaryl)methylamines by rhodium catalyzed addition of arylboroxines to cyclic N-sulfonyl ketimines. *J. Am. Chem. Soc.* **134**, 5056–5059 (2012).

56. Clayden, J., Donnard, M., Lefranc, J. & Tetlow, D. J. Quaternary centres bearing nitrogen (α-tertiary amines) as products of molecular rearrangements. *Chem. Commun.* **47**, 4624–4639 (2011).

57. Miura, T., Tanaka, T., Biyajima, T., Yada, A. & Murakami, M. One-pot procedure for the introduction of three different bonds onto terminal alkynes through N-sulfonyl-1,2,3-triazole intermediates. *Angew. Chem. Int. Ed.* **52**, 3883–3886 (2013).

58. Gheorghe, A., Quiclet-Sire, B., Vilaa, X. & Zard, S. Synthesis of 3-arylpiperidines by a radical 1,4-aryl migration. *Org. Lett.* **7**, 1653–1656 (2005).

59. Kong, W., Casimiro, M., Merino, E. & Nevado, C. Copper-catalyzed one-pot trifluoromethylation/aryl migration/desulfonylation and C(sp²) – N bond formation of conjugated tosyl amides. *J. Am. Chem. Soc.* **135**, 14480–14483 (2013).

60. Kong, W., Merino, E. & Nevado, C. Arylphosphonylation and arylazidation of activated alkenes. *Angew. Chem. Int. Ed.* **53**, 5078–5082 (2014).

61. Gao, P. et al. Copper-catalyzed one-pot trifluoromethylation/aryl migration/carbonyl formation with homopropargylic alcohols. *Angew. Chem. Int. Ed.* **53**, 7629–7633 (2014).

62. Song, Z.-L., Fan, C.-A. & Tu, Y.-Q. Semipinacol rearrangement in natural product synthesis. *Chem. Rev.* **111**, 7523–7556 (2011).

63. Paquette, L. A. & Hofferberth, J. E. *Org. React* **62**, 477–567 (2003).

64. Zhang, X., Staples, R. J., Rheingold, A. L. & Wulff, W. D. Catalytic asymmetric α-iminol rearrangement: new chiral platform. *J. Am. Chem. Soc.* **136**, 13971–13974 (2014).

## Acknowledgements

We acknowledge support for this work from the National NSF of China (21172033, 21372041 and 21302017).

## Author contributions

G.Z., J.H. and T.X. performed the experiments and analysed the data. Y.L. and Q.Z. designed and directed the project and wrote the manuscript. G.Z. and Y.L. contributed equally to this work. All the authors discussed the results and commented on the manuscript.

## Additional information

**Competing financial interests:** The authors declare no competing financial interests.

# 7

# Spacer-free BODIPY fluorogens in antimicrobial peptides for direct imaging of fungal infection in human tissue

Lorena Mendive-Tapia[1], Can Zhao[2], Ahsan R. Akram[3], Sara Preciado[1], Fernando Albericio[1,4,5,6], Martin Lee[7], Alan Serrels[7], Nicola Kielland[8], Nick D. Read[2], Rodolfo Lavilla[5,8] & Marc Vendrell[3]

Fluorescent antimicrobial peptides are promising structures for *in situ*, real-time imaging of fungal infection. Here we report a fluorogenic probe to image *Aspergillus fumigatus* directly in human pulmonary tissue. We have developed a fluorogenic Trp-BODIPY amino acid with a spacer-free C-C linkage between Trp and a BODIPY fluorogen, which shows remarkable fluorescence enhancement in hydrophobic microenvironments. The incorporation of our fluorogenic amino acid in short antimicrobial peptides does not impair their selectivity for fungal cells, and enables rapid and direct fungal imaging without any washing steps. We have optimized the stability of our probes in human samples to perform multi-photon imaging of *A. fumigatus* in *ex vivo* human tissue. The incorporation of our unique BODIPY fluorogen in biologically relevant peptides will accelerate the development of novel imaging probes with high sensitivity and specificity.

[1] Institute for Research in Biomedicine, Barcelona Science Park, Baldiri Reixac 10-12, Barcelona 08028, Spain. [2] Manchester Fungal Infection Group, Institute of Inflammation and Repair, University of Manchester, CTF Building, Grafton St, Manchester M13 9NT, UK. [3] MRC/UoE Centre for Inflammation Research, University of Edinburgh, 47 Little France Crescent, Edinburgh EH16 4TJ, UK. [4] Department Organic Chemistry, University of Barcelona, Martí i Franqués 1-11, Barcelona 08028, Spain. [5] CIBER-BBN, Networking Centre for Bioengineering, Biomaterials and Nanomedicine, Baldiri Reixac 10-12, Barcelona 08028, Spain. [6] School of Chemistry, University of KwaZulu-Natal, Durban 4001, South Africa. [7] Edinburgh Cancer Research Centre, University of Edinburgh, Crewe South Road, Edinburgh EH4 2XR, UK. [8] Laboratory of Organic Chemistry, Faculty of Pharmacy, University of Barcelona, Barcelona Science Park, Baldiri Reixac 10-12, Barcelona 08028, Spain. Correspondence and requests for materials should be addressed to R.L. (email: rlavilla@pcb.ub.es) or to M.V. (email: mvendrel@staffmail.ed.ac.uk).

Invasive pulmonary aspergillosis (IPA) is a highly fatal disease in immunocompromised patients. IPA results from the infection with the fungal pathogen *Aspergillus fumigatus*, and it is a frequent cause of fungal pneumonia with mortality rates up to 40% (ref. 1). Current diagnostic approaches for IPA rely on histological analysis, cultures from bronchoalveolar lavage fluid and sampling peripheral blood[2]. These methods are fraught with problems of upper airway contamination and diagnostic delays, by which time the disease may have progressed or been treated empirically with inappropriate drugs. Moreover, blood markers are unlikely to provide useful information about events deep in pulmonary tissue, especially in patients with multi-system disease, such as immunosuppressed patients affected by IPA. These limitations of current diagnostic tools have prompted the development of imaging probes that can provide *in situ* and real-time information on the progression of infection[3-6]. Fluorescent probes based on antibiotics and antimicrobial peptides are chemical entities with enormous potential for imaging infection sites due to their high selectivity for microbial cell structures over mammalian cells[7-11]. van Oosten *et al.*[12] recently reported a near-infrared fluorescently labelled vancomycin for real-time *in vivo* imaging of bacterial infections in a mouse myocitis model. Similarly, Thiberville and co-workers have described fluorescein-conjugated peptides to visualize fungal biofilms in immunosupressed rats using fibre-based microendoscopy[13]. These probes have been prepared by conjugating peptides of interest to suitable fluorophores via chemical spacers. While such approaches have been useful to functionalise long peptides or proteins[14], alternative strategies are needed for shorter peptides, where relevant modifications can compromise their specificity. Our group and others have studied the mechanism of action of Peptide AntiFungal 26 (PAF26), a synthetic antimicrobial hexapeptide with high affinity for fungal cells and selectivity over bacterial and mammalian cells[15,16]. We envisaged that fluorescent analogues of PAF26 would enable imaging of fungal infection sites provided that the main recognition features of PAF26 remained unaffected after labelling. However, the incorporation of fluorophores in short antimicrobial peptides is challenging as chemical modifications are likely to alter the distribution of positive charges as well as their amphipathic character. PAF26 has a highly conserved sequence with a C-terminal hydrophobic domain (Trp–Phe–Trp) and an N-terminal cationic domain (Arg–Lys–Lys) that are essential to exert its antifungal action. Site-specific peptide labelling can be achieved by incorporation of amino acids with bio-orthogonal[17-20] or fluorogenic groups[21-23]. Fluorogenic amino acids are advantageous in that they provide high signal-to-noise ratios without the need for washing or additional labelling steps. A number of fluorogenic amino acids have been reported[24-26], but most exhibit inherent limitations as fluorophores (for example, short emission wavelengths, low extinction coefficients and compromised cell permeability). We have developed a spacer-free fluorogenic amino acid based on the 4,4-difluoro-4-bora-3*a*,4*a*-diaza-*s*-indacene (BODIPY) scaffold, and incorporated it in the hydrophobic domain of PAF26 to maintain the recognition features of the peptide while providing an excellent reporter of the interaction with fungal cells. This innovative approach has rendered fluorogenic BODIPY-labelled antimicrobial peptides as highly stable probes to image *A. fumigatus* directly in *ex vivo* human tissue.

## Results

### Design and synthesis of a Trp-BODIPY fluorogenic amino acid.
BODIPY is a fluorescent structure with excellent cell permeability and photophysical properties[27,28]. Moreover, the BODIPY scaffold can be derivatized with radioisotopes to prepare multimodal agents for both optical imaging and positron emission tomography[29,30], enabling quantitative whole-body imaging with high sensitivity[31,32]. Multimodal agents, which are designed to be compatible with complementary imaging modalities, are excellent tools to achieve good spatial resolution and specificity without compromising high sensitivity[33]. Despite the numerous BODIPY derivatives described to date[34-38], there are no reports of BODIPY-based fluorogenic amino acids. Environmentally sensitive fluorogens can be prepared by direct conjugation of the BODIPY core to electron-rich groups leading to photo-induced electron transfer quenching[39-41]. We envisaged that the direct coupling of the indole group of Trp to the BODIPY core would render a fluorogenic amino acid with potential to replace Trp in the preparation of fluorogenic antimicrobial peptides. Our group has recently described some Pd-catalysed C-H activation[42-44] as an efficient way to arylate the indole $C_2$ position[45] of Trp and prepare Trp-derivatized peptides and peptidomimetics[46-48]. In this way, we synthesized two BODIPY iodide derivatives (**1** and **2**, Fig. 1a) in good yields using our recently developed procedures and assessed their reactivity in Pd-catalysed $C_2$-arylation of Fmoc-Trp-OH. Notably, only the conjugate **3** was obtained from the *m*-iodophenyl-BODIPY (**2**)[49], while the corresponding *p*-iodophenyl **1** was unreactive, reflecting electronic preferences (Fig. 1a and Supplementary Discussion). We further optimized the gram-scale synthesis of **3** using microwave-assisted irradiation to readily isolate the fluorogenic amino acid as a solid stable compound with 74% yield, suitably protected to be directly used in solid-phase peptide synthesis (SPPS).

### Synthesis and evaluation of fluorogenic antifungal peptides.
The amino acid **3** displayed characteristic absorption and emission wavelengths of BODIPY probes as well as very high extinction coefficients (Fig. 1a, Supplementary Figs 1,2). Next we evaluated the properties of **3** as a fluorogenic probe and its potential to report interactions of antimicrobial peptides with fungal cells. Many antimicrobial peptides, including PAF26, recognize molecular components of the microbial cell membrane and accumulate in lipophilic intracellular compartments. Therefore, we examined the fluorescence spectra of **3** in phospholipid bilayer membranes that mimic such microenvironments. As shown in Fig. 1b, the BODIPY core embedded in **3** displayed remarkable fluorogenic behaviour with strong fluorescence emission upon binding to phospholipid membranes. In view of the properties of **3** as a fluorogenic surrogate of Trp, we prepared fluorogenic derivatives of PAF26 by SPPS. Since the sequence of PAF26 (**4**, Fig. 2a) contains two Trp residues, we synthesized all three possible combinations (**5–7**, Fig. 2a) to assess the impact of the amino acid **3** at different positions of the antimicrobial peptide. The amino acid **3** proved to be fully compatible with SPPS as it tolerates standard Fmoc deprotection and coupling conditions as well as mildly acidic (that is, 1% trifluoroacetic acid) cleavage cocktails for acid-labile solid supports (for example, Sieber amide and chlorotrityl-based polystyrene resins) without observing any degradation (Supplementary Methods and Supplementary Fig. 20). Being mildly acidic conditions harmless to the BODIPY core[50], peptides **5–7** were prepared using conventional SPPS protocols in a Sieber amide polystyrene resin. Molecular simulation models of both labelled and non-labelled peptides corroborated that the introduction of BODIPY scaffolds in the hydrophobic domain of PAF26 did not disrupt the conformation and hydrogen bonding pattern of the original peptide (Supplementary Fig. 3). Next we determined the activity of the peptides **4–7** in *A. fumigatus* as well as in

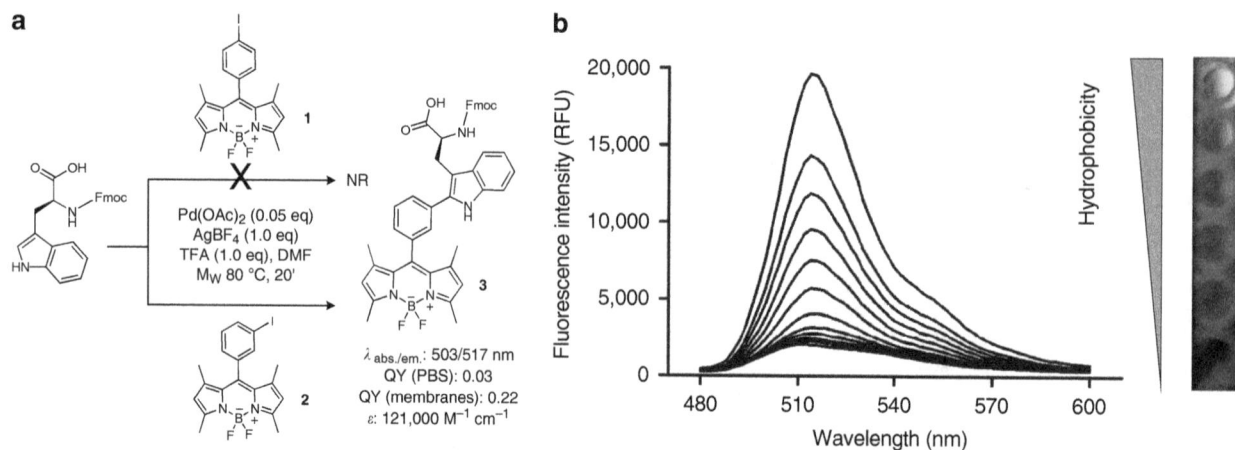

**Figure 1 | A Trp-BODIPY fluorogenic amino acid.** (a) Synthetic scheme and spectral properties of the Trp-BODIPY fluorogenic amino acid **3** (NR: no reaction). (b) The amino acid **3** displays strong fluorogenic behaviour in phospholipid membranes. Spectra of compound **3** (10 µM) were recorded after incubation with PC:cholesterol (7:1) liposome suspensions in PBS ranging from 3.75 to 0.004 mg ml$^{-1}$ of PC in two-fold serial dilutions, $\lambda_{exc}$: 450 nm. PBS alone was used as a negative control for a non-hydrophobic environment. On the right-hand side, pictures of the fluorescence emission of **3** under excitation with a 365 nm UV-lamp in PC:cholesterol liposome suspensions with increasing PC content (from top to bottom: 3.75, 1.88, 0.94, 0.47, 0.23 and 0 (only PBS) mg ml$^{-1}$ of PC).

bacterial strains and human RBCs as an indication of their affinity for both microbial and human cells. We included *Klebsiella pneumoniae*, *Escherichia coli* and *Pseudomonas aeruginosa* as clinically relevant bacterial strains commonly found in hospitalized pulmonary infections[51]. Likewise, we tested the activity of 4–7 in human RBCs, because positively charged peptides are potential haemolytic agents[52]. Remarkably, the incorporation of **3** in the hydrophobic domain of PAF26 rendered peptides (5–7) with slightly higher affinity for *A. fumigatus* than the non-labelled PAF26 peptide (4) (Fig. 2b and Supplementary Fig. 4). The marginal activity of PAF26 in bacterial and human cells was also maintained in all fluorogenic analogues (Fig. 2b, Supplementary Figs 5,6). Altogether, these results validate the direct C-C conjugation of BODIPY fluorogens to the C$_2$ position of the indole ring of Trp as a novel labelling approach with minimal interference in the molecular recognition properties of PAF26 while providing a suitable tag to report the interaction with *A. fumigatus*.

**Imaging *Aspergillus fumigatus* in co-culture with human cells.**
Peptides 5–7 exhibited similar spectral properties to **3** with an equally strong fluorogenic behaviour in phospholipid membranes (Fig. 2c and Supplementary Fig. 7). Double-labelled peptide **7** displayed a weaker fluorescence response than mono-labelled peptides (5, 6), partially due to the self-quenching derived from two neighbouring BODIPY fluorophores. In view of the excellent properties of 5-7 as fungi-targeting fluorogenic peptides, we evaluated them as live cell imaging agents of *A. fumigatus*. Peptides **5** and **6** brightly stained fungal cells, whereas **7** showed significantly weaker fluorescence, in accordance with its lower fluorogenicity (Fig. 2c). As negative controls, we assessed the activity and imaging properties of fluorogenic derivatives of PAF26 replacing some of the key residues for their interaction with fungal cells[53]. Peptide **5a**, which lacks the hydrophilic domain of PAF26, showed poor activity and staining in *A. fumigatus* (Supplementary Fig. 8). Similar results were obtained when we examined the activity and staining properties of the single BODIPY amino acid **3** (Supplementary Fig. 8). We also synthesized peptide **5b**, including less non-polar residues in the hydrophobic domain, which exhibited reduced activity and brightness in *A. fumigatus* (Supplementary Fig. 8). These observations confirmed the importance of embedding

the amino acid **3** within the full amphipathic sequence of PAF26 in order to efficiently interact with the cell membrane of *A. fumigatus*.

We further used peptide **5** to image live *A. fumigatus* in co-cultures with human lung epithelial cells. As shown in Fig. 2d, the fluorogenic properties of **5** enabled direct live fungal cell imaging without the need of any washing steps. Furthermore, we counterstained lung epithelial cells with the red fluorescent dye Syto82 and performed plot profile analysis to confirm that **5** specifically labelled *A. fumigatus* without staining human lung epithelial cells (Fig. 2d).

**Probe optimization for direct *ex vivo* tissue imaging.** Direct tissue imaging of infection sites is often hampered by the high concentration of proteolytic enzymes[54], which can compromise the integrity of imaging agents. Hence, we decided to examine the chemical stability of peptide **5** in human bronchoalveolar lavages from patients with acute respiratory dystress syndrome to assess the potential for *ex vivo* human tissue imaging. The linear peptide **5** was rapidly degraded in human lavages with a half-life shorter than 60 min (Fig. 3a, Supplementary Figs 9,10). To enhance the stability required for direct *ex vivo* imaging in human pulmonary tissue, we synthesized **8** as the corresponding BODIPY-labelled cyclic analogue (Fig. 3b). Cyclic peptides do not contain free N- and C-terminal groups, leading to increased resistance to degradation by proteases[55,56]. We synthesized compound **8** using 2-chlorotrityl polystyrene resin, which enabled the preparation and subsequent cleavage of the protected linear peptide under mild acidic conditions (Supplementary Fig. 11). Head-to-tail cyclization was performed in solution with 87% yield using HATU as the coupling reagent. We optimized the reaction conditions to remove all the protecting groups without affecting the BODIPY scaffold. Reduction of the protected peptide in H$_2$ atmosphere with Pd(OH)$_2$/C using mild acidic conditions led to the desired product with yields around 60% and purities over 90%. The peptide **8** showed around two-fold enhanced affinity for fungal cells compared with peptide **5**, and maintained very high selectivity over bacteria and human cells (Fig. 2b). A similar activity profile was observed for peptide **9**, the non-labelled analogue of peptide **8** (Supplementary Fig. 8). Peptide **9** showed slightly enhanced affinity for *A. fumigatus* when compared with

**a**

Hydrophilic domain    Hydrophobic domain

R¹, R²: H          (4)
R¹: H, R²: BODIPY  (5)
R¹: BODIPY, R²: H  (6)
R¹, R²: BODIPY     (7)

**b**

|   | Aspergillus fumigatus[1] | Klebsiella pneumoniae[2] | Escherichia coli[2] | Pseudomonas aeruginosa[2] | Human RBCs[3] |
|---|---|---|---|---|---|
| 4 | $7.9 \pm 0.2$ | 93% | >99% | >99% | 99% |
| 5 | $3.6 \pm 0.1$ | 87% | 94% | >99% | >99% |
| 6 | $3.0 \pm 0.1$ | 96% | 95% | >99% | >99% |
| 7 | $2.5 \pm 0.1$ | 93% | 95% | >99% | 96% |
| 8 | $2.0 \pm 0.1$ | 94% | 96% | >99% | >99% |

**c**

**d**

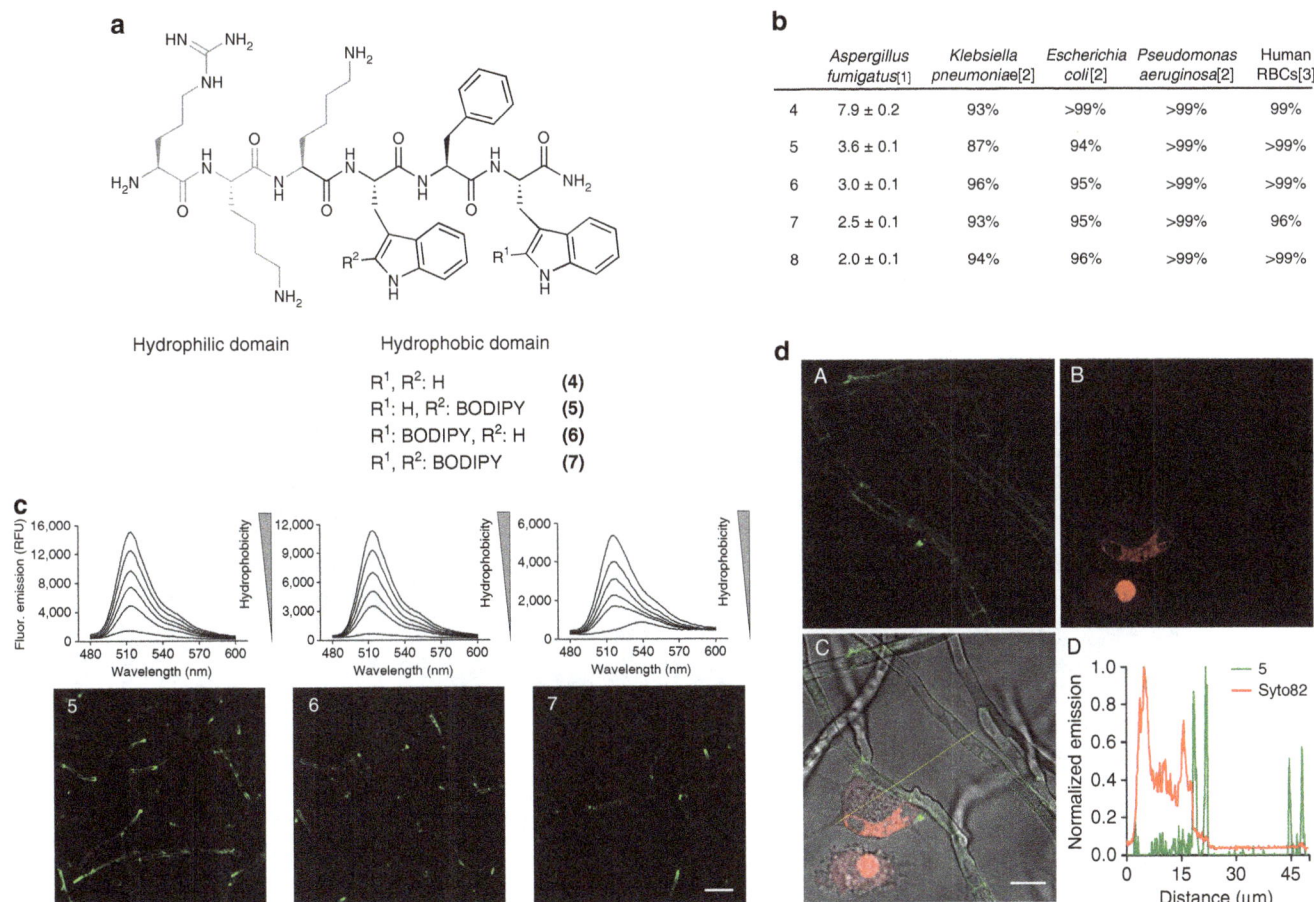

**Figure 2 | Fluorogenic peptides for live cell imaging of *A. fumigatus* in co-culture with human lung epithelial cells. (a)** Chemical structures of non-labelled and fluorogenic linear peptides (**4**-**7**), highlighting the two conserved hydrophilic (grey) and hydrophobic (black) domains of Peptide Antifungal 26 (PAF26). **(b)** Activity of antimicrobial peptides in *A. fumigatus*, several bacterial strains and in human RBCs.[1] $IC_{50}$ (μM) values represented as means ± s.e.m. from $n = 3$, [2] cell viability upon 16 h incubation with **4**-**8** at their respective $IC_{50}$ concentrations ($n = 3$), [3] cell viability upon 1 h incubation with **4**-**8** at their respective $IC_{50}$ concentrations ($n = 3$). **(c)** Fluorogenic behaviour of **5**-**7** (10 μM) in phosphatidylcoline (PC):cholesterol (7:1) liposome suspensions in PBS ranging from 3.75 to 0.004 mg ml$^{-1}$ of PC in two-fold serial dilutions ($\lambda_{exc}$: 450 nm), and wash-free live cell images of *A. fumigatus* at 37 °C using fluorescence confocal microscopy after incubation with peptides **5**-**7** (5 μM). Scale bar, 20 μm. **(d)** Peptide **5** (5 μM, green) and Syto82 (2.5 μM, red counterstain for lung epithelial cells) were incubated in co-cultures of *A. fumigatus* and human lung A549 epithelial cells and imaged under a fluorescence confocal microscope at 37 °C without any washing steps. Fluorescence staining of **5** (A), Syto82 (B), merged (C) and plot profile analysis (D) of peptide **5** (green) and Syto82 (red) from image C. Scale bar, 10 μm.

the linear PAF26 sequence (**4**), and maintained high selectivity over bacteria and human RBCs (Supplementary Fig. 8). These observations are in line with the fact that peptide cyclization can restrict conformational flexibility, which often leads to enhanced affinity and activity[57]. Preliminary NMR analysis of **8** showed no evidence of relevant structural modifications with respect to the non-labelled peptide **9**, in agreement with molecular simulations (Supplementary Fig. 3). Importantly, the peptide **8** remained intact after 24 h in human bronchoalveolar lavages from patients with acute respiratory dystress syndrome (Fig. 3a, Supplementary Figs 9,10). The peptide **8** also displayed stronger fluorogenic response than the linear peptides (**5**,**6**) and remarkable fluorescence emission in phospholipid membranes with quantum yields reaching 30% (Supplementary Figs 12,13). In addition to *A. fumigatus*, we examined the ability of peptide **8** to stain different fungal strains (Supplementary Fig. 14). While we observed slight differences in fluorescence intensity between strains, peptide **8** stained most fungal cells, indicating its potential as a probe for imaging fungal infection sites of variable origin. We also employed **8** to image *A. fumigatus* that had been pre-treated or not with an excess of non-labelled PAF26 (**4**) (Supplementary

Fig. 15 and Supplementary Movies 1,2). Cells that were pre-treated with compound **4** showed significantly lower staining when exposed to the same concentration of peptide **8**, confirming the specificity of our fluorogenic cyclic structure. We also confimed that the peptide **8** brightly stained *A. fumigatus* in co-cultures with human lung epithelial cells (Supplementary Fig. 16). All these observations assert the cyclic peptide **8** as a fluorogenic probe with high stability in lavage samples from patients with multi-system respiratory disease and potential for direct *ex vivo* imaging of *A. fumigatus* in human pulmonary tissue.

**Ex vivo imaging of *Aspergillus fumigatus* in human tissue.** Next we employed the peptide **8** for high-resolution imaging of *A. fumigatus*. Time-lapse imaging showed the fluorogenic response of **8** upon interaction with the fungal cell membrane and after being internalized and accumulated in lipid-rich intracellular compartments (Fig. 3d and Supplementary Movie 3). The kinetic analysis shows that the peptide **8** labelled fungal cells very rapidly, within few minutes after addition of the probe and

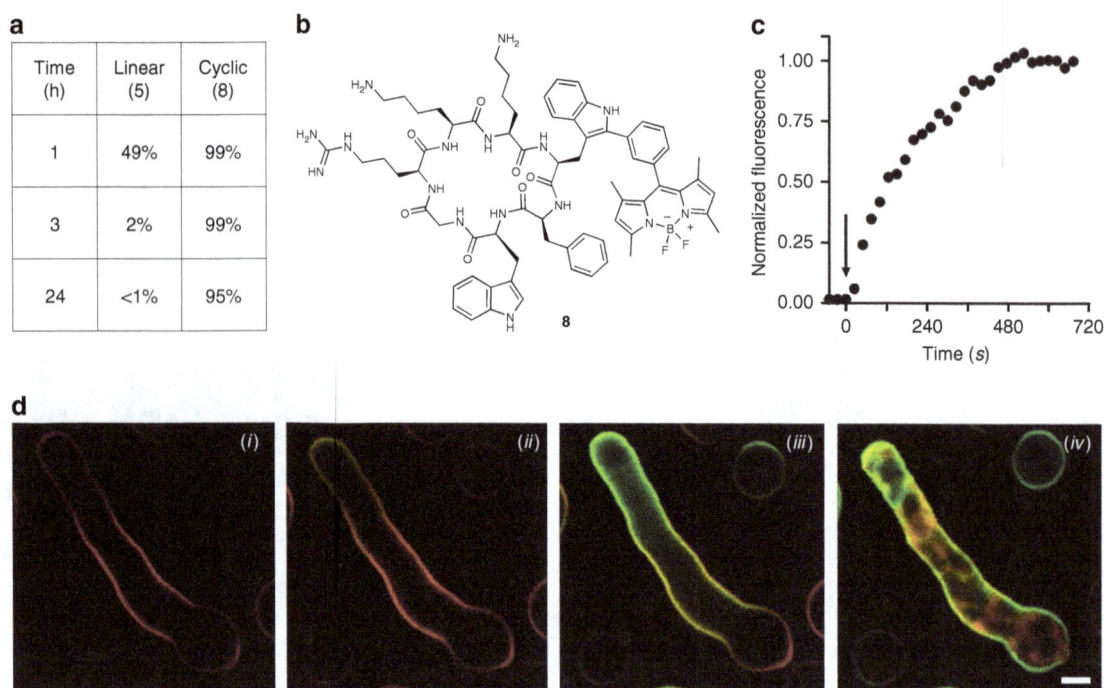

**Figure 3 | The cyclic peptide 8 is a highly stable fluorogenic agent for high-resolution imaging of *A. fumigatus*.** (**a**) Comparative chemical stability of mono-labelled BODIPY linear (**5**) and cyclic (**8**) PAF26 analogues in human bronchoalveolar lavage samples from patients with acute respiratory distress syndrome. (**b**) Chemical structure of the cyclic BODIPY-labelled peptide **8**. (**c**) Kinetic analysis (from time-lapse imaging in **d** of the fluorescence signal of compound **8** (2 μM) in the cell membrane of *A. fumigatus* (arrow points at the addition time for compound **8**). (**d**) Time-lapse high-resolution imaging of *A. fumigatus* upon incubation with a cell membrane counterstain (red) and compound **8** (2 μM, green) for 0 min (*i*), 1 min (*ii*), 3 min (*iii*) and 10 min (*iv*) (see Supplementary Movie 3). Scale bar, 2.5 μm.

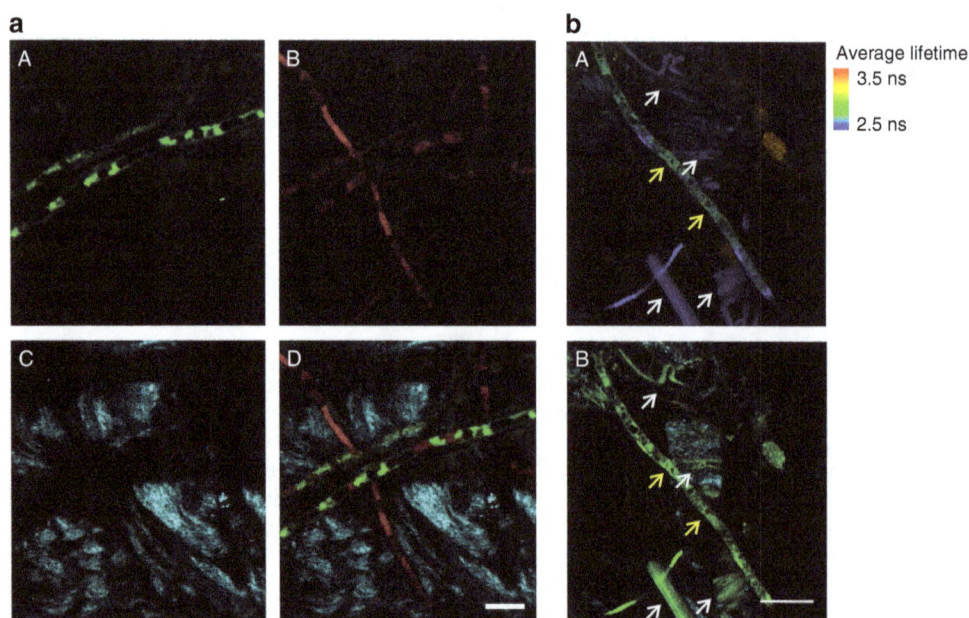

**Figure 4 | Multi-photon fluorescence microscopy of *ex vivo* human pulmonary tissue after incubation with RFP-expressing *A. fumigatus*.** (**a**) Multi-photon microscope images from peptide **8** (5 μM) (A), RFP-expressing *A. fumigatus* (B), second harmonic generation from collagen fibres (C) and merged (D) in *ex vivo* human lung tissue. Scale bar, 10 μm. (**b**) (A) Fluorescence lifetime image of **8**-stained *A. fumigatus* in *ex vivo* human lung tissue. White arrows point autofluorescent tissue structures and yellow arrows point **8**-stained fungal cells. (B) Corresponding fluorescence image of **8**-stained *A. fumigatus* (green) and collagen fibres (second harmonic generation, cyan) for the fluorescence lifetime image in A. Scale bar, 20 μm.

without requiring any washing steps (Fig. 3c). Moreover, the peptide **8** showed no cytotoxicity in lung epithelial cells, even at high concentrations (Supplementary Fig. 17). In view of these properties, we employed the peptide **8** for direct imaging of

*A. fumigatus* in human pulmonary tissue using multi-photon microscopy. In order to confirm the specific staning of **8**, we employed a transgenic strain of *A. fumigatus* expressing red fluorescent protein (RFP) in the cytoplasm. As shown in Fig. 4a, the peptide **8** (green) clearly stained RFP-expressing *A. fumigatus* (red), which confirmed the selectivity of our probe. Multi-photon excitation enabled the acquisition of second harmonic generation (cyan) from the collagen structures of the fibrilar network of human pulmonary tissue. Furthermore, the examination of these samples by fluorescence lifetime imaging revealed that autofluorescent human tissue structures (for example, collagen and elastin)[58], which could potentially overlap with the emission of BODIPY fluorogens, are readily distinguished from **8**-stained *A. fumigatus* by their fluorescence lifetimes (Fig. 4b). Altogether, these results validate our fluorogenic BODIPY-labelled cyclic peptide **8** as a highly stable imaging agent for direct and straightforward visualization of *A. fumigatus* in human tissue.

## Discussion

Peptides are excellent scaffolds for the development of imaging agents due to the highly specific molecular interactions with their respective targets. Since most peptides do not contain chemical groups that enable their direct visualization, they often need to be modified with reporters (for example, fluorophores) or reactive groups (for example, aldehydes, azides, alkynes and tetrazines) for further derivatization[59]. Unnatural amino acids containing bio-orthogonal tags can be incorporated at specific sites of peptide sequences by SPPS[60]. Likewise, the incorporation of genetically encoded unnatural amino acids in response to nonsense or frameshift codons has opened the possibility to synthesize protein and peptide structures with reactive groups for subsequent modification[61]. Bio-orthogonal approaches typically involve two-step labelling processes including a conjugation reaction (for example, 'click' chemistry) followed by the removal of excess labelling agent. Recent advances in bio-orthogonal chemistry have led to fluorogenic labelling agents that emit a signal only after conjugation, thus reducing background fluorescence and washing steps[62,63]. Alternatively, and most commonly, peptides are derivatized by incorporation of fluorophores into their sequence so they can be directly used for imaging. Since fluorophores are typically bulkier structures, it is imperative that they are introduced at specific positions of the sequence without impairing the molecular recognition properties of the peptide. Many conjugation methods to attach fluorophores to peptides involve a chemical spacer and rely on the reactivity of polar groups (that is, amines, carboxylic acids, thiols, alcohols); however, these modifications often disrupt the hydrogen bonding pattern of the original peptide, having a detrimental effect on its biological properties.

In the present work, we have engineered a methodology to prepare fluorogenic peptides that relies on a unique Trp-BODIPY derivative (**3**, Fig. 1), which mimics the molecular interactions of the native Trp. The incorporation of a BODIPY group into the $C_2$ position of Trp via a spacer-free C-C linkage does not affect the conformation and molecular interactions of the native amino acid, and introduces a fluorogenic tag that emits only in hydrophobic environments (Fig. 1). To assess the compatibility of our approach with SPPS and validate its utility to prepare peptide-based agents for imaging of fungal infection, we derivatized the antimicrobial hexapeptide PAF26, which shows high affinity for the membrane of fungal cells. PAF26 is an amphipathic peptide with highly conserved *C*-terminal hydrophobic and *N*-terminal cationic domains that are essential to exert its antifungal action[16]. Therefore, the derivatization of PAF26 is not straightforward since conventional labelling might alter the distribution of positive charges or its amphipathic character, resulting in a loss of activity and selectivity.

Analogues of PAF26 incorporating the fluorogenic amino acid **3** at specific sites in their sequence were prepared by SPPS and showed no impairment of the affinity and selectivity of the

original peptide for fungal cells (Fig. 2). Our fluorogenic peptides were used for real-time imaging of several fungal pathogens, namely *Fusarium oxysporum*, *Candida albicans*, *Cryptococcus neoformans* and *A. fumigatus*, suggesting a potential common target for different fungal strains (Supplementary Fig. 14). Given that *A. fumigatus* is the fungal pathogen responsible for IPA, a highly fatal disease in immunocompromised patients, we focused our imaging studies in this fungal strain.

Notably, the minimal fluorescence background in aqueous media and strong fluorogenic behaviour of our probes enabled their use for direct and wash-free imaging of *A. fumigatus* (Fig. 2). Competition experiments with the corresponding non-labelled analogues and comparative studies with non-antifungal negative controls—lacking key residues for the interaction at fungal cells—confirmed the specificity of our PAF26-derived fluorogenic peptides (Supplementary Figs 8,15).

A major advantage of our methodology is its wide applicability to bioconjugation and peptide chemistry. The fluorogenic amino acid **3** and its peptide derivatives are compatible with most Fmoc-based SPPS protocols as they tolerate standard deprotection and coupling conditions as well as mildly acidic (that is, 1% trifluoroacetic acid) cleavage cocktails without observing any degradation. Whereas the precise impact of the amino acid **3** in the molecular recognition properties of labelled sequences needs to be examined on a case-by-case basis, we observed similar activities for labelled and non-labelled peptides in a relatively broad range of short antimicrobial sequences, which confirms the ability of the Trp-BODIPY amino acid **3** to behave as a Trp surrogate (Fig. 2 and Supplementary Fig. 8).

With these peptides being promising imaging agents for *in situ* detection of fungal pathogens in clinically relevant samples, we optimized their chemical stability to image *A. fumigatus* in *ex vivo* human pulmonary tissue. Our optimization studies yielded peptide **8** as a highly fluorogenic cyclic structure with bright fluorescence emission in fungal cells and excellent chemical integrity in samples with high proteolytic activity (Fig. 3 and Supplementary Figs 9,16). The excellent properties of **8** enabled its use in multi-photon and lifetime imaging for the direct visualization of *A. fumigatus* in *ex vivo* human tissue and its discrimination from autofluorescent tissue structures (Fig. 4).

Given that the fluorogenic amino acid **3** can be readily incorporated and has general applicability to both linear and cyclic peptides, we envisage that the introduction of our spacer-free BODIPY fluorogen in relevant peptides will become a transformative methodology to develop peptide-based imaging probes with high sensitivity and specificity. Furthermore, the extension of our methodology to other aromatic amino acids will create numerous opportunities for minimally invasive peptide tagging using synthetically available building blocks.

## Methods

**Chemical synthesis and characterization.** Synthetic procedures and chemical characterization (NMR and high-performance liquid chromatography analysis) for all the probes are included in the Supplementary Information (Supplementary Figs 18–25).

***In vitro* spectral measurements.** Spectroscopic and quantum yield data were recorded on a Synergy HT spectrophotometer (Biotek). Compounds were dissolved at the indicated concentrations and spectra were recorded at room temperature. Spectra are represented as means from at least two independent experiments with $n = 3$. Quantum yields were calculated by measuring the integrated emission area of the fluorescence spectra and comparing it to the area measured for fluorescein in basic ethanol as reference (QY: 0.97). Phosphatidylcholine (PC)-based liposome suspensions were purchased from Clodronateliposomes (Netherlands) and were prepared as previously reported[64].

**IC$_{50}$ determination in *Aspergillus fumigatus*.** The *A. fumigatus* (strain CEA10, source: FGSC A1163) was grown on Vogel's medium at 37 °C for 5 days before the spores (conidia) were harvested. Peptides **4–8** were incubated at different concentrations with *A. fumigatus* conidia to reach a final volume of 100 µl per well. The final conidia concentration was $5 \times 10^5$ cells ml$^{-1}$ in 10% Vogel's medium. After 24 h incubation at 37 °C in 96 well-plates, fungal growth was determined by measuring $OD_{610nm}$ in a spectrophotometer. $IC_{50}$ values were determined using four parameter logistic regression. Data is represented as means ± s.e.m from at least three independent experiments with $n = 3$.

**Cell culture of fungal strains.** *Neurospora crassa* (strain *74-OR23-1VA*, source: FGSC 2489) was grown on standard Vogel's agar (Vogel, 1956) at 25 °C under constant artificial light for 5 days. Conidia were collected using sterile dH$_2$O and then diluted in 20% Vogel's liquid medium for imaging. *F. oxysporum* (strain *4287*, source: FGSC 9935) was grown in liquid potato dextrose broth (PDB) at 28 °C with shaking. Conidia were re-suspended in 20% Vogel's liquid medium and imaged after incubation for 12 h at 30 °C. *Can. albicans* (strain SC5314, source: ATCC MYA-2876) was grown on yeast peptone dextrose liquid medium at 30 °C for 12 h with shaking and then diluted using minimal medium (0.7% yeast nitrogen base plus 2% glucose) before imaging. *Cry. neoformans* (strain H99, source: FGSC 9487) was grown on yeast peptone dextrose agar at 30 °C for 3 days. To collect the cells for imaging, a single colony 1-2 mm in diameter was re-suspended in PBS and washed once with fresh PBS before imaging.

***In vitro* measurements of antimicrobial activity.** *P. aeruginosa* (ATCC 47085), *K. pneumoniae* (ATCC BAA1706) and *E. coli* (ATCC 25922) were grown on Lysogeny Broth (LB) agar plates and stored at 4 °C. For assays, a single colony of bacteria was taken into 10 mL liquid broth and incubated at 37 °C for 16 h. Cultures were centrifuged at 4,000 r.p.m. for 5 min and the pellet was re-suspended in 1 ml of fresh PBS and washed three times. Cultures were reconstituted to 1.0 $OD_{595nm}$, then diluted 1:1,000 and incubated with compounds **4–8** at the indicated concentrations (that is, concentrations matching the $IC_{50}$ values in *A. fumigatus* for all compounds, except for compounds **3** and **5a** where a top concentration of 20 µM was used). Cell viability was monitored over 16 h by measuring $OD_{600nm}$ in a spectrophotometer. Data is represented as % of cell viability as means from at least two independent experiments with $n = 3$.

**Determination of haemolytic activity.** Erythrocytes were isolated from freshly drawn, anticoagulated human blood and diluted in PBS (1:5). An amount of 50 µl of erythrocyte suspension was added to 50 µl of compounds **4–8** at the indicated concentrations (that is, concentrations matching the $IC_{50}$ values in *A. fumigatus* for all compounds, except for compounds **3** and **5a** where a top concentration of 20 µM was used). 0.2% Triton X-100 was used as positive control and PBS as negative control. The plate was incubated at 37 °C for 1 h, each well was diluted with 150 µl of PBS and the plate was centrifuged at 1,200g for 15 min. A total of 100 µl of the supernatant from each well was transferred to a fresh plate, and the absorbance at 350 nm was measured in a microplate reader. Data is represented as % of cell viability as means from three independent experiments with $n = 3$.

**Confocal microscopy of *Aspergillus fumigatus* and human cells.** Human lung A549 epithelial cells (ATCC CCL-185) were grown using DMEM supplemented with 10% fetal bovine serum (FBS), antibiotics (100 U ml$^{-1}$ penicillin and 100 mg ml$^{-1}$ streptomycin) and 2 mM L-glutamine in a humidified atmosphere at 37 °C with 5% CO$_2$. A549 cells were regularly passaged in T-75 cell culture flasks. *A. fumigatus* was grown on standard Vogel's agar at 37 °C for 5 days. Conidia were collected using 0.05% Tween 80, re-suspended in 20% Vogel's liquid medium and incubated for 12 h at 25 °C. For co-cultures, human lung epithelial cells were plated on glass chamber slides Lab-Tek II (Nunc) 2 days before imaging and incubated for 16 h with *A. fumigatus* conidia reaching 75–90% confluence on the day of the experiment. For imaging experiments, cells were incubated for 15 min at 37 °C with compounds **5–8** (5 µM for compounds **5–7** and 2 µM for compound **8**) and imaged without washing in phenol red-free DMEM under a Zeiss LSM 510 META fluorescence confocal microscope equipped with a live cell imaging stage. Fluorescence and bright-field images were acquired using × 40 or × 63 oil objectives. Fluorescent probes were excited with 488 nm (compounds **5–8**) or 543 nm (Syto82) lasers. Confocal microscopy images were analysed and processed with ImageJ. Quantitative analysis of mean fluorescence intensities in competition experiments was performed with Imaris by calculating the mean intensity of each hyphae as independent regions of interests. For competition assays, all images were acquired and analysed using exactly the same conditions.

**Chemical stability in human bronchoalveolar lavages.** Peptides **5** and **8** (20 µM) were dissolved in human bronchoalveolar lavage samples (total volume: 100 µl) and incubated at 37 °C for the indicated times. Samples were injected into an high-performance liquid chromatography Agilent 1100 separations module connected to a UV detector with a Discovery C$_{18}$ column (5 µm, 4.6 × 50 mm). Matrix-assisted laser desorption/ionization data was recorded on a Bruker Ultraflex mass spectrometer using sinapinic acid as the matrix.

**Multi-photon imaging in *ex vivo* human tissue.** *Ex vivo* human lung tissue experiments were approved by the NHS Lothian Tissue Governance Committee and Regional Ethics Committee (REC reference: ref. 13/ES/0126). Human lung tissue was obtained from the periphery (non-cancerous) region of patients undergoing resection for lung cancer. A 1 cm$^3$ tissue was inflated with optimum cutting temperature formulation and stored at − 80 °C. Embedded tissue was cryosectioned at 10 µm intervals and fixed onto glass slides for imaging. RFP-expressing *A. fumigatus* conidia were grown overnight at 37 °C the day before the experiments and incubated with human lung tissue sections for 2–3 h before imaging. For multi-photon imaging experiments, the cyclic peptide **8** was used at a concentration of 5 µM. A custom-built multi-photon microscope was used to acquire second harmonic generation (SHG) and two-photon fluorescence images. Briefly, a picoEmerald (APE) laser provided both a tunable pump laser (720–990 nm, 7 ps, 80 MHz repetition rate) and a spatially overlapped Stokes laser (1064 nm, 5–6 ps and 80 MHz repetition rate). GFP two-photon fluorescence signals were filtered using the following series of filters: FF520-Di02, FF483/639-Di01 and ET500/40m. RFP two-photon fluorescence signals were filtered using FF520-Di02, FF757-Di01 and FF01-609/181, and SHG signals were filtered using FF520-Di02, FF483/639-Di01 and FF01-466/40. Fluorescence lifetime images were acquired by connecting the relevant detector to a PicoHarp 300 (Picoquant, Berlin) and configuring the PMT for photon counting mode for TCSPC-FLIM. SHG and GFP images were taken with the laser tuned to 950 nm and RFP images were recorded using a 1,064 nm laser. Lifetime images were recorded at 20 mW with a 10 µs pixel dwell using the SymPhoTime software (Picoquant). All images were analysed and processed using ImageJ.

## References

1. Chai, L. Y. & Hsu, L. Y. Recent advances in invasive pulmonary aspergillosis. *Curr. Opin. Pulm. Med.* **17**, 160–166 (2011).
2. Azoulay, E. & Afessa, B. Diagnostic criteria for invasive pulmonary aspergillosis in critically ill patients. *Am. J. Respir. Crit.Care Med.* **186**, 8–10 (2012).
3. Zhao, M. *et al.* Spatial-temporal imaging of bacterial infection and antibiotic response in intact animals. *Proc. Natl Acad. Sci. USA* **98**, 9814–9818 (2001).
4. Leevy, W. M. *et al.* Optical imaging of bacterial infection in living mice using a fluorescent near-infrared molecular probe. *J. Am. Chem. Soc.* **128**, 16476–16477 (2006).
5. Leevy, W. M. *et al.* Noninvasive optical imaging of staphylococcus aureus bacterial infection in living mice using a Bis-dipicolylamine-Zinc(II) affinity group conjugated to a near-infrared fluorophore. *Bioconjug. Chem.* **19**, 686–692 (2008).
6. Xie, X. *et al.* Rapid point-of-care detection of the tuberculosis pathogen using a BlaC-specific fluorogenic probe. *Nat. Chem.* **4**, 802–809 (2012).
7. Panizzi, P. *et al.* In vivo detection of *Staphylococcus aureus* endocarditis by targeting pathogen-specific prothrombin activation. *Nat. Med.* **17**, 1142–1146 (2011).
8. Shi, H. *et al.* Engineering the stereochemistry of cephalosporin for specific detection of pathogenic carbapenemase-expressing bacteria. *Angew. Chem. Int. Ed. Engl.* **53**, 8113–8116 (2014).
9. Cheng, Y. *et al.* Fluorogenic probes with substitutions at the 2 and 7 positions of cephalosporin are highly BlaC-specific for rapid *Mycobacterium tuberculosis* detection. *Angew. Chem. Int. Ed. Engl.* **53**, 9360–9364 (2014).
10. Welling, M. M. *et al.* Development of a hybrid tracer for SPECT and optical imaging of bacterial infections. *Bioconjug. Chem.* **26**, 839–849 (2015).
11. Akram, A. R. *et al.* A labelled-ubiquicidin antimicrobial peptide for immediate in situ optical detection of live bacteria in human alveolar lung tissue. *Chem. Sci.* **6**, 6971–6979 (2015).
12. van Oosten, M. *et al.* Real-time in vivo imaging of invasive- and biomaterial-associated bacterial infections using fluorescently labelled vancomycin. *Nat. Commun.* **4**, 2584 (2013).
13. Morisse, H. *et al.* In vivo molecular microimaging of pulmonary aspergillosis. *Med. Mycol.* **51**, 352–360 (2013).
14. Zhou, Q. *et al.* Bioconjugation by native chemical tagging of C-H bonds. *J. Am. Chem. Soc.* **135**, 12994–12997 (2013).
15. Lopez-Garcia, B., Perez-Paya, E. & Marcos, J. F. Identification of novel hexapeptides bioactive against phytopathogenic fungi through screening of a synthetic peptide combinatorial library. *Appl. Environ. Microbiol.* **68**, 2453–2460 (2002).
16. Muñoz, A. *et al.* Two functional motifs define the interaction, internalization and toxicity of the cell-penetrating antifungal peptide PAF26 on fungal cells. *PLoS ONE* **8**, e54813 (2013).
17. Beatty, K. E., Xie, F., Wang, Q. & Tirrell, D. A. Selective dye-labeling of newly synthesized proteins in bacterial cells. *J. Am. Chem. Soc.* **127**, 14150–14151 (2005).
18. Carrico, I. S., Carlson, B. L. & Bertozzi, C. R. Introducing genetically encoded aldehydes into proteins. *Nat. Chem. Biol.* **3**, 321–322 (2007).
19. Brustad, E. M., Lemke, E. A., Schultz, P. G. & Deniz, A. A. A general and efficient method for the site-specific dual-labeling of proteins for single molecule fluorescence resonance energy transfer. *J. Am. Chem. Soc.* **130**, 17664–17665 (2008).
20. Lang, K. *et al.* Genetically encoded norbornene directs site-specific cellular protein labelling via a rapid bioorthogonal reaction. *Nat. Chem.* **4**, 298–304 (2012).

21. Venkatraman, P. et al. Fluorogenic probes for monitoring peptide binding to class II MHC proteins in living cells. Nat. Chem. Biol. 3, 222–228 (2007).
22. Lee, H. S., Guo, J., Lemke, E. A., Dimla, R. D. & Schultz, P. G. Genetic incorporation of a small, environmentally sensitive, fluorescent probe into proteins in Saccharomyces cerevisiae. J. Am. Chem. Soc. 131, 12921–12923 (2009).
23. Lukinavičius, G. et al. A near-infrared fluorophore for live-cell super-resolution microscopy of cellular proteins. Nat. Chem. 5, 132–139 (2013).
24. Vazquez, M. E., Blanco, J. B. & Imperiali, B. Photophysics and biological applications of the environment-sensitive fluorophore 6-N,N-dimethylamino-2,3-naphthalimide. J. Am. Chem. Soc. 127, 1300–1306 (2005).
25. Loving, G. & Imperiali, B. A versatile amino acid analogue of the solvatochromic fluorophore 4-N,N-dimethylamino-1,8-naphthalimide: a powerful tool for the study of dynamic protein interactions. J. Am. Chem. Soc. 130, 13630–13638 (2008).
26. Ge, J., Li, L. & Yao, S. Q. A self-immobilizing and fluorogenic unnatural amino acid that mimics phosphotyrosine. Chem. Commun. 47, 10939–10941 (2011).
27. Loudet, A. & Burgess, K. BODIPY dyes and their derivatives: syntheses and spectroscopic properties. Chem. Rev. 107, 4891–4932 (2007).
28. Boens, N., Leen, V. & Dehaen, W. Fluorescent indicators based on BODIPY. Chem. Soc. Rev. 41, 1130–1172 (2012).
29. Li, Z. et al. Rapid aqueous [$^{18}$F]-labeling of a BODIPY dye for positron emission tomography/fluorescence dual modality imaging. Chem. Commun. 47, 9324–9327 (2011).
30. Hendricks, J. A. et al. Synthesis of [$^{18}$F]-BODIPY: bifunctional reporter for hybrid optical/positron emission tomography imaging. Angew. Chem. Int. Ed. Engl. 51, 4603–4606 (2012).
31. Holland, J. P. et al. Annotating MYC status with 89Zr-transferrin imaging. Nat. Med. 18, 1586–1591 (2012).
32. Thorek, D. L., Ogirala, A., Beattie, B. J. & Grimm, J. Quantitative imaging of disease signatures through radioactive decay signal conversion. Nat. Med. 19, 1345–1350 (2013).
33. Kircher, M. F. et al. A brain tumor molecular imaging strategy using a new triple-modality MRI-photoacoustic-Raman nanoparticle. Nat. Med. 18, 829–834 (2012).
34. Miller, E. W., Zeng, L., Domaille, D. W. & Chang, C. J. Preparation and use of Coppersensor-1, a synthetic fluorophore for live-cell copper imaging. Nat. Protoc. 1, 824–827 (2006).
35. Lee, J. S. et al. Synthesis of a BODIPY library and its application to the development of live cell glucagon imaging probe. J. Am. Chem. Soc. 131, 10077–10082 (2009).
36. Zhai, D., Lee, S. C., Vendrell, M., Leong, L. P. & Chang, Y. T. Synthesis of a novel BODIPY library and its application in the discovery of a fructose sensor. ACS Comb. Sci. 14, 81–84 (2012).
37. Vazquez-Romero, A. et al. Multicomponent reactions for de novo synthesis of BODIPY probes: in vivo imaging of phagocytic macrophages. J. Am. Chem. Soc. 135, 16018–16021 (2013).
38. Hong-Hermesdorf, A. et al. Subcellular metal imaging identifies dynamic sites of Cu accumulation in Chlamydomonas. Nat. Chem. Biol. 10, 1034–1042 (2014).
39. Sunahara, H., Urano, Y., Kojima, H. & Nagano, T. Design and synthesis of a library of BODIPY-based environmental polarity sensors utilizing photoinduced electron-transfer-controlled fluorescence ON/OFF switching. J. Am. Chem. Soc. 129, 5597–5604 (2007).
40. Bura, T., Retailleau, P., Ulrich, G. & Ziessel, R. Highly substituted BODIPY dyes with spectroscopic features sensitive to the environment. J. Org. Chem. 76, 1109–1117 (2011).
41. Er, J. C. et al. MegaStokes BODIPY-triazoles as environmentally sensitive turn-on fluorescent dyes. Chem. Sci. 4, 2168–2176 (2013).
42. White, M. C. C-H bond functionalization & synthesis in the 21st century: a brief history and prospectus. Synlett 23, 2746–2748 (2012).
43. Ackermann, L. Carboxylate-assisted transition-metal-catalyzed C-H bond functionalizations: mechanism and scope. Chem. Rev. 111, 1315–1345 (2011).
44. Chen, X., Engle, K. M., Wang, D. H. & Yu, J. Q. Palladium(II)-catalyzed C-H activation/C-C cross-coupling reactions: versatility and practicality. Angew. Chem. Int. Ed. Engl. 48, 5094–5115 (2009).
45. Lebrasseur, N. & Larrosa, I. Room temperature and phosphine free palladium catalyzed direct C-2 arylation of indoles. J. Am. Chem. Soc. 130, 2926–2927 (2008).
46. Ruiz-Rodriguez, J., Albericio, F. & Lavilla, R. Postsynthetic modification of peptides: chemoselective C-arylation of tryptophan residues. Chem. Eur. J 16, 1124–1127 (2010).
47. Preciado, S., Mendive-Tapia, L., Albericio, F. & Lavilla, R. Synthesis of C-2 arylated tryptophan amino acids and related compounds through palladium-catalyzed C-H activation. J. Org. Chem. 78, 8129–8135 (2013).
48. Mendive-Tapia, L. et al. New peptide architectures through C–H activation stapling between tryptophan–phenylalanine/tyrosine residues. Nat. Commun. 6, 7160 (2015).
49. Li, L. et al. Influence of the number and substitution position of phenyl groups on the aggregation-enhanced emission of benzene-cored luminogens. Chem. Commun. 51, 4830–4833 (2015).
50. Vendrell, M. et al. Solid-phase synthesis of BODIPY dyes and development of an immunoglobulin fluorescent sensor. Chem. Commun. 47, 8424–8426 (2011).
51. Lynch, 3rd J. P. Hospital-acquired pneumonia: risk factors, microbiology, and treatment. Chest 119, 373S–384S (2001).
52. Hamuro, Y., Schneider, J. P. & DeGrado, W. F. De novo design of antibacterial β-peptides. J. Am. Chem. Soc. 121, 12200–12201 (1999).
53. Muñoz, A. et al. Understanding the mechanism of action of cell-penetrating antifungal peptides using the rationally designed hexapeptide PAF26 as a model. Fungal Biol. Rev. 26, 146–155 (2013).
54. Nishimura, J. et al. Potent antimycobacterial activity of mouse secretory leukocyte protease inhibitor. J. Immunol. 180, 4032–4039 (2008).
55. White, C. & Yudin, A. K. Contemporary strategies for peptide macrocyclization. Nat. Chem. 3, 509–524 (2011).
56. Hill, T. A., Shepherd, N. E., Diness, F. & Fairlie, D. P. Constraining cyclic peptides to mimic protein structure motifs. Angew. Chem. Int. Ed. Engl. 53, 13020–13041 (2014).
57. Glas, A. et al. Constrained peptides with target-adapted cross-links as inhibitors of a pathogenic protein-protein interaction. Angew. Chem. Int. Ed. Engl. 53, 2489–2493 (2014).
58. Newton, R. C. et al. Imaging parenchymal lung diseases with confocal endomicroscopy. Respir. Med. 106, 127–137 (2012).
59. Schumacher, D. & Hackenberger, C. P. R. More than add-on: chemoselective reactions for the synthesis of functional peptides and proteins. Curr. Opin. Chem. Biol 22, 62–69 (2014).
60. Zhao, L. et al. Synthesis of a cytotoxic amanitin for bioorthogonal conjugation. Chembiochem 16, 1420–1425 (2015).
61. Liu, W., Brock, A., Chen, S., Chen, S. & Schultz, P. G. Genetic incorporation of unnatural amino acids into proteins in mammalian cells. Nat. Methods 4, 239–244 (2007).
62. Shien, P. et al. CalFluors: a universal motif for fluorogenic azide probes across the visible spectrum. J. Am. Chem. Soc. 137, 7145–7151 (2015).
63. Meimetis, L. G., Carlson, J. C., Giedt, R. J., Kohler, R. H. & Weissleder, R. Ultrafluorogenic coumarin-tetrazine probes for real-time biological imaging. Angew. Chem. Int. Ed. Engl. 53, 7531–7534 (2014).
64. van Rooijen, N. & van Nieuwmegen, R. Elimination of phagocytic cells in the spleen after intravenous injection of liposome-encapsulated dichloromethylene disphosphonate. Cell Tissue Res. 238, 355–358 (1984).

## Acknowledgements

L.M.-T. acknowledges the support of MECD—Spain in the form of a FPU Scholarship. A.S. and M.L. acknowledge the support of CRUK (C157/A15703 (A.S.) and C10195/A18075 (M.L.)). R.L. acknowledges the support of DGICYT-Spain (CTQ2015-67870-P) and Generalitat de Catalunya (2014 SGR 137). M.V. acknowledges the support of the Medical Research Council and the FP7 Marie Curie Integration Grant (333847). We acknowledge Dr Andrew Conway-Morris (University of Cambridge) for providing samples of human bronchoalveolar lavage and Dr Kevin Dhaliwal (University of Edinburgh) for providing samples of ex vivo human lung tissue. All experiments employing human samples were conducted with approval from the NHS Lothian Regional Ethics Committee and the NHS Lothian SAHSC Bioresource. We dedicate this article to the memory of Prof. Enrique Pérez-Payá for his contribution to the discovery of antimicrobial peptides.

## Author contributions

L.M.-T., S.P. and N.K. performed all compound syntheses and chemical characterization; C.Z. and N.D.R. designed the in vitro experiments with fungal cells; C.Z., A.R.A. and M.V. performed in vitro spectral and biological characterization, imaging experiments and analysed the data; M.L. and A.S. set up multi-photon and lifetime imaging experiments in ex vivo human lung tissue; F.A., R.L. and M.V. designed the chemical syntheses of the probes. All authors discussed the results and commented on the manuscript. R.L. and M.V. conceived and co-supervised the overall project; M.V. wrote the paper.

## Additional information

Competing financial interests: The authors declare competing financial interests: University of Edinburgh has filed an invention disclosure form to protect part of the technology described in the study.

# Exponential self-replication enabled through a fibre elongation/breakage mechanism

Mathieu Colomb-Delsuc[1], Elio Mattia[1], Jan W. Sadownik[1] & Sijbren Otto[1]

Self-replicating molecules are likely to have played a central role in the origin of life. Most scenarios of Darwinian evolution at the molecular level require self-replicators capable of exponential growth, yet only very few exponential replicators have been reported to date and general design criteria for exponential replication are lacking. Here we show that a peptide-functionalized macrocyclic self-replicator exhibits exponential growth when subjected to mild agitation. The replicator self-assembles into elongated fibres of which the ends promote replication and fibre growth. Agitation results in breakage of the growing fibres, generating more fibre ends. Our data suggest a mechanism in which mechanical energy promotes the liberation of the replicator from the inactive self-assembled state, thereby overcoming self-inhibition that prevents the majority of self-replicating molecules developed to date from attaining exponential growth.

[1]Centre for Systems Chemistry, Stratingh Institute, University of Groningen, Nijenborgh 4, 9747 AG, Groningen, The Netherlands. Correspondence and requests for materials should be addressed to S.O. (email: s.otto@rug.nl).

Replication of information-containing molecules is of key importance in life as we know it. Although in biology replication is mediated by complex biomolecular machinery, in the prebiotic world this process must have occurred through much simpler mechanisms. This postulate has spurred the development of relatively simple self-replicating and cross-replicating molecules[1,2] based on nucleic acids[3,4], peptides[5,6] or fully synthetic structures[7–9]. The typical design of self-replicating systems is based on template-directed ligation of two halves of the replicator, to produce a noncovalent dimer of the autocatalyst. Subsequent dissociation of this duplex will liberate two replicators that can each mediate another round of replication, potentially enabling exponential growth of the replicator (Fig. 1a). However, such exponential replication is only rarely realized, because achieving sufficient duplex dissociation is in most cases problematic. Von Kiedrowski demonstrated that, when a significant proportion of the replicator resides in the inactive duplex state, replicator growth is typically parabolic; the reaction has an order of 0.5 in the autocatalyst[10–12]. Exponential growth occurs only when the order in autocatalyst is 1. This difference in replication kinetics has important consequences for evolutionary scenarios where several replicators compete for a common resource. Equations 1, 2 and 3 describe the kinetics of a simple competition[11–13]:

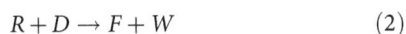

$$F \xrightarrow{R} R \qquad (1)$$

$$R + D \rightarrow F + W \qquad (2)$$

$$\frac{d[R]}{dt} = k_R[F]^f[R]^r - k_D[R]^d[D]^x \qquad (3)$$

where $F$ is a food molecule, $R$ is a replicator, $D$ is a destroying agent and $W$ a waste molecule. The rate of replicator formation is given in by equation 3, in which $k_R$ and $k_D$ are the rate constants for the replication and destruction reaction, respectively. The order of the replication process in food and replicator are given by $f$ and $r$, and the order of the destruction process in replicator and destroying agent are given by $d$ and $x$, respectively. In order for competition to result in destruction of the weakest replicator (a necessary but not sufficient requirement for Darwinian evolution), the order in replicator in the replication process has to be higher than or equal to the order in replicator in its destruction reaction (that is, $r \geq d$)[11–13]. In most plausible replicator destruction scenarios $d$ equals 1, for example, by a bimolecular reaction with a destroying agent shown in equation 2 or by removal of products through flowing part of the solution out of the system. Therefore, for most common competition/destruction scenarios, an order in replicator of at least 1 (that is, exponential replication) is required to achieve Darwinian evolution. Although systems have been reported that move from parabolic towards exponential replication[14–17], few systems have been reported which achieve $r \geq 1$ non-autonomously[18–20] or autonomously[4,21], but even in these examples no general mechanism for exponential growth was reported. The lack of design criteria for self-replicators capable of exponential growth constitutes a major problem that needs to be solved before approaches to Darwinian evolution of synthetic molecules can become mainstream.

We now report a new mechanism by which exponential replication can be achieved based on a fibre growth/breakage mechanism acting on self-assembling replicators. In principle, replication through this mechanism could give rise to exponential growth, as each time a fibre breaks this doubles the number of fibre ends. We now demonstrate experimentally that the replication process in our system is consistent with exponential

growth. We then offer experimental insights into the mechanism of replication: a key role for fibre breakage is evident from the fact that the stirring rate influences both the rate of replication and the concentration of fibre ends. We demonstrate a direct correlation between average fibre length and replication rate. Furthermore, we show that fibre length distributions remain constant during replication, which is necessary in order to achieve exponential replication. Finally, we also confirm computationally that our hypothesized fibre elongation/breakage mechanism results in exponential replication and that the fibre breakage step is required for exponential growth.

## Results

**Replicator emergence from a dynamic combinatorial library.** This work builds on our previous observations on the emergence of replicators from dynamic combinatorial libraries[22] arising from peptide building blocks, such as **1** (refs 23,24). Oxidation of an aqueous solution of **1** gives rise to a continuously exchanging pool of disulfide macrocycles (Fig. 1b). Initially, cyclic trimers and tetramers dominate but, upon agitation, self-replicating hexamers emerge, which self-assemble into fibres and eventually become the major product in solution (Fig. 1d). Note that monomers, trimers or tetramers do not assemble into fibres. Hexamer growth is sigmoidal and seeding experiments confirmed that the growth process is autocatalytic. Shear stress is crucial for replication: in a non-agitated library trimers and tetramers are the only significant products, and no fibres could be observed by transmission electron microscopy (TEM) analysis of libraries dominated by those two species. The hexamer replicator only emerges when libraries are agitated. A mechanism for the replication process was proposed that involves two steps: fibre elongation by growth of the fibres from their extremities and breakage due to mechanically induced shear stress (Fig. 1c). Fibre growth by elongation may take place by sequestration of hexamer macrocycles from the macrocycle equilibrium, or through a mechanism in which fibre ends catalyse the formation of the hexamer macrocycle. Parallels exist between this mechanism and the nucleation-growth mechanisms of amyloid fibres, in which mechanically induced breakage can also play a role[25–29]. The hierarchical self-assembly process (monomers organizing into hexamers that then stack into fibres) is somewhat reminiscent of the fully non-covalent assembly of rosettes made from melamine derivatives and cyanuric acid, which have been proposed as prebiotic precursors to RNA analogues[30,31]. However, neither amyloids nor melamine rosettes involve covalent bond formation or self-replication.

**Experimental determination of the order in replicator.** In the absence of replicator destruction, the rate law for replication (equation (3)) simplifies to:

$$\frac{d[R]}{dt} = k_R[F]^f[R]^r \qquad (4)$$

To determine the order in replicator $r$, equation 4 may be re-written as:

$$\log \frac{d[R]}{dt} = \log k_R + f \log[F] + r \log[R] \qquad (5)$$

During the initial phase of growth, the concentration of food molecules $F$ is approximately constant, hence $\log k_R + f \log[F]$ is constant. Following the methodology developed by von Kiedrowski, the replication order $r$ may now be determined from the slope of a plot of the initial rate of replication ($\log d[R]/dt$) versus $\log[R]$[3,10]. We determined the initial rate of replication for different replicator concentrations through a set of seeding experiments. A series of identical samples rich in monomers,

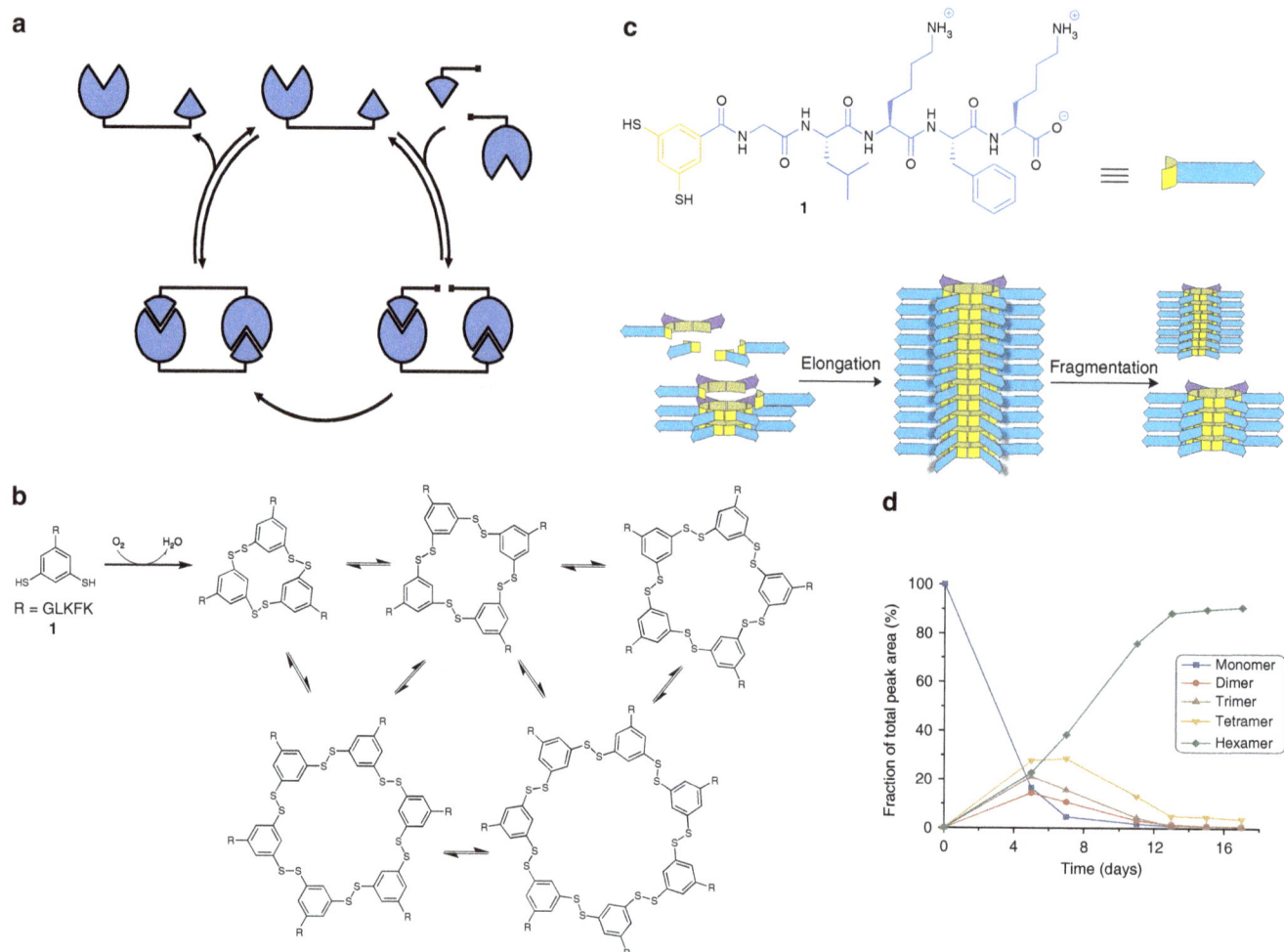

**Figure 1 | Replication mechanisms.** (**a**) Traditional mechanism based on template-directed ligation of two replicator precursors. (**b**) Oxidation of building block **1** containing two thiol functionalities leads to a mixture of interconverting macrocycles. (**c**) Self-assembly of hexamers **1**$_6$ results in the formation of fibres. Fibre fragmentation results in doubling of the number of fibre ends. (**d**) Species distribution as a function of time of a non-seeded dynamic combinatorial library made from 3.8 mM **1** in 50 mM borate buffer pH 8.1 stirred at 1,500 r.p.m.

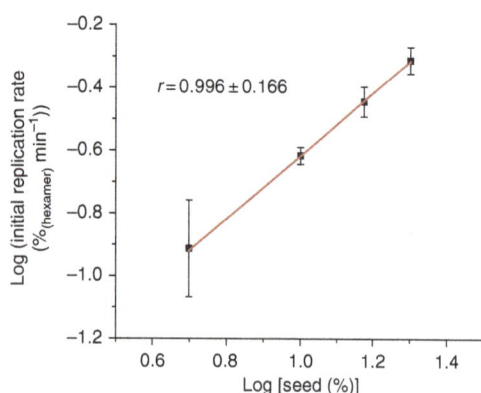

**Figure 2 | Experimental determination of the order in replicator.** Initial replication rates against initial replicator concentration. Data points correspond to seed concentrations (expressed as concentration of **1**$_6$) of 31, 63, 95 and 127 µM, respectively. The error bars on the data points correspond to one standard deviation for each seed concentration and the error on the slope is the standard deviation based on the complete set of measurements.

determined by ultra performance liquid chromatography (UPLC) analysis (under the chromatographic conditions, the fibres dissociate into their constituent hexamer macrocycles, which can be quantified from their ultraviolet–visible absorbance). The individual kinetic profiles are shown in Supplementary Fig. 1. The resulting plot of the initial rate of replication versus hexamer seed concentration is shown in Fig. 2, which allowed us to determine a value for the order in replicator $r$ of $0.996 \pm 0.166$. The error bars in Fig. 2 represent one standard deviation for each seed concentration (three to four measurements for each data point) and the error in $r$ is a standard deviation based on the complete set of seeding experiments. Thus, within the experimental error, this system appears to be capable of exponential replication. However, given the experimental error on the value of $r$, we cannot exclude that replication is sub-exponential based on these experiments alone. To obtain additional evidence for exponential replication, we performed a detailed investigation to establish the replication mechanism. We then calculated the exact value of $r$ associated with this replication mechanism through computer simulations (which are not subject to experimental errors).

**Experimental insights into the mechanism.** To ascertain the validity of our postulated mechanism, we further investigated our system experimentally. We probed the role of fibre breakage in

trimers and tetramers of **1** was prepared and seeded with different amounts of a stirred solution of pre-formed hexamer fibres. The initial rate of increase of the hexamer concentration was

replication and the relationship between fibre end concentration and replicator concentration.

First, the effects of stirring on the average fibre length and on the rate of replication were investigated. Fibre length distributions were determined using TEM for identical samples of pre-formed hexamer fibres that were subjected to different stirring rates. As expected, the average fibre length decreases substantially with increasing stirring rate, ranging from 745 nm at 200 r.p.m. to 97 nm at 1,500 r.p.m. (Fig. 3a; for details, see Supplementary Fig. 2 and Supplementary Table 1). Higher stirring rates also result in more efficient replication. Figure 3b shows the change in replicator concentration with time for different libraries stirred at rates ranging from 200 to 1,500 r.p.m. At higher stirring rates, the hexamers emerge faster than at lower rates of stirring.

The two previous results together imply that in systems with shorter fibres replication is more efficient; this is also evident from Fig. 3c where $t_{50}$, that is, the time it takes for the hexamer to represent 50% of the total library material, is plotted against the average fibre length (values obtained from the data in Fig. 3b). Average fibre length and $t_{50}$ do indeed appear to be correlated, as expected for a mechanism where fibre growth takes place from fibre ends; that is, when the same quantity of replicator is distributed over a larger number of shorter fibres (hence, more fibre ends), replication is more efficient than when it is distributed over a smaller number of longer fibres (hence, fewer fibre ends).

Finally, we also studied whether the average fibre length changed during the replication process. In fact, in a mechanism where replication takes place at the fibre ends, for replication to be exponential (that is, $r = 1$), the average fibre length should not vary with time during replication. The concentration of fibre ends is only directly proportional to the total replicator concentration if the average fibre length is constant during replication. Note that, at a given amount of replicator, the concentration of fibre ends is independent of the fibre length distribution. This fact follows directly from the definition of the average fibre length, which equals the sum of the lengths of all fibres divided by the number of fibres. The first term is constant and determined by the replicator concentration (assuming all replicator to be incorporated into the fibres), whereas the second term equals half the number of fibre ends. We monitored product distribution by UPLC (Fig. 4a) and the average fibre length by TEM (Fig. 4b) over the course of the replication process in a partially oxidized library containing mainly monomer, trimer and tetramer, which was seeded at $t = 0$ min with pre-formed hexamer fibres and stirred at 1,200 r.p.m.

Figure 4b shows that the average fibre length is essentially constant (see Supplementary Fig. 3 and Supplementary Table 2 for details), whereas the total fibre concentration increased approximately fourfold (21–75%). Thus, the system reaches a dynamic stationary state in fibre length in which fibre elongation is balanced by mechanically induced fibre breakage. Constant fibre length suggests a direct proportionality between replication rate and concentration of fibre ends, a prerequisite for exponential replication with $r = 1$.

**Computational modelling of the elongation/breakage mechanism.** To corroborate that our proposed elongation/breakage

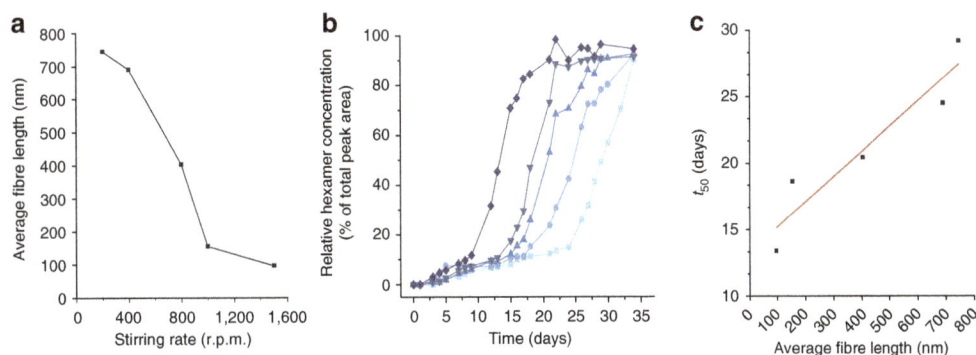

**Figure 3 | Influence of the stirring rate on fibre length and replication kinetics.** (**a**) Average fibre length for different stirring rates. (**b**) Kinetics of replicator growth at various stirring rates (lighter to darker blue: 200, 400, 800, 1,000 and 1,500 r.p.m.). (**c**) Time needed for the replicator to represent 50% of the library material ($t_{50}$), as a function of the average fibre length; the line represents a linear fit of the data.

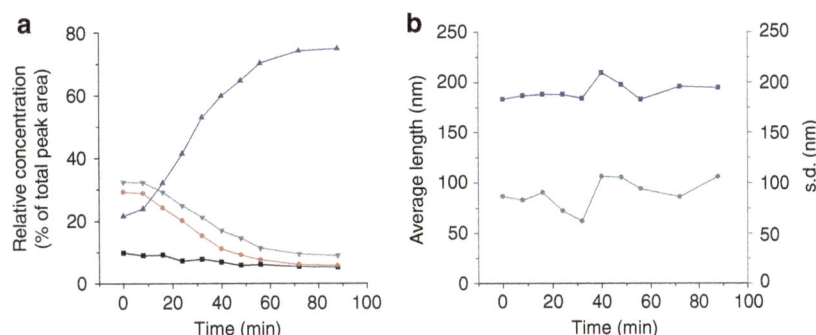

**Figure 4 | Change in replicator concentration and average fibre length during replication.** (**a**) Change in product distribution with time of a library made from building block **1** (3.8 mM) stirred at 1,200 r.p.m. composed of **1** and cyclic trimer and tetramer seeded at $t = 0$ min with 20% pre-formed hexamer fibres (monomer concentration as black squares, trimers as green triangles, tetramers as red circles, hexamers as blue triangles). (**b**) Average fibre length (blue squares) and associated standard deviation of the fibre length distribution (green circles) of this seeded library as a function of time (determined by TEM).

mechanism gives rise to exponential growth of hexamer fibres, the replicating system was also studied computationally and the order in replicator was determined through numerical simulations that are unencumbered by experimental errors. The exchange reactions in solution and the fundamental steps for replication according to our hypothesis, that is, fibre nucleation, elongation and breakage, were included in a kinetic model (Fig. 5, see Supplementary Note 1 for more details), which was studied under a set of parameters. We used this model to address two key questions: first, does the elongation/breakage mechanism result in exponential replication, and if so, can we obtain a more accurate estimate of the order in replicator $r$ than was possible experimentally? Second, is breakage necessary for exponential replication? Related to this second question we also set out to determine the value of $r$ in a model that lacks breakage, but is otherwise identical to the original model.

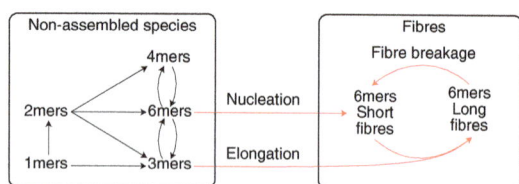

**Figure 5 | Computational model of the replicating system.** Fibre elongation and fibre breakage could be toggled on or off in order to study their role in replication.

The kinetic scheme, which has been employed for our numerical simulations, is depicted in Fig. 5 and the simulations are described in further detail in the Supplementary Note 2 (the full code is provided in Supplementary Method 1). It considers non-assembled species and self-assembled fibres of varying length, along with exchange processes and fibre elongation and breakage steps.

As shown in Fig. 6a, simulations with the complete elongation/breakage mechanism were able to qualitatively reproduce the behaviour of the experimental system, where monomers are oxidized to trimers and tetramers via the formation of dimers (not shown for simplicity), before replication takes over and consumes the smaller macrocycles through the sigmoidal growth of hexamer species (Fig. 1d).

We determined the order in replicator $r$ by using equation 5, which is only valid if the concentration of food molecules $F$ is constant. In the present model, the supply of food molecules was kept constant by fixing the concentrations of trimers and non-assembled hexamers to a constant value throughout the simulation (equivalent to a system in which any consumed food molecules are instantly replenished from the surroundings). Figure 6b shows the central region of a plot of the logarithm of replication rate against the logarithm of replicator concentration (see Supplementary Note 2 and Supplementary Fig. 4 for the complete plot and further details). A linear least squares fit of the data yielded an order in replicator of $r = 1.000$ confirming that a fibre elongation/breakage mechanism indeed results in exponential replication. Note that obtaining an observed order in

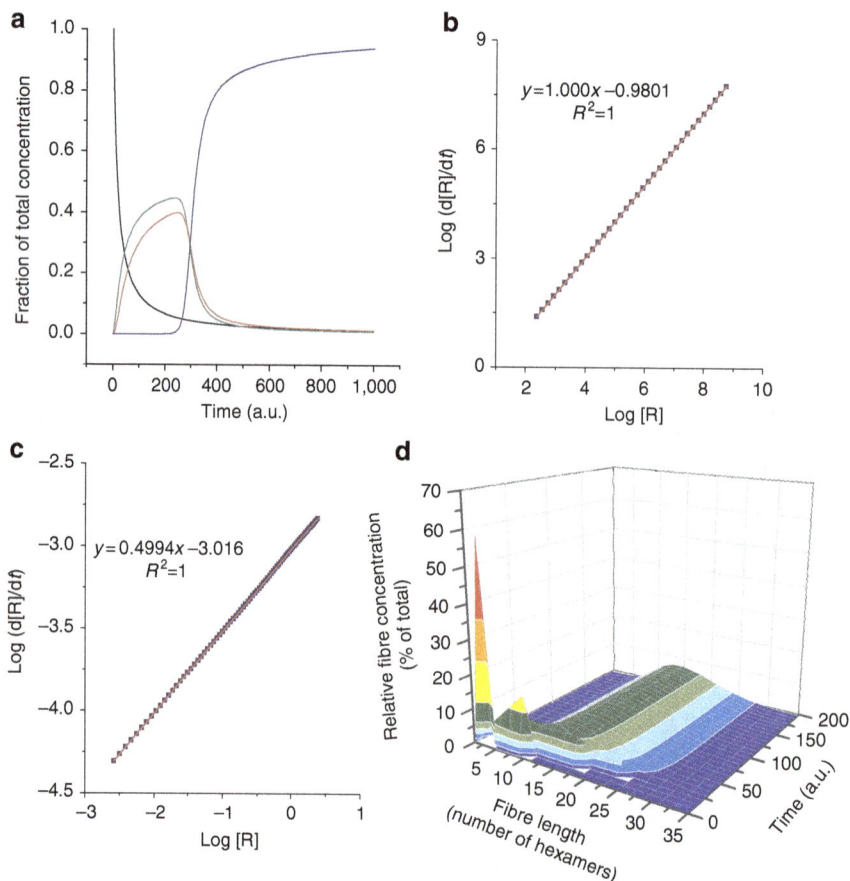

**Figure 6 | Computational studies on the growth/breakage mechanism.** The numerical simulations show distinct kinetics in the presence of the breakage mechanism and in the absence thereof. (**a**) Typical kinetics observed in the numerical simulations of the growth/breakage mechanism (monomers in black, trimers in green, tetramers in red and hexamers in blue). (**b**) Computational determination of the order in replicator $r$ in the case of a growth/breakage mechanism. (**c**) Computational determination of the replication order in the case of an elongation-only (breakage-free) mechanism. (**d**) Fibre length distribution as a function of time. After an initial transient regime, replication occurs at a steady-state length distribution.

replicator of **1** is only possible under conditions in which the production of replicator through the uncatalysed background reaction is negligible. For our family of self-replicators the background reaction is extremely slow: in the absence of agitation no significant amount of replicator $\mathbf{1}_6$ forms, even after 1 month.

To further ascertain the role of fibre breakage in the nature of the replication process, the system was also studied computationally under breakage-free conditions, that is, by deactivating the corresponding breakage pathway. Interestingly, the results, shown in Fig. 6c, display a replication order close to 0.5, that is, the value commonly associated to self-replicators that suffer from self-inhibition through duplex formation. Theoretical work by Stadler on replicators that form triplexes (while not considering mechanical breakage) concluded that also in these systems product inhibition leads to an order in replicator of 0.5 (ref. 32; see Supplementary Note 3 for further details). This resemblance in order in replicator between our system and the duplex or triplex systems is coincidental; it has a different origin than the 'square root law' of autocatalysis typically associated with replicator dimerization. It may be rationalized as follows: in a mechanism in which new fibres nucleate continuously at a constant rate, but no fibres break, the rate of replication increases linearly with time as it depends on the number of fibre ends ($d[R]/dt \sim t$). Furthermore, under these conditions it can be shown that the concentration of replicator increases proportionally to $t^2$ ($[R] \sim t^2$) (see Supplementary Figs 5 and 6 for further details). Therefore, without fibre breakage, the rate of replication is proportional to the square root of the replicator concentration ($d[R]/dt \sim [R]^{0.5}$). Thus, the simulations reveal that in the absence of fibre breakage no exponential growth is obtained, highlighting the crucial role of fragmentation in exponential replication.

Finally, we also monitored the fibre length distributions over time during the simulations. As shown in Fig. 6d, after a short initial transient phase, the length distribution remained constant throughout the replication process, similar to the behaviour we observed experimentally (see Supplementary Fig. 7).

## Discussion

We have demonstrated that replicators which, upon replication, assemble into large but fragile aggregates may be liberated by fracturing the aggregate through mild mechanical forces. Such fragmentation can overcome replicator self-inhibition and, most importantly, enables exponential replication, consistent with our experimental data and evident from numerical simulations. Notably, our simulations also indicate that, without breakage, the system reverts back to obeying the 'square root law of autocatalysis', albeit through a mechanism different from that postulated for duplex-forming replicators.

The growth/breakage mechanism may well provide a general solution to the auto-inhibition problem almost inherently associated with self-replication. We speculate that the replicators developed by Ashkenasy[21] might well work by a similar mechanism. Our mechanistic understanding gives clear guidance that may be employed to successfully design new exponential replicators, which opens up realistic prospects of achieving Darwinian evolution in a purely synthetic chemical system. The next step in this direction involves developing conditions in which replication and destruction operate in parallel, giving rise to a dynamic kinetic stability regime[33] that is characteristic for life. Work in this direction is currently taking place in our laboratory.

## Methods

**Library preparation and monitoring.** Dynamic combinatorial libraries were prepared by dissolving building block **1**, obtained from Cambridge Peptides, in a 50 mM pH 8.1 potassium borate buffer to a final concentration of 3.8 mM. The pH

of the resulting solution was adjusted to 8.1–8.2 by addition of small amounts of a 2.0 M KOH solution. All libraries were contained in HPLC vials ($12 \times 32$ mm) tightly closed with Teflon-lined snap caps. The libraries were stirred using a Teflon-coated magnetic stirrer bar ($5 \times 2$ mm, obtained from VWR), on an IKA RCT basic stirrer hotplate at 1,200 r.p.m. unless otherwise specified. Library compositions were monitored by quenching 2.0 µl samples of the library in 98 µl of a solution of doubly distilled $H_2O$ containing 0.6% trifluoroacetic acid, in a glass UPLC vial and injecting 5.0 µl of this sample on the UPLC. For samples that were monitored over time, it was confirmed that the total peak area in the UPLC chromatograms remained constant.

**UPLC-MS analysis.** UPLC analyses were performed on a Waters Acquity UPLC I-class system equipped with a photodiode array detector. All analyses were performed using a reversed-phase UPLC column (Aeris Widepore 3.6 µm XB-C18 $150 \times 2.10$ mm, purchased from Phenomenex). Ultraviolet absorbance was monitored at 254 nm. Column temperature was kept at 35 °C. UPLC-MS was performed using a Waters Acquity UPLC H-class system coupled to a Waters Xevo-G2 TOF. The mass spectrometer (MS) was operated in the positive electrospray ionization mode. Injection volume was 5 µl of a 3.8 mM library subjected to a 1:50 dilution in a solution of 0.6 v% of trifluoroacetic acid in doubly distilled water. Eluent flow was 0.3 ml min$^{-1}$; eluent A: UPLC grade water (0.1 v% trifluoroacetic acid); eluent B: UPLC grade acetonitrile (0.1 v% trifluoroacetic acid). For full method, see Supplementary Table 3. For UPLC-MS compound identification see Supplementary Table 4.

**Seeding experiments.** A library was prepared by dissolving 3.8 mM **1** in 50 mM borate buffer at pH 8.2. The library was then oxidized up to 70% using a freshly prepared solution of sodium perborate (38 mM, pH 8.0). The composition of the mixture was at this point: 25% monomer **1**, 5% linear dimer $(\mathbf{1})_2$, 35% cyclic trimer $(\mathbf{1})_3$ and 35% cyclic tetramer $(\mathbf{1})_4$. The resulting solution was split into four samples of 200 µl and each one was then seeded with a pre-formed library rich in the hexamer of **1** (which had been continuously stirred at 1,200 r.p.m.), by adding, respectively, 5.0, 10, 15 and 20 mol% of hexamer (the mol% was calculated as equivalents of **1** in the hexamer relative to equivalents of **1** in the library). The library was then stirred at 1,200 r.p.m. and the change in hexamer concentration was monitored by sampling every 8 min as described above.

**Experiments at different stirring rates.** A library was prepared by dissolving 3.8 mM **1** in 50 mM borate buffer at pH 8.2. It was then split into five samples of 200 µl, that were each placed in an HPLC vial ($12 \times 32$ mm), which was tightly closed with a Teflon-lined snap cap and which contained a Teflon-coated magnetic stir bar $5 \times 2$ mm. The libraries were continuously stirred, each of them at identical positions on different stirring plates, all of the same brand (IKA RCT basic stirrer hotplate), at rates of 200, 400, 800, 1,000 and 1,500 r.p.m. The libraries were regularly sampled for UPLC monitoring.

**Negative staining TEM.** An aliquot of a sample taken from a peptide library was diluted 40 times in doubly distilled water. Shortly thereafter, a small drop of the diluted sample was deposited on a 400 mesh copper grid covered with a thin carbon film. After 30 s, the droplet was blotted on filter paper. The sample was then stained with a solution of uranyl acetate deposited on the grid and blotted on filter paper after 30 s. The grids were observed in a Philips CM120 electron microscope operating at 120 kV. Images were recorded on a slow scan CCD camera.

**Fibre length measurements.** TEM micrographs were analysed using ImageJ. A scale was put on each micrograph according to its magnification, and an average length of each sample was determined by measuring fibres from the micrograph, using the measuring tool of ImageJ. The data were then transferred and analysed using MS Excel.

## References

1. Patzke, V. & von Kiedrowski, G. Self replicating systems. *ARKIVOC* **v**, 293–310 (2007).
2. Vidonne, A. & Philp, D. Making molecules make themselves - the chemistry of artificial replicators. *Eur. J. Org. Chem* 593–610 (2009).
3. von Kiedrowski, G. A self-replicating hexadeoxynucleotide. *Angew. Chem. Int. Ed. Engl.* **25**, 932–935 (1986).
4. Lincoln, T. A. & Joyce, G. F. Self-sustained replication of an RNA enzyme. *Science* **323**, 1229–1232 (2009).
5. Lee, D. H., Granja, J. R., Martinez, J. A., Severin, K. & Ghadiri, M. R. A self-replicating peptide. *Nature* **382**, 525–528 (1996).
6. Samiappan, M., Dadon, Z. & Ashkenasy, G. Replication NAND gate with light as input and output. *Chem. Commun.* **47**, 710–712 (2011).
7. Tjivikua, T., Ballester, P. & Rebek, J. A self-replicating system. *J. Am. Chem. Soc.* **112**, 1249–1250 (1990).
8. Kassianidis, E., Pearson, R. J., Wood, E. A. & Philp, D. Designing instructable networks using synthetic replicators. *Faraday Discuss.* **145**, 235–254 (2010).

9.  Dieckmann, A., Beniken, S., Lorenz, C., Doltsinis, N. L. & von Kiedrowski, G. Unravelling a fulvene based replicator: experiment and theory in interplay. *J. Syst. Chem* **1**, 10 (2010).

10. von Kiedrowski, G. Minimal replicator theory I: parabolic versus exponential growth. *Bioorg. Chem. Front* **3**, 113–146 (1993).

11. Szathmary, E. & Gladkih, I. Sub-exponential growth and coexistence of non-enzymatically replicating templates. *J. Theor. Biol.* **138**, 55–58 (1989).

12. Szathmary, E. Simple growth laws and selection consequences. *Trends Ecol. Evol.* **6**, 366–370 (1991).

13. Lifson, S. & Lifson, H. Coexistence and Darwinian selection among replicators: response to the preceding paper by Scheuring and Szathmáry. *J. Theor. Biol.* **212**, 107–109 (2001).

14. Wang, B. & Sutherland, I. O. Self-replication in a Diels-Alder reaction. *Chem. Commun.* 1495–1496 (1997).

15. Issac, R. & Chmielewski, J. Approaching exponential growth with a self-replicating peptide. *J. Am. Chem. Soc.* **124**, 6808–6809 (2002).

16. Li, X. Q. & Chmielewski, J. Peptide self-replication enhanced by a proline kink. *J. Am. Chem. Soc.* **125**, 11820–11821 (2003).

17. Kindermann, M., Stahl, I., Reimold, M., Pankau, W. M. & von Kiedrowski, G. Systems chemistry: kinetic and computational analysis of a nearly exponential organic replicator. *Angew. Chem. Int. Ed.* **44**, 6750–6755 (2005).

18. Luther, A., Brandsch, R. & von Kiedrowski, G. Surface-promoted replication and exponential amplification of DNA analogues. *Nature* **396**, 245–248 (1998).

19. Montagne, K., Plasson, R., Sakai, Y., Fujii, T. & Rondelez, Y. Programming an in vitro DNA oscillator using a molecular networking strategy. *Mol. Syst. Bio* **7**, 466 (2011).

20. Schulman, R., Yurke, B. & Winfree, E. Robust self-replication of combinatorial information via crystal growth and scission. *Proc. Natl Acad. Sci. USA* **109**, 176405–176410 (2012).

21. Rubinov, B., Wagner, N., Rapaport, A. & Ashkenasy, G. Self-replicating amphiphilic beta-sheet peptides. *Angew. Chem. Int. Ed.* **48**, 6683–6686 (2009).

22. Corbett, P. T. *et al.* Dynamic combinatorial chemistry. *Chem. Rev.* **106**, 3652–3711 (2006).

23. Carnall, J. M. A. *et al.* Mechanosensitive self-replication driven by self-organization. *Science* **327**, 1502–1506 (2010).

24. Malakoutikhah, M. *et al.* Uncovering the selection criteria for the emergence of multi-building-block replicators from dynamic combinatorial libraries. *J. Am. Chem. Soc.* **135**, 18406–18417 (2013).

25. Bolder, S. G., Sagis, L. M. C., Venema, P. & van der Linden, E. Effect of stirring and seeding on whey protein fibril formation. *J. Agric. Food Chem.* **55**, 5661–5669 (2007).

26. Buttstedt, A. *et al.* Different morphology of amyloid fibrils originating from agitated and non-agitated conditions. *Amyloid J. Protein Fold. Disord* **20**, 86–92 (2013).

27. Dunstan, D. E., Hamilton-Brown, P., Asimakis, P., Ducker, W. & Bertolini, J. Shear flow promotes amyloid-beta fibrilization. *Protein Eng. Des. Sel.* **22**, 741–746 (2009).

28. Hill, E. K., Krebs, B., Goodall, D. G., Howlett, G. J. & Dunstan, D. E. Shear flow induces amyloid fibril formation. *Biomacromolecules* **7**, 10–13 (2006).

29. Tiiman, A. *et al.* Effect of agitation on the peptide fibrillization: Alzheimer's amyloid-beta peptide 1-42 but not amylin and insulin fibrils can grow under quiescent conditions. *J. Pept. Sci.* **19**, 386–391 (2013).

30. Cafferty, B. J. *et al.* Efficient self-assembly in water of long noncovalent polymers by nucleobase analogues. *J. Am. Chem. Soc.* **135**, 2447–2450 (2013).

31. Chen, M. C. *et al.* Spontaneous prebiotic formation of a ribofuranoside that self-assembles with a complementary heterocycle. *J. Am. Chem. Soc.* **136**, 5640–5646 (2014).

32. Stadler, B., Stadler, P. F. & Schuster, P. Dynamics of autocatalytic replicator networks based on higher-order ligation reactions. *Bull. Math. Biol.* **62**, 1061–1086 (2000).

33. Pross, A. Toward a general theory of evolution: extending Darwinian theory to inanimate matter. *J. Syst. Chem* **2**, 1 (2011).

## Acknowledgements

This work was supported by the NWO, the ERC, the University of Groningen and the Dutch Ministry of Education, Culture and Science (Gravitation Program 024.001.035) and COST CM1005 and CM1304.

## Author contributions

All authors designed the experiments. M.C.-D. performed all the experiments, J.W.S. helped with developing the methodology for the seeding experiments. E.M. designed the model and carried out the simulations. S.O. guided the research and wrote the manuscript together with E.M. and M.C.-D. and J.W.S.

## Additional information

**Competing financial interests**: The authors declare no competing financial interests.

# 9

# Optically responsive supramolecular polymer glasses

Diederik W.R. Balkenende[1], Christophe A. Monnier[1], Gina L. Fiore[1] & Christoph Weder[1]

The reversible and dynamic nature of non-covalent interactions between the constituting building blocks renders many supramolecular polymers stimuli-responsive. This was previously exploited to create thermally and optically healable polymers, but it proved challenging to achieve high stiffness and good healability. Here we present a glass-forming supramolecular material that is based on a trifunctional low-molecular-weight monomer ((UPyU)$_3$TMP). Carrying three ureido-4-pyrimidinone (UPy) groups, (UPyU)$_3$TMP forms a dynamic supramolecular polymer network, whose properties are governed by its cross-linked architecture and the large content of the binding motif. This design promotes the formation of a disordered glass, which, in spite of the low molecular weight of the building block, displays typical polymeric behaviour. The material exhibits a high stiffness and offers excellent coating and adhesive properties. On account of reversible dissociation and the formation of a low-viscosity liquid upon irradiation with ultraviolet light, rapid optical healing as well as (de)bonding on demand is possible.

[1] Adolphe Merkle Institute, University of Fribourg, Chemin des Verdiers 4, CH-1700 Fribourg, Switzerland. Correspondence and requests for materials should be addressed to C.W. (email: christoph.weder@unifr.ch).

Supramolecular polymers are assembled from monomeric building blocks through non-covalent, directional interactions, such as H-bonding, π–π stacking and ligand–metal complexation[1]. The nature and strength of useful supramolecular interactions vary widely. The reversible and in many cases dynamic nature of supramolecular binding can be used to impart a wide range of stimuli-responsive characteristics[2,3], including mechanochromism[4,5], bioactivity[6] and mechanical morphing[7]. The possibility to temporarily reduce the molecular weight of the supramolecular assemblies by shifting the equilibrium to the monomer side upon exposure to an appropriate stimulus can also be used to create healable (or self-healing, if the system is sufficiently dynamic under ambient conditions) polymers. This is the result of an increased chain mobility and decreased viscosity upon disassembly, which enable the material to flow and fill cracks and gaps before the original material is reformed by shifting the equilibrium back to the assembled state[8-12]. The same approach can be used to create reversible adhesives that permit (de)bonding on demand[13]. Examples of thermally healable materials based on this general concept include hydrogen-bonded rubbers based on telechelic poly(ethylene-co-butylene) functionalized with ureido-4-pyrimidinone (UPy) units[8,14], elastomer networks based on fatty acids, ethylene diamine and urea[9], and phase-separated elastomers based on a polystyrene-polyacrylic acid brush copolymer[10]. Optically healable supramolecular polymers, which are advantageous because the stimulus can be applied in a targeted manner, have also been realized, for example, on the basis of telechelic poly(ethylene-co-butylene) that was chain-terminated with terdentate ligands and assembled into a polymer with stoichiometric amounts of $Zn^{2+}$ or $Eu^{3+}$ salts[15-18]. However, on account of the dynamic nature of the supramolecular motifs employed, and the use of building blocks with low glass transition temperature ($T_g$), virtually all known healable supramolecular polymers exhibit a low resistance to mechanical stress. This problem can to a certain extent be

overcome by the introduction of a reinforcing (nano)filler[19-21], but even with this improvement the stiffness (storage modulus of <250 MPa) and strength (tensile strength of <5 MPa) are rather low and stifle the exploitation of such materials as a replacement of glassy thermoset resins in coatings, adhesives and other applications[22]. In this study, we show that the combination of high stiffness imparted by supramolecular moieties and stimuli-responsive behaviour is accessible with supramolecular polymer glasses based on a low-molecular-weight, tri-functional supramolecular monomer, which assembles into a disordered supramolecular network. Although *molecular* glasses represent a well-investigated class of materials[23-25], *supramolecular polymer* glasses based on small molecules have been rarely observed and remain hardly explored[26-28]. Known examples of such materials show either a glass transition close to room temperature or the tendency to rapidly crystallize above the $T_g$. Here we report a supramolecular glass that exhibits a high $T_g$ and a low tendency to crystallize upon cooling from the melt. The design principles used here appear to be general and should permit easy access to other representatives of this interesting class of materials.

## Results

**Design and synthesis of a supramolecular glass.** The optically responsive supramolecular material studied here is based on a trifunctional low-molecular-weight monomer (UPy functionalized 1,1,1-tris(hydroxymethyl)propane ((UPyU)₃TMP)) that carries ureido-4-pyrimidinone (UPy) groups (Fig. 1a,b). The UPy motif, originally developed by Meijer and co-workers[29], was chosen because it forms strong self-complementary hydrogen-bonded dimers, is easy to synthesize and its dynamic binding is well investigated[1,30]. Much of the previous work on UPy-based supramolecular polymers has focused on telechelic monomers with two terminal binding motifs that promote linear chain extension, and macromolecular systems in which UPy side chains serve as reversible cross-links[14,31]; these approaches afford

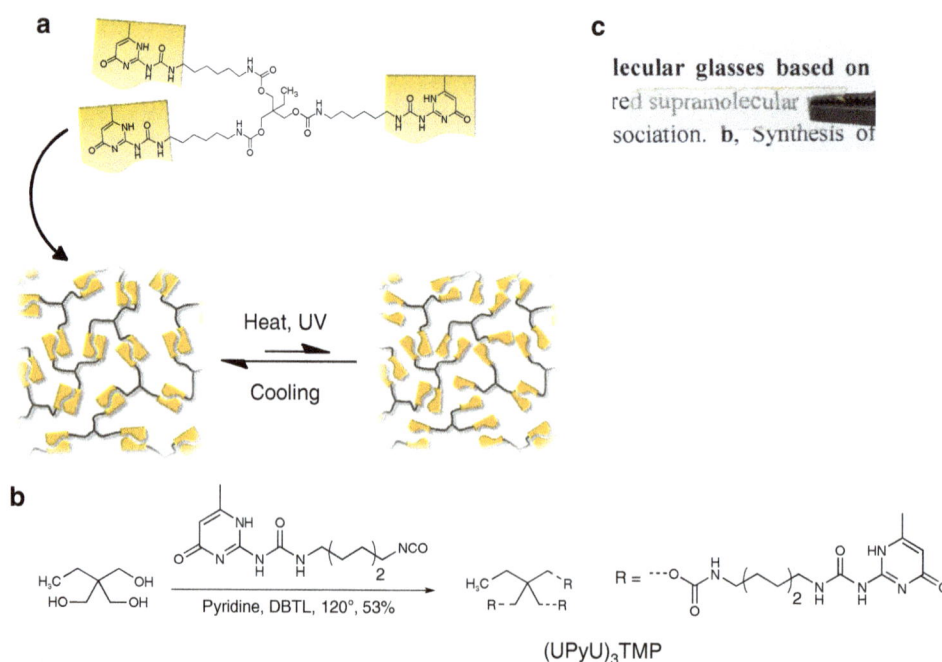

**Figure 1 | Supramolecular glasses based on (UPyU)₃TMP.** (**a**) Schematic representation of the formation of disordered supramolecular networks based on (UPyU)₃TMP and their reversible, heat- or light-induced dissociation. (**b**) Synthesis of (UPyU)₃TMP by the dibutyltindilaurate (DBTL) catalysed reaction of 1,1,1-tris(hydroxymethyl)propane with three equivalents of 2-(6-isocyanatohexylaminocarbonylamino)-6-methyl-4[1H]pyrimidinone in hot pyridine. (**c**) Picture of a self-supported (UPyU)₃TMP film prepared by compression molding at 145 °C.

supramolecular polymers whose properties are to a large extent governed by the nature of the telechelic or polymeric building blocks[32]. By contrast, the trifunctional (UPyU)$_3$TMP introduced here was designed to form a supramolecular network whose properties are dictated by the cross-linked nature and the large content of the binding motif. We further surmised that the high concentration of the supramolecular motif, which tilts the dynamic equilibrium towards the bound state (Fig. 1a), and the cross-linked nature, which reduces the molecular mobility of the monomers, would hamper crystallization and permit kinetic trapping of a disordered amorphous glass upon cooling the material from a dissociated melted state[1,6,30]. Finally, we have shown recently that, if used in a sufficiently high concentration, the UPy motif can serve as an efficient light-heat converter[13,21] and can be used to bestow polymers with optical responsiveness to permit features such as optical healing and (de)bonding on demand.

(UPyU)$_3$TMP was prepared by reacting 1,1,1-tris (hydroxymethyl)propane with three equivalents of 2-(6-isocya-natohexylaminocarbonylamino)-6-methyl-4[1H]pyrimidinone using isocyanate chemistry (Fig. 1b). This simple reaction, which was carried out in hot pyridine to prevent network formation during the reaction and co-crystallization of solvent, as was observed when dimethylformamide was employed, afforded the new monomer in good yield. Precipitation of the product from the reaction mixture upon cooling afforded the product as analytically pure semi-crystalline powder, which upon melting turned into a clear, glassy material (Fig. 1c).

**Properties of (UPyU)$_3$TMP.** Thermogravimetric analysis (TGA) and differential scanning calorimetry (DSC) were used to determine the thermal properties of (UPyU)$_3$TMP. The TGA of the as-prepared material (Supplementary Fig. 1) shows a 2% weight loss at 246 °C; above this temperature the decomposition rapidly accelerates. The DSC first heating trace of the as-prepared monomer shows a weak endothermic transition at 80 °C, which is associated with the glass transition, and an endothermic peak at 178 °C, corresponding to melting of the crystalline portion (Fig. 2a). The cooling scan reveals a glass transition around 100 °C and is void of any other transitions, even at a cooling rate as low as 1 °C min$^{-1}$ (Supplementary Fig. 2), demonstrating that upon melting and cooling, (UPyU)$_3$TMP forms a completely amorphous solid. This is confirmed by the second DSC heating trace, which also only shows a glass transition at 106 °C. Moreover, the DSC traces of (UPyU)$_3$TMP glasses that were kept at ambient temperature for up to 7 months did not show any signs of crystallization upon storage (Supplementary Fig. 3). The interpretation of the DSC experiments was confirmed by powder X-ray diffraction experiments. The diffractogram of the as-prepared (UPyU)$_3$TMP shows a superposition of an amorphous halo and well-defined reflections, whereas the diffractogram of a sample that had been heated to 200 °C and cooled to ambient temperature only displays diffuse diffraction (Fig. 2b). Taken together, these data indicate that the initially semicrystalline (UPyU)$_3$TMP does not readily crystallize after being heated above its melting point; instead, the material forms a (kinetically trapped) amorphous glass, even when cooled very slowly.

(UPyU)$_3$TMP can readily be melt-processed into solid supramolecular objects of various shapes by heating either the as-prepared crystalline monomer or material that had previously been converted into a glassy form to 200 °C (that is, above the melting temperature) to form a clear, slightly viscous liquid. Subsequent cooling to room temperature, optionally in a mold, affords a transparent hard material, for example, in the form of self-standing films (Fig. 1c) or coatings on substrates such as wood, glass or paper (Supplementary Fig. 4). Thin coatings on a

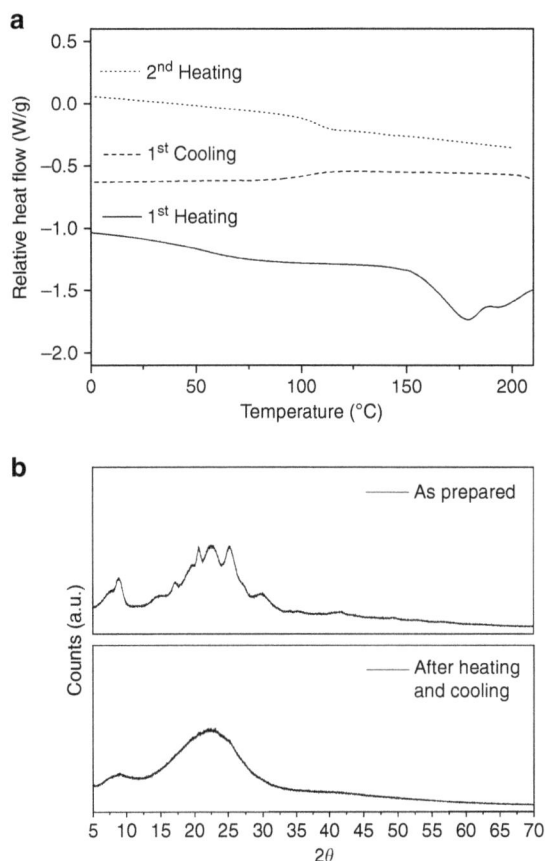

**Figure 2 | Thermal properties and morphology of (UPyU)$_3$TMP. (a)** Differential scanning calorimetry (DSC) traces (first heating ( — ), first cooling (- - -) and second heating (···)) of the as-prepared material. The experiment was conducted with heating and cooling rates of 10 °C min$^{-1}$ under N$_2$ atmosphere; traces with other heating and cooling rates are provided as Supplementary Information. **(b)** Powder X-ray diffractograms for the as-prepared (top) material and a sample that had been heated to 200 °C and cooled to ambient.

paper substrate remained intact upon flexing the substrate (Supplementary Movie 1). Compression molding at 145 °C (5 tons, 5 s) afforded self-supporting films that were cut into rectangular shape (Fig. 1c) and sufficiently robust to allow for dynamic mechanical analysis (DMA) and three-point bending tests. The DMA trace of (UPyU)$_3$TMP shows a glassy regime up to 125 °C with a room-temperature (25 °C) storage modulus of 3.65 ± 0.51 GPa (Fig. 3a and Table 1), reflecting a very high stiffness. Three-point bending tests conducted at the same temperature confirmed the high stiffness (flexural modulus = 3.04 ± 0.26 GPa) and revealed a strain at break of 0.26 ± 0.03% (Fig. 3b and Table 1) and a stress at break of 6.42 ± 0.20 MPa. These data reflect that (UPyU)$_3$TMP is quite brittle. We further probed the mechanical properties of the surface of a *ca.* 300-μm-thin (UPyU)$_3$TMP film that had been melt-deposited on a glass substrate by depth-sensing indentation and atomic force microscopy (AFM) in force spectroscopy mode, in the latter case also as function of temperature. The room temperature Young's moduli—3.7 ± 0.1 GPa measured by indentation (Supplementary Fig. 5) and 2.7 ± 0.1 GPa measured by AFM (Fig. 3c)—are comparable to the value determined by DMA. Temperature-dependent AFM data reveal a significant modulus decrease upon heating, with an onset around 105 °C (Fig. 3c). A comparison with the DSC and DMA data makes evident that this stiffness decrease is associated with the transition from the glassy

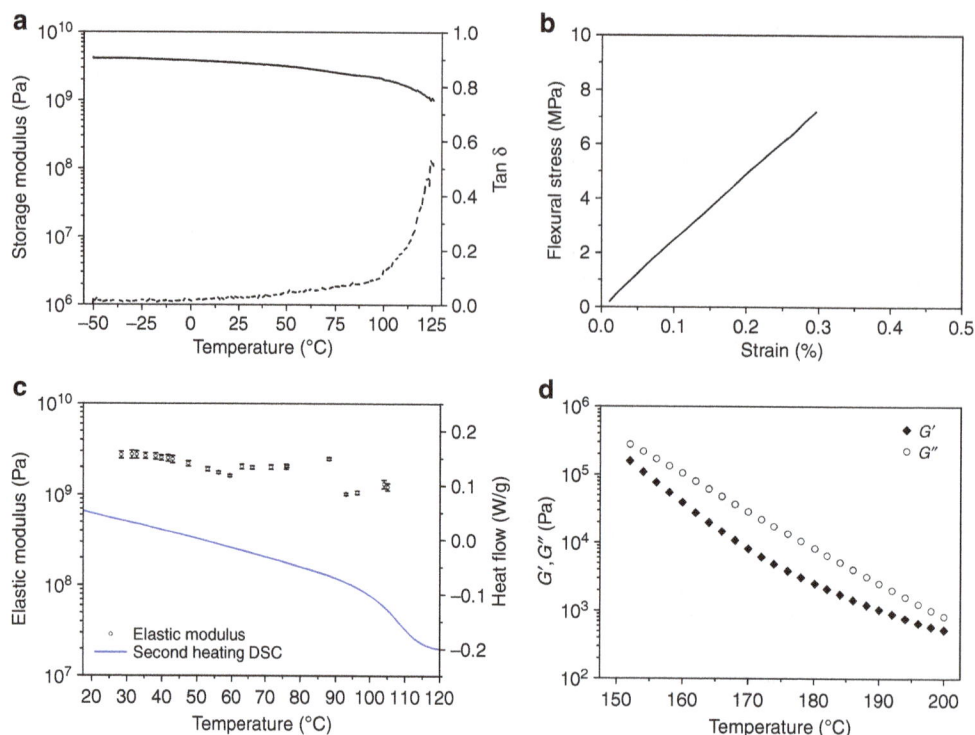

**Figure 3 | Mechanical and rheological properties of the (UPyU)₃TMP supramolecular glass.** (**a**) Representative dynamic mechanical analysis (DMA) trace of (UPyU)$_3$TMP; shown are storage modulus ( — ) and tan δ (···). (**b**) Representative flexural stress–strain curve of (UPyU)$_3$TMP at 25 °C established by a three-point bending test. (**c**) Average surface elastic moduli ($n = 20$) determined by AFM force spectroscopy as a function of temperature; also shown is the heat flow determined by DSC (second heating recorded at a rate of 10 °C min$^{-1}$). (**d**) Storage ($G'$) and loss ($G''$) moduli as function of temperature.

**Table 1 | Mechanical properties of (UPyU)₃TMP and (UPyU)₃TMP(UPy-C6-U-EH)ₓ.**

| Sample | Storage modulus (GPa)* | Flexural modulus (GPa)† | Flexural stress at break (MPa)† | Flexural strain at break (%)† |
|---|---|---|---|---|
| (UPyU)$_3$TMP | 3.65 ± 0.51 | 3.04 ± 0.26 | 6.42 ± 0.20 | 0.26 ± 0.03 |
| (UPyU)$_3$TMP(UPy-C6-U-EH)$_{0.33}$ | 2.87 ± 0.36 | 2.08 ± 0.28 | 6.41 ± 0.96 | 0.30 ± 0.03 |
| (UPyU)$_3$TMP(UPy-C6-U-EH)$_{1.0}$ | 2.21 ± 0.77 | 1.68 ± 0.14 | 5.58 ± 0.95 | 0.33 ± 0.09 |
| (UPyU)$_3$TMP(UPy-C6-U-EH)$_{1.5}$ | 1.47 ± 0.29 | 1.29 ± 0.12 | 4.44 ± 0.83 | 0.34 ± 0.06 |
| (UPyU)$_3$TMP healed‡ | NA | 2.80 ± 0.43 | 4.81 ± 2.74 | 0.27 ± 0.15 |

Abbreviation: NA, not applicable; UPy, ureido-4-pyrimidinone; (UPyU)$_3$TMP, UPy functionalized 1,1,1-tris(hydroxymethyl)propane; UPy-C6-U-EH, 2(6-(2-ethylhexyl)-8-hexylurea-aminocarbonylamino)-6-methyl-4[1H]pyrimidinone.
*Measured by DMA at 25 °C with $n = 3$ individual measurements.
†Measured by three-point bending stress-strain experiments at 25 °C, $n = 5$ (except for healed samples, where $n = 10$).
‡Samples were cut in half, partly overlapped and irradiated with ultraviolet light (320–390 nm, 650 mW cm$^{-2}$, 2 × 12 s).

into a rubbery state. AFM force spectroscopy measurements also reveal a significant increase in adhesion above the $T_g$ (Supplementary Fig. 6). The AFM data permit the conclusion that the UPy–UPy interactions are not simply 'switched off' when the (UPyU)$_3$TMP supramolecular glass is heated above $T_g$; instead, a dynamic equilibrium between bound and dissociated states exists, which is shifted to the monomer side as the temperature is increased. Rheological studies were conducted to confirm this interpretation, which is consistent with the results of previous studies in which UPy dimerization was investigated as a function of temperature[14]. Frequency sweep experiments performed in the linear viscoelastic regime show a frequency dependence of the storage ($G'$) and loss ($G''$) moduli between 150 and 200 °C (Supplementary Fig. 7), suggesting polymer-like viscoelastic properties. Further, a significant decrease of the $G'$ and $G''$ was observed upon heating the material (Fig. 3d),

suggesting a reduction of the virtual molecular weight on account of reducing the cross-link density by shifting the dynamic equilibrium towards the dissociated state.

The mechanical properties of (UPyU)$_3$TMP are largely a result of the high cross-link density of the material in the solid state as well as above $T_g$. We surmised that the introduction of a UPy-functionalized chain stopper would decrease the cross-link density and therefore influence the properties. To test this hypothesis, a mono-functional UPy-containing molecule was synthesized (2(6-(2-ethylhexyl)-8-hexylurea-aminocarbonyla-mino)-6-methyl-4[1H]pyrimidinone (UPy-C6-U-EH), Fig. 4a) and melt-mixed in various ratios with (UPyU)$_3$TMP to form a series of amorphous supramolecular glasses (Supplementary Fig. 8). DSC experiments show that the $T_g$ of these glasses decreases with increasing content of UPy-C6-U-EH (Fig. 4b and Supplementary Fig. 9).

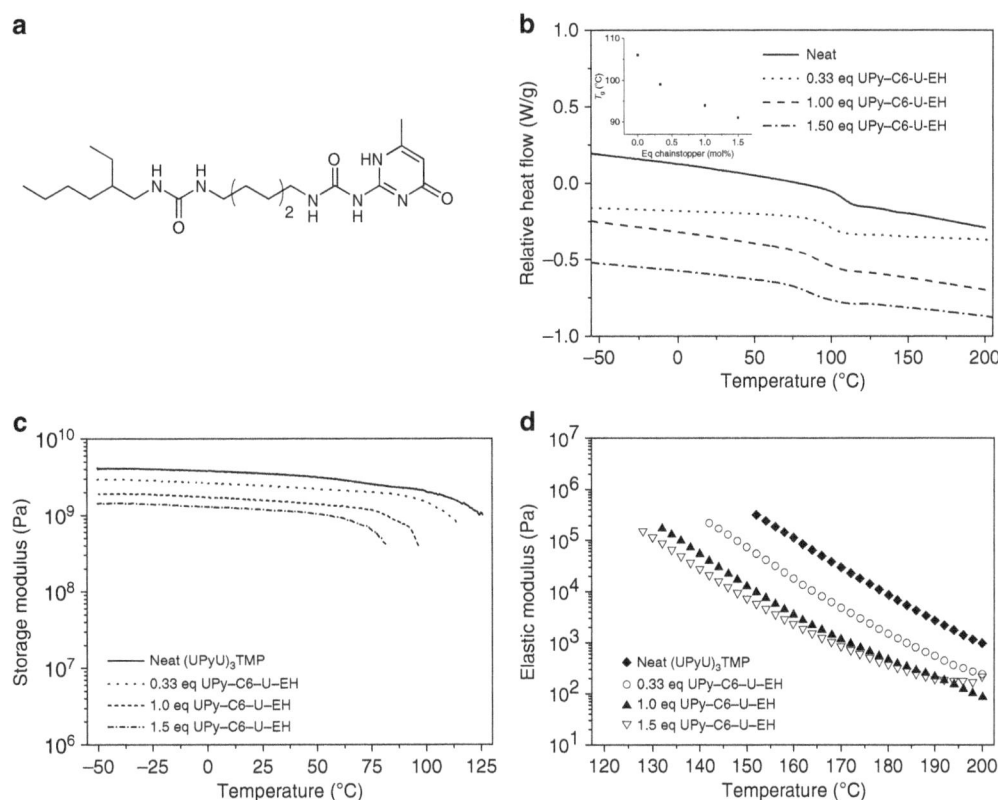

**Figure 4 | Influence of the addition of a chain stopper on the properties of (UPyU)$_3$TMP.** (**a**) Molecular structure of the chain stopper UPy-C6-U-EH. (**b**) Differential scanning calorimetry (DSC) traces. (**c**) Representative dynamic mechanical analysis (DMA) traces. (**d**) Elastic modulus as function of temperature. All samples were prepared by melt mixing (UPyU)$_3$TMP with UPy-C6-U-EH at 200 °C and cooling to ambient.

DMA and three-point bending tests showed a decrease of the storage modulus, flexural modulus and stress at break upon addition of UPy-C6-U-EH (Fig. 4c and Table 1). The strain at break remained largely the same (Table 1), showing that a reduction of the cross-link density does not directly lead to a decrease in brittleness, arguably due to the glassy nature of the materials. Temperature-dependent AFM force spectroscopy data show the same trend for the stiffness of thin film surfaces (Supplementary Fig. 10). As expected, rheological studies show that increasing amounts of UPy-C6-U-EH cause a shift of the storage and loss modulus traces to lower temperatures (Fig. 4d). The DMA and rheology results are characteristic of a reduction of the cross-link density and confirm clearly that the addition of the UPy-C6-U-EH chain stopper has the desired significant effect.

**Stimuli-responsive behaviour of (UPyU)$_3$TMP.** The targeted optically responsive nature of the new supramolecular glass relies on the conversion of (locally harnessed) optical energy into heat by non-radiative relaxation of the excited state. The rheological data presented show clearly that this causes the reversible dissociation of the network and formation of a melt. We first tested this by using (UPyU)$_3$TMP as a reversible supramolecular adhesive. Single lap joints were prepared by joining two glass substrates, of which one was coated with a 30-μm-thin film of the glassy material, bonding them by heating to 200 °C for 10 s and cooling to ambient temperature (Supplementary Fig. 11). The lap joints displayed a shear stress of 1.2 ± 0.2 MPa (Fig. 5a), which is comparable to that of other supramolecular adhesives[13].

The optical absorption spectrum of amorphous (UPyU)$_3$TMP shows an absorption band below 350 nm that can be attributed to the UPy motif (Fig. 5b)[21]. When the bonded lap joints were placed under load and exposed to ultraviolet light

($\lambda = 320$–390 nm, 1,000 mW cm$^{-2}$), the samples debonded within 30 s (Supplementary Movie 2). They could be re-bonded through exposure to light or heat, and the original adhesive properties were restored.

The high optical absorption imparted by the high UPy content (Fig. 5b), and the capability to dissociate into a low-viscosity melt should bestow the supramolecular (UPyU)$_3$TMP glass with excellent optical healing capabilities. To test this, a piece of wood was coated with a 300-μm-thin layer of amorphous (UPyU)$_3$TMP and the coating was intentionally damaged by cutting with a razor blade (Fig. 6a). The damaged area was subsequently exposed to ultraviolet irradiation (320–390 nm, 500 mW cm$^{-2}$), which led to disappearance of the cut in as little as 12 s (Fig. 6a and Supplementary Movie 3). We monitored the temperature increase of the material with the help of an infrared thermometer; the data show a rapid and localized temperature increase to 188 °C (Fig. 6b). AFM images (Fig. 6c) show that a ca 10-μm-wide cut vanishes after 12 s of ultraviolet exposure, although a very shallow scar remained (Fig. 6c and Supplementary Fig. 12), whereas AFM force spectroscopy confirmed that the original and healed samples are identical in regard to their mechanical behaviour (Supplementary Fig. 13). The high transparency of (UPyU)$_3$TMP also allowed optical welding of two films that were overlapped by irradiating such an assembly (320–390 nm, 650 mW cm$^{-2}$, $2 \times 12$ s, Fig. 6d). The results of the three-point bending tests show no statistically significant difference from the original samples (Fig. 6e and Table 1) and samples were observed to always break outside of the mended area. We note that the healing and welding time depends on the light intensity and thermal conductivity of the substrate, which serves as a heat sink. Healing was also possible on glass (which has a higher thermal conductivity than wood), even when the power density was reduced to 250 mW cm$^{-2}$ (Supplementary Fig. 14).

**a**

**b**

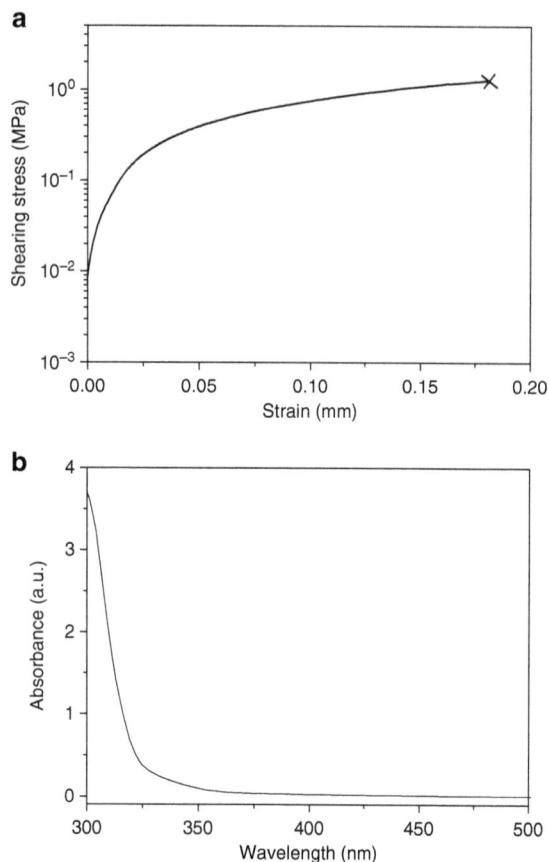

**Figure 5 | Adhesive and optical properties of (UPyU)₃TMP.** (**a**) Adhesion test of a lap joint formed by two glass slides that were bonded with (UPyU)₃TMP (30 μm). (**b**) Ultraviolet–visible absorption spectrum of a solid film of (UPyU)₃TMP (<30 μm) on a quartz substrate.

## Discussion

In summary, we have introduced a novel optically responsive glass-forming supramolecular material, which, in spite of the low-molecular-weight nature of the building block, displays typical polymeric behaviour, including high stiffness in the glassy state, viscoelastic behaviour in the melt and excellent coating and adhesive properties. Two specific characteristics appear to be particularly important in the context of the development of healable coatings. To our best knowledge, the supramolecular (UPyU)₃TMP glass is not only stiffer than any other healable supramolecular polymer reported to date, but the material also heals exceedingly fast. This attractive combination of properties is a direct result of the design principle applied, that is, the use of a low-molecular-weight multifunctional building block to form a dynamic, disordered supramolecular network, which can readily be frozen into a glassy solid. The concept appears to be broadly applicable to other supramolecular glasses made of multifunctional monomers with binding motifs exhibiting sufficiently dynamic supramolecular interactions and light/heat conversion abilities. The most important limitation of the materials studied here vis-à-vis conventional thermosetting polymers are its limited toughness and the high brittleness, which may be improved by the adaptation of conventional toughening approaches[33–35].

## Methods

**Reagents.** All reagents were used as received. Unless indicated otherwise, all chemicals and solvents were obtained from Sigma-Aldrich and used as received. 3-Aminopropyl silica gel (0.6–1.3 mmol g⁻¹) was acquired from TCI Japan. Anhydrous pyridine was purchased from Acros. Anhydrous dimethylformamide (DMF) was purified by passage through alumina columns. 2-(6-Isocyanatohexyl-

aminocarbonylamino)-6-methyl-4[1H]pyrimidinone was synthesized as previously reported[36].

**Characterization.** ¹H (360 MHz) and ¹³C (90 MHz) NMR spectra were recorded on a Bruker Avance III spectrometer in dimethylsulphoxide (DMSO)-$d6$.¹H NMR coupling constants are given in Hz. ¹H NMR spectra were referenced against the signal of residual DMSO at 2.50 p.p.m. and ¹³C NMR spectra were referenced against the DMSO-$d6$ signal at 39.52 p.p.m. TGAs were conducted under $N_2$ using a Mettler-Toledo STAR thermogravimetric analyser in the range of 25–500 °C with a heating rate of 10 °C min⁻¹. DSC measurements were performed under $N_2$ using a Mettler-Toledo STAR system operating at a heating/cooling rate of 10 °C min⁻¹ in the range −70 to 150 °C, unless indicated otherwise. Data from the second heating cycle and the reverse heat flow curve are reported unless indicated otherwise ($T_g$ = glass transition temperature). DMAs were performed on a TA Instruments DMA Q800 with a heating rate of 5 °C min⁻¹ and a frequency of 1 Hz in the range of −50 to 140 °C using a three-point bending setup. Three-point bending tests were conducted on the same instrument at 25 °C, with a displacement rate of 100 μm min⁻¹. All mechanical tests were conducted on rectangular samples (typical dimensions: 25 mm × 5 mm × 230 μm). Flexural moduli were determined from the entire curves, as all samples displayed a linear relation between stress and strain until failure. Load-sensing indentation measurements were performed using a CSM Ultra Nanoindenter equipped with a Berkovich tip (diamond). All experiments were performed using a loading and unloading rate of 100 μN min⁻¹, and 15 min of constant load (100 μN) before unloading to allow for creep deformation. AFM images and force spectroscopy measurements were performed on a JPK Nano Wizard II. AFM images were recorded with NanoWorld NCHR high-resonance frequency tips. Force spectroscopy tests were performed with a Bruker DNISP cantilever with a cube corner diamond tip (nominal sensitivity = 249 N m⁻¹). Temperature-dependent AFM force spectroscopy tests were performed on coatings with a thickness of 300 μm on a thin round microscopy glass slide and placed on a JPK HTHS high-temperature heating stage, and the sample surface temperature was continuously monitored using a hand-held infrared camera. All recorded unloading curves were fitted (upper 50% of the unloading curve) and analysed according to the Oliver and Pharr model to yield the elastic modulus, assuming a Poisson ratio of 0.3 and a perfect cube corner tip[37]. Rheological studies were performed on a TA instruments ARG2 Rheometer operating with a Peltier heating stage and parallel plate geometry. Ultraviolet–visible absorption spectra were measured on a Jasco V-670 spectrophotometer. Electrospray ionization mass spectrometry (ESI-MS) was measured by the Laboratory for Mass Spectroscopy of the University of Fribourg. Elemental analysis was performed by the Service d'analyses chimiques de l'Ecole d'ingénieurs et d'architectes of Fribourg. FT-IR spectra were measured on a Perkin-Elmer Spectrum 65 spectrometer using dried (crystalline) powder in attenuated total reflection (ATR) mode between 800 and 4,000 cm⁻¹ with five accumulated scans per sample. Ultraviolet irradiation of the samples to achieve healing was performed with a Hoenle Bluepoint 4 Ecocure lamp connected to an optical fibre. An optical filter was used to irradiate limit the output to the wavelength range of 320–390 nm. The temperature increase during ultraviolet irradiation was monitored using an Optris PI connect infrared camera. Adhesion tests were performed by adhering two regular glass slides (overlapping area = 10 × 25.7 mm²) with a thin layer of (UPyU)₃TMP. The adhesive properties were determined with a Zwick/Roell Z010 tensile tester at room temperature with a strain rate of 0.1 mm min⁻¹.

**Synthetic methods.** *Synthesis of (UPyU)₃TMP.* For the synthesis of (UPyU)₃TMP, a round bottom flask equipped with a reflux cooler was charged with 1,1,1-tris (hydroxymethyl)propane (635 mg, 4.7 mmol), 2-(6-isocyanatohexylaminocarbonylamino)-6-methyl-4[1H]pyrimidinone (5.0 g, 17 mmol), dibutyltin dilaurate (300 μl, ~3 mol%) and dry pyridine (300 ml) under an $N_2$ atmosphere. The reaction mixture was heated to reflux temperature (120 °C) and was stirred for 36 h under reflux. Aminopropyl functionalized silica (5 g) was added to eliminate any excess of the UPy isocyanate. The reaction mixture was stirred for an additional 30 min at reflux temperature to prevent precipitation of the product. The reaction mixture was cooled to 100 °C and solids were removed from the reaction mixture by vacuum filtration. The filtrate was cooled to room temperature and a white precipitate was observed. Acetone (200 ml) was added to the filtrate, and the solids were collected by filtration. The precipitate was washed with acetone (3 × 100 ml) and dried *in vacuo* at 70 °C for 12 h to yield the analytically pure product as white crystalline powder (3.7 g, 3.6 mmol, 53%). We note that the product has poor solubility in common solvents, therefore all NMR samples were prepared by heating to 100 °C in DMSO-$d6$ for 20 min to achieve complete dissolution ($c = 2$ mg ml⁻¹). Cooling to room temperature resulted in solution that was stable for up to 2 h. ¹H NMR (DMSO-$d6$, 360 MHz) $\delta$ = 11.53 (s, 3H), 9.67 (s, 3H), 7.39 (s, 3H), 7.09 (s, 3H), 5.76 (s, 3H), 3.85 (ds, 6H), 3.12 (t, 6H), 2.93 (s, 6H), 2.10 (s, 9H), 1.43–1.27 (bm, 26H), 0.8 (s, 3H). ¹³C NMR (CDCl₃, 91 MHz) $\delta$ = 173.67, 158.98, 156.99, 155.13, 148.73, 147.08, 107.05, 103.68, 41.11, 40.10, 36.59, 30.56, 29.85, 26.75, 19.35. ESI-MS: *m/z*: calculated: 1,014.1; found: 1,014.2. Analysis calculated for C₄₅H₇₁N₁₅O₁₂: C = 53.29, H = 7.06, N = 20.72. Found: C = 53.30, H = 7.46, N = 20.80.

*Synthesis of UPy-C6-U-EH.* For the synthesis of UPy-C6-U-EH, a round bottom flask was charged with 2-ethylhexyl-1-amine (559 μl, 3.4 mmol),

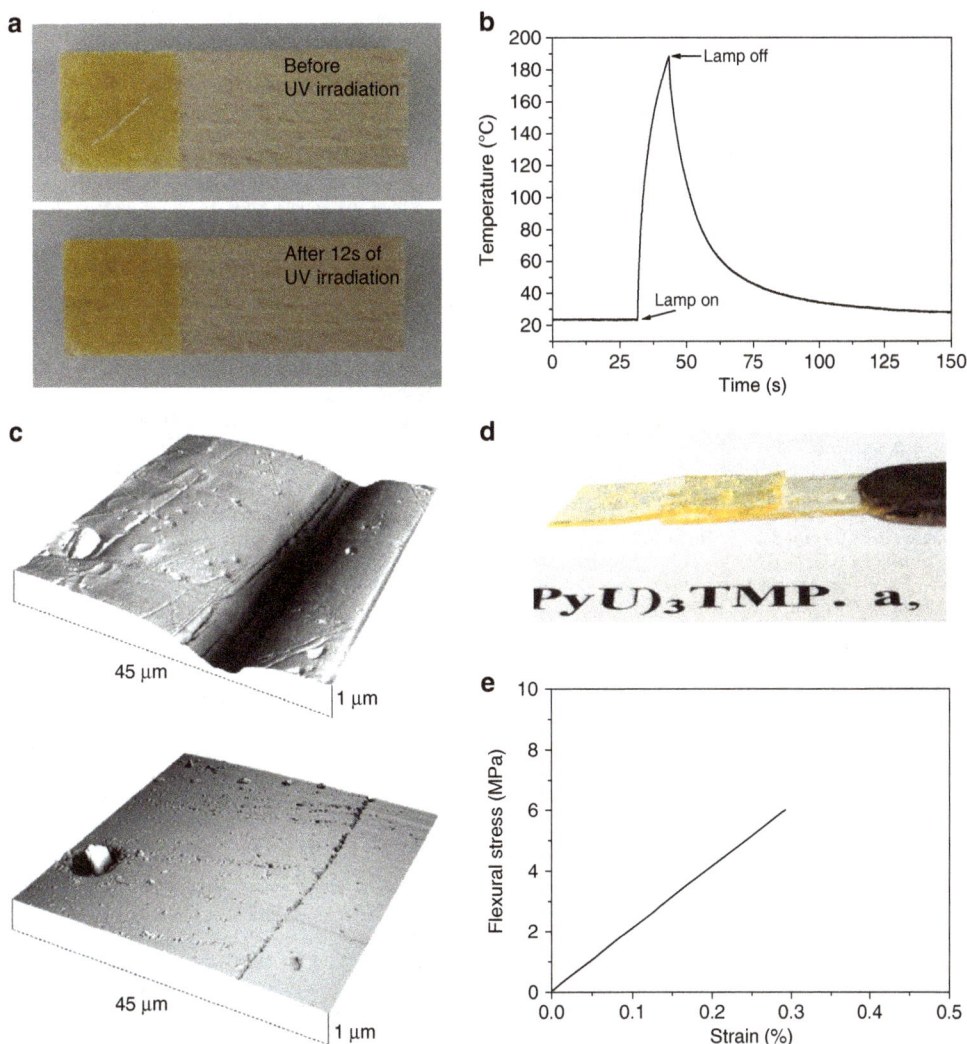

**Figure 6 | Optical healing of the (UPyU)₃TMP supramolecular glass.** (**a**) Pictures showing the optical healing of a damaged coating on wood. The 300-μm-thin coating was cut with a razor blade (top) and subsequently exposed to ultraviolet light for 12 s, which caused complete healing (bottom). (**b**) Surface temperature measured using an infrared camera upon irradiating the coating with ultraviolet light. (**c**) AFM images of damaged (top) and healed coating exposed to ultraviolet light for 12 s (bottom). (**d**) Picture of welded (UPyU)₃TMP films; a film sample was cut in half, overlapped and irradiated with ultraviolet light (2 × 12 s). (**e**) Representative flexural stress–strain curve of a film mended as in **c** at 25 °C. An ultraviolet light source emitting at 320–390 nm and having a power density of 500 mW cm⁻² (for experiments in **a,b,d**) or 650 mW cm⁻² (for experiments in **c,e**) was used.

2-(6-isocyanatohexylaminocarbonylamino)-6-methyl-4[1H]pyrimidinone (1.0 g, 3.4 mmol) and dry DMF (25 ml) under an $N_2$ atmosphere. The reaction mixture was heated to 90 °C for 12 h, and subsequently cooled to room temperature. The mixture was diluted with diethyl ether (100 ml), and a white precipitate was formed. The precipitate was filtered, washed with diethyl ether (2 × 50 ml) and dried *in vacuo* at 50 °C for 12 h to yield the pure product as white crystalline powder (1.3 g, 3.1 mmol, 90%). ¹H NMR (DMSO-*d6*, 360 MHz) $\delta = 11.56$ (s, 1H), 9.66 (s, 1H), 7.34 (s, 1H), 5.77 (s, 1H), 5.71 (q, 2H), 3.13 (q, 2H), 2.95 (dq, 4H), 2.10 (s, 3H), 1.44–1.25 (bm, 17H), 0.81 (dt, 6H). ¹³C NMR (101 MHz, DMSO) $\delta = 158.67$, 42.32, 30.88, 30.48, 29.58, 28.89, 26.48, 26.45, 24.10, 23.01, 14.44, 11.27. ESI-MS: *m/z*: calculated: 422.3; found: 423.3 $[M+1]^+$.

**General methods.** *Preparation of thin coatings and self-standing films.* As-prepared (UPyU)₃TMP was heated on a microscopy glass for 5 min at 200 °C to yield a clear viscous material. The material was transferred with a spatula onto a preheated glass slide and doctor-bladed to form a homogenous film. To form a coating on wood, the viscous liquid material was placed on a hot microscopy glass slide and doctor-bladed onto the wood surface to achieve a homogenous coating. This process was repeated to yield a coating of around 300 μm thickness. Self-standing films were prepared by first heating the (UPyU)₃TMP powder to 200 °C in a vial and cooling the resulting viscous liquid to room temperature. The resulting amorphous material was compression molded in a Carver press between two Kapton sheets that were separated by Teflon spacers (145 °C, 5 s at a pressure of 5 tons after the material had liquefied). Large crack formation was observed when samples were cooled to room temperature between the Kapton sheets. To produce films without macroscopic cracks, the assembly was removed from the hot press, and as soon as the (UPyU)₃TMP had reached the temperature at which it had solidified into a clear, slightly yellow film, the Kapton liners were removed. The samples were subsequently cooled to room temperature. To prepare rectangular samples for mechanical testing, (UPyU)₃TMP films were placed on a hotstage that was heated to 145 °C and lined with a Kapton film, before they were cut using a sharp razor blade. To minimize the influence of environmental conditions, mechanical tests were conducted within a day after producing the samples.

*Indentation measurements.* For indentation measurements, a glassy (UPyU)₃TMP coating with a thickness of around 300 μm on a thin round glass substrate was prepared by melt deposition at 200 °C onto a thin glass substrate. The coating thickness was chosen with respect to Buckle's one-tenth law to avoid any influence of the substrates on the measurement[38], and maximum indentation depths did not exceed 3 μm (<1%). Furthermore, as a control experiment to improve the reliability of the fitted data, AFM images of indents (for indentation forces of 150 and 300 μN) were acquired. Perfect cube corner indents with only limited pile-up (Supplementary Fig. 15) are seen, especially for indents with a force of 150 μN, which was used for all AFM force spectroscopy experiments. We note that fitted data from nano indentation measurements yield a comparative impression of the order magnitude of stiffness, and do not give absolute values for the stiffness. Depth-sensing indentation measurements were performed by indentation at room temperature with a maximum force of 100 μN at a loading and unloading rate of 100 μN min⁻¹. Before each measurement, a height calibration of the local sample surface was performed. Unloading curves were used to determine the elastic modulus according to the Oliver and Pharr model using CSM nanoindentation software[37]. AFM force spectroscopy measurements were carried out by using a cube corner diamond tip cantilever. An indentation force of $150\,\mu N$ was chosen to respect the indentation depth limitation dictated by Buckle's Law regarding effects of the substrate. Data were then analysed using the Oliver and Pharr model with a Poisson ratio of 0.3 and the slope of the initial part (that is, upper 50%) of the unloading indentation curve.

*Rheology.* All rheological studies were carried out using parallel plate geometry and controlled normal force (between $-0.5$ and $0.5\,N$). Samples were measured by heating $(UPyU)_3TMP$ or mixtures of $(UPyU)_3TMP$ and (UPy-C6-U-EH) to $200\,^{\circ}C$ for 2 min before the rotating parallel plate was lowered until the material was observed to flowed out on all sides of the upper plate; subsequently, excess material was removed to prevent edge effects on the data. Frequency sweep experiments were conducted at fixed temperature intervals between 200 and $90\,^{\circ}C$ at a strain of 0.2% and a frequency of $20\,rad\,s^{-1}$. The data were analysed with TA instruments software, and further processed with graphing software. Data were rejected when the strain on the sample was $<75\%$ of the strain applied.

*Chain stopper experiment.* $(UPyU)_3TMP$ and varying amounts of UPy-C6-U-EH (0.33, 1.0 and 1.5 molecular equivalents) were mixed as powders and then heated to $200\,^{\circ}C$ for 15 min to yield a clear homogenous melt. Coatings and films for AFM force spectroscopy measurements, DMA and three-point bending tests were prepared as described above for the neat $(UPyU)_3TMP$.

*Adhesive properties and optical debonding.* The adhesive properties of $(UPyU)_3TMP$ on robust glass substrates were determined with a tensile tester at room temperature with a strain rate of $0.1\,mm\,min^{-1}$. To do that, two glass slides were joined by a $30\text{-}\mu m$-thick layer of $(UPyU)_3TMP$ at $200\,^{\circ}C$ and subsequent cooling to room temperature. For optical debonding experiments, glass slides (thickness $= 140\,\mu m$) were joined together with $(UPyU)_3TMP$ in the same manner, and one side of the resulting lap joint was mounted into a holder, whereas a weight of 500 g was attached to the glass slide. Debonding was achieved by irradiation with ultraviolet light ($320$–$390\,nm$ $1,000\,mW\,cm^{-2}$). This experiment was conducted with thin glass slides to limit heat dissipation.

*Optical healing.* Optical healing of scratches was performed on coatings of $(UPyU)_3TMP$ (thickness $= 300\,\mu m$) that were damaged with a razor blade. Healing was achieved by irradiation with ultraviolet light ($320$–$390\,nm$, $500\,mW\,cm^{-2}$, 12 s) and the temperature was monitored by an infrared camera. After 12 s, a small scar is still visible, and can be completely removed by another 12 s irradiation with the same intensity. For optical mending rectangular pieces of $(UPyU)_3TMP$ films were cut into two with a razor blade. The samples were overlapped and without external pressure irradiated with ultraviolet light ($320$–$390\,nm$, $650\,mW\,cm^{-2}$, $2\times 12\,s$ with 5 min between the exposures to allow for cooling to room temperature).

# References

1. Brunsveld, L., Folmer, B. J. B., Meijer, E. W. & Sijbesma, R. P. Supramolecular polymers. *Chem. Rev.* **101**, 4071–4098 (2001).
2. Wojtecki, R. J., Meador, M. A. & Rowan, S. J. Using the dynamic bond to access macroscopically responsive structurally dynamic polymers. *Nat. Mater.* **10**, 14–27 (2011).
3. Yan, X., Wang, F., Zheng, B. & Huang, F. Stimuli-responsive supramolecular polymeric materials. *Chem. Soc. Rev.* **41**, 6042–6065 (2012).
4. Balkenende, D. W. et al. Mechanochemistry with metallosupramolecular polymers. *J. Am. Chem. Soc.* **136**, 10493–10498 (2014).
5. Haehnel, A., Sagara, Y., Simon, Y. & Weder, C. *Topics in Current Chemistry* Ch. 640, 1–31 (Springer, 2015).
6. Aida, T., Meijer, E. W. & Stupp, S. I. Functional Supramolecular Polymers. *Science* **335**, 813–817 (2012).
7. Capadona, J. R., Shanmuganathan, K., Tyler, D. J., Rowan, S. J. & Weder, C. Stimuli-responsive polymer nanocomposites inspired by the sea cucumber dermis. *Science* **319**, 1370–1374 (2008).
8. Bosman, A. W., Sijbesma, R. P. & Meijer, E. W. Supramolecular polymers at work. *Mater. Today* **7**, 34–39 (2004).
9. Cordier, P., Tournilhac, F., Soulie-Ziakovic, C. & Leibler, L. Self-healing and thermoreversible rubber from supramolecular assembly. *Nature* **451**, 977–980 (2008).
10. Chen, Y., Kushner, A. M., Williams, G. A. & Guan, Z. Multiphase design of autonomic self-healing thermoplastic elastomers. *Nat. Chem.* **4**, 467–472 (2012).
11. van Gemert, G. M. L., Peeters, J. W., Söntjens, S. H. M., Janssen, H. M. & Bosman, A. W. Self-Healing Supramolecular Polymers In Action. *Macromol. Chem. Phys.* **213**, 234–242 (2012).
12. de Espinosa, L. M., Fiore, G. L., Weder, C., Johan Foster, E. & Simon, Y. C. Healable supramolecular polymer solids. *Prog. Polym. Sci.* 60–78 (2015).
13. Heinzmann, C., Coulibaly, S., Roulin, A., Fiore, G. L. & Weder, C. Light-induced bonding and debonding with supramolecular adhesives. *ACS Appl. Mater. Interfaces* **6**, 4713–4719 (2014).
14. Folmer, B. J. B., Sijbesma, R. P., Versteegen, R. M., van der Rijt, J. A. J. & Meijer, E. W. Supramolecular polymer materials: chain extension of telechelic polymers using a reactive hydrogen-bonding synthon. *Adv. Mater.* **12**, 874–878 (2000).
15. Burnworth, M. et al. Optically healable supramolecular polymers. *Nature* **472**, 334–337 (2011).
16. Fiore, G. L., Rowan, S. J. & Weder, C. Optically healable polymers. *Chem. Soc. Rev.* **42**, 7278–7288 (2013).
17. Hong, G. et al. Mechanoresponsive healable metallosupramolecular polymers. *Macromolecules* **46**, 8649–8656 (2013).
18. Yang, B. et al. Self-healing metallo-supramolecular polymers from a ligand macromolecule synthesized via copper-catalyzed azide-alkyne cycloaddition and thiol-ene double "click" reactions. *Polym. Chem.* **5**, 1945–1953 (2014).
19. Michal, B. T., Jaye, C. A., Spencer, E. J. & Rowan, S. J. Inherently photohealable and thermal shape-memory polydisulfide networks. *ACS Macro Lett.* **2**, 694–699 (2013).
20. Coulibaly, S. et al. Reinforcement of optically healable supramolecular polymers with cellulose nanocrystals. *Macromolecules* **47**, 152–160 (2014).
21. Biyani, M. V., Foster, E. J. & Weder, C. Light-healable supramolecular nanocomposites based on modified cellulose nanocrystals. *ACS Macro Lett.* **2**, 236–240 (2013).
22. Hoogenboom, R. Hard Autonomous self-healing supramolecular materials—a contradiction in terms? *Angew. Chem. Int. Ed.* **51**, 11942–11944 (2012).
23. Hancock, B. C. & Zografi, G. Characteristics and significance of the amorphous state in pharmaceutical systems. *J. Pharm. Sci.* **86**, 1–12 (1997).
24. Shirota, Y. Photo- and electroactive amorphous molecular materials-molecular design, syntheses, reactions, properties, and applications. *J. Mater. Chem.* **15**, 75–93 (2005).
25. Vadrucci, R., Weder, C. & Simon, Y. C. Low-power photon upconversion in organic glasses. *J. Mater. Chem. C* **2**, 2837–2841 (2014).
26. Hirschberg, J. H. K. K. et al. Supramolecular polymers from linear telechelic siloxanes with quadruple-hydrogen-bonded units. *Macromolecules* **32**, 2696–2705 (1999).
27. St. Pourcain, C. B. & Griffin, A. C. Thermoreversible supramolecular networks with polymeric properties. *Macromolecules* **28**, 4116–4121 (1995).
28. Boils, D. et al. Molecular tectonics. Disruption of self-association in melts derived from hydrogen-bonded solids. *Macromolecules* **37**, 7351–7357 (2004).
29. Sijbesma, R. P. Reversible polymers formed from self-complementary monomers using quadruple hydrogen bonding. *Science* **278**, 1601–1604 (1997).
30. De Greef, T. F. A. et al. Supramolecular polymerization. *Chem. Rev.* **109**, 5687–5754 (2009).
31. Dankers, P. Y. W., Harmsen, M. C., Brouwer, L. A., Van Luyn, M. J. A. & Meijer, E. W. A modular and supramolecular approach to bioactive scaffolds for tissue engineering. *Nat. Mater.* **4**, 568–574 (2005).
32. van Beek, D. J. M., Gillissen, M. A. J., van As, B. A. C., Palmans, A. R. A. & Sijbesma, R. P. Supramolecular copolyesters with tunable properties. *Macromolecules* **40**, 6340–6348 (2007).
33. Meng, Q. S. et al. Toughening polymer adhesives using nanosized elastomeric particles. *J. Mater. Res.* **29**, 665–674 (2014).
34. Arends, C. B. *Polymer Toughening* (CRC, 1996).
35. Pearson, R. A., Sue, H.-J. & Yee, A. F. *Toughening of Plastics: Advances in Modeling and Experiments* (American Chemical Society, 2000).
36. Keizer, H. M., van Kessel, R., Sijbesma, R. P. & Meijer, E. W. Scale-up of the synthesis of ureidopyrimidinone functionalized telechelic poly(ethylenebutylene). *Polymer* **44**, 5505–5511 (2003).
37. Oliver, W. C. & Pharr, G. M. Measurement of hardness and elastic modulus by instrumented indentation: Advances in understanding and refinements to methodology. *J. Mater. Res.* **19**, 3–20 (2004).
38. Westbrook, J. H. J. H., Conrad, H. & Metals, A. S. f. *The Science of Hardness Testing and its Research Applications* (American Society for Metals, 1973).

# Acknowledgements

This work has been supported by the US Army Research Office (W911NF-12-1-0339), and the Adolphe Merkle Foundation. We thank Professor Bernard Grobéty, Professor Stefan Hengsberger and Adam Luczak for their assistance with the powder X-ray diffraction (B.G.) and depth-sensing indentation (S.H., A.L.) measurements.

# Author contributions

C.W., G.L.F. and D.W.R.B. developed the original concept for the study and designed the materials and experiments. D.W.R.B. synthesized and characterized all materials and performed the experiments. C.A.M. carried out and interpreted the AFM indentation and imaging experiments. All authors discussed the results and contributed to the interpretation of data. C.W. and D.W.R.B. wrote the paper. All authors contributed to editing the manuscript.

# Additional information

**Competing financial interests:** The Adolphe Merkle Institute has filed a patent application related to healable supramolecular polymers with high stiffness, which lists some of the authors as inventors (C.W., D.W.R.B., G.F.).

# 10

# Total synthesis of tetraacylated phosphatidylinositol hexamannoside and evaluation of its immunomodulatory activity

Pratap S. Patil[1], Ting-Jen Rachel Cheng[1], Medel Manuel L. Zulueta[1], Shih-Ting Yang[1], Larry S. Lico[1] & Shang-Cheng Hung[1]

Tuberculosis, aggravated by drug-resistant strains and HIV co-infection of the causative agent *Mycobacterium tuberculosis*, is a global problem that affects millions of people. With essential immunoregulatory roles, phosphatidylinositol mannosides are among the cell-envelope components critical to the pathogenesis and survival of *M. tuberculosis* inside its host. Here we report the first synthesis of the highly complex tetraacylated phosphatidylinositol hexamannoside ($Ac_2PIM_6$), having stearic and tuberculostearic acids as lipid components. Our effort makes use of stereoelectronic and steric effects to control the regioselective and stereoselective outcomes and minimize the synthetic steps, particularly in the key desymmetrization and functionalization of *myo*-inositol. A short synthesis of tuberculostearic acid in six steps from the Roche ester is also described. Mice exposed to the synthesized $Ac_2PIM_6$ exhibit increased production of interleukin-4 and interferon-$\gamma$, and the corresponding adjuvant effect is shown by the induction of ovalbumin- and tetanus toxoid-specific antibodies.

[1] Genomics Research Center, Academia Sinica, No. 128, Section 2, Academia Road, Taipei 115, Taiwan. Correspondence and requests for materials should be addressed to S.-C.H. (email: schung@gate.sinica.edu.tw).

*M*ycobacterium tuberculosis is a dreaded pathogen that causes tuberculosis, one of the leading causes of death in the world. Although the disease becomes active for only 5–10% of infected individuals, 1.5 million people died of tuberculosis in 2013 alone despite progress in the global effort for diagnosis, treatment and prevention[1]. Moreover, it is estimated that one-third of the human population is latently infected with *M. tuberculosis* and is highly vulnerable if immunocompromised. The antituberculosis vaccine bacillus Calmette-Guérin, made by using an attenuated strain of *M. bovis*, only gives protection to children but is highly variable in adults[2]. Co-infection with HIV and the rising cases of multidrug and extensive drug resistance also add to the high morbidity and mortality of the disease[3]. Clearly, fresh insights on the character of the causative agent and its pathogenesis are needed to help alleviate this human condition[4].

The thick glycolipid-containing cell envelope[5] of *M. tuberculosis* is critical for bacterial survival and growth. It is involved in sabotaging immunoregulatory responses[6–8] and it forms a protective barrier for various drugs[9,10]. Among the vital cell-envelope components, phosphatidylinositol mannosides (PIMs) and their hypermannosylated structural relatives (lipomannans and lipoarabinomannans) are found noncovalently anchored to the plasma membrane and the outer capsule through palmitate, stearate and tuberculostearate lipid chains[11]. PIMs, in particular, dictate the intercellular fate of mycobacteria by binding to macrophages[12], regulate cytokines and reactive radical species and stimulate early endosomal fusion by acting as ligands to Toll-like receptors, C-type lectins and DC-SIGN[13]. PIMs can also act as CD1d antigen to activate natural killer T cells for the production of interferon-γ (ref. 14), indicating their potential as vaccine or adjuvant candidates. In addition, PIMs interact with $\alpha_5\beta_1$ integrin on CD4$^+$ lymphocytes, which can either promote granuloma formation and enhance host immune response or help in bacterial survival[15].

Structurally, *myo*-inositol is the central support unit of PIMs with a diacylated glycerophospholipid moiety at O1 and α-mannosylation sites at O2 and O6 (ref. 16). Additional lipid chains may be linked at the primary hydroxyl of the 2-*O*-mannosyl unit and at the O3 position of *myo*-inositol to form triacylated PIMs (AcPIMs) and tetraacylated PIMs (Ac₂PIMs), respectively.

Higher PIMs (for example, $Ac_nPIM_3 - Ac_nPIM_6$) are formed by elongation at the mannose residue linked at O6 of *myo*-inositol. The number of mannose residues and the degree and type of the fatty acyl groups present in the PIM molecules determine their unique role in immunoregulation[11]. As a result, elegant synthetic strategies have been developed for PIMs and their related compounds[17–34]. Nevertheless, the synthesis of a tetraacylated phosphatidylinositol hexamannoside ($Ac_2PIM_6$), the most complex among this class of compounds, is yet to be reported. Thus far, previous disclosures explored the synthesis of $Ac_2PIM_2$ (ref. 34), $PIM_4$ (ref. 24), $PIM_6$ (ref. 32) and $AcPIM_6$ (ref. 25) to name a few.

We describe herein the first synthesis of $Ac_2PIM_6$, using stearic and tuberculostearic acids as the lipid components. The immunomodulatory activity of the synthesized $Ac_2PIM_6$ was also evaluated.

## Results

### Synthetic strategy
Compound **1** possesses multiple components and functionalizations. To arrive at this molecule, one rational synthetic design would be to fragment this sizeable structure into separate segments, which could later be assembled in a convergent manner. For this purpose, we conceived the pseudo-trisaccharide **2**, tetramannoside donor **3** and phosphonate **4** as the primary targets (Fig. 1). The readily perceptible synthetic issues include the transformation of the ordinarily meso *myo*-inositol into the unsymmetrical counterpart in **2** as well as the regioselective protection to afford the mannosyl-building blocks useful enough to deliver the necessary $\alpha 1 \rightarrow 2$ and $\alpha 1 \rightarrow 6$ linkages and the acylation of one mannosyl unit. Accordingly, along with benzyl groups for the global protection of hydroxyls that would be free in the desired product, we selected two additional orthogonal protecting groups for the primary positions of the mannosyl residues in intermediates leading to compound **2**. The *tert*-butyldiphenylsilyl (TBDPS) group should allow, on deprotection, the subsequent coupling with the tetramannoside **3**, whereas the 2-naphthylmethyl (2-NAP) group protects the position that would later be acylated. Being a core constituent of inositol phosphates and other phosphatidyl lipoglycans, various methods have been published for the *myo*-inositol resolution and

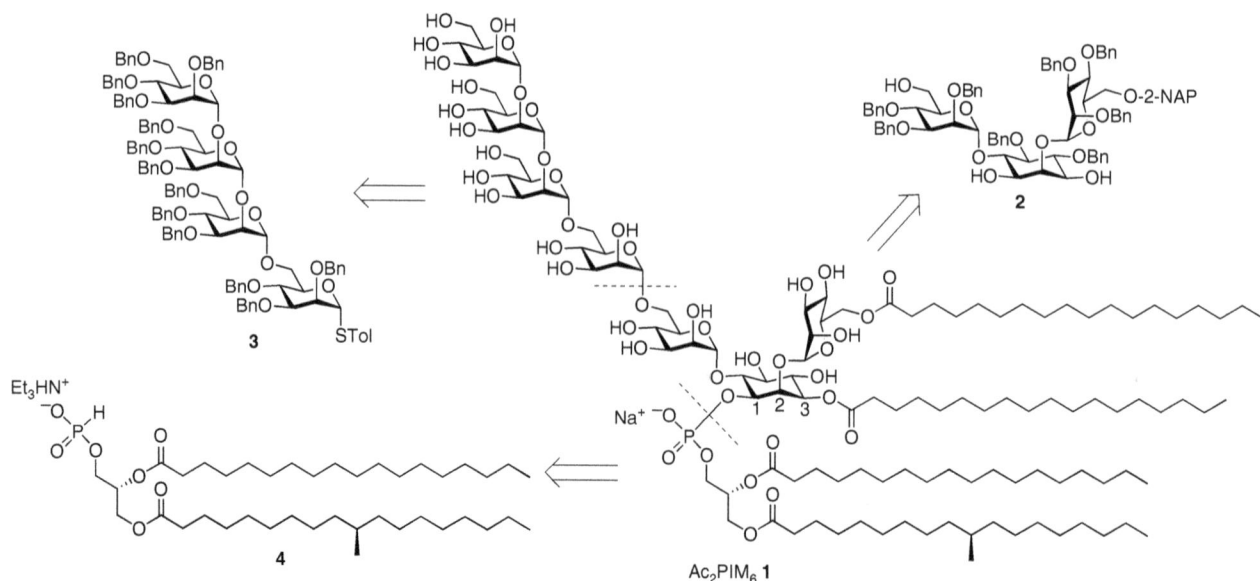

**Figure 1 | Our target tetraacylated phosphatidylinositol hexamannoside (Ac₂PIM₆) and the main blocks designed to represent each segment.** 2-NAP, 2-naphthylmethyl; Bn, benzyl; Tol, *p*-tolyl.

desymmetrization[35,36]. However, the suitably protected chiral *myo*-inositol derivative required for PIMs and glycosyl phosphatidylinositide synthesis were mainly achieved through multistep synthesis from D-glucose via the Ferrier reaction[37]. Recently, we have shown that a mannosyl donor can directly act as chiral auxiliary to differentiate the enantiotopic hydroxyls of meso *myo*-inositol derivatives[30,31]. This direct approach bypasses many transformation steps in affording an appropriately mannosylated and desymmetrized *myo*-inositol. We intend to apply this capacity towards diol 2 while also relying on the steric hindrance created by the installed mannosyl residues to strategically position the 4,5-di-*O*-benzyl moieties on the *myo*-inositol unit and sufficiently favour regioselective acylation at the free 3-OH. An extra temporary protecting group and participating moiety is also needed to permit the construction of the tetramannoside 3, in which case we chose a 2-*O*-benzoyl group. We planned to assemble 3 through linear glycosylation from the reducing end to the nonreducing end using a single elongation unit. For the phosphonate 4, access to the rare tuberculostearic acid is the main concern, and this fatty acid should be synthesized to complete the desired phospholipid moiety.

**Mannosyl-building blocks.** Considering the stability of the thiotolyl leaving group on various functional group interconversions, we selected the thiomannoside 5 (ref. 38) as a starting point in our transformations towards several mannosyl building blocks (Fig. 2). In general, the building blocks needed for the assembly of our target structure required differentiation at either O6 or O2. With a bulky functionality such as the TBDPS group, the protection sequence aimed regioselectively at the primary O6 position seems clear-cut. Thus, the 6-*O*-silylation of 5 using *tert*-butyldiphenylchlorosilane, triethylamine and 4-(*N,N*-dimethylamino)pyridine gave compound 6 in excellent yield. Subsequent benzylation under Williamson condition supplied the necessary thioglycoside 7. The 6-alcohol 8, intended as the starting acceptor in generating the tetramannoside 3, was also readily acquired from 7 by acidic desilylation.

In contrast to the route above, traditional approaches concerning the effective acquisitions of the 6-*O*-naphthylmethylated thiomannoside 13 and the 2-benzoate 14 do not appear to be straightforward. The complexity arises from the desire to carry out the fully regioselective installations of the vital ether groups. Apparently, the regioselective one-pot protection strategy that we introduced[39,40] and further expanded to other sugars[41–44] could simplify such preparations. Our recent work on the stereoselective dioxolane-type benzylidene formation on thiomannosides[43] should provide a convenient gateway to the 2,6-diol 11, a potential common intermediate towards compounds 13 and 14. It was envisioned that, with benzyl groups permanently protecting O3 and O4, the

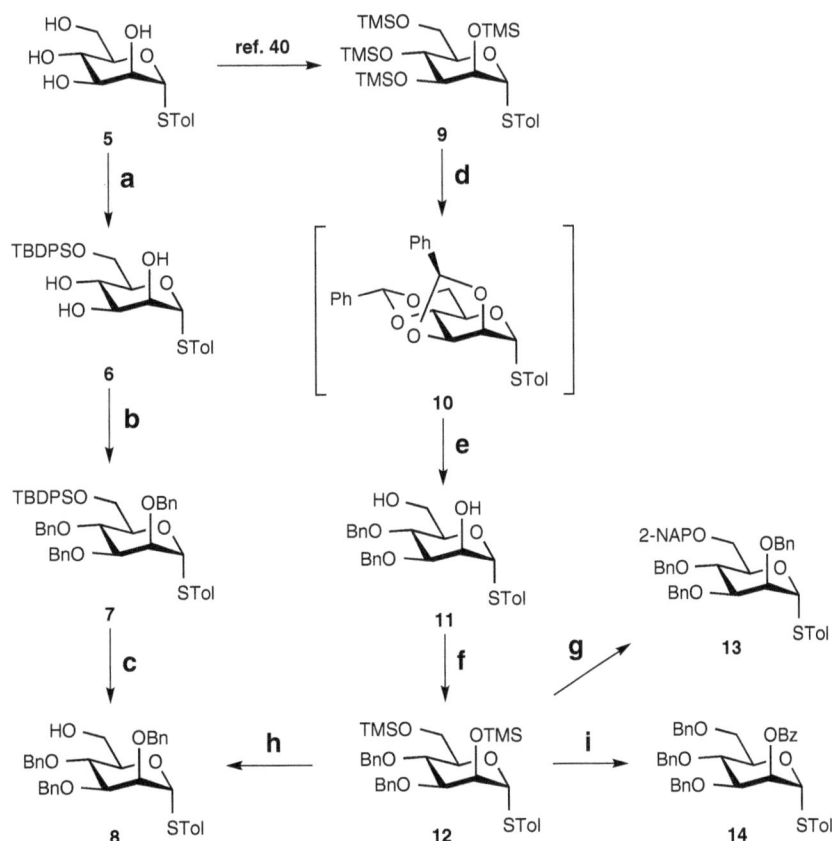

**Figure 2 | Preparations of the mannosyl-building blocks.** Reagents and conditions: (**a**) *tert*-butyldiphenylchlorosilane, Et₃N, DMAP, 93%; (**b**) NaH, BnBr, DMF, 94%; (**c**) PTSA, MeOH, CH₂Cl₂, 92%; (**d**) benzaldehyde (2.1 equivalents), TMSOTf, MeCN, 0 °C, 30 min; (**e**) BH₃·THF, Cu(OTf)₂, CH₂Cl₂, 87% (one pot from **9**); (**f**) trimethylchlorosilane, Et₃N, quantitative; (**g**) 2-naphthaldehyde, Et₃SiH, TMSOTf, CH₂Cl₂, − 78 to − 40 °C, 2 h, then, NaH, BnBr, DMF, 81% (one pot); (**h**) 2-naphthaldehyde, Et₃SiH, TMSOTf, CH₂Cl₂, − 78 to − 40 °C, 2 h, then, NaH, BnBr, DMF, then, DDQ, H₂O, 73% (one pot); (**i**) benzaldehyde, Et₃SiH, TMSOTf, − 78 °C, 1.5 h, then, BF₃·Et₂O, MeCN, − 78 to − 20 °C, 30 min, then, Bz₂O, Et₃N, 93% (one pot). Bz, benzoyl; Bz₂O, benzoic anhydride; Cu(OTf)₂, copper(II) trifluoromethanesulfonate; DDQ, 2,3-dichloro-5,6-dicyano-1,4-benzoquinone; DMAP, 4-(*N,N*-dimethylamino)pyridine; DMF, *N,N*-dimethylformamide; Ph, phenyl; PTSA, *p*-toluenesulfonic acid; TBDPS, *tert*-butyldiphenylsilyl; THF, tetrahydrofuran; TMS, trimethylsilyl; TMSOTf, trimethylsilyl trifluoromethanesulfonate.

primary 6-hydroxyl could be easily differentiated from the secondary and axial 2-hydroxyl. We were also keen to check whether stereoselective dibenzylidenation and simultaneous regioselective ring opening could be achieved in one pot.

Starting from the tetrakis-trimethylsilyl ether 9 acquired in one step from 5 (ref. 43), treatment with 2.1 equivalents of benzaldehyde along with catalytic trimethylsilyl trifluoromethanesulfonate (TMSOTf) in acetonitrile at 0 °C exclusively delivered the exo-product 10 as evidenced by NMR spectroscopy and X-ray crystallography (Supplementary Fig. 1, Supplementary Data 1). This fully stereoselective transformation is beneficial because unlike the regioselectivity in the 4,6-O-benzylidene ring opening, which is determined by the choice of reducing agent[45], the opening of the dioxolane-type 2,3-O-benzylidine moiety is guided by the orientation of the phenyl group (that is, exo-isomers generally open at the axial position). Delightfully, the subsequent exposure of 10 to BH₃·tetrahydrofuran in the same vessel provided the diol 11 in a two-stage one-pot yield of 87%. Anticipating a smooth regioselective reductive etherification at O6 (ref. 46), we subjected the diol 11 to trimethylsilylation. Consequently, treatment of the so-formed 12 with 2-naphthaldehyde, triethylsilane and TMSOTf in subzero temperature supplied the intermediate that was benzylated in one pot using the typical etherification method to afford compound 13. An X-ray single crystal analysis fully supported the desired structure (Supplementary Fig. 1,

Supplementary Data 2). Continuing further, addition of 2,3-dichloro-5,6-dicyano-1,4-benzoquinone to the in situ-generated 13 successfully delivered the same acceptor 8, thus providing an alternative pathway for its acquisition. Similarly, the corresponding reductive 6-O-benzylation of 12 was carried out, followed by desilylation at O2 with tetrabutylammonium fluoride (TBAF) and basic benzoylation in the same flask. Unfortunately, the yield for compound 14 was less than satisfactory even when benzoic acid and TBAF were used together[41] to perform the desilylation, probably due to the interference of TBAF to the benzoylation stage. Desilylation with BF₃·Et₂O apparently solved this issue and furnished 14 in an excellent 93% yield from 12.

**Synthesis of the pseudotrisaccharide 2.** For the desymmetrization of myo-inositol, we evaluated the coupling of the mannosyl donor 7 with the meso diol 16 (Fig. 3), which can be easily prepared in one step[30] from the commercially available Kishi's triol. The asymmetric nature of the mannosyl donor itself should provide certain preferences between the axial hydroxyls of 16 as we have demonstrated previously[30] but without the TBDPS group on the sugar. Because of the wider opening available for a nucleophilic attack on the half-chair mannosyl oxocarbenium ion intermediate by O6 as compared with O4 (Supplementary Fig. 2), it is expected that the required 6-O-mannosylation would be more favoured. Our attempts at coupling of 7 and 16 using

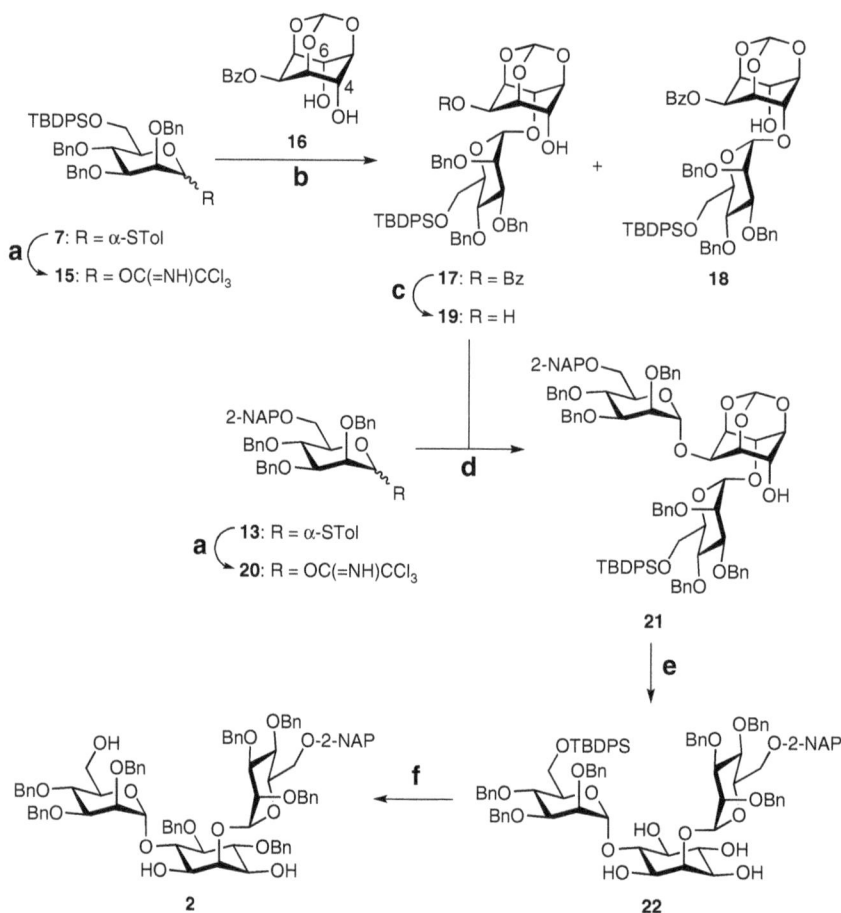

**Figure 3 | Preparation of the pseudotrisaccharide 2.** Reagents and conditions: (**a**) (1) NBS, acetone, H₂O; (2) K₂CO₃, CCl₃CN, **15**: 94% (two steps), **20**: 90% (two steps); (**b**) silver trifluoromethanesulfonate, 1,4-dioxane, CH₂Cl₂, **17**: 68%, **18**: 20%; (**c**) NaOMe, MeOH, CH₂Cl₂, quantitative; (**d**) BF₃·Et₂O, CH₂Cl₂, − 60 to − 20 °C, 72%; (**e**) (1) PTSA, MeOH, CH₂Cl₂, 84%; (2) tert-butyldiphenylchlorosilane, Et₃N, DMAP, 82%; (**f**) (1) trimethylchlorosilane, Et₃N, quantitative; (2) benzaldehyde (3 equivalent), Et₃SiH, TMSOTf, CH₂Cl₂, − 40 °C, then, tetrabutylammonium fluoride, 72%. NBS, N-bromosuccinimide.

N-iodosuccinimide and TMSOTf in $CH_2Cl_2$ and 1,4-dioxane to improve the solubility of diol **16**, unfortunately, led only to donor hydrolysis and full recovery of the acceptor. We suspected that the axial hydroxyls are too unreactive for mannosyl thioglycoside to foster productive couplings. With strong activators avoided to maintain the acid-sensitive orthoformate group, focus was shifted to the imidate versions of the donor. After some optimization (see the Supplementary Table 1), silver trifluoromethanesulfonate promoted the glycosylation step at room temperature, supplying the desired pseudodisaccharide **17** at 68% yield, along with its regioisomer **18** (20%). Here and in the succeeding glycosylations, we verified the α-orientations of the mannosidic bonds through the coupling constants of the anomeric carbons and protons (~170 Hz, see Supplementary Methods)[47,48]. Distinguishing the structures of **17** and **18** with confidence is not possible with NMR analysis alone. We, therefore, resorted to exchange the primary TBDPS with benzyl group (Supplementary Fig. 3) and compare agreement with the NMR spectra from previously published data[30].

For the preparation of the key intermediate **21**, the pseudodisaccharide **17** was subjected to Zemplén deacylation to generate the diol **19**. Regioselective mannosylation at the equatorial hydroxy group should be more likely because of steric reasons. A thorough evaluation of the subsequent coupling also made us consider the application of the imidate **20** over the thioglycoside **13**. Under $BF_3 \cdot Et_2O$ promotion, compound **21** was, therefore, acquired in 72% yield with complete regioselectivity and stereoselectivity. The orthoformate group was cleaved using p-toluenesulfonic acid, which also removed the TBDPS group. The tetraol **22** was obtained after re-installation of the silyl group. With **22** in hand, regioselective benzylation at O4 and O5 of the inositol unit is the next challenge. Williamson condition and acidic benzylation using benzyl imidate are not sufficiently selective, whereas the reductive benzylation of the trimethylsilylated substrate showed greater promise. True enough, excellent selectivity was achieved by using 3 equivalents of benzaldehyde, furnishing, after further full desilylation with TBAF, the desired compound **2** in 72% yield from **22**.

It should be stated that other less successful means in acquiring the pseudotrisaccharide backbone have been studied. Our effort at condensation of the imidate donor **20** with the diol **16** led to mixtures of inseparable regioisomers and stereoisomers, a demonstration of the known potential of the bulky 6-O-TBDPS group at enhancing α-selectivity[49]. A participating moiety at

O2 of the mannosyl donor was ruled out to avoid complications that may be encountered in later reactions. Sequential dimannosylation of Kishi's triol also seemed feasible under our synthetic design, with glycosylation at the more reactive O2 using donor **20** followed by asymmetric 6-O-mannosylation with donor **15**. Unfortunately, poor yields for both couplings were observed. Another procedure we have tried included the 2-O-mannosylation of **19** with a donor already carrying the fatty acyl functionality at the primary position. While the glycosylation step worked as intended, the acyl moiety was also removed along with the TBDPS group on acid treatment intended to cleave the orthoformate function.

**Synthesis of tuberculostearic acid and the H-phosphonate 4.** Tuberculostearic acid was first isolated from *M. tuberculosis* in 1927 (ref. 50) and several methods for its synthesis have been reported[25,26,51–53]. Nevertheless, an updated, shorter and more effective method for accessing this important fatty acid is still desirable. We decided to acquire the chiral carbon of tuberculostearic acid from the commercially available Roche ester (**23**). Tosylation of **23** to afford compound **24**, followed by reduction with diisobutylaluminium hydride and methylene insertion by Wittig reaction furnished the olefin **25** (ref. 54; Fig. 4). The first long-chain elongation of **25** towards compound **26** was achieved by Grignard reaction under catalytic $Li_2CuCl_2$. Grubbs metathesis of olefin **26** with the olefinic acid **27** provided the E/Z olefin mixture, which was exposed to palladium-catalysed hydrogenation to finally secure tuberculostearic acid (**28**). Accomplished in just six steps, this acquisition is the shortest synthetic preparation reported, thus far, for this compound.

Elaborations of the commercially available 3-O-benzyl-sn-glycerol were performed next. Under dicyclohexylcarbodiimide and 4-(N,N-dimethylamino)pyridine, the fatty acid **28** was first condensed with the primary hydroxyl followed by stearic acid esterification at the secondary position in good yields. Cleavage of the benzyl group was achieved through hydrogenolysis and the generated alcohol was phosphorylated using $PCl_3$, imidazole and $Et_3N$ to afford the H-phosphonate **2**.

**$Ac_2PIM_6$ assembly and final transformations.** Our planned sugar assembly towards the tetramannoside **3** hinges on the chemoselective activation of a trichloroacetimidate donor in the presence of a thioglycoside acceptor[55] (Fig. 5). Accordingly,

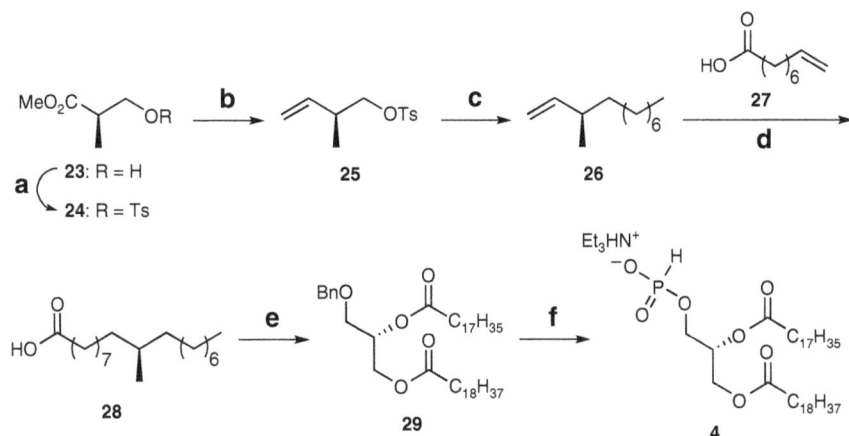

**Figure 4 | Preparation of tuberculostearic acid (28) and the H-phosphonate 4.** Reagents and conditions: (**a**) TsCl, $Et_3N$, DMAP, 94%; (**b**) (1) diisobutylaluminium hydride, $-78\,°C$; (2) $Ph_3P=CH_2$, 72% (two steps); (**c**) $C_7H_{15}MgBr$, $Li_2CuCl_4$, $-78$ to $0\,°C$, 92%; (**d**) (1) Grubbs second-generation catalyst, $CH_2Cl_2$, reflux; (2) $H_2$, Pd/C, 75% (two steps); (**e**) (1) 3-O-benzyl-sn-glycerol, DCC, DMAP, 75%; (2) stearic acid, DCC, DMAP, 88%; (**f**) (1) $H_2$, Pd/C, 92%; (2) $PCl_3$, imidazole, $Et_3N$, $-10\,°C$, 69%. DCC, dicyclohexylcarbodiimide; Ts, Tosyl.

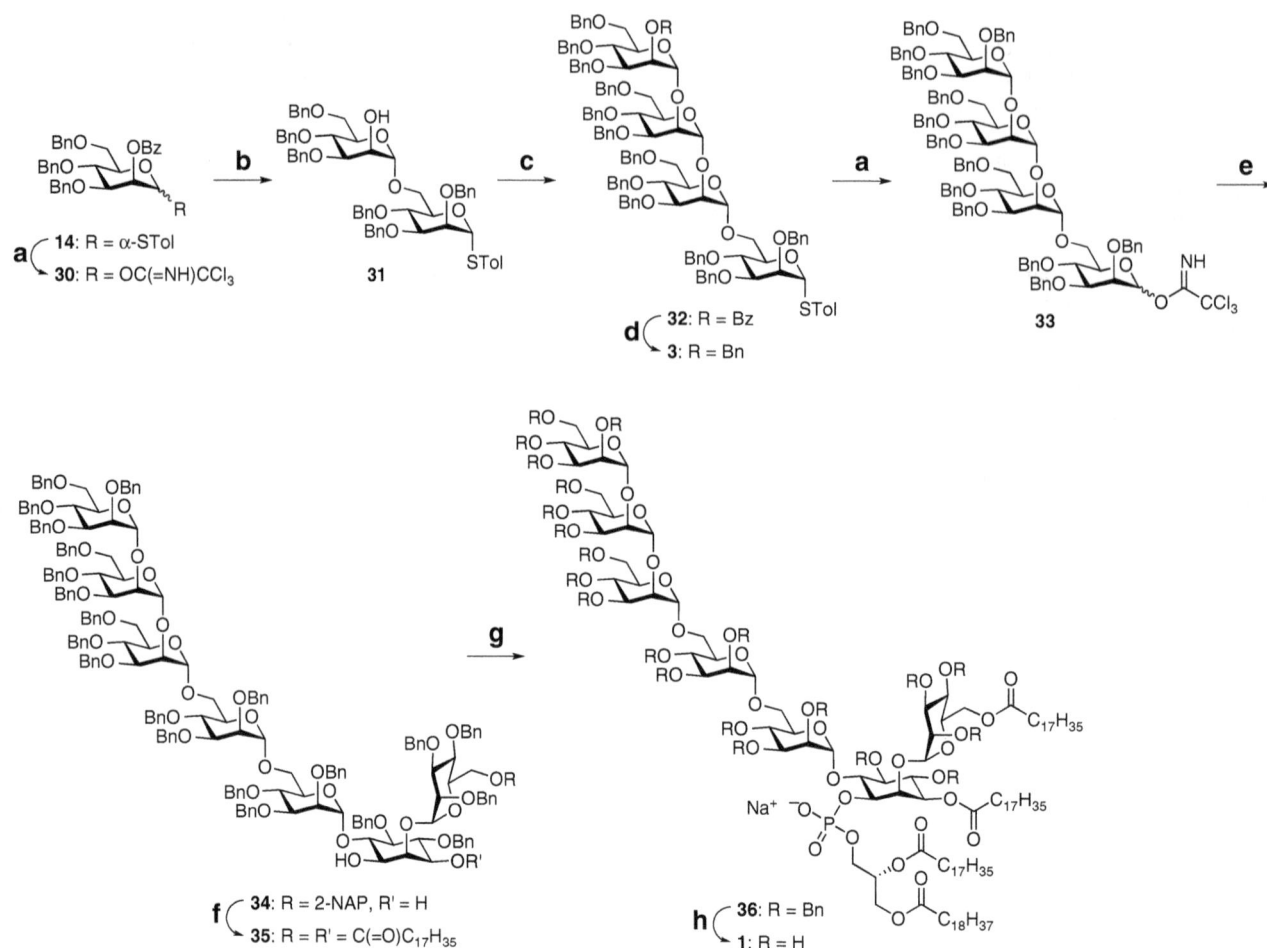

**Figure 5 | Synthesis of compound 1.** Reagents and conditions: (**a**) (1) NBS, acetone, H$_2$O; (2) CCl$_3$CN, 1,8-diazabicyclo[5.4.0]undec-7-ene, **30**: 96% (two steps), **33**: 86% (two steps); (**b**) **8**, TMSOTf, CH$_2$Cl$_2$, $-78\,°$C, then, NaOMe, MeOH, 87% (one pot); (**c**) (1) **30**, TfOH, CH$_2$Cl$_2$, $-60$ to $-40\,°$C, then, NaOMe, MeOH, 70% (one pot); (2) **30**, TfOH, CH$_2$Cl$_2$, $-60$ to $-40\,°$C, 74%; (**d**) BnBr, NaH, 98%; (**e**) **2**, TMSOTf, Et$_2$O, $-40\,°$C, 52% (89% yield based on the recovered **2**); (**f**) (1) DDQ, CH$_2$Cl$_2$, H$_2$O, 71%; (2) stearic acid, DCC, DMAP, 86%; (**g**) (1) **4**, pivaloyl chloride, pyridine; (2) I$_2$, pyridine, H$_2$O; (3) DOWEX 50WX8 Na$^+$ form, 77% from **35**; (**h**) H$_2$, Pd/C, 88%. TfOH, trifluoromethanesulfonic acid.

the elongation unit is formed by converting compound **14** to **30** under the usual procedures. Glycosylation of the thioglycoside **8** with **30** followed by debenzoylation in the same flask supplied the disaccharide acceptor **31** in 87% yield. Two more elongation cycles easily formed the tetramannoside **32**. Knowing that the benzoate group is base-sensitive, the benzoyl-to-benzyl exchange was achieved in one step by NaH and benzyl bromide treatment, smoothly offering the target compound **3** in 98% yield.

With all segment backbones available, we moved forward in putting all these pieces together. The glycosylation of the pseudotrisaccharide acceptor **2** by the thioglycoside **3**, however, produced only a meager 10% yield for compound **34** despite our best efforts, prompting us to use the more reactive imidate counterpart **33** instead (see Supplementary Table 2). With Et$_2$O as solvent and TMSOTf as activator, we eventually obtained the desired **34** in 52% yield (89%, if the recovered acceptor is considered). The 2-NAP ether was then cleaved, which paved the way for the concurrent installation of two stearate esters at the mannosyl and the inositol units, leading to the alcohol **35**. This reaction exhibited no regioselectivity issues, with O1 of the inositol unit spared because it experiences the highest steric hindrance among the three free hydroxyls. The attachment of the *H*-phosphonate **4** and the pseudo-heptasaccharide **35** was carried out by using pivaloyl chloride,

followed by iodine-mediated *in situ* oxidation and cation exchange, delivering the derivative **36**. Global hydrogenolysis of the benzyl ethers provided the Ac$_2$PIM$_6$ construct **1** in 82% yield.

**Evaluation of immunomodulatory activity.** The adjuvant effects of compound **1** were examined through co-administration with ovalbumin (Fig. 6a) or tetanus toxoid (Fig. 6b) antigen in BALB/c mice. PIMs isolated from *M. tuberculosis* strain H37Rv (iPIM$_{1,2}$ and iPIM$_6$) and alum were also investigated in parallel for comparison. It was observed that compound **1** induced an approximately two to fourfold increase in the level of antigen-specific antibodies. The adjuvant activity of **1** is similar to the bacteria-derived PIMs and slightly lower than alum.

Furthermore, we evaluated the cytokine-producing activity of compound **1** as well as iPIM$_{1,2}$ and iPIM$_6$ (Fig. 6c,d). The level of interleukin-4 and interferon-$\gamma$ was not detectable in mouse sera at 1 h after injection of Ac$_2$PIM **1** and the bacteria-derived PIMs. At 18 h after injection, the cytokine levels increased. Lipid and glycolipid molecules derived from *M. tuberculosis* are presented to T cells by CD1 antigen-presenting molecules, specifically CD1d[14,56]. Compared with the well-known CD1d-targeting $\alpha$-galactosylceramide, which can activate the invariant natural killer T cells and induce high levels of interleukin-4 and

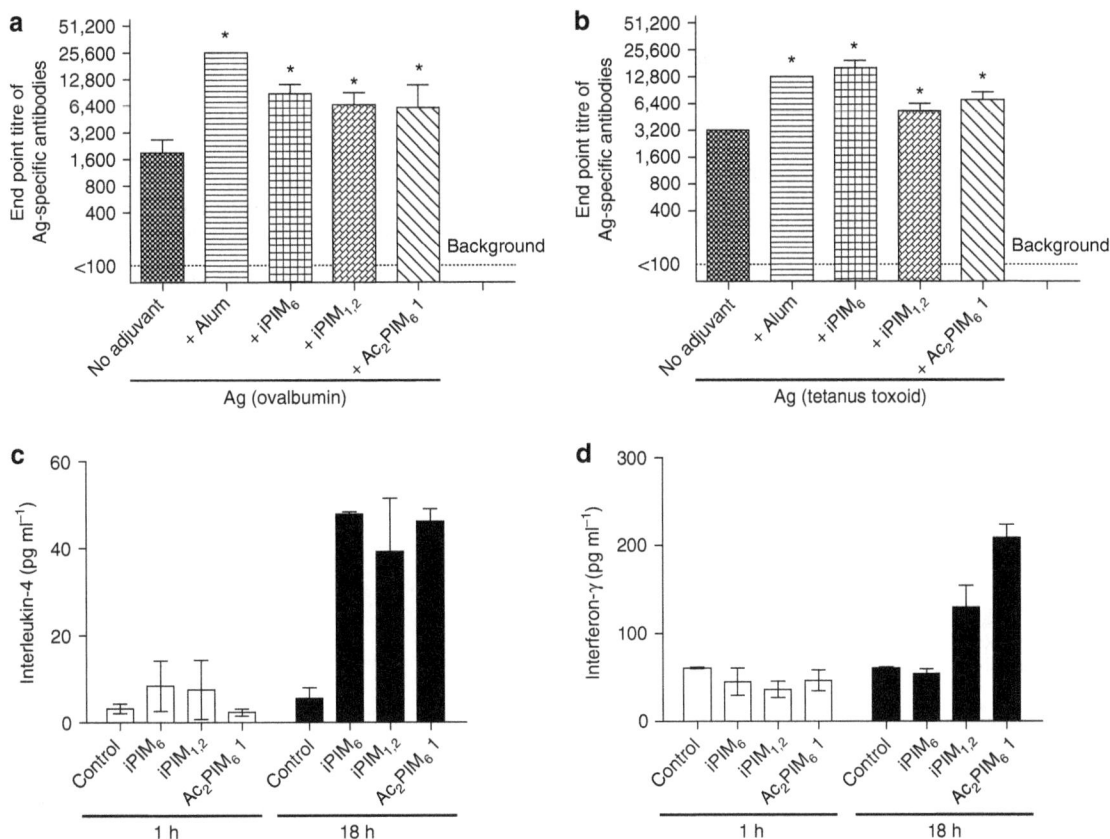

**Figure 6 | Immunological evaluation in BALB/c mice. (a,b)** Induction of antigen (Ag)-specific antibodies in mice immunized with ovalbumin or tetanus toxoid adjuvanted with alum or various PIMs; **(c,d)** secreted cytokines (interleukin-4 and interferon-γ) in mice 1 and 18 h after injection with various PIMs (control represents injection only with PBS). Both end point antibody titres and the cytokine levels were measured by using enzyme-linked immunosorbent assay. The results displayed represent the mean + s.d.'s ($n = 5$). Data were analysed by using one-way analysis of variance, and differences were considered significant at $*P < 0.05$. iPIM$_6$ and iPIM$_{1,2}$, isolated PIM$_6$ and mixture of PIM$_1$ and PIM$_2$, respectively, from *M. tuberculosis* strain H37Rv.

interferon-γ within 24 h (ref. 57), Ac$_2$PIM$_6$ **1** appeared to have moderate effects.

## Discussion

We have successfully developed a convenient route to synthesize an Ac$_2$PIM$_6$ construct in the form of compound **1** containing tuberculostearic acid and stearic acid as the fatty acid components. This is the first time that an Ac$_2$PIM$_6$ molecule was synthesized. Further, a novel and short synthetic route towards tuberculostearic acid was developed, with only six synthetic steps from the commercially available Roche ester and four purification stages. Our synthetic approach benefitted from the use of shared mannoside-building blocks, the carefully chosen orthogonal protecting groups and the features of the regioselective one-pot transformations from trimethylsilylated starting materials previously developed by us. The trichloroacetimidate donor types[58] are vital factors in achieving the successful assembly processes. Regioselectivity and stereoselectivity were achieved through the aid of steric and stereoelectronic effects. Steric effects were also exploited in the direct desymmetrization of *myo*-inositol by mannosyl donors and in minimizing the number of protecting groups used in the synthesis. With practical access and functional group flexibility, the key intermediates such as the pseudotrisaccharide **2** possess good potential in supplying PIMs of different mannosylation and lipidation patterns as well. Our synthesized Ac$_2$PIM$_6$ has comparable adjuvant activity with the natural PIMs against ovalbumin and tetanus toxoid antigens and induced the production of interleukin-4 and interferon-γ, thus,

validating the immunological qualities of PIM molecules and its value in vaccine research.

## Methods

**Chemical synthesis.** The complete experimental details and compound characterization data can be found in the Supplementary Methods. For the NMR spectra of the compounds in this article, see Supplementary Figs 4–120. The mass spectrum of the synthesized Ac$_2$PIM$_6$ **1** is shown in Supplementary Fig. 121.

**Materials for immunological evaluation.** All BALB/c mice were housed at the animal facility in the Institute of Cell Biology, Academia Sinica, Taiwan in accordance with the Institutional Animal Care Committee guidelines. Purified iPIM$_{1,2}$ (NR-14846) and iPIM$_6$ (NR-14847) were obtained through BEI Resources, National Institute of Allergy and Infectious Diseases, National Institutes of Health (USA). Ovalbumin and tetanus toxoid were purchased from InvivoGen (San Diego, CA, USA) and Adimmune Inc. (Taichung, Taiwan), respectively.

**Evaluation of adjuvant activity.** Five- to six-week-old female BALB/c mice were immunized with ovalbumin (100 μg) or tetanus toxoid (2 μg) adjuvanted with 10 μg of PIM compounds (Ac$_2$PIM$_6$ **1**, iPIM$_{1,2}$ or iPIM$_6$) or alum in PBS for three times at 2-week intervals by intramuscular injection. Two weeks after the third immunization, the immunized mice were bled for antigen-specific antibody analysis.

Ovalbumin- and tetanus toxoid-specific antibodies in heat-inactivated serum were monitored with direct enzyme-linked immunosorbent assay (ELISA). The ovalbumin- or tetanus toxoid-coated plates were incubated with mouse serum in twofold serial dilutions for 1 h. Antigen-specific IgG was monitored by using horseradish peroxidase-conjugated anti-mouse antibodies and 3,3',5,5'-tetramethylbenzidine substrate (Thermo Scientific Inc). After colour development, absorbance at 450 nm was recorded by using a plate reader (SpectraMax M5, Molecular Device). The end point antibody titre was defined as the highest dilution of serum to produce an absorbance 2.5 times higher than the optical absorbance produced by the pre-immune serum. The background end point antibody titre was assigned as < 1:100.

**Evaluation of cytokine-producing activity.** Five- to six-week-old female BALB/c mice were intramuscularly injected with $10\,\mu g$ of the PIM compounds ($Ac_2PIM_6$ **1**, $iPIM_{1,2}$ or $iPIM_6$) in PBS and bled at 1 or 18 h after injection (five mice per group). The cytokines in the sera were measured with sandwich ELISA using paired anti-interleukin-4 and anti-interferon-$\gamma$ monoclonal antibodies (R&D Systems).

**Statistical analysis.** The response of each mouse was counted as an individual data point for statistical analysis. Data obtained from animal studies were analysed by using one-way analysis of variance from Graphpad and differences were considered significant at $P < 0.05$.

# References

1. WHO. Tuberculosis factsheet N°104 http://www.who.int/mediacentre/factsheets/fs104/en/ (2014).
2. Orme, I. M. The search for new vaccines against tuberculosis. *J. Leukoc. Biol.* **70**, 1–10 (2001).
3. Zumla, A. I. *et al.* New antituberculosis drugs, regimens, and adjunct therapies: needs, advances, and future prospects. *Lancet Infect. Dis.* **14**, 327–340 (2014).
4. Dye, C. & Williams, B. G. The population dynamics and control of tuberculosis. *Science* **328**, 856–861 (2010).
5. Karakousis, P. C., Bishai, W. R. & Dorman, S. E. *Mycobacterium tuberculosis* cell envelope lipids and the host immune response. *Cell Microbiol.* **6**, 105–116 (2004).
6. Flynn, J. L. & Chan, J. What's good for the host is good for the bug. *Trends Microbiol.* **13**, 98–102 (2005).
7. Russell, D. G. Who puts the tubercle in tuberculosis? *Nat. Rev. Microbiol.* **5**, 39–47 (2007).
8. Hestvik, A. L. K., Hmama, Z. & Av-Gay, Y. Mycobacterial manipulation of the host cell. *FEMS Microbiol. Rev.* **29**, 1041–1050 (2005).
9. Bhowruth, V., Alderwick, L. J., Brown, A. K., Bhatt, A. & Besra, G. S. Tuberculosis: a balanced diet of lipids and carbohydrates. *Biochem. Soc. Trans.* **36**, 555–565 (2008).
10. Józefowski, S., Sobota, A. & Kwiatkowska, K. How *Mycobacterium tuberculosis* subverts host immune responses. *BioEssays* **30**, 943–954 (2008).
11. Cao, B. & Williams, S. J. Chemical approaches for the study of the mycobacterial glycolipids phosphatidylinositol mannosides, lipomannan and lipoarabinomannan. *Nat. Prod. Rep.* **27**, 919–947 (2010).
12. Gilleron, M. *et al.* Acylation state of the phosphatidylinositol mannosides from *Mycobacterium bovis* Bacillus Calmette Guérin and ability to induce granuloma and recruit natural killer T cells. *J. Biol. Chem.* **276**, 34896–34904 (2001).
13. Torrelles, J. B., Azad, A. K. & Schlesinger, L. S. Fine discrimination in the recognition of individual species of phosphatidyl-*myo*-inositol mannosides from *Mycobacterium tuberculosis* by C-type lectin pattern recognition receptors. *J. Immunol.* **177**, 1805–1816 (2006).
14. Fischer, K. *et al.* Mycobacterial phosphatidylinositol mannoside is a natural antigen for CD1d-restricted T cells. *Proc. Natl Acad. Sci. USA* **101**, 10685–10690 (2004).
15. Rojas, R. E. *et al.* Phosphatidylinositol mannoside from *Mycobacterium tuberculosis* binds $\alpha_5\beta_1$ integrin (VLA-5) on CD4$^+$ T cells and induces adhesion to fibronectin. *J. Immunol.* **177**, 2959–2968 (2006).
16. Guerin, M. E., Korduláková, J., Alzari, P. M., Brennan, P. J. & Jackson, M. Molecular basis of phosphatidyl-*myo*-inositol mannoside biosynthesis and regulation in mycobacteria. *J. Biol. Chem.* **285**, 33577–33583 (2010).
17. Elie, C. J. J., Verduyn, R., Dreef, C. E., van der Marel, G. A. & van Boom, J. H. Iodinium ion-mediated mannosylations of *myo*-inositol: synthesis of a mycobacteria phospholipid fragment. *J. Carbohydr. Chem.* **11**, 715–739 (1992).
18. Watanabe, Y., Yamamoto, T. & Okazaki, T. Synthesis of 2,6-di-O-α-D-mannopyranosylphosphatidyl-D-*myo*-inositol. Utilization of glycosylation and phosphorylation based on phosphite chemistry. *Tetrahedron* **53**, 903–918 (1997).
19. Stadelmaier, A. & Schmidt, R. R. Synthesis of phosphatidylinositol mannosides (PIMs). *Carbohydr. Res.* **338**, 2557–2569 (2003).
20. Stadelmaier, A., Biskup, M. B. & Schmidt, R. R. Synthesis of serine-linked phosphatidylinositol mannosides (PIMs). *Eur. J. Org. Chem.* **2004**, 3292–3303 (2004).
21. Jayaprakash, K. N., Lu, J. & Fraser-Reid, B. Synthesis of a key *Mycobacterium tuberculosis* biosynthetic phosphoinositide intermediate. *Bioorg. Med. Chem. Lett.* **14**, 3815–3819 (2004).
22. Jayaprakash, K. N., Lu, J. & Fraser-Reid, B. Synthesis of a lipomannan component of the cell-wall complex of *Mycobacterium tuberculosis* is based on Paulsen's concept of donor/acceptor "match". *Angew. Chem. Int. Ed.* **44**, 5894–5898 (2005).
23. Ainge, G. D. *et al.* Phosphatidylinositol mannosides: synthesis and suppression of allergic airway disease. *Bioorg. Med. Chem.* **14**, 5632–5642 (2006).
24. Ainge, G. D. *et al.* Phosphatidylinositol mannosides: synthesis and adjuvant properties of phosphatidylinositol di- and tetramannosides. *Bioorg. Med. Chem.* **14**, 7615–7624 (2006).
25. Liu, X., Stocker, B. L. & Seeberger, P. H. Total synthesis of phosphatidylinositol mannosides of *Mycobacterium tuberculosis*. *J. Am. Chem. Soc.* **128**, 3638–3648 (2006).
26. Dyer, B. S. *et al.* Synthesis and structure of phosphatidylinositol dimannoside. *J. Org. Chem.* **72**, 3282–3288 (2007).
27. Ainge, G. D. *et al.* Phosphatidylinositol mannoside ether analogues: syntheses and interleukin-12-inducing properties. *J. Org. Chem.* **72**, 5291–5296 (2007).
28. Boonyarattanakalin, S., Liu, X., Michieletti, M., Lepenies, B. & Seeberger, P. H. Chemical synthesis of all phosphatidylinositol mannoside (PIM) glycans from *Mycobacterium tuberculosis*. *J. Am. Chem. Soc.* **130**, 16791–16799 (2008).
29. Ali, A., Wenk, M. R. & Lear, M. J. Total synthesis of a fully lipidated form of phosphatidyl-*myo*-inositol dimannoside (PIM-2) of *Mycobacterium tuberculosis*. *Tetrahedron Lett.* **50**, 5664–5666 (2009).
30. Patil, P. S. & Hung, S.-C. Total synthesis of phosphatidylinositol dimannoside: a cell-envelope component of *Mycobacterium tuberculosis*. *Chem. Eur. J.* **15**, 1091–1094 (2009).
31. Patil, P. S. & Hung, S.-C. Synthesis of mycobacterial triacylated phosphatidylinositol dimannoside containing an acyl lipid chain at 3-O of inositol. *Org. Lett.* **12**, 2618–2621 (2010).
32. Ainge, G. D. *et al.* Chemical synthesis and immunosuppressive activity of dipalmitoyl phosphatidylinositol hexamannoside. *J. Org. Chem.* **76**, 4941–4951 (2011).
33. Ainge, G. D. *et al.* Synthesis and Toll-like receptor 4 (TLR4) activity of phosphatidylinositol dimannoside analogues. *J. Med. Chem.* **54**, 7268–7279 (2011).
34. Rankin, G. M. *et al.* Synthesis and mass spectral characterization of mycobacterial phosphatidylinositol and its dimannosides. *J. Org. Chem.* **77**, 6743–6759 (2012).
35. Sureshan, K. M., Shashidhar, M. S., Praveen, T. & Das, T. Regioselective protection and deprotection of inositol hydroxyl groups. *Chem. Rev.* **103**, 4477–4504 (2003).
36. Patil, P. S., Zulueta, M. M. L. & Hung, S.-C. Synthesis of phosphatidylinositol mannosides. *J. Chin. Chem. Soc.* **61**, 151–162 (2014).
37. Takahashi, H., Kittaka, H. & Ikegami, S. Novel synthesis of enantiomerically pure natural inositols and their diastereoisomers. *J. Org. Chem.* **66**, 2705–2716 (2001).
38. Watt, J. A. & Williams, S. J. Rapid, iterative assembly of octyl α-1,6-oligomannosides and their 6-deoxy equivalents. *Org. Biomol. Chem.* **3**, 1982–1992 (2005).
39. Wang, C.-C. *et al.* Regioselective one-pot protection of carbohydrates. *Nature* **446**, 896–899 (2007).
40. Wang, C.-C., Kulkarni, S. S., Lee, J.-C., Luo, S.-Y. & Hung, S.-C. Regioselective one-pot protection of glucose. *Nat. Protoc.* **3**, 97–113 (2008).
41. Chang, K.-L., Zulueta, M. M. L., Lu, X.-A., Zhong, Y.-Q. & Hung, S.-C. Regioselective one-pot protection of D-glucosamine. *J. Org. Chem.* **75**, 7424–7427 (2010).
42. Huang, T.-Y., Zulueta, M. M. L. & Hung, S.-C. One-pot strategies for the synthesis of the tetrasaccharide linkage region of proteoglycans. *Org. Lett.* **13**, 1506–1509 (2011).
43. Patil, P. S., Lee, C.-C., Huang, Y.-W., Zulueta, M. M. L. & Hung, S.-C. Regioselective and stereoselective benzylidene installation and one-pot protection of D-mannose. *Org. Biomol. Chem.* **11**, 2605–2612 (2013).
44. Huang, T.-Y., Zulueta, M. M. L. & Hung, S.-C. Regioselective one-pot protection, protection-glycosylation and protection-glycosylation-glycosylation of carbohydrates: a case study with D-glucose. *Org. Biomol. Chem.* **12**, 376–382 (2014).
45. Shie, C.-R. *et al.* Cu(OTf)₂ as an efficient and dual-purpose catalyst in the regioselective reductive ring opening of benzylidene acetals. *Angew. Chem. Int. Ed.* **44**, 1665–1668 (2005).
46. Wang, C.-C. *et al.* Synthesis of biologically potent α1 → 2-linked disaccharide derivatives via regioselective one-pot protection-glycosylation. *Angew. Chem. Int. Ed.* **41**, 2360–2362 (2002).
47. Bock, K. & Pedersen, C. A study of $^{13}$CH coupling constants in hexopyranoses. *J. Chem. Soc. Perkin. Trans.* **2**, 293–297 (1974).
48. Duus, J. Ø., Gotfredsen, C. H. & Bock, K. Carbohydrate structural determination by NMR spectroscopy: modern methods and limitations. *Chem. Rev.* **100**, 4589–4614 (2000).
49. Zulueta, M. M. L. *et al.* α-Glycosylation by D-glucosamine-derived donors: synthesis of heparosan and heparin analogues that interact with mycobacterial heparin-binding hemagglutinin. *J. Am. Chem. Soc.* **134**, 8988–8995 (2012).
50. French, G. L. *et al.* Diagnosis of tuberculous meningitis by detection of tuberculostearic acid in cerebrospinal fluid. *Lancet* **330**, 117–119 (1987).
51. Prout, F. S., Cason, J. & Ingersoll, A. W. The synthesis of tuberculostearic acid. *J. Am. Chem. Soc.* **69**, 1233–1233 (1947).

52. Schmidt, G. A. & Shirley, D. A. A new synthesis of tuberculostearic acid. *J. Am. Chem. Soc.* **71**, 3804–3806 (1949).

53. Roberts, I. O. & Baird, M. S. A new short synthesis of 10-(*R*)-tuberculostearic acid and its enantiomer. *Chem. Phys. Lipids* **142**, 111–117 (2006).

54. Li, H., Wu, J., Luo, J. & Dai, W.-M. A concise total synthesis of amphidinolide T2. *Chem. Eur. J.* **16**, 11530–11534 (2010).

55. Kaeothip, S. & Demchenko, A. V. Expeditious oligosaccharide synthesis via selective, semi-orthogonal, and orthogonal activation. *Carbohydr. Res.* **346**, 1371–1388 (2011).

56. De Libero, G. & Mori, L. The T-Cell response to lipid antigens of *Mycobacterium tuberculosis*. *Front. Immunol.* **5**, 219 (2014).

57. Shiozaki, M. *et al.* Synthesis of RCAI-172 (C6 epimer of RCAI-147) and its biological activity. *Bioorg. Med. Chem.* **22**, 827–833 (2014).

58. Schmidt, R. R. & Kinzy, W. Anomeric-oxygen activation for glycoside synthesis: the trichloroacetimidate method. *Adv. Carbohydr. Chem. Biochem.* **50**, 21–123 (1994).

## Acknowledgements

This work was supported by the Ministry of Science and Technology (MOST 100-2113-M-001-019-MY3 and MOST 102-2628-M-001-001), National Health Research Institute (NHRI-EX103-10146NI) and Academia Sinica.

## Author contributions

S.-C.H. designed the study, supervised students and staffs, and finalized the manuscript preparation. P.S.P. performed the synthesis of Ac$_2$PIM$_6$ **1**. T.-J.R.C. and S.-T.Y. carried out the evaluation of immunological activity. M.M.L.Z. prepared the figures and wrote the manuscript. L.S.L. is involved in the early stages of manuscript preparation and assisted in compiling the Supplementary Information. All authors discussed the results and commented on the manuscript.

## Additional information

**Competing financial interests:** The authors declare no competing financial interests.

# 11

# Supramolecular block copolymers by kinetically controlled co-self-assembly of planar and core-twisted perylene bisimides

Daniel Görl[1], Xin Zhang[1], Vladimir Stepanenko[1] & Frank Würthner[1]

New synthetic methodologies for the formation of block copolymers have revolutionized polymer science within the last two decades. However, the formation of supramolecular block copolymers composed of alternating sequences of larger block segments has not been realized yet. Here we show by transmission electron microscopy (TEM), 2D NMR and optical spectroscopy that two different perylene bisimide dyes bearing either a flat (A) or a twisted (B) core self-assemble in water into supramolecular block copolymers with an alternating sequence of $(A_mBB)_n$. The highly defined ultralong nanowire structure of these supramolecular copolymers is entirely different from those formed upon self-assembly of the individual counterparts, that is, stiff nanorods (A) and irregular nanoworms (B), respectively. Our studies further reveal that the as-formed supramolecular block copolymer constitutes a kinetic self-assembly product that transforms into thermodynamically more stable self-sorted homopolymers upon heating.

[1] Institut für Organische Chemie and Center for Nanosystems Chemistry, Universität Würzburg, Am Hubland, Würzburg 97074, Germany. Correspondence and requests for materials should be addressed to F.W. (email: wuerthner@chemie.uni-wuerzburg.de).

Self-assembly has become a key technology for the creation of nanostructured materials with desirable properties for application, for example, in organic electronics, molecular sensing and bionanotechnology[1-3]. However, major focus in this field being so far on single-component self-assembly to elucidate structure–property relationships. In contrast, multicomponent assembly is by far less explored, despite its high promises for the construction of novel nanosized composite materials. With his pioneering work on the elucidation of mixtures of established small-sized self-assembling systems, in particular based on hydrogen-bonding-recognition patterns, Isaacs established the research field of self-sorting phenomena[4] that has gained increasingly more importance during the last decade. In this seminal work as well as in most of the subsequent contributions, however, typically only small-sized finite and analytically assignable supramolecular architectures are addressed[5-9].

On the other hand, for large[10,11] and non-finite self-assembled systems[12-14] few studies on multicomponent self-assembly with detailed insights into the self-assembly sequence exist. Remarkably, perfect self-sorting into $A_nB_m$ copolymers has been afforded for unidirectionally growing systems by sequential assembly[15-19], while perfect mixing into supramolecular copolymers with alternating sequences $(AB)_n$ has been realized by well-directed molecular design for mutual interactions of the individual components[20-25]. However, unravelling constitutional aspects of supramolecular polymer architectures formed by bottom-up self-assembly of multicomponent mixtures remains still challenging, particularly in water. One reason for the reluctance of researchers to enter this field is given by the difficulties encountered in the analysis of the large-sized and polydisperse self-assembly products. Another reason might be the additional challenges encountered by pathway complexity that arises when the growth of a larger nano- or even micrometre-sized structure operates either under kinetic or thermodynamic control, depending on the experimental conditions[26-28]. As pointed out by Rybtchinski and by Fernández in two recent reviews, pathways towards kinetic products are in particular to be considered for the self-assembly of larger aromatic π-systems in water because of the hydrophobic effect[29,30]. Our previous achievements[31] in elucidation of the hierarchical growth of a planar perylene bisimide (PBI) amphiphile (PBI 1) in water into nanorods and nanolamellae by transmission electron microscopy (TEM) encouraged us to elucidate the co-assembly of planar PBI 1 with core-twisted and conformationally more flexible ('soft') PBI amphiphile 2.

Here we report our unique findings on the bicomponent system PBI 1/PBI 2 that hydrophobic interactions lead to the formation of kinetically controlled supramolecular copolymers in water as explored by TEM and two-dimensional (2D) nuclear magnetic resonance (NMR) studies as well as optical spectroscopy. These results provide novel insights into the evolution of π-stacked supramolecular structures in aqueous media[32-36].

## Results

**Molecular design.** The molecular structures of amphiphilic PBI 1 (ref. 31) and PBI 2 are shown in Fig. 1a and the synthesis of the new PBI 2, along with its characterization data (Supplementary Fig. 1), is reported in the Supplementary Methods. The perylene core of PBI 1 is almost flat, while that of PBI 2 is enforced to twist by ∼30° due to the presence of four bulky aryloxy substituents in the bay region[37-39]. These bay substituents also strongly influence the optical properties of PBIs (Fig. 1b). Thus, core-twisted PBI 2 molecularly dissolved in tetrahydrofuran (THF) displays a 46-nm red-shifted absorption maximum compared with that of planar PBI 1 and a second absorption band at a shorter wavelength of 444 nm, which can be attributed to a partially symmetry allowed $S_0$–$S_2$ transition with a transition dipole moment perpendicular to the molecular long axis[40]. Consequently, PBI 1 shows a green fluorescence, while PBI 2 a red one.

**Self-assembly of unimolecular PBI systems.** We have first studied the self-assembly behaviour of the new PBI 2 in aqueous medium and compared it with that of the previously investigated PBI 1. For PBI 1, we had observed a very slow (ca. 2 days) dissolution process in water, leading to either one-dimensional (1D) nanorods (at a concentration of ∼0.05 mM) or nanolamellae (at concentrations >0.1 mM)[31]. Because the dissolution process is such slow and the conversion between nanorods and nanolamellae proved to be reversible, it is reasonable to consider that these aggregate structures are thermodynamically equilibrated. This view is further supported by the fact that these structures are formed pathway independently. Thus, the same structures are also generated upon injection of water to molecularly dissolved PBI 1 in THF and subsequent evaporation of THF ensuring a slow gradual increase of water content in the solvent mixture. Different from PBI 1, PBI 2 does not dissolve in pure water under ambient conditions due to its increased hydrophobicity imparted by the aryloxy substituents at bay area. However, a stable suspension of aggregates was obtained by

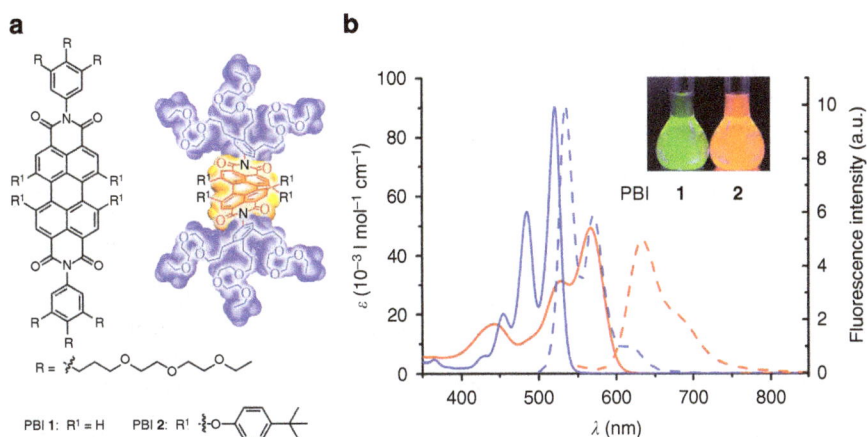

**Figure 1 | Molecular design and optical properties of PBIs 1 and 2.** (**a**) Chemical structures and space-filling (CPK) model of PBIs **1** and **2**. (**b**) Ultraviolet-visible absorption spectra (solid lines) and corresponding fluorescence spectra (dashed lines) of PBI **1** (blue) and PBI **2** (red) in THF; concentration, ∼1 × 10$^{-5}$ M of each sample. Excitation wavelength is 470 nm. Inset, photograph of PBI **1** and PBI **2** solutions in THF under ultraviolet light.

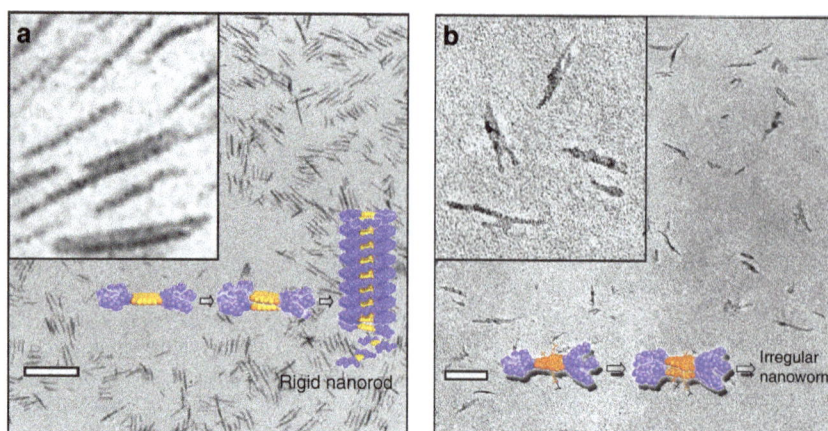

**Figure 2 | Morphology of unimolecular PBI self-assemblies. (a)** TEM image of PBI **1** aggregates in water, [PBI **1**] = 0.05 mM (0.077 mg ml$^{-1}$) and schematic space-filling models illustrating the self-assembly of PBI **1**. Scale bar, 100 nm. Inset, magnified TEM image of the same sample. **(b)** TEM image of PBI **2** aggregates in water, [PBI **2**] = 0.03 mM (0.077 mg ml$^{-1}$) and schematic space-filling models illustrating the self-assembly of PBI **2**. Scale bar, 50 nm. Inset, magnified TEM image of the same sample.

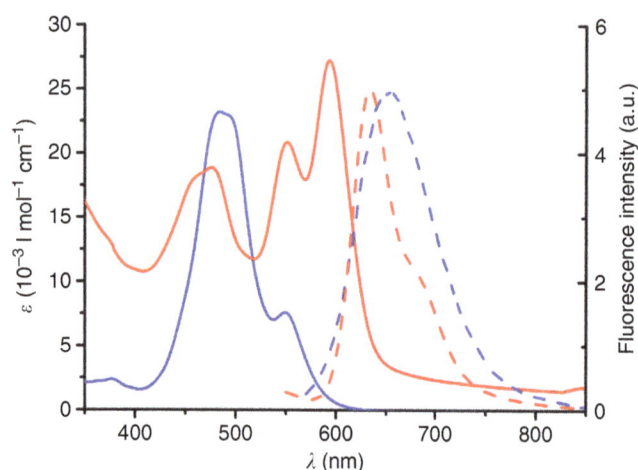

**Figure 3 | Optical properties of unimolecular PBI self-assemblies.** Ultraviolet–visible absorption spectra (solid lines) and corresponding normalized fluorescence spectra (dashed lines) of PBI **1** (blue) and PBI **2** (red) in water; concentration, ~$1 \times 10^{-5}$ M of each sample. Excitation wavelength is 470 nm. Magic-angle setup was applied.

adding water to molecularly dissolved PBI **2** in THF and subsequent slow evaporation of THF from the THF/H$_2$O solution of PBI **2**. Very different from the homogeneous and rigid segmented nanorod structures observed for PBI **1** (Fig. 2a), TEM analysis of the aggregates of PBI **2** revealed more irregular and flexible worm-like aggregates at low concentration of $3 \times 10^{-5}$ M in water (Fig. 2b and Supplementary Fig. 2). The formation of these nanoworm structures is apparently evoked by the twisting of the perylene core of PBI **2**, leading to a different and less regular packing arrangement compared with PBI **1** aggregates.

To understand the formation of these nanoworms upon removal of THF from THF/H$_2$O mixtures, we first performed concentration-dependent ultraviolet–visible studies of PBI **2** in various THF/H$_2$O mixtures. These studies reveal a transition from monomeric to aggregated species upon increasing concentration, which is accompanied by a decrease and a small bathochromic shift (<5 nm) of the absorption maximum (Supplementary Fig. 3). The corresponding plot of extinction versus concentration of PBI **2** could be fitted best to the dimer model[41] (see Supplementary Fig. 3, insets). This is reasonable because the bulky bay substituents prevent the π-surfaces in PBI **2**

dimer from further stacking interactions as already described in our earlier work on the aggregation of related core-distorted PBIs in organic solvents[37]. Upon increasing the water content in THF/H$_2$O mixtures from 7:3 to 5:5 (v/v) the binding constant $K$ increases only slightly from $K$(PBI **2**) = $5 \times 10^1$ M$^{-1}$ (7:3) to $K$(PBI **2**) = $4 \times 10^3$ M$^{-1}$ (5:5), but then distinctly to $K$(PBI **2**) = $7 \times 10^4$ M$^{-1}$ when reaching a specific solvent composition of THF/H$_2$O = 4:6. A similar transition from monomers to dimers is also observed upon increasing the water content and concomitantly the hydrophobic effect in dilute solutions of PBI **2** in THF/H$_2$O mixtures (Supplementary Fig. 4). Driven by the hydrophobic effect, these PBI **2** dimers are then further aggregated into structurally less-defined nanoworms in pure water as revealed by our TEM studies. The ultraviolet–visible spectrum of PBI **2** in pure water (Fig. 3) shows a more pronounced bathochromic shift (27 nm) compared with those in the THF/H$_2$O mixtures as well as some scattering effects (baseline does not go to zero), pinpointing the suspension character of the sample.

The binding constants for PBI **1** self-assembly in THF/H$_2$O mixtures were determined by the isodesmic model that assumes equal binding constants for both PBI π-surfaces in accordance with our earlier work[42] and the observation of defined 1D aggregates by TEM (Fig. 2). In contrast to PBI **2**, a pronounced hypsochromic shift of the absorption maximum is observed upon aggregation of PBI **1** (Supplementary Fig. 5). As expected, the binding constant of PBI **1** increases with increasing amount of water from $K$(PBI **1**) = $1 \times 10^3$ M$^{-1}$ in THF/H$_2$O 7:3 to $K$(PBI **1**) = $4 \times 10^4$ M$^{-1}$ in THF/H$_2$O 4:6. Interestingly, for higher THF contents (>50%) PBI **1** exhibits significantly higher binding constants than PBI **2** whilst for H$_2$O-enriched mixtures, the dimerization constant for PBI **2** surpasses the aggregation constant of PBI **1** due to the pronounced hydrophobicity of PBI **2** (Supplementary Fig. 6). Unfortunately, for more water-enriched mixtures (>60% H$_2$O), binding constants for both PBI **1** and PBI **2** could not be determined reliably due to very strong aggregation that prohibits analysis of concentration-dependent ultraviolet–visible experiments.

As shown in Fig. 3, the aggregates of PBI **1** and PBI **2** in water exhibit indeed very different ultraviolet–visible spectra. For PBI **1** the major band is hypsochromically displaced compared with that of the monomers (in THF), while for PBI **2** a bathochromic shift compared with the monomer band is observed. Moreover, owing to pronounced excitonic coupling between the closely stacked dyes[43], spectra of PBI **1** aggregate do not show the vibronic

coupling pattern of the monomeric dyes. However, such vibronic transitions are clearly observed for the aggregates of PBI **2** as expected for less-tightly stacked dyes owing to the twisted core and the bulky phenoxy substituents. Compared with the PBI **1** aggregates, also quite different fluorescence bands are observed for PBI **2** aggregates (Fig. 3 and Supplementary Fig. 7). The former display typical excimer-type broad emission band[44] with a large Stokes shift (emission maximum at 650 nm), whereas the fluorescence spectra of the latter show only a moderate Stokes shift and have similar shapes as the fluorescence bands of the molecularly dissolved dyes. These features suggest that the core-twisted PBI dyes **2** are only weakly coupled in aggregates, which corroborates our earlier conclusions that the main driving force for self-assembly of core-twisted PBI **2** is the hydrophobic effect whilst additional strong π–π interactions are operative in aggregates of planar PBI **1**.

**Co-assembly of PBI 1 and PBI 2.** Since both types of aggregates, namely well-dissolved PBI **1** nanorods and suspension-forming PBI **2** nanoworms, respectively, can be generated by the same preparation method, that is, by injection of water to the respective PBI dye monomer solution in THF and subsequent slow removal of THF under ambient conditions, we were able to address the effect of water on the self-assembly of a mixture of both PBI dyes and to elucidate the structural features of the aggregates formed. Thus, we prepared a solution of a 2:1 mixture of PBI **1** and PBI **2** in THF, injected water to this solution and slowly removed the organic solvent under ambient conditions. Microscopic analyses by TEM (Fig. 4 and Supplementary Fig. 8) and atomic force microscopy (Supplementary Fig. 9) revealed that the assemblies of this mixed system in water are quite different from those of the individual components, as the co-assemblies grow along 1D axis into long-range, structurally ordered nanowires. These nanowires are ultrathin with uniform diameter of 3–4 nm (Fig. 4), and are among the thinnest organic nanowires ever revealed by electron microscopy[45,46]. These 'single-molecule'-thin nanowires are also extraordinarily long up to 5 μm and display high aspect ratio. Segmented structures within the nanowires could be resolved by TEM (Fig. 4, inset). Cryo-TEM images (Supplementary Fig. 10) further corroborate their 1D constitution, which significantly differs from previously reported PBI **1** nanorod and nanolamellae self-assemblies[31], as well as from those of PBI **2** (Fig. 2).

The different nanostructures observed by TEM for pure PBI **1**, pure PBI **2** and PBI **1**/PBI **2** mixtures are a strong indication for a co-assembly process between PBI **1** and PBI **2** molecules. However, on this macroscopic level, we cannot get insight into the local contacts and distinguish between, for example, alternating (AB)$_n$- and block A$_n$B$_m$-type co-aggregates. Accordingly, the initial steps for the formation of supramolecular structures of this PBI **1**/PBI **2** mixed system were further analysed by NMR spectroscopy. For this purpose, D$_2$O was gradually added to a solution of PBI **1** and PBI **2** in a 2:1 molar ratio in deuterated THF and the chemical shifts of the perylene protons were monitored by NMR spectroscopy (Supplementary Fig. 11). With increasing D$_2$O content, the chemical shifts of the perylene protons of both planar PBI **1** (H$^1$ and H$^2$) and core-twisted PBI **2** (H$^a$ and H$^b$) move to higher field as a consequence of π–π stacking induced by water (D$_2$O) addition. Two-dimensional $^1$H–$^1$H-ROESY NMR spectroscopy was used to elucidate three-dimensional structures and conformations of PBI **1**/PBI **2** mixtures by through-space proton correlations. In ROESY NMR spectra (Fig. 4b) of 2:1 mixture of **1** and **2** cross-peaks appear between H$^a$ (8.10 p.p.m.) and H$^1$ (8.45 p.p.m.), H$^a$ (8.10 p.p.m.) and H$^2$ (8.32 p.p.m.), H$^b$ (7.28 p.p.m.) and H$^1$ (8.45 p.p.m.), and H$^b$ (7.28 p.p.m.) and H$^2$ (8.32 p.p.m.). This suggests that the

protons H$^1$ and H$^2$ of PBI **1** have through-space correlation with the protons H$^a$ and H$^b$ of PBI **2**, implying an intermolecular face-to-face co-assembly of planar and core-twisted PBI molecules. In addition, the cross-peaks between the protons of the same kind of molecules, particularly PBI **1**, appear between H$^3$ (7.10 p.p.m.) and H$^1$ (8.45 p.p.m.), and H$^3$ (7.10 p.p.m.) and H$^2$ (8.32 p.p.m.). To distinguish the intramolecular (through-bond, shown in blue arrow) and intermolecular (through-space, shown in red arrow) correlation signals, we compared 2D $^1$H–$^1$H TOCSY (total correlated spectroscopy) NMR spectra, which only show the through-bond cross-peak signals (Supplementary Figs 12 and 13). The cross-peak between H$^3$ (7.10 p.p.m.) and H$^1$ (8.36 p.p.m.) observed in ROESY spectra also appear in $^1$H–$^1$H TOCSY spectra, which confirms that this cross-peak signal arose from the intramolecular (through-bond) correlation. Accordingly, this NMR analysis provided evidence for the evolvement of mixed dye assemblies, that is, contact of π-surfaces between PBI **1** and PBI **2**, during the formation process of the nanowires in THF/water mixtures.

The nanowires of PBI **1**/PBI **2** (2:1) mixed system were further analysed by optical spectroscopy. The absorption spectrum (Fig. 5) shows three maxima (at 483, 552 and 601 nm) and is again affected by scattering effects. Since both molecules PBI **1** (ref. 31) and PBI **2** do not show concentration-dependent absorption changes in water, we compared the spectrum of the mixed system with a calculated spectrum based on the spectra of the single components in water. This comparison reveals same spectral characteristics, suggesting that the mixed PBI **1**/PBI **2** system is composed of blocks of PBI **1** aggregates, exhibiting same optical properties like the previously reported PBI **1** aggregates, and blocks of PBI **2** aggregates. Accordingly, this result suggests that the PBI **1** and PBI **2** molecules are not alternately stacked in (AB)$_n$ co-assemblies, which would exhibit different spectral properties. This outcome is in apparent contradiction with the previous insights gained by NMR spectroscopy. However, both findings can be reconciled in a structural model of these nanowires based on blocks of short homoaggregates that are dominated by the excitonic coupling pattern arising from same neighbour molecules, which further assemble into the ultrathin nanowires. Indeed, as revealed from our earlier work[37], PBI **2** has a strong preference for dimerization, but is reluctant to grow into more extended supramolecular polymers (anti-cooperative self-assembly) and, accordingly, dimeric blocks of PBI **2** constitute a very reasonable option for the size of the 'soft' block in these mixed aggregate nanowires.

To assess the effect of the molar ratio of PBI **1** and PBI **2** on the co-assemblies of these PBIs and thus on the optical properties of the resulted aggregates, we have investigated the mixtures of these PBIs in 4:1, 8:1 and 10:1 molar ratios in water by ultraviolet–visible absorption spectroscopy. The absorption profiles of these mixtures were elucidated similarly as for the 2:1 mixture shown in Fig. 5 and are displayed in Supplementary Fig. 14. The spectra of the various mixtures are all in good agreement with the calculated ones, suggesting that the mixed aggregates are in all cases composed of individual self-assembled PBI **1** and PBI **2** blocks, and thus exhibit the same optical properties as the respective unimolecular aggregates in water. Furthermore, we have measured TEM for the 10:1 mixture. The obtained image (Supplementary Fig. 15) differs considerably from that observed for the 2:1 mixture, but resembles that obtained from pure PBI **1** in water (Fig. 2a). Thus, we can conclude that minor amounts of PBI **2** cannot transform the morphology from stiff segmented nanorods (as observed for pure PBI **1**) into flexible elongated nanowires (as observed for 2:1 mixed aggregates). On the other hand, the use of >50% of PBI **2** in the mixture results in colloidal samples, whose suspensious character is clearly visible by the eyes.

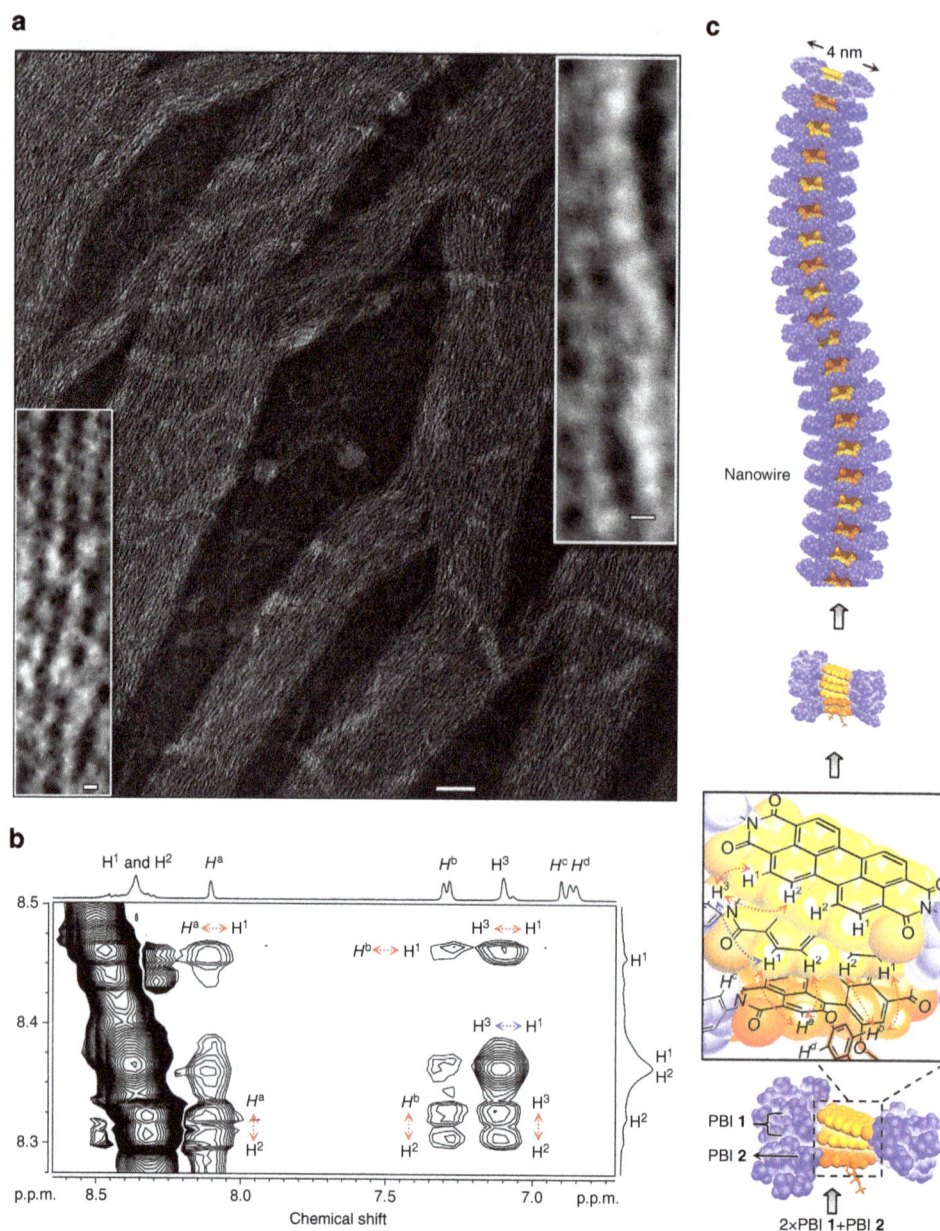

**Figure 4 | Co-assembly of planar and core-twisted PBIs.** (**a**) TEM images of segmented nanowires formed by co-assembly of PBI **1**/PBI **2** in water. Scale bar, 50 nm. Inset, magnified TEM images of the single-molecule-thin nanowires. Scale bars, 2 nm. [PBI **1**]:[PBI **2**] = 2:1 in molar ratio, [PBI **1**] = $3.2 \times 10^{-4}$ M (0.5 mg ml$^{-1}$). (**b**) $^1$H, $^1$H-ROESY spectroscopy for PBI **1**/PBI **2** co-assemblies in D$_2$O/[D$_8$]THF (400/600 µl). [PBI **1**]:[PBI **2**] = 2:1, [PBI **1**] = $4.9 \times 10^{-3}$ M. (**c**) Schematic illustration based on space-filling model for PBI **1**/PBI **2** co-assembly. For clarity, only representative protons are shown in the structural model of one co-assembling segment from semiempirical AM1 calculations.

Clearly, such large amounts of PBI **2** cannot be embedded homogeneously in PBI **1** aggregates and accordingly similar aggregates like those of pure PBI **2** are formed. From these studies we can deduce that homoaggregated blocks of PBI **1** and PBI **2**, respectively, are given for all accessible mixing ratios according to ultraviolet–visible spectroscopy, but that morphologically distinct macroscopic nanowires as shown in Fig. 4 are only possible in a narrow mixing regime around 2:1 molar ratio.

Further insight into the most interesting 2:1 composition of the mixed PBI **1**/PBI **2** nanowire system is obtained by fluorescence spectroscopy. The nanowires formed by co-assembly of PBI **1** and PBI **2** display an intense fluorescence (Fig. 5, Supplementary Figs 16 and 17) with fluorescence quantum yield (Φ) of 6.4%, which is higher than the value for both PBI **1** nanorods (Φ = 1.7%; ref. 27) and PBI **2** nanoworms in water (Φ = 2.2%). Excitation-dependent fluorescence spectra

(Supplementary Fig. 18) show that the fluorescence spectrum obtained upon excitation at the shortest wavelength absorption maximum (480 nm) affords the bathochromically displaced excimer emission at 650 nm that undoubtedly originates from PBI **1** aggregates, whereas the fluorescence spectrum obtained upon excitation at the red-shifted maximum (600 nm) shows a rather similar emission as observed for pure PBI **2** nanoworms (Fig. 3). Accordingly, as for the absorption spectra, also the emission spectra corroborate the idea that the nanowires are composed of alternating segments of PBI **1** and PBI **2** aggregates, that is, self-sorting on the nanoscopic scale is involved in the co-assembly process due to the very distinctive structural features of PBI **1** (flat aromatic core) and PBI **2** (twisted aromatic core)[47,48]. Furthermore, the emission spectra indicate a lack of energy transfer between the respective blocks. At a first glance this is a very surprising result for heteroaggregates composed of short

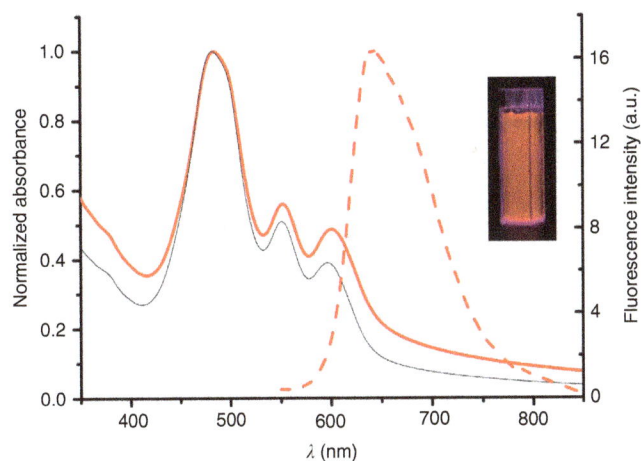

**Figure 5 | Optical properties of PBI 1/PBI 2 nanowires.** Normalized absorption spectrum (red solid) and corresponding fluorescence spectrum (red dashed line, $\lambda_{exc} = 480$ nm, magic-angle setup) of nanowires formed by co-assembly of PBI **1**/PBI **2** in a molar ratio of 2:1 ([PBI] = $7.5 \times 10^{-5}$ M in water) and calculated absorption spectrum (black solid) for a 2:1 mixture of PBI **1** and PBI **2** based on the linear superposition of normalized absorption spectra of PBI **1** and PBI **2** (see Fig. 3), respectively, in water. Inset, photograph of a solution of nanowires in water under ultraviolet light.

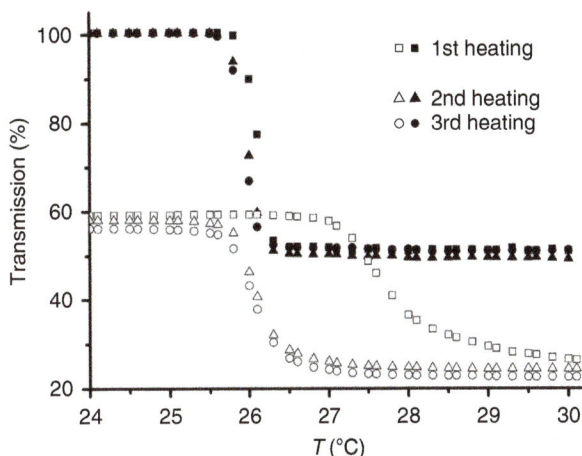

**Figure 6 | LCST behaviour of co-assembled PBI 1/PBI 2 nanowires and unimolecular PBI 1 self-assemblies in water.** Determination of LCSTs for nanowires (open symbols) formed by co-assembly of PBI **1** and PBI **2** in a molar ratio of 2:1 ([PBI **1**] = $5 \times 10^{-5}$ M in water) and for unimolecular PBI **1** aggregates (closed symbols, [PBI **1**] = $5 \times 10^{-5}$ M). Transmission was monitored at 800 nm. Heating rate was 0.1 °C min$^{-1}$.

blocks, that is, close neighbourhood of PBI **1** and PBI **2** segments. However, we have to consider the following unusual situation in these heteroaggregates: PBI **1** aggregate has the higher-lying $S_1$ state than PBI **2** aggregate, but the emission of PBI **1** aggregates takes place from a lower-lying excimer state[44], as evidenced by a characteristic broad emission band at longer wavelength, that is, lower energy. Owing to the ultrafast relaxation into this excimer state[44], photoexcited PBI **1** segments will accordingly not transfer their energy to neighbouring segments of PBI **2**. On the other hand, if the latter are photoexcited, they can also not transfer their energy to PBI **1** segments because there is no overlap of the emission band of PBI **2** segments with the absorption band of PBI **1** segments due to the more hypsochromic shift of the latter.

**Self-sorting of PBI 1 and PBI 2.** Our studies so far indicate that the mixed PBI **1**/PBI **2** system is composed of alternating small-sized segments of homoaggregated PBI **1** and homoaggregated PBI **2** that form ultralong nanowires composed of thousands of stacked molecules. If we assume the presence of the smallest possible segment size for PBI **2**, that is, dimer, we can calculate a block size of four PBI **1** homoaggregates at the chosen 2:1 molecular ratio (for model see Fig. 4c). The remaining question is now whether these nanowires constitute a thermodynamically stable state or they were formed under kinetic control upon evaporation of THF from the THF/H$_2$O solvent mixture. To address this question we subjected the 2:1 mixed nanowire system to heating–cooling cycles (Fig. 6). Upon first heating, a phase transition for the mixed system occurred at about 27.5 °C, whereas further cooling–heating cycles showed a phase transition at a lower temperature of 26 °C, the latter being almost identical to that of pure PBI **1** aggregates in water. These phase transitions originate from a pronounced conformational change of the oligoethylene glycol (OEG) chains and concomitant partial desolvation[49]. This phenomenon leads to the unexpected precipitation upon heating of OEG-functionalized hydrophobic molecules and is characterized by the respective phase-transition temperature that is called the lower critical solution temperature (LCST). Due to the heating above its LCST of 27.5 °C, PBI **1**/PBI **2** mixed nanowires are not preserved during the first phase

transition. Instead, PBI **2** sediments upon re-cooling, whereas PBI **1** is re-dissolved as confirmed by subsequent ultraviolet–visible measurements that show merely the absorbance of PBI **1** aggregates. A similar separation of PBI **1** and PBI **2** was observed when a different preparation method was applied, namely the addition of water to a perfectly mixed solid material consisting of a 2:1 mixture of PBI **1** and PBI **2**. In this case only PBI **1** is dissolved. Thus, these results unambiguously prove that the mixed PBI **1**/PBI **2** nanowire system obtained upon evaporation of THF from a THF/H$_2$O solvent mixture constitutes a kinetically trapped self-assembly product. Notably, pure PBI **2** in water is not subject to LCST behaviour because it does not dissolve in pure water as discussed above. Thus, after the first heating cycle, the co-assembled system is transformed into dissolved PBI **1** aggregates and dehydrated suspension-forming PBI **2** aggregates. Because the transmission of the mixed system at 800 nm can be completely regenerated during the heating cycles, this change in transmission can only be attributed to the PBI **1** homoaggregate system, which undergoes phase transition.

The integration of PBI **2** dimers in 1D PBI **1** aggregates during the co-self-assembly process prohibits a pronounced elongation of the PBI **1** homostacks into stiff segmented nanorods. Statistically, four PBI **1** molecules interact with each other at the 2:1 ratio of PBI **1** with dimers of PBI **2**, which might be the reason for the higher LCST of the mixed system compared with that of PBI **1** homoaggregates. The presence of PBI **2** further prevents the fusion of 1D PBI **1** nanorods into lamellar structures as it is the case for pure PBI **1** (ref. 31). This kinetically stabilized state can be transformed into the thermodynamic one in water by heating, leading to self-sorted and well-dissolved PBI **1** homoaggregates and PBI **2** precipitate.

Since LCST is an entropic phenomenon, the higher LCST of the mixed PBI **1**/PBI **2** system implies that the nanowires created from **1** and **2** are of different order compared with the homoaggregates of PBI **1**. These differences may be attributed to the mutual interaction[50] of planar and core-twisted PBI molecules in the nanowires. The slow evaporation of THF during nanowire preparation enforces both PBIs to mix with each other. This is presumably supported by the similarity in molecular structure, especially with regard to the OEG-bearing imide substituents, that guarantee solubility at the given experimental conditions. The trend of the binding constants (Supplementary Fig. 6) suggests that with increasing water content PBI **2** reaches a fully aggregated but only dimeric state, while PBI **1** aggregation into elongated columnar structures is by far not complete. This state is supposed to be very close to the end point of nanofiber formation. As mentioned before, PBI **2** dimers obviate further interactions with each other via π–π stacking because of steric hindrance of the bulky aryloxy bay substituents. To overcome the unfavourable exposure of π-surfaces of such dimers to the aqueous environment, co-assembly with sterically less encumbered PBI **1** takes place as evidenced by 2D NMR spectroscopy (Fig. 4). Thus, 1D block copolymer structures are formed, where PBI **2** dimers and PBI **1** assemblies of variable size (depending on the excess of PBI **1**) are alternately distributed.

Because PBI **2** molecules are not soluble in pure water at any temperature, it is reasonable to assume that PBI **1** replaced THF as a solubilizing environment upon evaporation of THF from THF/H$_2$O mixture during the co-assembly process. Finally, at higher water, content kinetic trapping occurs mainly because of the much stronger hydrophobic interactions constraining the reversibility[46]. Because of its pronounced hydrophobicity, PBI **2** looses its solvation shell along the way from monomeric state to being part of the nanowires. Thus, the mixed structure remains stable even in pure water due to strong cohesive forces between planar and core-twisted PBIs, although PBI **2** alone cannot exist in the given solvent and prevails in a non-hydrated 'precipitated' state.

## Discussion

This study has elucidated the co-self-assembly of two amphiphilic PBI dyes that bear identical OEG side chains for solubilization in water but are distinguished by the π-conjugated core that is planar for PBI **1** and twisted (as well as more hydrophobic) for PBI **2**. Owing to the different structural features and hydrophobicity, PBI **1** exhibits remarkable solubility at room temperature in water even at quite high concentrations, whereas the core-twisted PBI **2** can only form suspension-forming self-assemblies of small size in water. In the presence of PBI **1**, however, co-assembly with **2** becomes possible upon evaporation of the co-solvent THF, providing ultrathin and micrometre-long nanowires with uniform diameter of 3–4 nm. Our studies showed that these co-assemblies consist of an alternating sequence of respective self-assembled blocks of PBI **1** and PBI **2** that exhibit unusual fluorescence properties, that is, depending on the excitation wavelength emission originates from either PBI **1** or PBI **2** segments, respectively. Such mixed dye assemblies can only be formed by means of a kinetically controlled pathway during passing through the THF/water gradient upon THF evaporation. After formation, these co-assemblies are persistent in pure water due to strong hydrophobic interactions among PBI **1** and PBI **2** molecules that are co-assembled in ultrathin nanowires. However, desolvation of PBI **1** upon heating (LCST behaviour) destabilizes the nanowires, thus ending in the thermodynamically favoured self-sorting of both PBI components into their respective homoaggregates. Our studies demonstrate that mixed supramolecular block copolymers can be generated by applying appropriate sample preparation conditions, that is, strong cohesive forces in water. On the basis of our results we can conclude that co-assembly of two-component systems can follow different pathways than that of their individual components, facilitating the creation of nanosystems with different morphology and properties. This principle may be applied to other multicomponent self-assembled nanosystems whose functions may ultimately bridge the gap to the amazing examples found in nature.

## Methods

**Preparation of PBI aggregates.** PBI **2** aggregates: deionized water (1.0 ml) was added dropwise into a THF solution of PBI **2** (50 µl, 1.0 mg ml$^{-1}$) to induce the aggregate formation. The aggregate suspension was exposed to air to remove THF at room temperature for several days until it is condensed to 0.65 ml, leading to a final concentration of 0.077 mg ml$^{-1}$ for PBI **2** in deionized water.

PBI **1**/PBI **2** co-aggregates: deionized water (0.5 ml) was added dropwise into the mixed THF solution of PBI **1** (25 µl, 1.0 mg ml$^{-1}$) and PBI **2** (17 µl, 1.0 mg ml$^{-1}$); [PBI **1**]:[PBI **2**] = 2:1 in molar ratio. The aggregate suspension was exposed to air to remove THF at room temperature for several days until it is condensed to 0.05 ml, leading to a final concentration of 0.5 mg ml$^{-1}$ for PBI **1** in deionized water.

For further information on Methods, see the Supplementary Methods.

## References

1. Aida, T., Meijer, E. W. & Stupp, S. I. Functional supramolecular polymers. *Science* **335**, 813–817 (2012).
2. Kim, F. S., Ren, G. & Jenekhe, S. A. One-dimensional nanostructures of π-conjugated molecular systems: assembly, properties, and applications from photovoltaics, sensors, and nanophotonics to nanoelectronics. *Chem. Mater.* **23**, 682–732 (2011).
3. Adler-Abramovich, L. & Gazit, E. The physical properties of supramolecular peptide assemblies: from building block association to technological applications. *Chem. Soc. Rev.* **43**, 6881–6893 (2014).
4. Wu, A. & Isaacs, L. Self-sorting: the exception or the rule? *J. Am. Chem. Soc.* **125**, 4831–4835 (2003).
5. Safont-Sempere, M. M., Fernández, G. & Würthner, F. Self-sorting phenomena in complex supramolecular systems. *Chem. Rev.* **111**, 5784–5814 (2011).
6. Lin, N.-T. *et al.* Enantioselective self-sorting on planar, π-acidic surfaces of chiral anion-π transporters. *Chem. Sci.* **3**, 1121–1127 (2012).
7. Ke, C. *et al.* Quantitative emergence of hetero[4]rotaxanes by template-directed click chemistry. *Angew. Chem. Int. Ed.* **52**, 381–387 (2013).
8. Talotta, C., Gaeta, C., Qi, Z., Schalley, C. A. & Neri, P. Pseudorotaxanes with self-sorted sequence and stereochemical orientation. *Angew. Chem. Int. Ed.* **52**, 7437–7441 (2013).
9. Cera, L. & Schalley, C. A. Stimuli-induced folding cascade of a linear oligomeric guest chain programmed through cucurbit[n]uril self-sorting (n = 6, 7, 8). *Chem. Sci.* **5**, 2560–2567 (2014).
10. Kaeser, A. *et al.* Side chains control dynamics and self-sorting in fluorescent organic nanoparticles. *ACS Nano* **7**, 408–416 (2013).
11. Pal, A., Besenius, P. & Sijbesma, R. P. Self-sorting in rodlike micelles of chiral bisurea bolaamphiphiles. *J. Am. Chem. Soc.* **133**, 12987–12989 (2011).
12. Morris, K. L. *et al.* Chemically programmed self-sorting of gelator networks. *Nat. Commun.* **4**, 1480 (2013).
13. Ajayaghosh, A., Vijayakumar, C., Praveen, V. K., Babu, S. S. & Varghese, R. Self-location of acceptors as "isolated" or "stacked" energy traps in a supramolecular donor self-assembly: a strategy to wavelength tunable FRET emission. *J. Am. Chem. Soc.* **128**, 7174–7175 (2006).
14. Samanta, S. K. & Bhattacharya, S. Wide-range light-harvesting donor-acceptor assemblies through specific intergelator interactions via self-assembly. *Chem. Eur. J.* **18**, 15875–15885 (2012).
15. Zhang, W. *et al.* Supramolecular linear heterojunction composed of graphite-like semiconducting nanotubular segments. *Science* **334**, 340–343 (2011).
16. Qiu, H. *et al.* Tunable supermicelle architectures from the hierarchical self-assembly of amphiphilic cylindrical B-A-B triblock co-micelles. *Angew. Chem. Int. Ed.* **51**, 11882–11885 (2012).
17. Rupar, P. A., Chabanne, L., Winnik, M. A. & Manners, I. Non-centrosymmetric cylindrical micelles by unidirectional growth. *Science* **337**, 559–562 (2012).
18. Finnegan, J. R. *et al.* Gradient crystallization-driven self-assembly: cylindrical micelles with "patchy" segmented coronas via the coassembly of linear and brush block copolymers. *J. Am. Chem. Soc.* **136**, 13835–13844 (2014).
19. Ogi, S., Fukui, T., Jue, M. L., Takeuchi, M. & Sugiyasu, K. Kinetic control over pathway complexity in supramolecular polymerization through modulating the energy landscape by rational molecular design. *Angew. Chem. Int. Ed.* **53**, 14363–14367 (2014).
20. Huang, Z. *et al.* Supramolecular polymerization promoted and controlled through self-sorting. *Angew. Chem. Int. Ed.* **53**, 5351–5355 (2014).
21. Wang, F. *et al.* Self-sorting organization of two heteroditopic monomers to supramolecular alternating copolymers. *J. Am. Chem. Soc.* **130**, 11254–11255 (2008).
22. Venkata Rao, K. & George, S. J. Supramolecular alternate co-assembly through a non-covalent amphiphilic design: conducting nanotubes with a mixed D-A structure. *Chem. Eur. J.* **18**, 14286–14291 (2012).
23. Felder, T., de Greef, T. F. A., Nieuwenhuizen, M. M. L. & Sijbesma, R. P. Alternation and tunable composition in hydrogen bonded supramolecular copolymers. *Chem. Commun.* **50**, 2455–2457 (2014).
24. Frisch, H. *et al.* pH-Switchable ampholytic supramolecular copolymers. *Angew. Chem. Int. Ed.* **52**, 10097–10101 (2013).

25. Frisch, H., Nie, Y., Raunser, S. & Besenius, P. pH-Regulated selectivity in supramolecular polymerizations: switching between co- and homopolymers. *Chem. Eur. J.* **21**, 3304–3309 (2015).
26. de Greef, T. F. A. *et al.* Supramolecular polymerization. *Chem. Rev.* **109**, 5687–5754 (2009).
27. Lohr, A. & Würthner, F. Chiral amplification, kinetic pathways, and morphogenesis of helical nanorods upon self-assembly of dipolar merocyanine dyes. *Isr. J. Chem.* **51**, 1052–1066 (2011).
28. Shahar, C. *et al.* Self-assembly of light-harvesting crystalline nanosheets in aqueous media. *ACS Nano* **7**, 3547–3556 (2013).
29. Weissman, H. & Rybtchinski, B. Noncovalent self-assembly in aqueous medium: mechanistic insights from time-resolved cryogenic electron microscopy. *Curr. Opin. Colloid Interface Sci.* **17**, 330–342 (2012).
30. Rest, C., Mayoral, M. J. & Fernández, G. Aqueous self-sorting in extended supramolecular aggregates. *Int. J. Mol. Sci.* **14**, 1541–1565 (2013).
31. Zhang, X., Görl, D., Stepanenko, V. & Würthner, F. Hierarchical growth of fluorescent dye aggregates in water by fusion of segmented nanostructures. *Angew. Chem. Int. Ed.* **53**, 1270–1274 (2014).
32. Kim, H.-J., Kim, T. & Lee, M. Responsive nanostructures from aqueous assembly of rigid-flexible block molecules. *Acc. Chem. Res.* **44**, 72–82 (2011).
33. Görl, D., Zhang, X. & Würthner, F. Molecular assemblies of perylene bisimide dyes in water. *Angew. Chem. Int. Ed.* **51**, 6328–6348 (2012).
34. Rao, K. V., Jayaramulu, K., Maji, T. K. & George, S. J. Supramolecular hydrogels and high-aspect-ratio nanofibers through charge-transfer-induced alternate coassembly. *Angew. Chem. Int. Ed.* **49**, 4218–4222 (2010).
35. García, F. & Sánchez, L. Structural rules for the chiral supramolecular organization of OPE-based discotics: induction of helicity and amplification of chirality. *J. Am. Chem. Soc.* **134**, 734–742 (2012).
36. Tidhar, Y., Weissman, H., Wolf, S. G., Gulino, A. & Rybtchinski, B. Pathway-dependent self-assembly of perylene diimide/peptide conjugates in aqueous medium. *Chem. Eur. J.* **17**, 6068–6075 (2011).
37. Chen, Z., Baumeister, U., Tschierske, C. & Würthner, F. Effect of core twisting on self-assembly and optical properties of perylene bisimide dyes in solution and columnar liquid crystalline phases. *Chem. Eur. J.* **13**, 450–465 (2007).
38. Kohl, C., Weil, T., Qu, J. & Müllen, K. Towards highly fluorescent and water-soluble perylene dyes. *Chem. Eur. J.* **10**, 5297–5310 (2004).
39. Yang, S. K. *et al.* Monovalent, clickable, uncharged, water-soluble perylenediimide-cored dendrimers for target-specific fluorescent biolabeling. *J. Am. Chem. Soc.* **133**, 9964–9967 (2011).
40. Sadrai, M. *et al.* Lasing action in a family of perylene derivatives: singlet absorption and emission spectra, triplet absorption and oxygen quenching constants, and molecular mechanics and semiempirical molecular orbital calculations. *J. Phys. Chem.* **96**, 7988–7996 (1992).
41. Martin, R. B. Comparisons of indefinite self-association models. *Chem. Rev.* **96**, 3043–3064 (1996).
42. Chen, Z., Fimmel, B. & Würthner, F. Solvent and substituent effects on aggregation constants of perylene bisimide π-stacks – a linear free energy relationship analysis. *Org. Biomol. Chem.* **10**, 5845–5855 (2012).
43. Seibt, J. *et al.* On the geometry dependence of molecular dimer spectra with an application to aggregates of perylene bisimide. *Chem. Phys.* **328**, 354–362 (2006).
44. Lim, J. L. *et al.* Exciton delocalization and dynamics in helical π-stacks of self-assembled perylene bisimides. *Chem. Sci.* **4**, 388–397 (2013).
45. Yamamoto, Y. Programmed self-assembly of large π-conjugated molecules into electroactive one-dimensional nanostructures. *Sci. Technol. Adv. Mater.* **13**, 033001 (2012).
46. Baram, J., Weissman, H., Tidhar, Y., Pinkas, I. & Rybtchinski, B. Hydrophobic self-assembly affords robust noncovalent polymer isomers. *Angew. Chem. Int. Ed.* **53**, 4123–4126 (2014).
47. Shaller, A. D., Wang, W., Gan, H. Y. & Li, A. D. Q. Tunable molecular assembly codes direct reaction pathways. *Angew. Chem. Int. Ed.* **47**, 7705–7709 (2008).
48. Safont-Sempere, M. M. *et al.* Impact of molecular flexibility on binding strength and self-sorting of chiral π-surfaces. *J. Am. Chem. Soc.* **133**, 9580–9591 (2011).
49. Weber, C., Hoogenboom, R. & Schubert, U. S. Temperature responsive bio-compatible polymers based on poly(ethylene oxide) and poly(2-oxazoline)s. *Prog. Polym. Sci.* **37**, 686–714 (2012).
50. Ajayaghosh, A., Varghese, R., Mahesh, S. & Praveen, V. K. From vesicles to helical nanotubes: a sergeant-and-soldiers effect in the self-assembly of oligo(p-phenyleneethynylene). *Angew. Chem. Int. Ed.* **45**, 7729–7732 (2006).

## Acknowledgements

We thank the DFG (grant project: Wu 317/11) for financial support.

## Author contributions

F.W. initiated and guided the research programme; D.G. synthesized the molecules; D.G. and X.Z. performed the spectroscopy and TEM investigations; V.S. performed the atomic force microscopy and cryo-TEM experiments; X.Z. designed the graphical artwork; D.G. and F.W. wrote the manuscript.

## Additional information

Competing financial interests: The authors declare no competing financial interests.

# 12

# Iterative protecting group-free cross-coupling leading to chiral multiply arylated structures

Cathleen M. Crudden[1,2], Christopher Ziebenhaus[1], Jason P.G. Rygus[1], Kazem Ghozati[1], Phillip J. Unsworth[1], Masakazu Nambo[2], Samantha Voth[1], Marieke Hutchinson[1], Veronique S. Laberge[1], Yuuki Maekawa[1,3] & Daisuke Imao[1]

The Suzuki–Miyaura cross-coupling is one of the most often utilized reactions in the synthesis of pharmaceutical compounds and conjugated materials. In its most common form, the reaction joins two $sp^2$-functionalized carbon atoms to make a biaryl or diene/polyene unit. These substructures are widely found in natural products and small molecules and thus the Suzuki–Miyaura cross-coupling has been proposed as the key reaction for the automated assembly of such molecules, using protecting group chemistry to affect iterative coupling. We present herein, a significant advance in this approach, in which multiply functionalized cross-coupling partners can be employed in iterative coupling without the use of protecting groups. To accomplish this, the orthogonal reactivity of different boron substituents towards the boron-to-palladium transmetalation reaction is exploited. The approach is illustrated in the preparation of chiral enantioenriched compounds, which are known to be privileged structures in active pharmaceutical compounds.

[1] Queen's University, Department of Chemistry, Chernoff Hall, Kingston, Ontario, Canada K7L 3N6. [2] Institute of Transformative Bio-Molecules (WPI-ITbM), Nagoya University, Chikusa, Nagoya 464-8602, Japan. [3] Department of Chemistry and Biomolecular Science, Faculty of Engineering, Gifu University, 1-1 Yanagido, Gifu 501-1193, Japan. Correspondence and requests for materials should be addressed to C.M.C. (email: cruddenc@chem.queensu.ca).

he Suzuki–Miyaura cross-coupling, in which organoboranes are coupled with organic halides or their equivalents (Fig. 1a), has changed the way organic molecules are assembled[1,2]. This reaction is the method of choice for the preparation of biaryl or polyene units in pharmaceutical[3] and materials industries[4]. Because of the ease with which biaryl molecules are prepared using the Suzuki–Miyaura reaction, these substructures are now widely found in pharmaceutical products and precursors, perhaps even to the detriment of structural diversity. Indeed, evidence is emerging that 'flat' molecules lacking stereochemistry are suboptimal drug candidates when compared with chiral molecules, which have a more complex and tunable three-dimensional shape and improved pharmacokinetic properties[5,6]. Molecules with chiral centres are found with significantly greater frequency (up to 30%) in final drugs than in discovery compounds, a fact attributed to improved drug performance compared with flat structures composed of sp[2] centres[5]. Despite these compelling facts, the creation of C–C bonds with stereochemistry using the Suzuki–Miyaura reaction has only been demonstrated in the last few years, enabling the preparation of molecules with considerable diversity (Fig. 1b)[7–23].

The development of methods to include the construction of chiral centres with control of stereochemistry is clearly critical to access more complex and valuable structures, and doing this in an iterative manner is a key for efficient, potentially automated applications. In fact, Burke has proposed the use of the Suzuki–Miyaura reaction as the key reaction with which the majority of non-peptidic organic molecules can be assembled in an automated manner (Fig. 1c)[24,25]. One critical component in any iterative synthesis involving organoboranes is the ability to modulate the reactivity of B–C bonds between 'off' states, where no reaction takes place; and 'on' states, from which coupling can occur. At present, this can only be accomplished through the use of blocking ligands on boron that deactivate the substrate towards coupling, which is followed by chemical removal of these ligands to generate an active-coupling partner[26,27].

In a significant divergence from existing approaches, we describe herein the first example the iterative coupling of up to three B–C bonds within the same molecule without employing protecting group chemistry. This is accomplished by taking advantage of inherent differences in the transmetalation efficiency of closely related C–Bpin bonds (Fig. 1d). The orthogonal reactivity of B–C bonds in different positions in a single molecule permits the chemoselective, sequential coupling of aromatic, aliphatic and stereochemistry-bearing B–C bonds. This method permits the rapid generation of multiply arylated, chiral organic

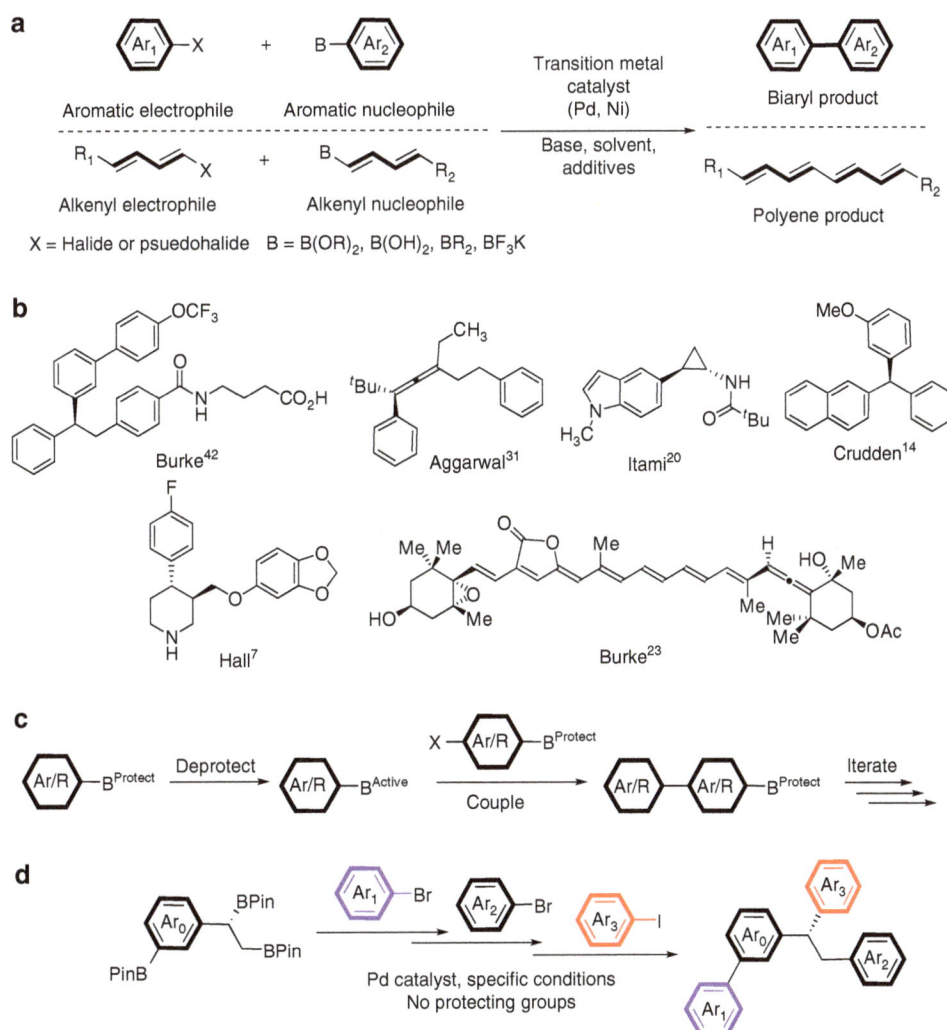

**Figure 1 | Advances in the Suzuki–Miyaura cross-coupling reaction.** (**a**) Classical Suzuki–Miyaura reaction for the formation of C–C bonds between aryl or alkenyl electrophiles and aryl or alkenyl organoboron nucleophiles. (**b**) Selected examples of molecules prepared by enantiospecific Suzuki–Miyaura cross-coupling reactions. (**c**) Iterative coupling concept employing the Suzuki–Miyaura reaction as the key structure-building component. (**d**) This work: chemoselective, protecting group-free cross-coupling of multiply borylated organic compounds, including the coupling of chiral, enantioenriched B–C bonds.

molecules with control of stereochemistry, without the need for protection/deprotection sequences. The method is simple, straightforward and holds considerable promise for the facile synthesis of new classes of pharmaceutical structures with improved properties. The Molander group has reported examples of protecting group-free iterative couplings accomplished employing the differential ability of benzylic RBF$_3$K salts to undergo radical-based transmetallation[19,28,29]. Although this has not been reported on molecules containing more than one identical boron substituent, such reactivity can be envisioned.

## Results

### Orthogonal coupling of benzylic and non-benzylic C–B bonds.

In 2009, we reported the first example of the cross-coupling of chiral benzylic boronic esters (1) that occurred with retention of stereochemistry (Fig. 2)[9]. This method has been extended to include the cross-coupling of allylic[11,30], propargylic[31] and doubly benzylic boronic esters[14]. Using different substrates, Suginome[10], Hall[13], Molander[16] and Biscoe[15] have reported groundbreaking invertive couplings under alternative conditions.

In our 2009 report, two key observations were critical for the current work. First, we noted that when bases other than silver oxide were employed with benzylic boronic ester 1, the starting boronic ester was recovered untouched, indicating that silver oxide was required for transmetalation. Second, under optimized conditions for the coupling of 1, namely Pd(0)/PPh$_3$ and Ag$_2$O, its linear achiral isomer 3 was completely unreactive, such that 3 could be recovered from the reaction mixture with the B–C bond intact (Fig. 2a)[9]. Thus, the mild silver oxide-promoted conditions employed for branched substrate 1 were clearly insufficient to promote transmetalation of isomer 3. This orthogonal reactivity offered the exciting opportunity to control the arylation of multiborylated organic compounds based on inherent differences in transmetalation efficiencies of different types of B–C bonds.

The critical question, then, was whether the relative reactivity described in Fig. 2 would translate to distinct B–C bonds in diborylated molecules. We first chose to examine the reactivity of 1,2 diboronates (5), which contain both a linear and a branched benzylic boron substituent. These compounds can be easily prepared with high enantioselectivity by diboration of the corresponding styrene derivative 4 (refs 32–34). Pinacol substituents were chosen on boron, as these are among the most common and stable ancillary groups for boron[8,35]. The

coupling of one B–C(OR)$_2$ unit in diborylated compounds followed by oxidation or, less commonly, amination or homologation of a remaining B–C bond has been amply demonstrated by Fernández[36], Morken[37–39] and Suginome[40]. However, with the exception of one example of a hydroxyl-promoted coupling[39], there have been no reports describing the use of unprotected multiborylated compounds in iterative cross-couplings.

Initial attempts at reacting the benzylic, secondary B–C bond selectively in the presence of the adjacent linear B–C bond were unsuccessful, leading to a mixture of products, including those resulting from protodeboronation. However, the linear B–C bond in 5 could be enticed to undergo transmetalation and cross-coupling with aryl bromides in the presence of Pd(OAc)$_2$ with RuPhos (2-dicyclohexylphosphino-2′,6′-diisopropoxybiphenyl) or SPhos (2-dicyclo-hexylphosphino-2′,6′-dimethoxybiphenyl) as ligand, and K$_2$CO$_3$ as base, leaving the branched B–C bond intact (Fig. 3a). Although counterintuitive, substrates showing sensitivity to deboronation benefited from the use of a significantly higher proportion of water (as seen in substrates 6aC, 6bC, 6gA, 6gE and 6hA, and later in the linear B–C$^{(sp3)}$ coupling of trisborylated substrate 11a (Fig. 4b)). On the basis of the work of the Lloyd-Jones group[41], this could be due to a decrease in the effective pH of the reaction medium as the proportion of water is increased. In some cases, lower yields were obtained due to losses upon chromatography, as illustrated with compound 6hA (Fig. 3a). Under these relatively simple coupling conditions, a variety of coupling partners for the linear position were tolerated, including electron-rich, electron-poor and π-extended substrates. Heteroaromatic aryl bromides were less successful in this position, with 2-thienyl, 3-bromopyridyl and 3-bromoquinoline giving suboptimal yields, although 4-pyridylbromide could be effectively coupled as shown in Fig. 4c. In all cases studied, this reaction occurred without compromising the stereochemistry at the benzylic B–C bond.

For the next step in the iterative sequence, the branched B–C bond was coupled using our previously described silver oxide-promoted conditions (Fig. 3b)[9]. Again, electron-rich, electron-poor, π-extended and heteroaromatic aryl iodides were well tolerated as coupling partners under the conditions shown in Fig. 3b. Enantiospecificities of this reaction are on the order of 90%, which is slightly lower than those observed in couplings of simpler 1-phenethyl boronic esters[9,42], possibly indicating a slightly increased susceptibility to stereochemical erosion, because of the conjugated nature of any presumed stilbene intermediate resulting from β-hydride elimination; however, no regioisomeric products were observed, and thus the possibility of loss of stereochemistry during transmetalation must also be considered (enantiospecificity, e.s. = (e.e. product/e.e. starting material) × 100%). The electronics of the aryl iodide did not appear to play any effect.

### Orthogonal coupling of benzylic sp$^3$ versus sp$^2$ C–B bonds.

Having demonstrated that two adjacent aliphatic B–C bonds can be sequentially coupled with high chemoselectivity, we next probed whether the same approach could be used in molecules containing other types of B–C bonds with different propensities for transmetalation. Thus, we chose to pit B–C$^{(sp2)}$ bonds against secondary B–C$^{(sp3)}$ bonds, as in substrate 8, which was prepared by hydroboration of 4-pinacol boronato styrene. This reaction took place with 95:5 e.r. (ref. 43). In this case, the B–C$^{(sp2)}$ bond was targeted for initial coupling since it should clearly undergo transmetalation, the most readily based on >30 years of precedent. Although a number of conditions were effective for cross-coupling of this B–C bond and indeed left the secondary

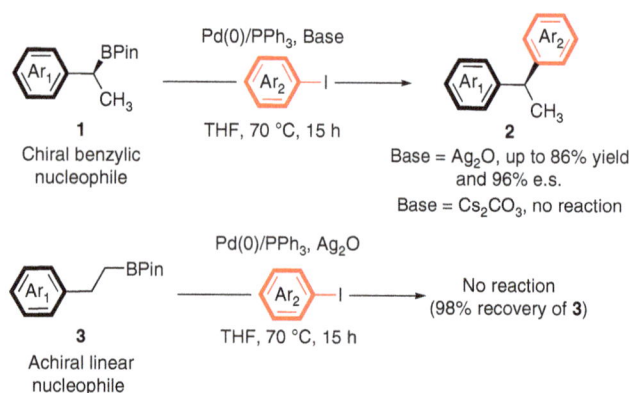

**Figure 2 | Orthogonal reactivity of branched benzylic and linear aliphatic boronic esters.** Observation of complete inversion of reactivity such that branched benzylic boronic esters (1) react in the presence of Ag$_2$O, but not carbonate bases, while linear aliphatic isomers (3) are recovered unchanged when treated with Ag$_2$O.

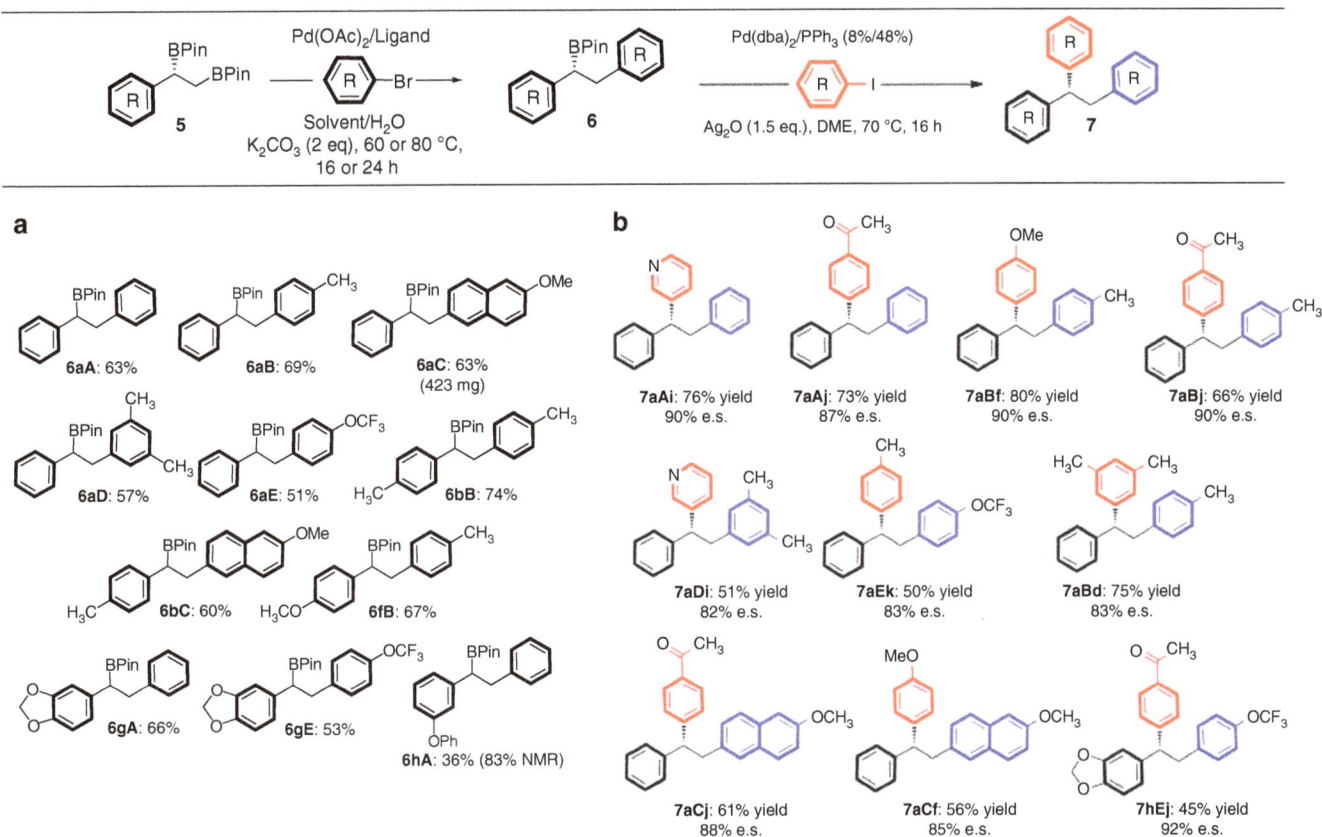

Figure 3 | Chemoselective Suzuki–Miyaura cross-coupling of chiral 1,2-diboronic esters. (a) Demonstration of selective coupling of linear boronic ester, while the branched B–C bond remains intact. (b) Stereoretentive coupling of remaining secondary benzylic B–C bond in enantioenriched boronic esters, leading to enantiomerically enriched unsymmetrical 1,1′,2-triarylethanes with up to 92% enantiospecificity.

benzylic B–C bond untouched, small impurities derived from the incorporation of phenyl rather than aryl substituents were observed consistently. These side products originated from the use of triphenylphosphine as the ligand via well-documented P–Ph activation[44]. After several unsuccessful attempts to circumvent this unwanted side reaction, the use of P$^t$Bu$_3$ proved successful[45], giving the desired arylated products 9aX in high isolated yields, with the benzylic B–C bond intact for the next coupling (Fig. 4a). The reaction was readily scalable and could be routinely run on several hundred milligrams scale. Comparison of the enantiopurity of products 9aX, after coupling at the B–C$^{(sp2)}$ bond indicated, as expected, that there was no detectable loss of enantiopurity during this coupling.

With these products in hand, we next examined the scope of the Ag$_2$O-promoted benzylic cross-coupling reaction (Fig. 4a). The enantiospecific cross-coupling took place as expected, except that π-extended biphenyl boronic esters 9aX displayed higher sensitivity to the loss of enantiomeric purity during coupling than previously observed for simple phenylated boronic esters[9]. In particular, we noted that electron-withdrawing substituents on the biphenyl derivative as in 9aJ led to a greater erosion in enantiospecificity (for example, 76% e.s. for coupling with 4-iodotrifluoromethyltoluene yielding 10aJk and 54% e.s. for coupling of the same boronic ester with 3-iodopyridine yielding 10aJi).

To confirm that this effect was due to the starting boronic ester, 3-iodopyridine and 4-iodotrifluoromethyltoluene were reacted with coupling partner 9aF, bearing a methoxyphenyl instead of phenacyl substituent. As shown in Fig. 4a, these reactions occurred with very high enantiospecificities (96% and 89%, respectively), confirming that it is the electron-withdrawing

substituent on the biphenyl that leads to higher loss of enantiopurity in the coupling. Bulky aryl iodides such as 1-napthyliodide were effective, but also resulted in decreased enantiospecificity (that is, 10aDm). Comparing simple 1-phenethyl pinacol boronate with the π-extended systems illustrates the sensitivity of the latter systems: π-extended 9aD reacts with 1-naphthyliodide with 76% e.s., while PhCH(BPin)CH$_3$ reacts with 85% e.s. when coupled with the same aryl iodide[46]. These results will provide important information in ongoing mechanistic studies of this reaction.

Protecting group-free iterative coupling of trisborylated species. To fully illustrate how the orthogonal reactivity of the various B–C bonds can be applied in iterative couplings, we prepared compound 11a containing three different types of B–C bonds: a B–C$^{(sp2)}$ bond, a primary B–C$^{(sp3)}$ bond and a secondary B–C$^{(sp3)}$ bond, and we attempted all three couplings in the same pot. For the first B–C$^{(sp2)}$ coupling, Fu-type conditions[45], previously shown to be optimal for the coupling of diborylated substrate 8, were also effective with trisborylated substrate 11a, giving the desired product without loss of either aliphatic pinacol boronate. The yield of this first arylation was determined to be 92–95% by NMR analysis. Optimized conditions for the second-coupling reaction involved the use of solvent with a high proportion of water (1:2 ratio of organic solvent:water) as described above. In test studies, these conditions led to NMR yields on the order of 70%, boding well for a fully iterative coupling of 11. Indeed, we found the reaction proceeded with good yield, and importantly, leaving the benzylic B–C bond intact. Although the final benzylic cross-coupling worked well with isolated starting materials, when run in

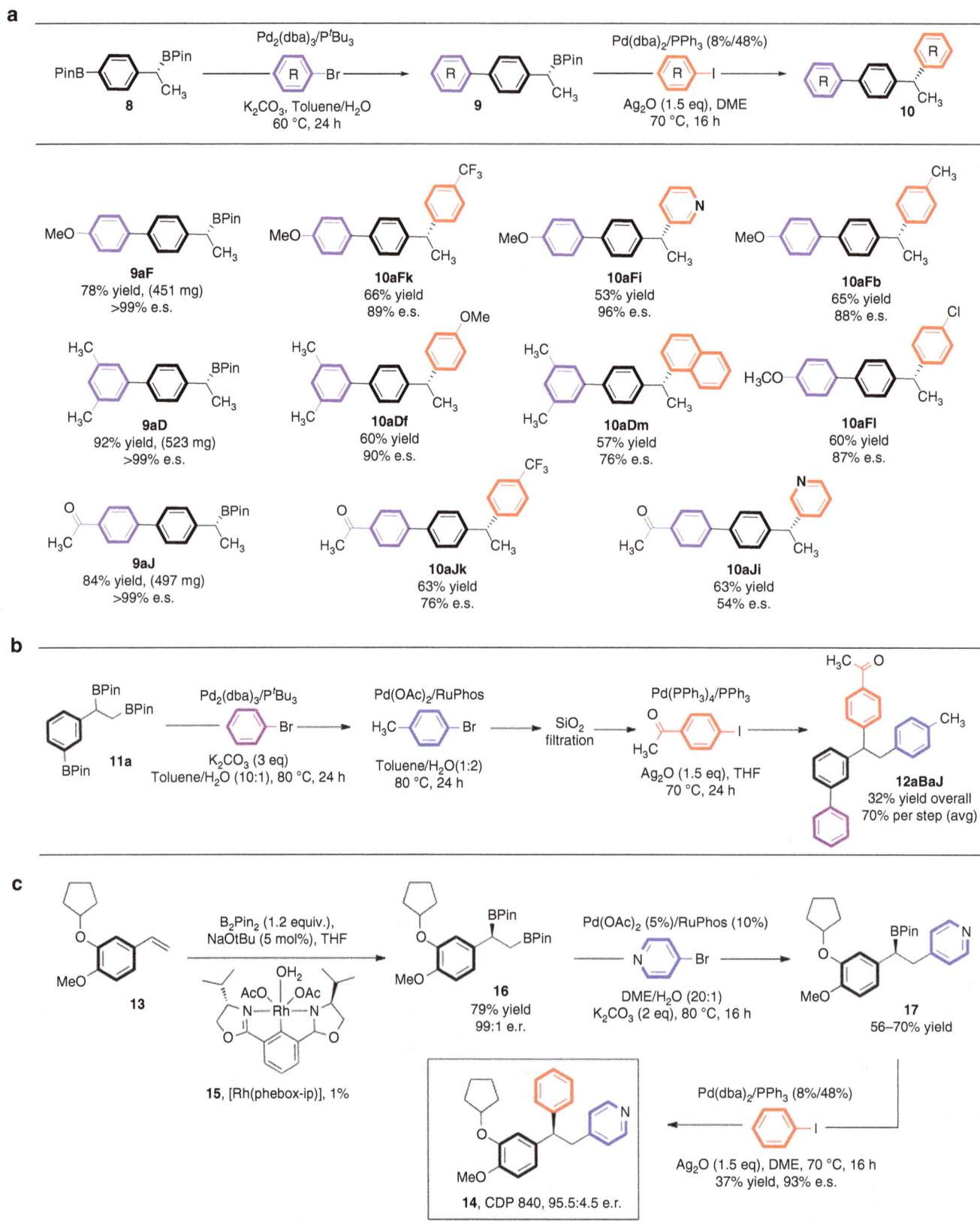

**a**

**b**

**c**

**Figure 4 | Chemoselective cross-couplings of bis- and tris-borylated chiral aromatics.** (**a**) Initial selective coupling of aryl B–C bonds using Pd/P*t*Bu₃ as the catalyst and carbonate bases, followed by second, stereospecific coupling of chiral secondary benzylic B–C bond leading to biphenyl-substituted 1,1–diaryl alkanes. (**b**) The introduction of three unique aryl groups sequentially by coupling at (1) the aryl B–C bond, (2) the linear achiral aliphatic B–C bond, and (3), the chiral benzylic B–C bond in a single pot for the first two steps, and the use of filtration through a short silica plug before the final step. (**c**) Illustration of the orthogonal coupling concept in the synthesis of phosphodiesterase inhibitor CDP 840 (**14**).

sequence, the one-pot procedure required filtration of the reaction mixture through a small silica plug, likely to deal with excess halides remaining from previous coupling reactions (Fig. 4b). In this manner, triarylated product **12aBaJ** could be obtained in 32% overall yield, representing an average yield per step of just over 70%.

**Application to compounds of medicinal importance.** Finally, to illustrate the effectiveness of our orthogonal coupling methodology, we carried out the synthesis of a triarylated compound of pharmaceutical importance. Thus, compound **16** was prepared by diborylation of styrene **13** as shown in Fig. 4c, and, in only two steps, converted into biologically active derivative **14**, otherwise known as CDP 840. This compound is a prototype orally active anti-inflammatory phosphodiesterase with potential inhibitory effects against phosphodiesterase-4 (ref. 47). The key coupling step occurred with 93% e.s. giving the final product in 95.5:4.5 e.r. Although outside the scope of this paper, we envisage that our iterative coupling approach will be highly effective for the rapid preparation of derivatives of active compounds by the introduction of a series of different aryl groups at various borylated positions of a central core molecule.

In conclusion, we have shown that the resistance of certain types of B–C bonds to transmetalation, one of the key steps in the Suzuki–Miyaura reaction[48], can be capitalized upon to develop a protecting group-free sequential cross-coupling of multiply borylated organic compounds. This approach is complementary to other approaches that involve the manipulation of the substituents on boron to control reactivity of the B–C bond. In all cases, we employed pinacol esters since these are the most versatile and readily employed type of organoboron substituent[8,35]. Chiral substrates are compatible with the method, leading to products with significant complexity, which are likely to provide interesting leads for pharmaceutical and medicinal applications.

## Methods

**Coupling of linear B–C$^{(sp3)}$ bond in 1,2-diborylated compounds 5x.** In a nitrogen-filled glovebox, diboronate **5x** (1 equiv.), bromoarene (1.2 equiv.), Pd(OAc)$_2$ (0.1 equiv.), RuPhos (0.25 equiv.) and K$_2$CO$_3$ (1.9 equiv.) were weighed into a vial and tetrahydrofuran was added. The reaction was sealed with a septum and removed from the glovebox, and placed under a flow of argon. Degassed water was added (20:1, organic:water) and the septum was replaced with a teflon cap. The reaction mixture was sonicated for about 2 min before being stirred at 80 °C for 15 h. The reaction mixture was cooled and filtered through a plug of silica, washed through with EtOAc and concentrated *in vacuo*. Purification by column chromatography was affected as the final purification.

**Coupling of branched B–C$^{(sp3)}$ bond in 1-boryl,1,2-diaryl species 6xX.** In a nitrogen-filled glovebox, boronic ester **6xX** (1 equiv.), iodoarene (1.5 equiv.), Pd(dba)$_2$ (0.08 equiv.), PPh$_3$ (0.64 equiv.) and Ag$_2$O (1.5 equiv.) were weighed into a 1 dram vial and dimethoxyethane (DME) was added. The reaction vessel was sealed, removed from the glovebox and heated at 70 °C for 16 h. The reaction mixture was cooled and filtered through a plug of silica, eluted with EtOAc and concentrated *in vacuo*. Purification by column chromatography was affected as the final purification.

**Coupling of aromatic B–C$^{(sp2)}$ bond in diboryl species 8x.** An oven-dried pressure tube with a stir bar in a glovebox was charged with diboronate **8x** (1 equiv.), Pd$_2$(dba)$_3$ (0.05 equiv.), [($t$Bu)$_3$PH]BF$_4$ (0.2 equiv.), K$_2$CO$_3$ (3 equiv.), bromoarene (1.2 equiv.) and toluene (3.4 equiv.). The pressure tube was sealed with a rubber septum, removed from the glovebox and placed under argon. Water degassed (34:1 organic to water) was added and the rubber septum was replaced with a lid. The reaction was heated to 60 °C for 24 h. After cooling to room temperature, the reaction mixture was filtered through a plug of silica gel (ca. 2 mL) and a PROMAX 0.22 μm polytetrafluoroethylene (PTFE) syringe filter using copious ethyl acetate. The resulting product was purified by chromatography.

For NMR and super critical fluid chromatography (SFC) analysis of the compounds in this article, see Supplementary Figs 1–99.

## References

1. Miyaura, N., Yamada, K. & Suzuki, A. New stereospecific cross-coupling by the palladium-catalyzed reaction of 1-alkenylboranes with 1-alkenyl or 1-alkynyl halides. *Tetrahedron Lett.* **20**, 3437–3440 (1979).
2. Suzuki, A. Cross-coupling reactions of organoboranes: an easy way to construct C-C bonds (nobel lecture). *Angew. Chem. Int. Ed.* **50**, 6722–6737 (2011).
3. Carey, J. S., Laffan, D., Thomson, C. & Williams, M. T. Analysis of the reactions used for the preparation of drug candidate molecules. *Org. Biomol. Chem.* **4**, 2337–2347 (2006).
4. Schlüter, A. D. The tenth anniversary of Suzuki polycondensation (spc). *J. Polym. Sci.* **39**, 1533–1556 (2001).
5. Lovering, F., Bikker, J. & Humblet, C. Escape from flatland: Increasing saturation as an approach to improving clinical success. *J. Med. Chem.* **52**, 6752–6756 (2009).
6. Over, B. *et al.* Natural-product-derived fragments for fragment-based ligand discovery. *Nature Chem* **5**, 21–28 (2013).
7. Ding, J., Rybak, T. & Hall, D. G. Synthesis of chiral heterocycles by ligand-controlled regiodivergent and enantiospecific suzuki miyaura cross-coupling. *Nature Commun.* **5**, 5474 (2015).
8. Leonori, D. & Aggarwal, V. K. Stereospecific couplings of secondary and tertiary boronic esters. *Angew. Chem. Int. Ed.* **54**, 1082–1096 (2015).
9. Imao, D., Glasspoole, B. W., Laberge, V. S. & Crudden, C. M. Cross coupling reactions of chiral secondary organoboronic esters with retention of configuration. *J. Am. Chem. Soc.* **131**, 5024–5025 (2009).
10. Ohmura, T., Awano, T. & Suginome, M. Stereospecific suzuki-miyaura coupling of chiral alpha-(acylamino)benzylboronic esters with inversion of configuration. *J. Am. Chem. Soc.* **132**, 13191–13193 (2010).
11. Chausset-Boissarie, L. *et al.* Enantiospecific, regioselective cross-coupling reactions of secondary allylic boronic esters. *Chem. Eur. J.* **19**, 17698–17701 (2013).
12. Molander, G. A. & Wisniewski, S. R. Stereospecific cross-coupling of secondary organotrifluoroborates: potassium 1-(benzyloxy)alkyltrifluoroborates. *J. Am. Chem. Soc.* **134**, 16856–16868 (2012).
13. Lee, J. C. H., MacDonald, R. & Hall, D. G. Enantioselective preparation and chemoselective cross-coupling of 1,1-diboron compounds. *Nature Chem.* **3**, 894–899 (2011).
14. Matthew, S. C., Glasspoole, B. W., Eisenberger, P. & Crudden, C. M. Synthesis of enantiomerically enriched triarylmethanes by enantiospecific suzuki-miyaura cross coupling reactions. *J. Am. Chem. Soc.* **136**, 5828–5831 (2014).
15. Li, L., Zhao, S., Joshi-Pangu, A., Diane, M. & Biscoe, M. R. Stereospecific Pd-catalyzed cross-coupling reactions of secondary alkylboron nucleophiles and aryl chlorides. *J. Am. Chem. Soc.* **136**, 14027–14030 (2014).
16. Sandrock, D. L., Jean-Gerard, L., Chen, C.-Y., Dreher, S. D. & Molander, G. A. Stereospecific cross-coupling of secondary alkyl beta-trifluoroboratoamides. *J. Am. Chem. Soc.* **132**, 17108–17110 (2010).
17. Maity, P., Shacklady-McAtee, D. M., Yap, G. P. A., Sirianni, E. R. & Watson, M. P. Nickel-catalyzed cross couplings of benzylic ammonium salts and boronic acids: Stereospecific formation of diarylethanes via C-N bond activation. *J. Am. Chem. Soc.* **135**, 280–285 (2013).
18. Wilsily, A., Tramutola, F., Owston, N. A. & Fu, G. C. New directing groups for metal-catalyzed asymmetric carbon-carbon bond-forming processes: Stereoconvergent alkyl-alkyl Suzuki cross-couplings of unactivated electrophiles. *J. Am. Chem. Soc.* **134**, 5794–5797 (2012).
19. Tellis, J. C., Primer, D. N. & Molander, G. A. Single-electron transmetalation in organoboron cross-coupling by photoredox/nickel dual catalysis. *Science* **345**, 433–436 (2014).
20. Miyamura, S., Araki, M., Suzuki, T., Yamaguchi, J. & Itami, K. Stereodivergent synthesis of arylcyclopropylamines by sequential C-H borylation and Suzuki-Miyaura coupling. *Angew. Chem. Int. Ed.* **54**, 846–851 (2015).
21. Harris, M. R., Hanna, L. E., Greene, M. A., Moore, C. E. & Jarvo, E. R. Retention or inversion in stereospecific nickel-catalyzed cross- coupling of benzylic carbamates with arylboronic esters: control of absolute stereochemistry with an achiral catalyst. *J. Am. Chem. Soc.* **135**, 3303–3306 (2013).
22. Awano, T., Ohmura, T. & Suginome, M. Inversion or retention? Effects of acidic additives on the stereochemical course in enantiospecific suzuki-miyaura coupling of alpha-(acetylamino)benzylboronic esters. *J. Am. Chem. Soc.* **133**, 20738–20741 (2011).
23. Woerly, E. M., Cherney, A. H., Davis, E. K. & Burke, M. D. Stereoretentive suzuki-miyaura coupling of haloallenes enables fully stereocontrolled access to (-)-peridinin. *J. Am. Chem. Soc.* **132**, 6941–6943 (2010).
24. Woerly, E. M., Roy, J. & Burke, M. D. Synthesis of most polyene natural product motifs using just twelve building blocks and one coupling reaction. *Nature Chem.* **6**, 484–491 (2014).
25. Li, J. *et al.* Synthesis of many different types of organic small molecules using one automated process. *Science* **34**, 1221–1226 (2015).
26. Noguchi, H., Hojo, K. & Suginome, M. Boron-masking strategy for the selective synthesis of oligoarenes via iterative Suzuki-Miyaura coupling. *J. Am. Chem. Soc.* **129**, 758–759 (2007).

27. Gillis, E. P. & Burke, M. D. A simple and modular strategy for small molecule synthesis: Iterative Suzuki-Miyaura coupling of B-protected haloboronic acid building blocks. *J. Am. Chem. Soc.* **129**, 6716–6717 (2007).

28. Primer, D. N., Karakaya, I., Tellis, J. C. & Molander, G. A. Single-electron transmetalation: an enabling technology for secondary alkylboron cross-coupling. *J. Am. Chem. Soc.* **137**, 2195–2198 (2015).

29. Yamashita, Y., Tellis, J. C. & Molander, G. A. Protecting group-free, selective cross-coupling of alkyltrifluoroborates with borylated aryl bromides via photoredox/nickel dual catalysis. *Proc. Natl Acad. Sci. USA* **112**, 12026–12029 (2015).

30. Glasspoole, B. W., Ghozati, K., Moir, J. & Crudden, C. M. Suzuki-Miyaura cross couplings of secondary allylic boronic esters. *Chem. Commun.* **48**, 1230–1232 (2012).

31. Partridge, B. M., Chausset-Boissarie, L., Burns, M., Pulis, A. P. & Aggarwal, V. K. Enantioselective synthesis and cross-coupling of tertiary propargylic boronic esters using lithiation-borylation of propargylic carbamates. *Angew. Chem. Int. Ed.* **51**, 11795–11799 (2012).

32. Kliman, L. T., Mlynarski, S. N. & Morken, J. P. Pt-catalyzed enantioselective diboration of terminal alkenes with $B_2(pin)_2$. *J. Am. Chem. Soc.* **131**, 13210–13211 (2009).

33. Bonet, A., Pubill-Ulldemolins, C., Bo, C., Gulyas, H. & Fernandez, E. Transition-metal-free diboration reaction by activation of diboron compounds with simple Lewis bases. *Angew. Chem. Int. Ed.* **50**, 7158–7161 (2011).

34. Toribatake, K. & Nishiyama, H. Asymmetric diboration of terminal alkenes with a rhodium catalyst and subsequent oxidation: Enantioselective synthesis of optically active 1,2-diols. *Angew. Chem. Int. Ed.* **52**, 11011–11015 (2013).

35. Crudden, C. M., Hleba, Y. B. & Chen, A. C. Regio and enantiocontrol in the room temperature hydroboration of vinyl arenes. *J. Am. Chem. Soc.* **125**, 9200–9201 (2004).

36. Penno, D. *et al.* Multifaceted palladium catalysts towards the tandem diboration-arylation reactions of alkenes. *Chem. Eur. J.* **14**, 10648–10655 (2008).

37. Miller, S. P., Morgan, J. B., Nepveux, F. J. & Morken, J. P. Catalytic asymmetric carbohydroxylation of alkenes by a tandem diboration/Suzuki cross-coupling/oxidation reaction. *Org. Lett.* **6**, 131–133 (2004).

38. Mlynarski, S. N., Schuster, C. H. & Morken, J. P. Asymmetric synthesis from terminal alkenes by cascades of diboration and cross-coupling. *Nature* **505**, 386–390 (2014).

39. Blaisdell, T. P. & Morken, J. P. Hydroxyl-directed cross-coupling: a scalable synthesis of debromohamigeran E and other targets of interest. *J. Am. Chem. Soc.* **137**, 8712–8715 (2015).

40. Daini, M. & Suginome, M. Palladium-catalyzed, stereoselective, cyclizative alkenylboration of carbon-carbon double bonds through activation of a boron-chlorine bond. *J. Am. Chem. Soc.* **133**, 4758–4761 (2011).

41. Lennox, A. J. J. & Lloyd-Jones, G. C. Organotrifluoroborate hydrolysis: boronic acid release mechanism and an acid – base paradox in cross-coupling. *J. Am. Chem. Soc.* **134**, 7431–7441 (2012).

42. Li, J. & Burke, M. D. Pinene-derived iminodiacetic acid (pida): a powerful ligand for stereoselective synthesis and iterative cross-coupling of $c(sp^3)$ boronate building blocks. *J. Am. Chem. Soc.* **133**, 13774–13777 (2011).

43. Hayashi, T., Matsumoto, Y. & Ito, Y. Catalytic asymmetric hydroboration of styrenes. *J. Am. Chem. Soc.* **111**, 3426–3428 (1989).

44. Grushin, V. V. Thermal stability, decomposition paths, and ph/ph exchange reactions of $[(Ph_3P)_2Pd(Ph)X]$ (X = I, Br, Cl, F, and $HF_2$). *Organometallics* **19**, 1888–1900 (2000).

45. Littke, A. F. & Fu, G. C. A convenient and general method for Pd-catalyzed Suzuki cross-couplings of aryl chlorides and arylboronic acids. *Angew. Chem. Int. Ed.* **37**, 3387–3388 (1998).

46. Glasspoole, B. W. *The Chemoselective, Enantiospecific Cross-Coupling of Secondary Boronic Esters and the Stability of Mesoporous Silica Supports for Pd Catalysis.* (PhD thesis, Queen's Univ., 2011).

47. Alexander, R. P. *et al.* CDP840. A prototype of a novel class of orally active anti-inflammatory phosphodiesterase 4 inhibitors. *Bioorg. Med. Chem. Lett.* **12**, 1451–1456 (2002).

48. Lennox, A. J. J. & Lloyd-Jones, G. C. Transmetalation in the Suzuki-Miyaura coupling: the fork in the trail. *Angew. Chem. Int. Ed.* **52**, 7362–7370 (2013).

## Acknowledgements

C.M.C. acknowledges the Natural Sciences and Engineering Research Council of Canada (NSERC) for funding in terms of Discovery, Accelerator and RTI grants, and the Canada Foundation for Innovation (CFI) for infrastructure funding. M.N. and C.M.C. acknowledge the Japan Society for the Promotion of Science and the Institute for Transformative Bio-Molecules/Nagoya University for funding. J.R. acknowledges NSERC for a postgraduate scholarship. C.Z., S.V., J.R., M.H. and V.L. acknowledge Queen's department of Chemistry for Queen's Graduate Awards.

## Authors contributions

C.M.C. and D.I. conceived the overall concepts; C.Z., J.P.G.R, K.G., P.J.U., N.M., S.V., M.H., V.S.L. and Y.M. performed the experiments, compound characterization and data analysis. All authors contributed to the overall experiment design, discussions and manuscript preparation.

## Additional information

**Competing financial interests:** The authors declare no competing financial interests.

# 13

# Tandem C–H oxidation/cyclization/rearrangement and its application to asymmetric syntheses of ( − )-brussonol and ( − )-przewalskine *E*

Zhi-Wei Jiao[1], Yong-Qiang Tu[1,2], Qing Zhang[1], Wen-Xing Liu[1], Shu-Yu Zhang[1], Shao-Hua Wang[1], Fu-Min Zhang[1] & Sen Jiang[1]

Natural products are a vital source of lead compounds in drug discovery. Development of efficient tandem reactions to build useful compounds and apply them to the synthesis of natural products is not only a significant challenge but also an important goal for chemists. Here we describe a tandem C–H oxidation/cyclization/rearrangement of isochroman-derived allylic silylethers, promoted by DDQ and InCl$_3$. This method allows the efficient construction of tricyclic benzoxa[3.2.1]octanes with a wide substrate scope. We employ this tandem reaction to achieve the asymmetric total syntheses of ( − )-brussonol and ( − )-przewalskine *E*.

[1] State Key Laboratory of Applied Organic Chemistry & College of Chemistry and Chemical Engineering, Lanzhou University, Lanzhou 730000, P. R. China. [2] Collaborative Innovation Center of Chemical Science and Engineering, Tianjin 300071, P. R. China. Correspondence and requests for materials should be addressed to Y.-Q.T. (email: tuyq@lzu.edu.cn) or to F.-M.Z. (email: zhangfm@lzu.edu.cn).

The strained tricyclic benzoxa[3.2.1]octane skeleton exists in numerous bioactive and pharmaceutical molecules such as przewalskone[1] and brussonol[2], and it is a useful building block in organic synthesis such as for producing platensimycin[3] (Fig. 1a). In recent decades, the significant biological activity and potential pharmaceutical value of molecules with this skeleton have driven chemists to devise several methods for constructing it (Fig. 1b)[3-14]. Nevertheless, more efficient and practical methods are needed.

One possible approach is the C–C bond formation via direct ($sp^3$) α-C–H bond functionalization, which is increasingly being used to synthesize complex N- or O-containing molecules[15-18]. While several reactions have been described to achieve C–C formation at the α-position in amines, only a handful of reactions have been reported for C–C formation in ethers[19-29]. In connection with our long-standing interest in α-C–H bond functionalization of ethers[30,31] and in carbon–carbon rearrangement of allylic alcohol/silylether[32-34], we speculated that it might be possible to construct the benzoxa[3.2.1]octane unit via a tandem reaction that is initiated by benzylic C–H oxidation of an isochroman-derived allylic silylether and triggered by C–C bond rearrangement.

Here we use 2,3-dichloro-5,6-dicyanobenzoquinone (DDQ) and $InCl_3$ to promote tandem C–H oxidation/cyclization/ rearrangement of isochroman-derived allylic silylether. We then apply this efficient tandem reaction to asymmetric syntheses of the bioactive natural products ( − )-brussonol and ( − )-przewalskine E.

## Results

**Reaction optimization.** We began our efforts to generate the benzoxa[3.2.1]octane unit using the model substrate **1a**, prepared as a single diastereoisomer in a general procedure

**a** Selected natural products

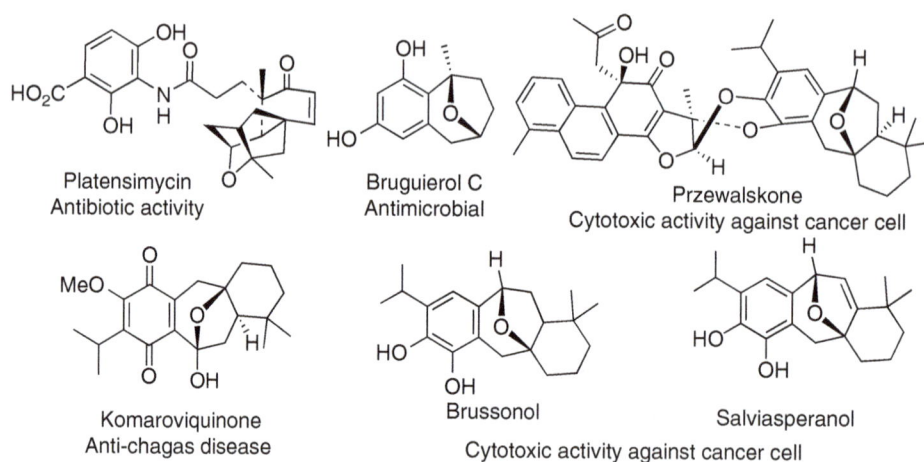

Platensimycin
Antibiotic activity

Bruguierol C
Antimicrobial

Przewalskone
Cytotoxic activity against cancer cell

Komaroviquinone
Anti-chagas disease

Brussonol
Cytotoxic activity against cancer cell

Salviasperanol

**b** Previous work

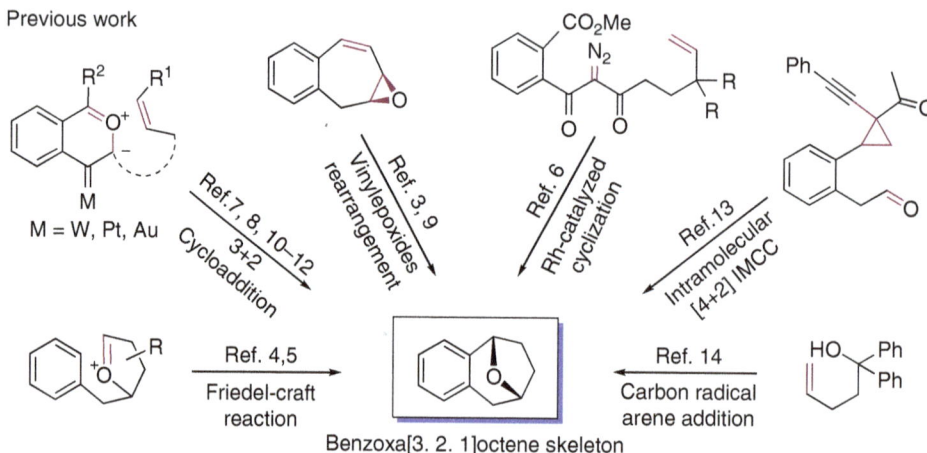

M = W, Pt, Au

Ref.7, 8, 10–12
3+2
Cycloaddition

Ref. 3, 9
Vinylepoxides rearrangement

Ref. 6
Rh-catalyzed cyclization

Ref.13
Intramolecular [4+2] IMCC

Ref. 4,5
Friedel-craft reaction

Benzoxa[3. 2. 1]octene skeleton

Ref. 14
Carbon radical arene addition

**c** Our tandem reaction (this work)

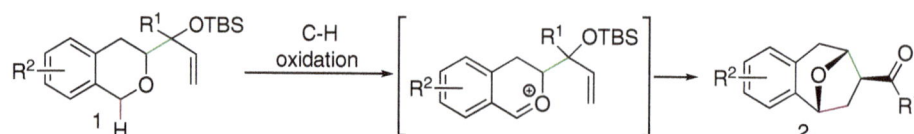

C–H oxidation

**Figure 1 | Representative natural products containing benzoxa[3.2.1]octane skeleton and approaches to it.** (**a**) Selected bioactive natural products having benzoxa[3.2.1]octane skeleton. (**b**) Previous methods for construction of benzoxa[3.2.1]octane skeleton. (**c**) Our synthetic proposal via a tandem C–H oxidation/cyclization/rearrangement reaction.

**Table 1 | Conditions optimization\*.**

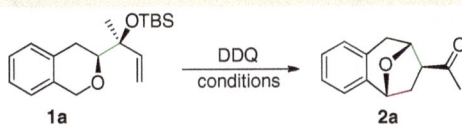

| Entry | Cat. (equiv.) | DDQ (equiv.) | Base (equiv.) | Time (h) | Yield[†] |
|---|---|---|---|---|---|
| 1 | — | 2.0 | — | 12 | NR |
| 2 | $FeCl_3$/1.0 | 1.2 | — | 0.5 | Dec |
| 3 | $SnCl_4$/1.0 | 1.2 | — | 0.5 | Dec |
| 4 | $SnBr_4$/1.0 | 1.2 | — | 8 | 33% |
| 5 | $Cu(OTf)_2$/1.0 | 1.2 | — | 12 | 23% |
| 6 | $LiClO_4$ (1.0) | 1.2 | — | 12 | 29% |
| 7 | $InCl_3$ (1.0) | 1.2 | — | 24 | 36% |
| 8 | $InCl_3$ (1.0) | 2.0 | — | 4 | 56% |
| 9 | $InCl_3$ (0.1) | 2.0 | — | 12 | 69% |
| 10[‡] | $InCl_3$ (0.1) | 2.0 | — | 12 | 30% |
| 11 | $InCl_3$ (0.1) | 2.0 | $Na_2CO_3$ | 12 | 44% |
| 12 | $InCl_3$ (0.1) | 2.0 | $Et_3N$ | 12 | NR |
| 13[§] | $InCl_3$ (0.1) | 2.0 | DTBP | 12 | Trace |
| 14[§] | $InCl_3$ (0.1) | 2.0 | DBP | 12 | 81% |

Dec, decomposed; DTBP, 2, 6-di-*tert*-butylpyridine; NR, no reaction.
\*All reactions unless specifically notified were performed with 0.3 mmol **1a** and 150 mg 4 Å MS in 3.0 ml $CH_2Cl_2$ at RT.
†Isolated yield.
‡No 4 Å MS used.
§5.0 equiv. base used.

(see Supplementary Methods 1 and Supplementary Data 1 and 4). Initial experiment with the common oxidant DDQ (2.0 equiv.) as the sole promoter failed to give the desired product and resulted in fully recovery of **1a** (Table 1, entry 1). Next we tried combinations of DDQ (1.2 equiv.) and Lewis acids in the presence of 4-Å molecular sieve. Using $FeCl_3$ or $SnCl_4$ led to the decomposition of **1a** (entries 2 and 3), whereas using $SnBr_4$ led to the consumption of **1a** in 8 h and the desired product **2a** in 33% yield (entry 4). $Cu(OTf)_2$ and $LiClO_4$ also promoted this transformation, albeit in lower yield (entries 5 and 6). When 1.0 equiv. $InCl_3$ was used, **2a** was obtained in 36% yield after 24 h and 15% of **1a** was recovered (entry 7). Increasing the load of DDQ to 2.0 equiv. gave higher yield (56%) in shorter time (4 h). Interestingly, decreasing the load of $InCl_3$ to 0.1 equiv. further increased the yield of **2a** to 69% (entry 9). The molecular sieve was essential to this reaction, since omitting it led to only 30% yield (entry 10). With $InCl_3$ as the catalyst, a side reaction in which **1a** was partially desilylated to give free allylic alcohol was always observed. To inhibit this, we screened several weakly basic additives (entries 11–14). We were pleased to find that using 2,6-dibromopyridine (DBP, 5.0 equiv.) significantly improved yield to 81% (entry 14). In addition, the reaction was also carried out in some other solvents ($C_2H_4Cl_2$, $CH_3NO_2$, $CH_3CN$, toluene, tetrahydrofuran (THF)) and oxidants (TEMPO, benzoquinone), but results were not better than with $CH_2Cl_2$ and DDQ. Therefore, the optimal conditions were defined to be DDQ (2.0 equiv.), $InCl_3$ (0.1 equiv.), DBP (5.0 equiv.) and 4-Å molecular sieve in $CH_2Cl_2$ (entry 14, see Supplementary Methods 2).

**Substrate scope.** Using these optimal reaction conditions, we tested the substrate scope of this transformation extensively (Fig. 2), starting with the allylic substituents at $R^1$. Substrates **1b** and **1c** with *n*-butyl and *i*-propyl groups at this position reacted well and gave the desired products **2b** and **2c** in respective yields of 80 and 83%. Benzyl-substituted **1d** also afforded the desired product **2d** in good 72% yield. In contrast, substrates with more electron-rich groups at $R^1$ gave only medium to good yields. For example, substrates **1e** and **1f** with phenyl and vinyl substituents gave the corresponding products **2e** and **2f** in respective yields of

74 and 63% yield. The substrate **1g** with an acetylenyl substituent generated the product **2g** in a lower yield of 23%. Using substrate **1h** with a methyl substitution at the ally C=C led to smooth production of **2h** with a quaternary stereocenter in 83% yield. The product's relative configuration was confirmed by X-ray diffraction analysis. Next, the substituent effects on the aromatic ring of the isochroman moiety were investigated—compounds **1i** and **1j** fully substituted with MeO or Me reacted well, giving products **2i** and **2j** in respective yields of 83 and 87%. These relatively high yields may mean that the electron-donating methoxyl group stabilizes the benzylic oxonium carbocation in the transition state in Fig. 1c better than hydrogen does. Compound **1k** with a bromo substituent at the 7th position in the isochroman moiety led to the desired product **2k** in 63% yield, extending the flexibility of our approach for further derivatization.

We also expanded the substrate scope to acyclic systems to allow the synthesis of multi-substituted THF derivatives, which exist in numerous bioactive and pharmaceutical molecules[35–37]. Four representative allylic siylethers **1l–1o** with terminal benzylic or allylic ethers as the reaction trigger generated the corresponding products **2l–2o** in 37–53% yields under the same optimal conditions. Notably, the reaction was triggered efficiently by either benzylic ethers (**1l**, **1m**, **1o**) or cinnamylic ethers. (**1n**) When a TBDPS ethylene ether was created in **1o** to compete with benzyl ether during the expected reaction, only the benzyl ether underwent reaction, affording the product **2o** in 42% yield. Under the same optimal conditions, yields were generally lower with acyclic substrates than with cyclic ones, which may be due to hyperoxidation of THF products that produced furan-type byproducts (Supplementary Methods 2.2).

**Asymmetric syntheses of ( – )-brussonol and ( – )-przewalskine E.** To further demonstrate the utility of this novel method for synthesizing polycyclic molecules, we applied it to the total synthesis of two important bioactive natural products—tetracyclic ( – )-przewalskine E (**3a**) and ( – )-brussonol (**3b**)[2,38–40]. Two diastereoselective routes to **3b** have been reported by Sarpong[41] and Jennings[42], while its asymmetric synthesis has been achieved by Majetich[43]. We are unaware of reports of the synthesis of **3a**.

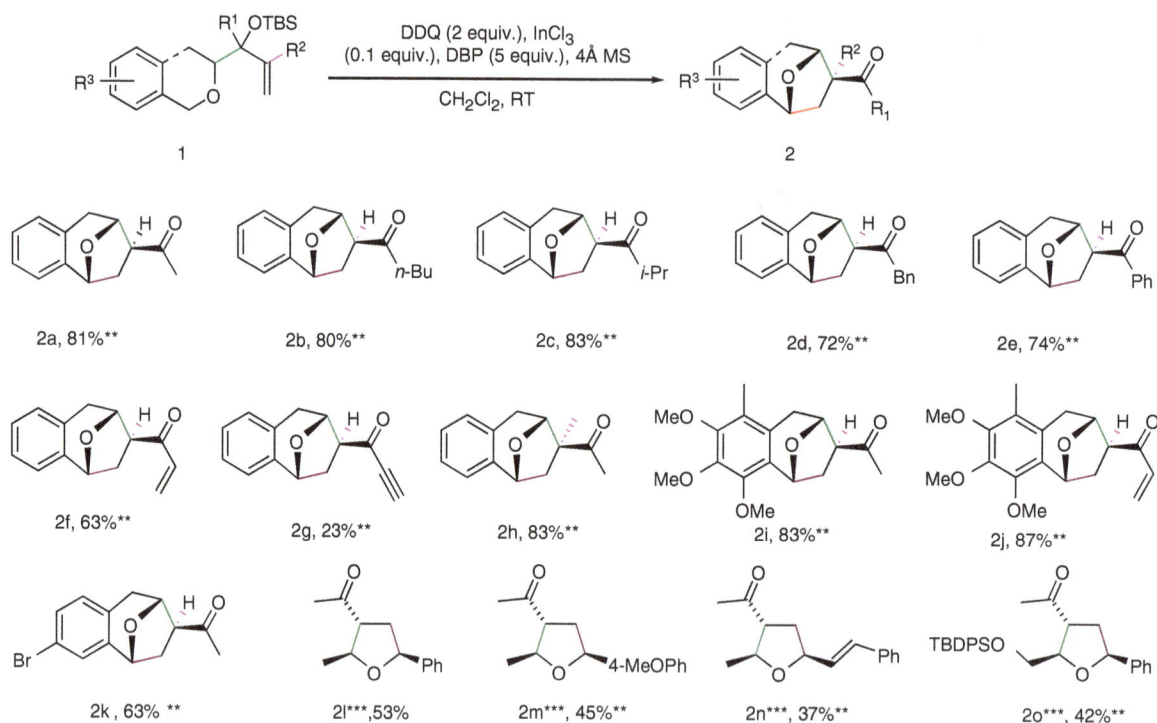

**Figure 2 | Reaction scope*.** *All reactions unless notified were performed in experimental section procedure on a 0.1–1.5 mmol substrate scale in CH₂Cl₂ (0.1 mmol ml⁻¹), 0.1 equiv. InCl₃, 4 Å MS (50 mg per 0.1 mmol), 2.0 equiv. DDQ, 5.0 equiv. DBP at RT. Relative configuration of the products were assigned based on X-ray structure of **2h** and **2j** (CCDC 1000811, CCDC 1000827, See Supplementary Data 2 and 3 for more details). **Isolated yields. ***1.1 equiv. DDQ was used.

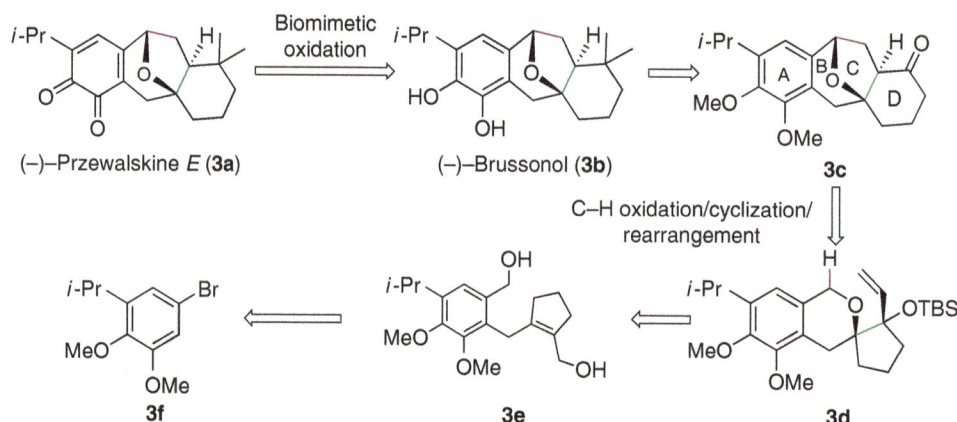

**Figure 3 | Retrosynthetic analysis of 3a,b.** The current method was used to construct the core framework.

In our retrosynthetic analysis (Fig. 3), we hypothesized that we could generate **3a** from **3b** via biomimetic oxidation, and that **3b** could be obtained from ketone intermediate **3c**. We planned to construct rings C and D in **3c** by applying our novel method to the tricyclic allylic silyether **3d**, which could be obtained from achiral allylic alcohol **3e** via a challenging tandem Sharpless asymmetric epoxidation/epoxy opening. The precursor **3e** could be prepared from the simple starting material **3f**[44] in a few short steps.

On the basis of this strategy, we started our synthesis by preparing allylic alcohol **3e** from bromide **3f** (Fig. 4), which was formalized and protected as a 1,3-dioxane **3g** to survive subsequent metallization[45,46]. Deprotonation of **3g** with *n*-BuLi followed by quenching with formaldehyde and finally bromination of the resulting hydroxyl group with Ph₃P/CBr₄ gave benzyl bromide **3h** in 41% yield over two steps. Coupling bromide **3h** with vinyl triflate **3i**[47] and then removing the

1,3-dioxane protecting group in aqueous HCl afforded aldehyde **3j** in 69% yield over two steps. Excess diisobutylaluminum hydride (DIBAL-H) was used to reduce **3j** in a single step to diol **3e** in 93% yield. Then we investigated the key tandem Sharpless asymmetric epoxidation/epoxy opening of **3e**. The expected tricyclic species **3k** was obtained in 90% yield and 83% ee in the presence of classic Sharpless catalyst (1.5 equiv.) at −50 °C. While this enantioselectivity is not ideal, it appears to be a rare example of successful *tetra*-substituted olefin epoxidation[48]. Selective oxidation of the primary hydroxyl of **3k** followed by methylenenation afforded the desired tertiary allylic alcohol **3l**, which was protected to give the precursor **3d**. Fortunately, our optimal conditions of oxidative cyclization/ring enlargement gave the expected tetracyclic skeleton **3c** in high yield (81%) and excellent stereoselectivity, no other isomer was detected.

At this stage, only the installation of a gem-dimethyl group remained to complete the synthesis of **3b**. Initial attempts to

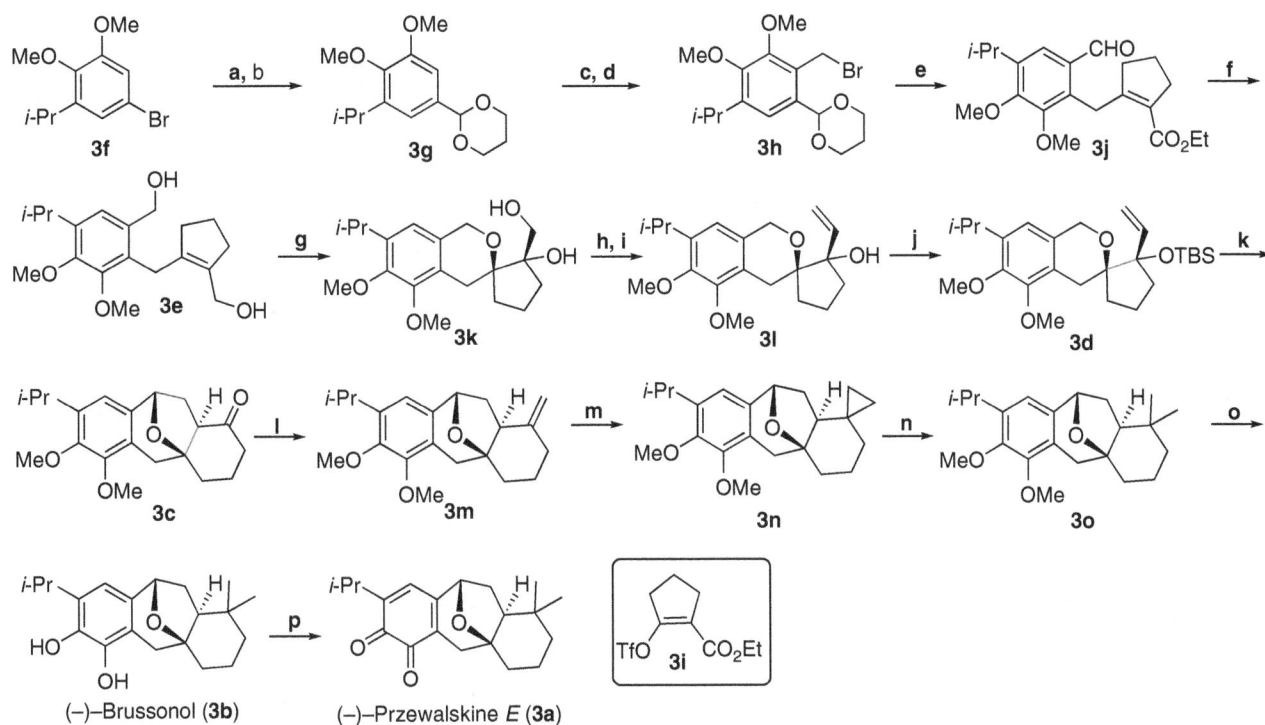

**Figure 4 | Asymmetric total synthesis of ( − )-przewalskine E (3a) and ( − )-brussonol (3b).** Reagent and conditions: (**a**) n-BuLi, Et$_2$O, − 78 °C, then DMF; (**b**) 1,3-propanediol, CH(OEt)$_3$, (n-Bu)$_4$N$^+$Br$_3^-$, 65 °C; 77% (2 steps); (**c**) n-BuLi, Hexane/Et$_2$O, RT, then CH$_2$O in THF, − 78 °C; (**d**) Ph$_3$P, CBr$_4$, CH$_2$Cl$_2$, 0 °C, 41% (2 steps); (**e**) (i) Zn, THF, 0 °C to RT, then Pd(Ph$_3$P)$_2$Cl$_2$, **3i** in DMF, 90 °C; (ii) 4 mol l$^{-1}$ HCl, THF/H$_2$O (4:1), 69% (2 steps); (**f**) DIBAL-H, CH$_2$Cl$_2$, − 78 °C to RT, 93%; (**g**) ( − )-DET, Ti(i-PrO)$_4$, t-BuO$_2$H, − 25 °C to − 50 °C, CH$_2$Cl$_2$, 90% yield, 83% ee; (**h**) DMSO, DIPEA, SO$_3$·Py, CH$_2$Cl$_2$;(**i**) Ph$_3$PCH$_3$Br, t-BuOK, Tol, 72% (2 steps); (**j**) TBSOTf, Et$_3$N, CH$_2$Cl$_2$, 0 °C to 40 °C, 98%; (**k**) 4 Å MS, 2,6-DBP, InCl$_3$, DDQ, CH$_2$Cl$_2$, RT, 82%; (**l**) Ph$_3$PCH$_3$Br, t-BuOK, Toluene, 88%; (**m**) Et$_2$Zn, CH$_2$I$_2$, Tol., 58%; (**n**) PtO$_2$, H$_2$ (1 atm), AcOH, 65 °C, 89%; (**o**) EtSH, NaH, DMF, 150 °C, 76%; (**p**) Ag$_2$O, Et$_2$O, RT, 71%. DIPEA, N,N-diisopropylethylamine.

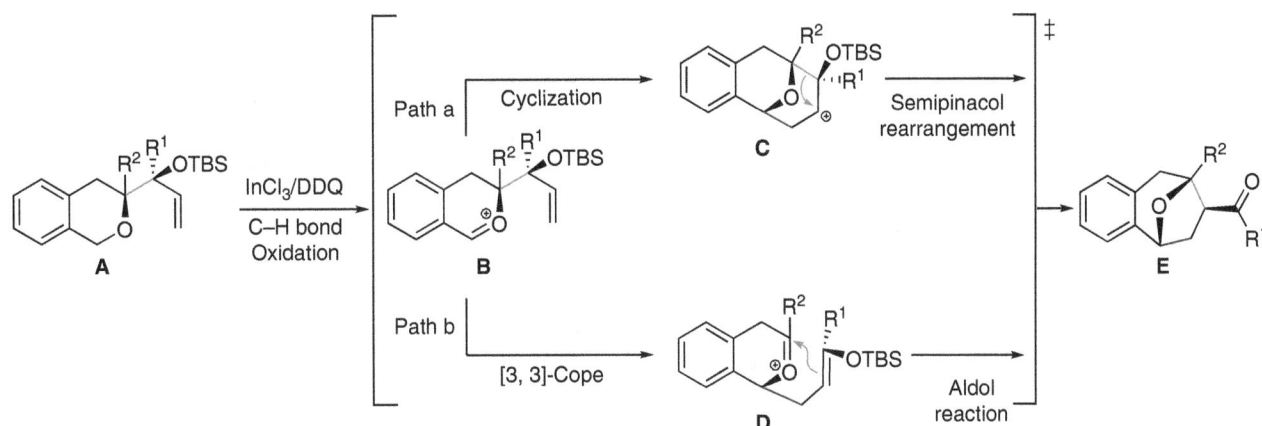

**Figure 5 | Proposed mechanism.** (**a**) Tandem C–H bond oxidation/cyclization/semipinacol rearrangement reaction; (**b**) tandem C–H bond oxidation/ [3,3]-Cope rearrangement/aldol reaction.

transform the carbonyl in ketone **3c** directly into gem-dimethyl using TiMe$_2$Cl$_2$ reagent[49] gave the desired product **3o** in low yield. Therefore, we adopted a three-step protocol involving Wittig reaction, cyclopropanation and reduction. Treating **3c** with Ph$_3$PCH$_3$Br/t-BuOK followed by Simmons–Smith cyclopropanation of the resulting exo-cyclic olefin gave the cyclopropane compound **3n**, which was hydrogenated with H$_2$ (1 atm)/PtO$_2$ to afford the desired **3o** in 45% yield over three steps[50]. Demethylating **3o** using with EtSNa/DMF provided natural product **3b** in 76% yield, and the spectral data were identical to those reported by Sarpong[41] (Supplementary Tables 1 and 2). We screened various oxidants for the biomimetic

oxidation of **3b** to **3a**; Ag$_2$O proved to be the best, affording the natural product **3a** in 71% yield. Its spectral data were identical to those reported by Zhao[40] (Supplementary Tables 3 and 4).

## Discussion

The tandem C–H oxidation/cyclization/rearrangement of iso-chroman-derived allylic silylether described here shows good chemo- and stereoselectivity as well as good product yield. We demonstrated the usefulness of this approach by using it to achieve the asymmetric total syntheses of ( − )-brussonol and

(−)-przewalskine *E*. We expect that this approach will find additional applications in organic synthesis.

This approach involves simpler initiation and substrate preparation as well as milder reaction conditions than a similar acetal rearrangement under acidic conditions via alkoxy release reported by Overman's group[51-58]. Notably, we nearly always obtained the desired product with a benzoxa[3.2.1]octane framework as a single diastereomer (Fig. 2). In term of the mechanism of this transformation, there are two possible pathways whereby intermediate **B** may react with a benzylic oxocarbenium cation under oxidative conditions (Fig. 5). One is a tandem cyclization/semipinacol rearrangement to give **E** (pathway a) and the other is a tandem [3,3]-Cope rearrangement/aldol reaction (pathway b)[52]. Our results suggest that pathway b is more likely, since we did not obtain products with a benzoxa[3.3.1] skeleton when we used substrates **1e**, **1f**, **1g** or **1j**, all of which should preferentially undergo migration of an electron-rich $R^1$ group when a semipinacol rearrangement process is involved. In addition, both substrates **1h** and **1h′**, which are epimers at the allylic position, gave a single diastereomeric product. Whether pathway b is the true one under our optimal conditions needs to be confirmed.

## Methods

**Materials.** For NMR analysis, see Supplementary Figs 1–136 and 139–154. For high-performance liquid chromatography traces, see Supplementary Figs 137 and 138. For X-ray structures of the compounds, see Supplementary Figs 155–158.

**General.** All reactions under standard conditions were monitored by thin-layer chromatography. Column chromatography was performed on silica gel (200–300 mesh). Reaction solvents were distilled before use, and all air- or moisture-sensitive reactions were conducted under an argon atmosphere. Melting points were measured using a micro-melting point apparatus. Optical rotations were measured using a 0.1-ml cell with a 1-cm path length. $^1$H NMR and $^{13}$C NMR spectra were recorded in CDCl$_3$ on a 400-MHz instrument ($^1$H NMR) and a 100-MHz instrument ($^{13}$C NMR); spectral data were reported in p.p.m. relative to trimethylsilane as the internal standard. Infrared spectra were recorded on a Fourier transform infrared spectrometer. High-resolution mass spectral analysis data were measured using the electrospray ionization technique on a Fourier transform ion cyclotron resonance mass analyser.

**General procedure for this tandem reaction.** To a solution of **1a** (249 mg, 0.78 mmol) in CH$_2$Cl$_2$ (8.0 ml), we successively added a 4-Å molecular sieve (400 mg), 2,6-DBP (924 mg, 3.90 mmol, 5.0 equiv.) and InCl$_3$ (18 mg, 0.08 mmol, 0.1 equiv.) at room temperature under an argon atmosphere. The mixture was stirred for 15 min, then DDQ (361 mg, 98%, 1.56 mmol, 2.0 equiv.) was added. The resulting brown mixture was stirred for 12 h at room temperature before it was filtered via an SiO$_2$ column with petroleum ether/EtOAc (4:1) as eluent to remove the molecular sieve and 2,6-DBP. The filtrate was concentrated under vacuum and chromatographed on an SiO$_2$ column (petroleum ether/EtOAc, 10:1) to give product **2a** as a colourless oil (128 mg, 0.63 mmol, 81% yield).

## References

1. Xu, G. *et al.* Przewalskone: a cytotoxic adduct of a danshenol type terpenoid and an icetexane diterpenoid via hetero-Diels–Alder reaction from *Salvia przewalskii*. *Chem. Commun.* **48**, 4438–4440 (2012).
2. Aoyagi, Y. *et al.* Semisynthesis of isetexane diterpenoid analogues and their cytotoxic activity. *Chem. Pharm. Bull.* **54**, 1602–1604 (2006).
3. McGrath, N. A., Bartlett, E. S., Sittihan, S. & Njardarson, J. T. A concise ring-expansion route to the compact core of platensimycin. *Angew. Chem. Int. Ed.* **48**, 8543–8546 (2009).
4. Marson, C. M., Campbell, J., Hursthouse, M. B. & Abdul Malik, K. M. Stereocontrolled routes to bridged ethers by tandem cyclizations. *Angew. Chem. Int. Ed.* **37**, 1122–1124 (1998).
5. Wu, Y., Li, Y. & Wu, Y.-L. First examples of Friedel–Crafts alkylation using ketals as alkylating agents: an expeditious access to the benzene-fused 8-oxabicyclo[3.2.1]octane ring system. *J. Chem. Soc. Perkin Trans. 1* 1189–1192 (1999).
6. Padwa, A., Boonsombat, J., Rashatasakhon, P. & Wills, J. Efficient construction of the oxatricyclo[6.3.1.0$^{0,0}$]dodecane core of komaroviquinone using a cyclization/cycloaddition cascade of a rhodium carbenoid intermediate. *Org. Lett.* **7**, 3725–3727 (2005).
7. Kusama, H., Funami, H., Shido, M. & Iwasawa, N. Generation and reaction of tungsten-containing carbonyl ylides: [3 + 2]-cycloaddition reaction with electron-rich alkenes. *J. Am. Chem. Soc.* **127**, 2709–2716 (2005).
8. Ito, K., Hara, Y., Mori, S., Kusama, H. & Iwasawa, N. Theoretical study of the cycloaddition reaction of a tungsten-containing carbonyl ylide. *Chem. Eur. J.* **15**, 12408–12416 (2009).
9. Simmons, E. M. & Sarpong, R. Ga(III)-catalyzed cycloisomerization strategy for the synthesis of icetexane diterpenoids: total synthesis of (±)-salviasperanol. *Org. Lett.* **8**, 2883–2886 (2006).
10. Oh, C. H., Lee, J. H., Lee, S. J., Kim, J. I. & Hong, C. S. Intramolecular Huisgen-type cyclization of platinum-bound pyrylium ions with alkenes and subsequent insertion into a benzylic C–H bond. *Angew. Chem. Int. Ed.* **47**, 7505–7507 (2008).
11. Oh, C. H., Lee, J. H., Lee, S. M., Yi, H. J. & Hong, C. S. Divergent insertion reactions of Pt-carbenes generated from [3 + 2] cyclization of platinum-bound pyrylliums. *Chem. Eur. J.* **15**, 71–74 (2009).
12. Song, X.-R. *et al.* Gold-catalyzed tandem cyclization/cycloaddition reaction of enynones: highly regioselective synthesis of oxabicyclic compounds and naphthyl ketones. *Asian J. Org. Chem.* **2**, 755–762 (2013).
13. Xing, S., Pan, W., Liu, C., Ren, J. & Wang, Z. Efficient construction of Oxa- and Aza-[n.2.1] skeletons: lewis acid catalyzed intramolecular [3 + 2] cycloaddition of cyclopropane 1,1-diesters with carbonyls and imines. *Angew. Chem. Int. Ed.* **49**, 3215–3218 (2010).
14. Miller, Y., Miao, L., Hosseini, A. S. & Chemler, S. R. Copper-catalyzed intramolecular alkene carboetherification: synthesis of fused-ring and bridged-ring tetrahydrofurans. *J. Am. Chem. Soc.* **134**, 12149–12156 (2012).
15. Doye, S. Catalytic C-H activation of sp$^3$ C-H bonds in α-position to a nitrogen atom-two new approaches. *Angew. Chem. Int. Ed.* **40**, 3351–3353 (2001).
16. Campos, K. R. Direct sp$^3$ C–H bond activation adjacent to nitrogen in heterocycles. *Chem. Soc. Rev.* **36**, 1069–1084 (2007).
17. Poesky, P. W. Catalytic hydroaminoalkylation. *Angew. Chem. Int. Ed.* **48**, 4892–4894 (2009).
18. Li, Z., Bohle, D. S. & Li, C.-J. Cu-catalyzed cross-dehydrogenative coupling: a versatile strategy for C–C bond formations via the oxidative activation of sp$^3$ C–H bonds. *Proc. Natl Acad. Sci. USA* **103**, 8928–8933 (2006).
19. Zhang, S.-Y., Zhang, F.-M. & Tu, Y.-Q. Direct Sp$^3$ α-C-H activation and functionalization of alcohol and ether. *Chem. Soc. Rev.* **40**, 1937–1949 (2011).
20. Zhang, Y. & Li, C.-J. Highly efficient cross-dehydrogenative-coupling between ethers and active methylene compounds. *Angew. Chem. Int. Ed.* **45**, 1949–1952 (2006).
21. Li, Z. & Li, C.-J. Catalytic allylic alkylation via the cross-dehydrogenative-coupling reaction between allylic sp$^3$ C – H and methylenic sp$^3$ C – H bonds. *J. Am. Chem. Soc.* **128**, 56–57 (2006).
22. Zhang, Y. & Li, C.-J. DDQ-mediated direct cross-dehydrogenative-coupling (CDC) between benzyl ethers and simple ketones. *J. Am. Chem. Soc.* **128**, 4242–4243 (2006).
23. Tu, W., Liu, L. & Floreancig, P. E. Diastereoselective tetrahydropyrone synthesis through transition-metal-free oxidative carbon-hydrogen bond activation. *Angew. Chem. Int. Ed.* **47**, 4184–4187 (2008).
24. Tu, W. & Floreancig, P. E. Oxidative carbocation formation in macrocycles: synthesis of the neopeltolide macrocycle. *Angew. Chem. Int. Ed.* **48**, 4567–4571 (2009).
25. Liu, L. & Floreancig, P. E. Cyclization reactions through DDQ-mediated vinyl oxazolidinone oxidation. *Org. Lett.* **11**, 3152–3155 (2009).
26. Yu, B. *et al.* A novel prins cyclization through benzylic/allylic C – H activation. *Org. Lett.* **11**, 3442–3445 (2009).
27. Liu, L. & Floreancig, P. E. Structurally and stereochemically diverse tetrahydropyran synthesis through oxidative C- H bond activation. *Angew. Chem., Int. Ed.* **49**, 3069–3072 (2010).
28. Liu, L. & Floreancig, P. E. Stereoselective synthesis of tertiary ethers through geometric control of highly substituted oxocarbenium ions. *Angew. Chem. Int. Ed.* **49**, 5894–5897 (2010).
29. Ghosh, A. K. & Cheng, X. Synthesis of functionalized 4-methylenetetrahydropyrans by oxidative activation of cinnamyl or benzyl ethers. *Tetrahedron Lett.* **53**, 2568–2570 (2012).
30. Cao, K. *et al.* A coupling reaction between tetrahydrofuran and olefins by Rh-catalyzed/Lewis acid-promoted C–H activation. *Tetrahedron Lett.* **49**, 4652–4654 (2008).
31. Jiao, Z.-W. *et al.* Organocatalytic asymmetric direct Csp$^3$-H functionalization of ethers: a highly efficient approach to chiral spiroethers. *Angew. Chem. Int. Ed.* **51**, 8811–8815 (2012).
32. Song, Z.-L., Fan, C.-A. & Tu, Y.-Q. Semipinacol rearrangement in natural product synthesis. *Chem. Rev.* **111**, 7523–7556 (2011).
33. Wang, B. & Tu, Y.-Q. Stereoselective construction of quaternary carbon stereocenters via a semipinacol rearrangement strategy. *Acc. Chem. Res.* **44**, 1207–1222 (2011).
34. Wang, S. H., Li, B. S. & Tu, Y.-Q. Catalytic asymmetric semipinacol rearrangements. *Chem. Commun.* **50**, 2393–2408 (2014).

35. Wesley, J. W. *Polyether Antibiotics: Naturally Occurring AcidIonophores.* Vol. I and II (Marcel Dekker, 1982).

36. Norcross, R. D. & Paterson, I. Total synthesis of bioactive marine macrolides. *Chem. Rev.* **95**, 2041–2114 (1995).

37. Alali, F. Q., Liu, X.-X. & McLaughlin, J. L. Annonaceous acetogenins: recent progress. *J. Nat. Prod.* **62**, 504–540 (1999).

38. Simmons, E. M. & Sarpong, R. Structure, biosynthetic relationships and chemical synthesis of the icetexane diterpenoids. *Nat. Prod. Rep.* **26**, 1195–1217 (2009).

39. Wu, Y.-B. *et al.* Constituents from Salvia species and their biological activities. *Chem. Rev.* **112**, 5967–6026 (2012).

40. Xu, G. *et al.* Three new diterpenoids from salvia przewalskiiMaxim. *Helv. Chim. Acta* **92**, 409–413 (2009).

41. Simmons, E. M., Yen, J. R. & Sarpong, R. Reconciling icetexane biosynthetic connections with their chemical synthesis: total synthesis of ( ± )-5, 6-dihydro-6α-hydroxysalviasperanol,( ± )-Brussonol and ( ± )-Abrotanone. *Org. Lett.* **9**, 2705–2708 (2007).

42. Martinez-Solorio, D. & Jennings, M. P. Convergent formal syntheses of ( ± )-Brussonol and ( ± )-Abrotanone via an intramolecular Marson-type cyclization. *Org. Lett.* **11**, 189–192 (2009).

43. Majetich, G. & Zou, G. Total synthesis of (-)-Barbatusol, ( + )-Demethylsalvicanol, (-)-Brussonol and ( + )-Grandione. *Org. Lett.* **10**, 81–83 (2008).

44. Chin, C.-L., Tran, D. D.-P., Shia, K.-S. & Liu, H.-J. The total synthesis of pygmaeocin C. *Synlett* 417–420 (2005).

45. Snieckus, V. Directed ortho metalation. Tertiary amide and O-carbamate directors in synthetic strategies for polysubstituted aromatics. *Chem. Rev.* **90**, 879–933 (1990).

46. Li, C., Lobkovsky, E. & Porco, Jr. J. A. Total synthesis of ( ± )-torreyanic acid. *J. Am. Chem. Soc.* **122**, 10484–10485 (2002).

47. Horst, M. A. V. T. *et al.* Locked chromophore analogs reveal that photoactive yellow protein regulates biofilm formation in the deep sea bacterium *Idiomarina loihiensis*. *J. Am. Chem. Soc.* **131**, 17443–17451 (2009).

48. Brandes, B. D. & Jacobsen, E. N. Enantioselective catalytic epoxidation of tetrasubstituted olefins. *Tetrahedron Lett.* **36**, 5123–5126 (1995).

49. Reetz, M. T., Westermann, J. & Kyung, S.-H. Direct geminal dimethylation of ketones and exhaustive methylation of carboxylic acid chlorides using dichlorodimethyltitanium. *Chem. Ber.* **118**, 1050–1057 (1985).

50. Taber, D. F., Nakajima, K., Xu, M. & Reheingold, A. L. Lactone-directed intramolecular Diels-Alder cyclization: synthesis of trans-dihydroconfertifolin. *J. Org. Chem.* **67**, 4501–4504 (2002).

51. Hopkins, M. H. & Overman, L. E. Stereocontrolled preparation of tetrahydrofurans by acid-catalyzed rearrangement of allylic acetals. *J. Am. Chem. Soc.* **109**, 4748–4749 (1987).

52. Hopkins, M. H., Overman, L. E. & Rishton, G. M. Stereocontrolled preparation of tetrahydrofurans from acid-promoted rearrangements of allylic acetals. *J. Am. Chem. Soc.* **113**, 5354–5365 (1991).

53. Ando, S., Minor, K. P. & Overman, L. E. Ring-enlarging cyclohexane annulations. *J. Org. Chem.* **62**, 6379–6387 (1997).

54. Cohen, F., MacMillan, D. W. C., Overman, L. E. & Romero, A. Stereoselection in the Prins-pinacol synthesis of acyltetrahydrofurans. *Org. Lett.* **3**, 1225–1228 (2001).

55. Overman, L. E. & Pennington, L. D. Strategic use of pinacol-terminated Prins cyclizations in target-oriented total synthesis. *J. Org. Chem.* **68**, 7143–7154 (2003).

56. Overman, L. E. & Velthuisen, E. J. Stereocontrolled construction of either stereoisomer of 12-Oxatricyclo [6.3.1.0$^{2, 7}$] dodecanes using Prins-Pinacol reactions. *Org. Lett.* **6**, 3853–3856 (2004).

57. Overman, L. E. & Velthuisen, E. J. Scope and facial selectivity of the prins-pinacol synthesis of attached rings. *J. Org. Chem.* **71**, 1581–1587 (2006).

58. Overman, L. E. & Tanis, P. S. Origin of stereocontrol in the construction of the 12-Oxatricyclo [6.3.1.0$^{2, 7}$] dodecane ring system by Prins − Pinacol Reactions. *J. Org. Chem.* **75**, 455–463 (2010).

## Acknowledgements

This work was supported by the NSFC (No.: 21202073, 21290180, 21272097, 21372104 and 21472077), the '111' Program of MOE, the Project of MOST (2012ZX 09201101-003) and the lzujbky-2013-236.

## Author contributions

Z.-W.J. conducted most of the experiments; Q.Z. prepared substrates for reaction scope evaluation and synthesis of ( − )-brussonol and ( − )-przewalskine E; W.-X.L. and S.J. (Wuyi University) prepared some substrates for reaction scope evaluation; Y.-Q.T., F.-M.Z., S.-H.W. and S.-Y.Z. conceptualized and directed the project, and prepared the manuscript with the assistance from all co-authors.

## Additional information

**Competing financial interests:** The authors declare no competing financial interests.

# 14

# Photoswitchable fatty acids enable optical control of TRPV1

James Allen Frank[1], Mirko Moroni[2], Rabih Moshourab[2,3], Martin Sumser[1], Gary R. Lewin[2] & Dirk Trauner[1]

Fatty acids (FAs) are not only essential components of cellular energy storage and structure, but play crucial roles in signalling. Here we present a toolkit of photoswitchable FA analogues (FAAzos) that incorporate an azobenzene photoswitch along the FA chain. By modifying the FAAzos to resemble capsaicin, we prepare a series of photolipids targeting the Vanilloid Receptor 1 (TRPV1), a non-selective cation channel known for its role in nociception. Several azo-capsaicin derivatives (AzCAs) emerge as photoswitchable agonists of TRPV1 that are relatively inactive in the dark and become active on irradiation with ultraviolet-A light. This effect can be rapidly reversed by irradiation with blue light and permits the robust optical control of dorsal root ganglion neurons and C-fibre nociceptors with precision timing and kinetics not available with any other technique. More generally, we expect that photolipids will find many applications in controlling biological pathways that rely on protein–lipid interactions.

[1] Department of Chemistry and Center for Integrated Protein Science, Ludwig Maximilians University Munich, Butenandtstrasse 5–13, Munich 81377, Germany. [2] Molecular Physiology of Somatic Sensation, Max Delbrück Center for Molecular Medicine, Berlin 13125, Germany. [3] Department of Anesthesiology, Campus Charité Mitte und Virchow Klinikum, Charité Universitätsmedizin Berlin, Augustburgerplatz 1, Berlin 13353, Germany. Correspondence and requests for materials should be addressed to D.T. (email: dirk.trauner@lmu.de).

Lipids serve not only as sources of energy and integral components of membranes but are also involved in cellular communication through participation in a variety of signalling cascades and the modulation of transmembrane proteins[1]. Over the past several decades, interest in lipid chemistry has been overshadowed by advancements in proteomic and genomic technologies. However, recent developments in lipid research, including analysis of the lipidome[2], have shed new light on the roles of these molecules at all levels of biology. Many lipids consist of fatty acids (FAs). These ancient molecular building blocks typically feature a long linear carbon chain (up to 28 carbons)[3] that often contains one or several cis- double bonds.

The Vanilloid Receptor 1 (TRPV1) is the most studied of the transient receptor potential ion channels[4,5]. This family of non-selective cation channels is renowned for its ability to respond to a wide variety of chemical and physical inputs[6,7]. TRPV1 is involved in the regulation of body temperature[8] and the transduction of painful stimuli from the periphery towards the central nervous system[9]. It is expressed in sub-populations of sensory nerve fibres within the dorsal root and trigeminal ganglia[10], where it responds to temperatures greater than 43 °C[11], protons[12], as well as environmental toxins and poisons[13,14]. Importantly, TRPV1 is modulated by a plethora of FA amides, including the endogenous arachidonic acid derivatives anandamide[15] and N-arachidonoyl dopamine[16]. Its most famous exogenous agonist is the vanilloid capsaicin (CAP), the pungent component of chilli peppers[11]. Synthetic TRPV1 agonists include olvanil[17] and arvanil[18], which are FA-derived vanilloids developed as non-pungent CAP analogues.

TRPV1 is not only involved in responses to noxious stimuli, but is also believed to initiate the neurogenic inflammatory response[19], which has made it an attractive target for novel analgesics[11]. However, these attempts have proven more challenging than anticipated, as TRPV1 is involved in a variety of other biological pathways that can lead to unwanted side effects. An agonist or antagonist that could be applied globally but activated only locally could offer a solution to this problem. In addition to this, such a tool would be highly valuable for untangling the complex interactions that TRPV1 has with other proteins, such as the serotonin (5-HT), bradykinin (BK) and recently GABA_B receptors[5,20,21].

Precision control can be achieved through photoswitchable small molecules that act as transducers between a light stimulus and protein function[22–24]. In 2013, our group was the first to place TRPV1 under reversible optical control when we developed a series of photoswitchable antagonists which could optically control TRPV1 in the presence of CAP to activate the channel[25]. Based on the structure of CAP and other TRPV1 agonists, we saw the opportunity to develop a photoswitchable TRPV1 agonist which would permit optical control over the ion channel without the use of a second factor.

In this study, we present a toolkit of photolipids that allow us to place lipid-modulated biological targets, such as TRPV1, under the precise spatial and temporal control of light. We show that AzCA4 permits optical control over TRPV1 in complex neural systems with a higher degree of spatiotemporal precision than is currently possible via other methods. More generally, this work represents the first example of the fusion of photopharmacology with lipid signalling and consequently sets the groundwork for future research in this field.

## Results

**Photolipid syntheses.** In an effort to mimic FAs with a chain length of 18 carbons, we prepared a series of 8 photoswitchable

FA derivatives, FAAzo1–8. Each of these compounds contained an azobenzene photoswitch, which allowed for controllable cis/trans-isomerization along the length of the chain (Fig. 1a). FAAzo1–8 were prepared in between two and six steps in moderate yield (Supplementary Fig. 1)[26]. In their dark-adapted state, the FAAzos existed predominantly in the trans-configuration. Ultraviolet–visible spectroscopy showed that isomerization from trans- to cis- could be achieved by irradiation at $\lambda = 365$ nm and this process could be reversed by $\lambda = 460$ nm light.

Supported by structure–activity relationships and recent structural data[27], we envisioned a unique opportunity to create a series of photoswitchable vanilloids for the optical control of TRPV1. Our approach combined the vanilloid head group of CAP with the FAAzos, which would act as a photoswitchable tail mimicking that of N-arachidonoyl dopamine, anandamide, olvanil or arvanil (Fig. 1b). The preparation of these compounds (Fig. 1c and Supplementary Fig. 2) required only a peptide coupling between vanillamine and the appropriate FAAzo to afford eight photoswitchable vanilloids, AzCA1–8, in good yields (Fig. 1b). Photoswitching of AzCA1–8 was also achieved using $\lambda = 365/460$ nm and they showed similar photoswitching properties when compared with the FAAzos (Supplementary Fig. 3). As such, both FAAzos and AzCAs could be classified as 'regular' azobenzenes that require ultraviolet-A light for isomerization to their thermally unstable cis-form.

**Optical control over TRPV1 in HEK293T cells.** We evaluated the photopharmacology of AzCA1–8 using whole-cell electrophysiology in human embryonic kidney (HEK) 293T cells transiently expressing the yellow fluorescent protein (YFP)-tagged ion channel (TRPV1-YFP)[28]. Each compound (1 μM) was continuously applied while alternating between irradiation at $\lambda = 350$ nm and at $\lambda = 450$ nm, until a steady state of activity/desensitization was achieved on photoswitching. Their relative light-dependant activities were then assessed by performing voltage ramps ( − 100 to + 100 mV over 5 s) under irradiation at both wavelengths (Fig. 2a and Supplementary Fig. 4a). Among the eight derivatives tested, three compounds—AzCA2, AzCA3 and AzCA4 (at 1 μM)—showed the most profound TRPV1 photoswitching effect (Fig. 2b and Supplementary Fig. 4b). For all three derivatives, larger currents were observed under irradiation at $\lambda = 350$ nm, indicating that these compounds had higher efficacies towards TRPV1 in their cis-configuration. At higher concentrations ( > 300 nM, bath application), smaller cellular currents were observed on application of the AzCAs in their dark-adapted states. This showed that in both configurations, the AzCAs were TRPV1 agonists; however, in all cases a larger current was observed under $\lambda = 350$ nm irradiation. Working at an optimized concentration (1 μM by puff pipette), the AzCAs could be applied to cells in the dark without any observable effect and this allowed immediate TRPV1 activation on irradiation with ultraviolet-A light (Supplementary Fig. 5). Photoswitching could be repeated over multiple cycles and only minor channel desensitization was observed (Fig. 2c).

The washout of AzCA4 with buffer was very slow and photoswitching currents persisted for minutes under constant perfusion. However, application of capsazepine (5 μM), a TRPV1 antagonist known to bind competitively against CAP, was capable of displacing AzCA4 and rapidly abolished inward currents on ultraviolet stimulation (Supplementary Fig. 6)[29,30]. In control experiments, no light-induced activity was observed in cells lacking TRPV1, before the application of an AzCA derivative (Supplementary Fig. 7a) or after the application of CAP (1 μM) (Supplementary Fig. 7b). The application of FAAzo2, FAAzo3

**a**

**b**

**c**

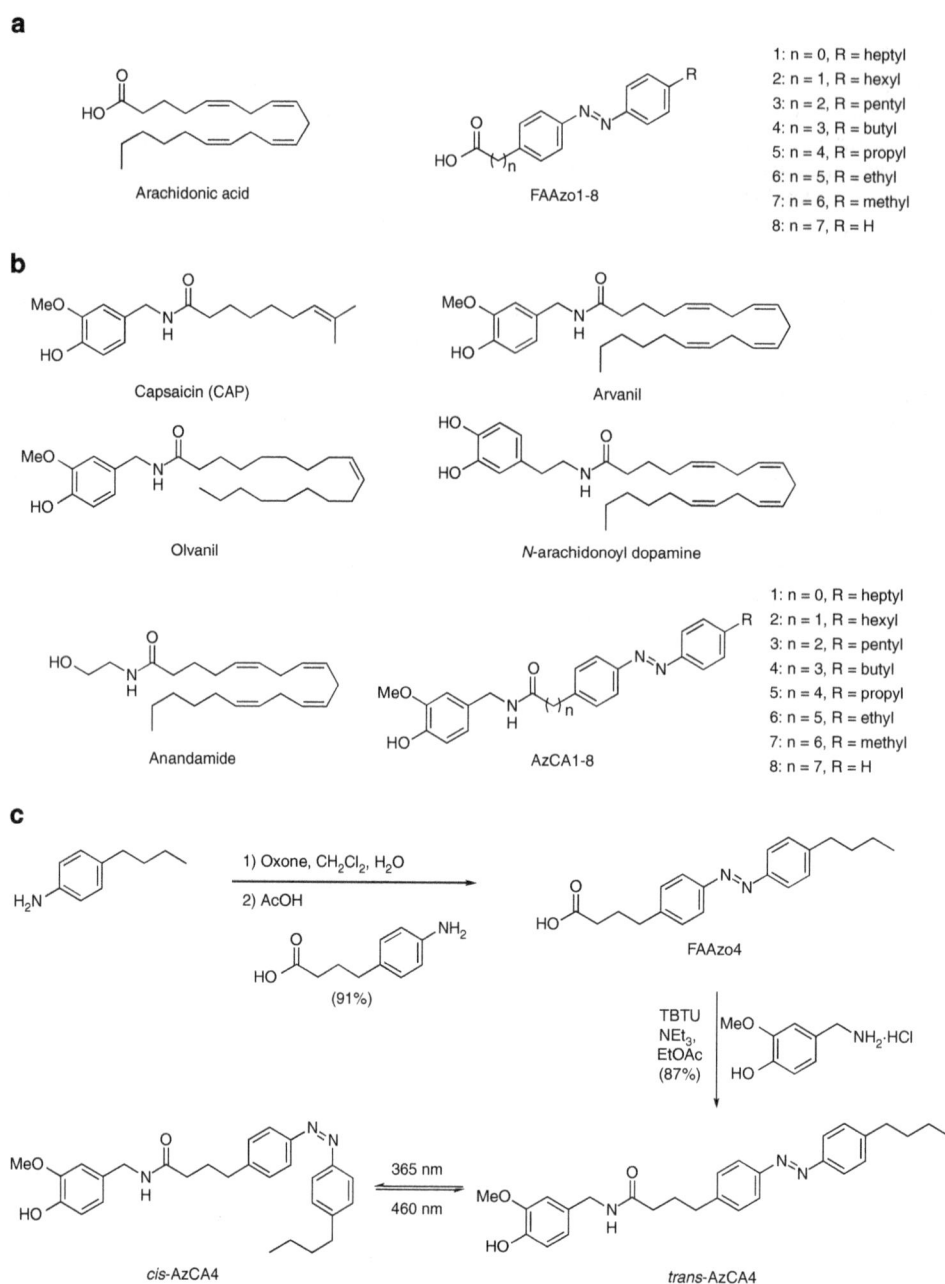

**Figure 1 | Photolipids for the optical control of TRPV1.** (**a**) Chemical structures of arachidonic acid and FAAzo1-8. (**b**) Chemical structures of TRPV1 agonists capsaicin (CAP), arvanil, olvanil, N-arachidonoyl dopamine and anandamide alongside photoswitchable vanilloids, AzCA1-8. (**c**) Chemical synthesis of AzCA4. Isomerization between cis- and trans-AzCA4 could be induced by irradiation with $\lambda = 365$ nm (trans to cis) and $\lambda = 460$ nm (cis to trans), respectively.

and FAAzo4 caused no effect (5 µM, $n \geq 3$ for each) in TRPV1-responding cells.

In voltage clamp experiments, the magnitude of the cellular currents could be controlled by adjusting the ON wavelength between $\lambda = 350$–$390$ nm. As shown by an action spectrum, the largest currents were observed under $\lambda = 350$ nm and they became smaller towards $\lambda = 390$ nm (Fig. 3a and Supplementary Fig. 8). Current clamp experiments revealed that the cellular membrane potential could be controlled in a similar manner, with $\lambda = 360$ nm yielding the largest depolarization (Fig. 3b). Through exponential curve fitting, we evaluated the effects of irradiation on the TRPV1 activation kinetics. The fastest ON response was observed at $\lambda = 360$ nm and all other $\tau_{on}$ values were normalized to this $\tau$-value for comparison (Fig. 3c). At longer wavelengths, a slower response was observed. Taken together, these results indicate that the AzCAs permit precise, optical control over the activity of TRPV1 that cannot be achieved with other lipophilic agonists, such as CAP, alone. As AzCA4 could be prepared in high yield at low cost, we decided to focus our future investigations on this compound.

**AzCA4 allowed optical control over cultured DRG neurons.** Having demonstrated that AzCA4 acts as a photoswitchable TRPV1 agonist that is relatively inactive in the dark, we next evaluated its activity in isolated wild-type (wt) mouse dorsal root ganglia (DRG) neurons using both electrophysiology and intracellular $Ca^{2+}$ imaging.

Whole-cell patch clamp electrophysiology showed that AzCA4 (200 nM) enabled reversible optical control over DRG neuronal

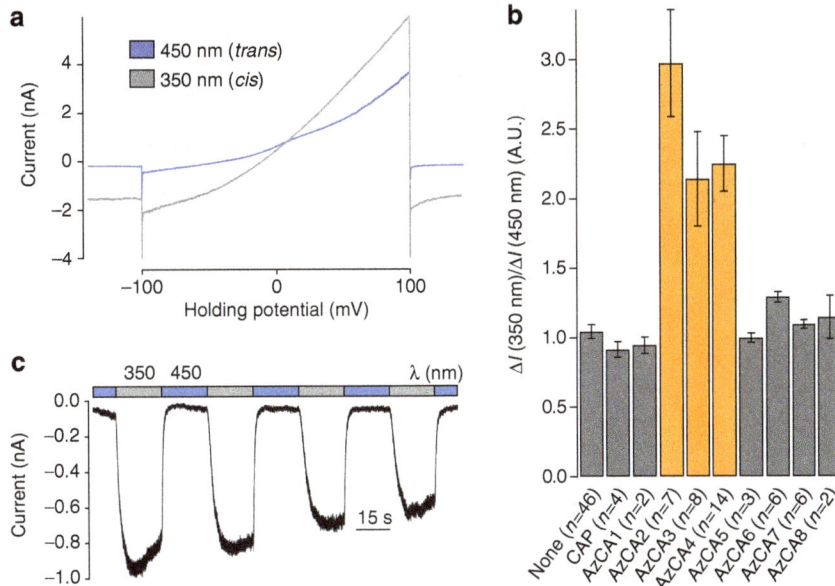

**Figure 2 | AzCAs permit optical control of TRPV1 in HEK293T cells.** HEK293T cells expressing TRPV1–YFP were observed using whole-cell patch clamp electrophysiology after the application of an AzCA derivative (1 µM). Error bars were calculated as s.e.m. (**a**) Voltage ramps were applied under both $\lambda = 350$ nm and $\lambda = 450$ nm irradiation. Larger currents were observed under $\lambda = 350$ nm irradiation (AzCA4). (**b**) AzCA2, AzCA3 and AzCA4 emerged as photoswitchable TRPV1 agonists. (**c**) When voltage clamped, photoswitching could be repeated over many cycles (AzCA3).

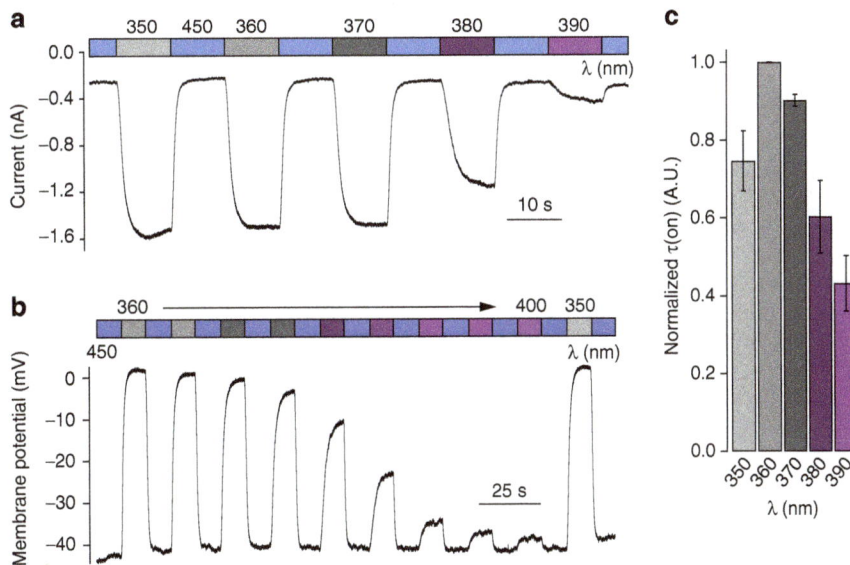

**Figure 3 | TRPV1 activation can be precisely controlled with light.** Cellular currents were observed using whole-cell patch clamp electrophysiology in HEK293T cells expressing TRPV1–YFP after the application of an AzCA derivative (1 µM). Error bars were calculated as s.e.m. (**a**) The current magnitude could be controlled by adjusting the ON wavelength between $\lambda = 350$ and 390 nm (AzCA3). (**b**) When current clamped, the membrane potential could be controlled by adjusting the ON wavelength between $\lambda = 350$ and 400 nm (AzCA4). (**c**) The activation rate could be controlled by adjusting the ON wavelength between $\lambda = 350$ and 390 nm (AzCA3, $n = 3$).

activity. Switching between $\lambda = 365$ nm and $\lambda = 460$ nm in the voltage clamp configuration at a holding potential of $-60$ mV resulted in reversible control over the transmembrane currents (Fig. 4a). Accordingly, the membrane potential and excitability of DRG neurons could be optically controlled as well. Action potential (AP) firing was reversibly switched ON and OFF by irradiation with $\lambda = 365$ nm and $\lambda = 460$ nm, respectively (Fig. 4b).

Bath application of AzCA4 (100 nM) and irradiation at $\lambda = 365$ nm (5 s) caused an increase in intracellular $Ca^{2+}$ concentration in ~30% of DRG neurons cultured from 2- to

3-week-old mice (Fig. 4c and Supplementary Fig. 9). The remaining 70% of the DRG neurons did not respond to AzCA4 but still responded to a high potassium (Hi-K$^+$) solution (100 mM) (Fig. 4d). The percentage of AzCA4 responding neurons was consistent with the percentage of TRPV1 containing DRG neurons at this stage of mouse development ($33.6 \pm 1.3\%$ at P15, $30.5 \pm 3\%$ at P22)[31].

**AzCA4 is selective for TRPV1-expressing neurons in the DRG.** TRPV1 knockout mutant mice ($Trpv1^{-/-}$) have proven to be an

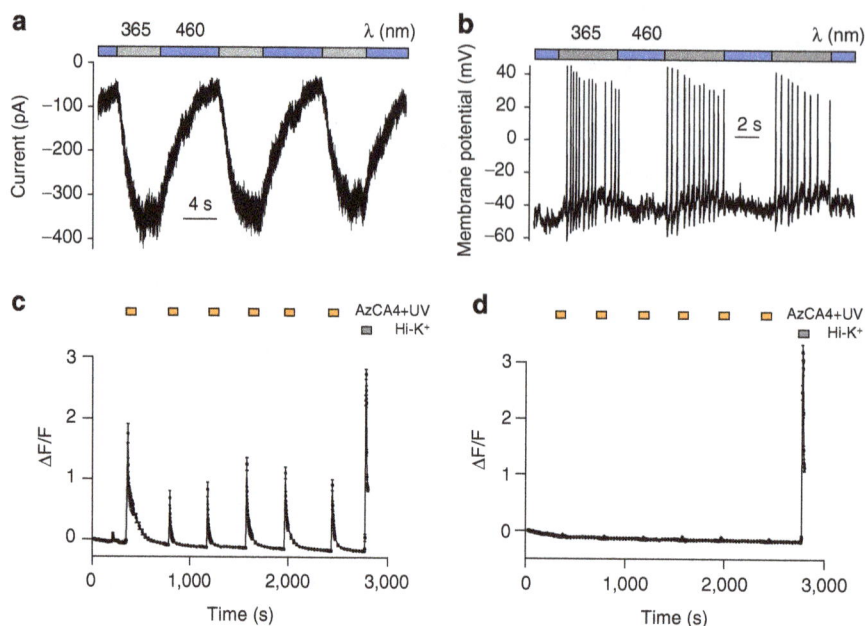

Figure 4 | AzCA4 enabled optical control over cultured wt DRG neurons. (a) After the bath application of AzCA4 (200 nM), neuronal currents could be modulated by monochromatic irradiation. When clamped at a holding potential of $-60$ mV, an inward current was observed on irradiation with $\lambda = 365$ nm and this effect could be reversed with irradiation at $\lambda = 460$ nm. (b) When clamped at a current of 0 pA and after the application of AzCA4 (200 nM), the neuronal membrane potential could be controlled. AP firing could be induced by irradiation at $\lambda = 365$ nm and halted with $\lambda = 460$ nm. This process could be repeated over many cycles. (c) Intracellular $Ca^{2+}$ imaging showed that AzCA4 (100 nM) significantly increased the level of intracellular $Ca^{2+}$ in TRPV1-positive neurons only after irradiation at $\lambda = 365$ nm (5 s). (d) Cells that did not respond to AzCA4 showed no increase in intracellular $Ca^{2+}$ after the bath application of AzCA4 and an ultraviolet pulse (6 pulses, 100 nM, 5 s at $\lambda = 365$ nm). These cells still responded to a Hi-K$^+$ solution (100 mM), suggesting that they did not possess TRPV1. Error bars were calculated as s.e.m.

excellent negative control to study the functional aspects of TRPV1 and the role of CAP-sensitive afferent neurons in the mammalian nervous system[32]. We used this mouse line to assay the selectivity of AzCA4 in DRG neurons. As expected, $Ca^{2+}$ imaging in $Trpv1^{-/-}$ mouse DRG neurons showed no neural activity in response to AzCA4 (300 nM) or CAP (300 nM). Control experiments showed that these neurons still responded to menthol (100 µM, data not shown), ATP (20 µM) and Hi-K$^+$ (100 mM) solutions (Supplementary Fig. 10). In combination, these results indicate that AzCA4 acted solely on TRPV1-positive neurons in the DRG.

**Optical control of C-fibre nociceptors.** We next evaluated the effects of AzCA4 on heat-sensitive C-fibre nociceptors (C-MH) in the *ex vivo* skin nerve preparation of the saphenous nerve. In wt mice, 19 out of 35 C-MH responded to AzCA4 photoswitching compared with none of the 11 C-MH examined in $Trpv1^{-/-}$ mice ($P < 0.001$, $\chi^2$-test; Fig. 5a). We compared the activation of AzCA4 with that of CAP on C-MH in terms of the latency time to the first AP spike, as well as the AP discharge rates. Peak discharge rates during photostimulation with $\lambda = 365$ nm surged rapidly to 9.6 spikes per second and lasted on average $11.0 \pm 6.9$ s (Fig. 5b). The average peak discharge rates of C-MH due to AzCA4 photoswitching were not significantly different from CAP stimulation (Fig. 5c). However, the mean latency time to the first AP spike on irradiation with $\lambda = 365$ nm in the presence of AzCA4 was significantly shorter ($1.9 \pm 0.5$ s; mean $\pm$ s.e.m.) when compared with the latencies to the first spike after CAP application ($9.5 \pm 3.5$ s) (Fig. 5d, Mann–Whitney test, $P < 0.05$). The discharges of C-MH in response to AzCA4 were usually transient and adapted during the 20 s photostimulation period. These responses were completely reproducible after a 5 min recovery interval, during which AzCA4 was kept on the receptive field with

no signs of desensitization (Fig. 5e). Therefore, the photoswitching of AzCA4 could selectively activate TRPV1 receptors on the peripheral terminals of cutaneous C-MH to a similar magnitude as CAP. Advantageously, AzCA4 had the ability to trigger a more rapid neuronal response and did not require washout of the drug.

**Serotonin and BK sensitized DRG neurons to AzCA4.** TRPV1-mediated hyperalgesia is a complex process, which underlies inflammatory pain[19,33]. At sites of tissue injury, a number of chemical agents are released, which cause an inflammatory response resulting in an increased thermal pain sensation in response to normally non-painful stimuli. Several G protein-coupled receptors (GPCRs) mediate TRPV1 sensitization[33,34]. BK- and 5-HT-triggered GPCR cascades are able to decrease the threshold for TRPV1 activation and increase the number of receptors at the cell surface[5,33,35,36].

We used intracellular $Ca^{2+}$ imaging in cultured wt mouse DRG neurons to determine whether these endogenous inflammatory agents still sensitized TRPV1 towards AzCA4. Our experiments showed that TRPV1 became sensitized to AzCA4 (200 nM) after the application of both BK (200 nM) and 5-HT (100 µM) (Fig. 6a,b). After 5 pulses of AzCA4 and ultraviolet irradiation, a steady state of TRPV1 desensitization was reached and the cells were washed with the sensitizing agent for 5 min (Fig. 6a). In both cases, a final pulse of AzCA4 and ultraviolet irradiation resulted in increased levels of intracellular $Ca^{2+}$ when compared with the previous pulse. Experiments with both inflammatory agents produced similar results to those achieved when using CAP as the channel agonist (Fig. 6b and Supplementary Fig. 11). These results suggest that AzCA4 could be used as a tool to study the involvement of TRPV1 in inflammatory pain.

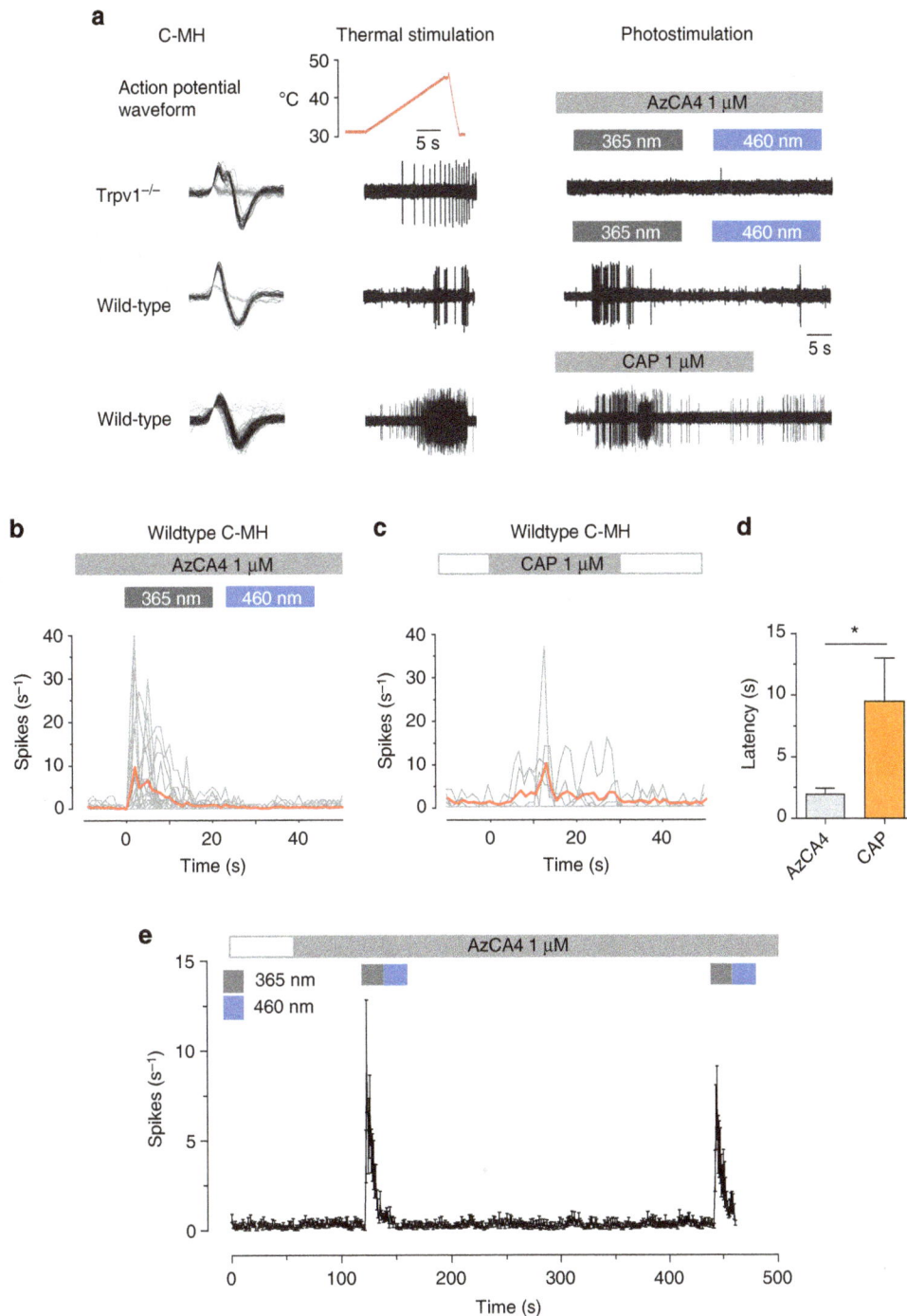

**Figure 5 | Optical activation of TRPV1 in C-fibre nociceptors.** (**a**) A typical sample trace showing the activation of a heat-sensitive C-fibre nociceptor (C-MH) by optical switching of AzCA4 (1 μM) with $\lambda = 365$ nm light and CAP (1 μM) application in wt mice. No photostimulation occurred in $Trpv1^{-/-}$ mice. (**b,c**) The average AP spiking rate (red) and the spiking rate for individual C-MHs (grey) in response to photostimulation in the presence of AzCA4 (1 μM, $n = 19$), and activation by CAP (1 μM, $n = 6$). (**d**) The latencies to the occurrence of the first AP spikes after photostimulation and after the application of CAP (*$P < 0.05$, Mann–Whitney test). (**e**) A second photostimulation in the presence of AzCA4 by $\lambda = 365$ nm light can be triggered 5 min after the first photostimulation; data are represented as the mean and s.d. of the number of spikes per 1s bins of all activated C-MH.

**AzCA4 is compatible with genetic tools**. Genetically encoded $Ca^{2+}$ indicators, such as GCaMP3 (ref. 37), have proven very useful for the study of neuronal activity *in vitro* and *in vivo*. We therefore tested the ability of AzCA4 to work in combination with this genetic tool. We crossed a mouse line that expresses the Cre recombinase under the promoter of TRPV1 (ref. 38), with a reporter mouse line that expresses GCaMP3 in a Cre-dependant manner ($Trpv1^{Cre/GCaMP3}$). As the majority of DRG neurons

express TRPV1 at embryonic stage E14.5 (ref. 31), we observed that 90% of the DRG neurons in culture showed a faint basal fluorescence, confirming that recombination had occurred in most neurons at an earlier developmental stage.

However, on application of CAP or AzCA4 alongside ultraviolet irradiation, only 30% of the neurons showed a $Ca^{2+}$-dependent increase in fluorescence. This confirmed not only the correct function and localization of GCaMP3 in

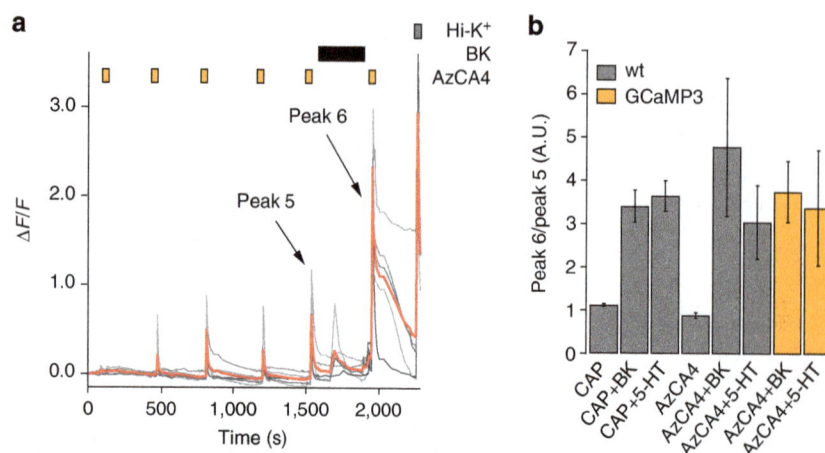

**Figure 6 | Serotonin and BK sensitized TRPV1 to AzCA4.** Intracellular $Ca^{2+}$ imaging showed that both BK and 5-HT sensitized TRPV1 to CAP (100 nM) and AzCA4 (200 nM), with ultraviolet irradiation ($\lambda = 365$ nm, 5 s) in cultured wt DRG neurons. (**a**) After the application of BK (200 nM), an increased intensity and duration of $Ca^{2+}$ influx was observed on application of AzCA-4 with ultraviolet irradiation when compared with the previous pulse. The neurons responded to a Hi-$K^+$ solution (100 mM). Shown here are five representative traces (grey) and average $\Delta F/F$ value (red). (**b**) TRPV1 sensitization experiments in wt mouse DRG neurons (grey) and the $Trpv1^{Cre/GCaMP3}$ mouse line (orange). The results were plotted as the ratio of peak heights for Peak 6/Peak 5 for the wt mouse as follows: CAP with no sensitization agent ($n = 520$); CAP with BK ($n = 210$); CAP with 5-HT ($n = 175$); AzCA4 with no sensitizing agent ($n = 36$); AzCA4 with BK ($n = 15$); and AzCA4 with 5-HT ($n = 12$). For the $Trpv1^{Cre/GCaMP3}$ mouse line, the results are plotted for the following: AzCA4 with BK ($n = 28$) and AzCA4 with 5-HT ($n = 10$). Error bars were calculated as s.e.m.

TRPV1-positive DRG neurons, but also that this small molecular photoswitch, AzCA4, can be used in combination with a genetic tool.

To further confirm the applicability of AzCA4 and the use of genetically encoded $Ca^{2+}$ indicators to address physiologically relevant issues, we repeated the sensitization experiments on the $Trpv1^{Cre/GCaMP3}$ mouse line (Fig. 6b, orange). These results again confirmed that TRPV1 could be sensitized to AzCA4 by BK or 5-HT, even in the $Trpv1^{Cre/GCaMP3}$ mouse line.

**QX-314 can be selectively transported into DRG neurons.** QX-314 is a permanently charged lidocaine derivative that blocks $Na^+$ channels from the intracellular side but is unable to penetrate the plasma membrane due to its charged nature[39]. It has been shown that QX-314, and its photoswitchable derivative QAQ[40], could be shuttled into cells via TRPV1 when co-applied with CAP, to open the channel[41]. We hypothesized that intracellular $Ca^{2+}$ build-up is not only caused by $Ca^{2+}$ influx through TRPV1 but is a consequence of AP firing leading to $Ca^{2+}$ release from intracellular stores. We showed AzCA4 (200 nM) in combination with $\lambda = 365$ nm irradiation, opened TRPV1 and allowed QX-314 to enter TRPV1-positive neurons in the DRG. The cells were first washed with a Hi-$K^+$ solution (40 mM) causing massive electrical activity. After the cells recovered, four pulses of AzCA4 (200 nM, 5 s at $\lambda = 365$ nm) reached a steady state of TRPV1 desensitization. Next, AzCA4 was co-applied with QX-314 (5 mM), followed by another pulse of AzCA4 and a final pulse of the Hi-$K^+$ solution. By comparing the peak heights of the first and final Hi-$K^+$ pulses (HK$_1$ and HK$_2$, respectively), we observed that the cells that were responsive to AzCA4 showed a lower HK$_2$/HK$_1$ ratio when compared with cells that did not respond to AzCA4 (Fig. 7). These results indicate that in combination with AzCA4, TRPV1 could be used as an import channel to localize a charged anaesthetic in TRPV1-positive cells, only.

## Discussion

In this study, we present eight photoswitchable FAs, the FAAzos. We expect that these lipophilic modules will allow

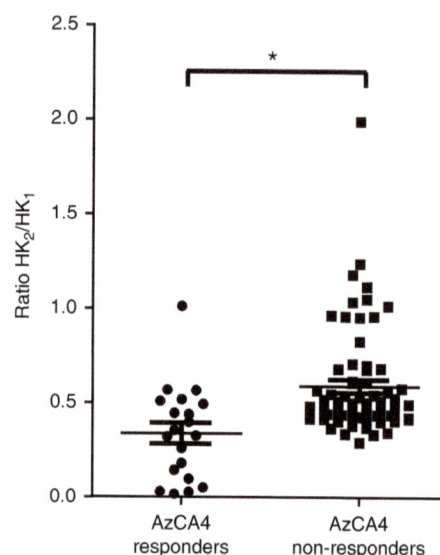

**Figure 7 | QX-314 could be shuttled into TRPV1-positive DRG neurons using AzCA4.** After the co-application of AzCA4 (200 nM) and QX-314 (5 mM), the peak heights of the $Ca^{2+}$ signals of the Hi-$K^+$ (40 mM) pulses before (HK$_1$) and after (HK$_2$) drug application were compared. The results were plotted as the peak height ratio of two pulses, HK$_2$/HK$_1$, calculated for both AzCA4 responding and AzCA4 non-responding neurons. The error bars were calculated as s.e.m.

us to place a variety of lipid-modulated biological targets under photopharmacological control. The FAAzos are FAs mimetics, resembling in particular highly unsaturated FAs such as arachidonic acid. These compounds could be used as molecular building blocks for the construction of more complicated photolipids, which could facilitate the optical control of a wide range of ion channels, GPCRs and enzymes associated with FA signalling.

We hypothesized that an azobenzene photoswitch would be best suited towards integration into the FA backbone, as its hydrophobic nature may cause only a minimal disruption to the properties of a natural aliphatic chain. The FAAzos allow the position of the switch to be fine-tuned to complement structure–activity relationships between the ligand and its target. As a first installment illustrating this concept, we showed that the FAAzos could be incorporated into other photolipids through a simple amide coupling reaction. In HEK293T cells, three AzCAs stood out as the most promising candidates to enable the optical control of TRPV1 and showed significant, light-dependant efficacy at concentrations as low as 100 nM. All three compounds were more potent in their *cis*-configuration. Owing to the structural similarities between the AzCAs and other vanilloid TRPV1 agonists, we presume that the AzCAs act on the same vanilloid binding site as CAP, olvanil and arvanil. This is further supported by the requirement of the vanilloid headgroup, as was shown by the inactivity of FAAzo2–4. Channel activation by AzCA4 can also be completely blocked by the application of capsazepine, a known competitive antagonist for CAP, suggesting a common binding site and mode of activation.

We then demonstrated that AzCA4 was a powerful modulator of DRG neurons in wt mice and the $Trpv1^{Cre/GCaMP3}$ mouse line. Control experiments in untransfected HEK293T cells and $Trpv1^{-/-}$ mice showed no response to AzCA4, even on light stimulation. These results rule out the possibility of action through off-target mechanisms. This characteristic is essential for the study of signal transduction in the nociceptive neurons involved in inflammatory pain. To this end, we envision that AzCA4 could be used to study TRPV1-mediated hyperalgesia and the physiological processes involved with the inflammatory state that occurs at sites of tissue injury. We showed that application of components of the so-called 'inflammatory soup', such as BK or 5-HT, could sensitize TRPV1 to AzCA4 in DRG neurons. Furthermore, we proved that AzCA4 could be used in conjunction with genetic tools such as the GCaMP3 $Ca^{2+}$ indicator selectively expressed in TRPV1-positive mouse neurons.

Importantly, AzCA4 allowed for greater temporal control over TRPV1 than that which can be achieved by other small-molecule agonists such as CAP, olvanil and arvanil. By working at an optimized concentration, AzCA4 showed no activity towards TRPV1 in its *trans*-configuration, but rapid TRPV1 activation was observed when it was isomerized to its *cis*-configuration. This ON and OFF effect became even more pronounced in neuronal systems when the nonlinear, 'all or nothing,' nature of the AP took effect. Previous studies have suggested that the rate of TRPV1 activation determines the balance between agonist potency and pungency[42–45]. Molecular weight and lipophilicity normally define the pharmacodynamics of TRPV1 agonists[42,43]. In the case of AzCAs, light provides another level of control. We showed that the magnitude and rate of cellular activation could be precisely tuned by adjusting the ON wavelength between $\lambda = 350$ and 390 nm. Therefore, AzCAs could provide a platform for the further understanding of hyperalgesia and could lead to the development of new anaesthetics.

When using CAP to stimulate C-fibres in the saphenous nerve, a relatively slow increase in AP firing was observed (Fig. 5b,c). This effect was probably caused by the time required for CAP to diffuse through the skin and plasma membrane to reach the vanilloid binding site[46–48]. We showed that AzCA4 could be applied to neurons as its dark-adapted, relatively inactive configuration. On isomerization, TRPV1 was activated and AP firing was immediately observed. It is this characteristic that makes AzCA4 a useful tool for the study of nociception, a process that relies on the rapid transmission of noxious stimuli from the periphery towards the coordinating centres of the central nervous system. AzCA4 also possesses significant advantages when

compared with other small molecules that have been used to place TRPV1 under the control of light. Caged CAP is a useful tool that has increased the level of control with which we are able to activate TRPV1 (ref. 49). However, compound uncaging and TRPV1 activation is a non-reversible, one-shot process. Repeated activation by uncaging relies on fast-acting transporters or deactivating enzymes to clear the synapse of the free ligand[50]. The fact that CAP and its analogues exhibit long-lasting effects suggests that they do not undergo transporter-mediated reuptake or significant enzymatic hydrolysis. These aspects are circumvented by the use of AzCA4, which allows for successive rounds of activation/inactivation without the requirement for washout of the drug. Previously, we showed that photoswitchable TRPV1 antagonists can optically control the activity of a constitutively active agonist on photoswitching[25]. In comparison, AzCA4 greatly simplifies this system and is capable of activating TRPV1 directly. This permits the optical control of neuronal excitability without the use of a second factor. Advantageously, AzCA4 is relatively inactive in the dark; therefore, a more rapid and reproducible initiation of activity can be achieved after it has distributed itself uniformly within more complex tissues.

In conclusion, this study provides the first application of photopharmacology to lipid signalling. Given the ubiquitous distribution of FA derivatives at all levels of nature, we envision that the FAAzos and their conjugates will emerge as broadly applicable tools for the optical control of biological functions which rely on protein–lipid interactions.

## Methods

**Whole-cell electrophysiology in HEK293T cells.** HEK293T cells (obtained from the Leibniz-Institute DSMZ: 305) were incubated at 37 °C (10% $CO_2$) in DMEM medium + 10% fetal bovine serum and were split at 80%–90% confluency. For cell detachment, the medium was removed and the cells were washed with calcium-free PBS buffer and treated with trypsin for 2 min at 37 °C. The detached cells were diluted in growth medium and plated on acid-etched coverslips coated with poly-L-lysine in a 24-well plate. Cells (50,000) were added to each well in 500 μl standard growth medium along with the DNA (per coverslip: 500 ng TRPV1-YFP[28]) and JetPRIME transfection reagents, according to the manufacturer's instructions (per coverslip: 50 μl JetPRIME buffer, 0.5 μl JetPRIME transfection reagent). The transfection medium was exchanged for normal growth media 4 h after transfection and electrophysiological experiments were carried out 20–40 h later. Whole-cell patch clamp experiments were performed using a standard electrophysiology setup equipped with a HEKA Patch Clamp EPC10 USB amplifier and PatchMaster software (HEKA Electronik). Micropipettes were generated from 'Science Products GB200-F-8P with filament' pipettes using a Narishige PC-10 vertical puller. The patch pipette resistance varied between 5 and 9 MΩ. The bath solution contained the following: Solution A (in mM: 150 NaCl, 6.0 CsCl, 1.0 $MgCl_2$, 1.5 $CaCl_2$, 10 HEPES and 10 glucose (adjusted to pH 7.4 with 3 M NaOH))[51] or Solution B (in mM: 140 NaCl, 5 KCl, 5 HEPES, 1 MgCl and 5 glucose (adjusted to pH 7.4 with 3 M NaOH)). The pipette solution contained th following: Solution A' (in mM: 150 NaCl, 3 $MgCl_2$, 10 HEPES and 5 EGTA (adjusted to pH 7.2 with 3 M NaOH)) or solution B' (in mM: 100 K-gluconate, 40 KCl, 5 HEPES, 5 MgATP and 1 MgCl (adjusted to pH 7.2 with 1 M KOH)). The cells were first visualized to contain TRPV1-YFP by irradiation at $\lambda = 480$ nm using a Polychrome V (Till Photonics) monochromator. All cells had a leak current between $-15$ to $-300$ pA on breakin at $-60$ mV. All voltage clamp measurements were carried out at a holding potential of $-60$ mV. The cells were held at 0 pA for current clamp measurements. The compounds were applied by puff pipette using a 'Toohey Spritzer pressure system IIe' at 25 psi. The puff pipette resistance varied between 3 and 5 MΩ. All experiments were performed at room temperature.

**Determination of AzCA photoswitching properties.** The photoswitching properties of the prepared compounds were assessed using whole-cell voltage clamp electrophysiology in HEK293T cells transiently expressing TRPV1-YFP[28]. The compounds were dissolved as stock solutions in dimethylsulfoxide (DMSO; 2–6 mM) and then diluted into warmed extracellular solution at a concentration of 1 μM. The cell was held at $-60$ mV and voltage ramps ($-100$ to $+100$ mV over 5 s) were applied under both $\lambda = 450$ nm and $\lambda = 350$ nm irradiation. The AzCA derivative (1 μM) was then constantly applied (puff pipette application) while switching between the two wavelengths until a steady state of activity/desensitization was observed on photoswitching (Supplementary Fig. 4a). The voltage ramps were applied again under each wavelength and the current change ($\Delta I$) between the baseline ($-60\,mV$) and the ramp maximum ($+100\,mV$) was recorded (Supplementary Fig. 4b). $\Delta I(350\,nm)$ was normalized to $\Delta I(450\,nm)$ and this potentiation factor was averaged over multiple cells and plotted in Fig. 2b. TRPV1 activation and inactivation kinetics were determined by exponential curve fitting in Igor Pro.

**Dorsal root ganglion neuronal culture.** DRG were quickly dissected, collected in ice-cold DRG medium and digested in $1\,mg\,ml^{-1}$ Collagenase IV (Gibco) at $37\,^{\circ}C$ for 50 min, to dissociate the tissue, followed by incubation in 0.05% trypsin (Gibco) in PBS at $37\,^{\circ}C$ for 15 min. The trypsin was removed and cells were re-suspended in 1 ml of DRG medium. After gentle trituration, the DRGs were loaded onto a 2-ml BSA pillow and centrifuged at $250g$ for 10 min, to separate the myelin and debris. The resulting cell pellet was suspended in $50\,\mu l$ fresh DRG medium and plated onto poly-D-lysine ($100\,\mu g\,ml^{-1}$) and laminin ($10\,\mu g\,ml^{-1}$)-coated coverslips. The cells were flooded with medium 30 min after plating.

**Whole-cell electrophysiology in cultured DRG neurons.** Electrophysiological recordings of DRG neurons from 14- to 21-day-old mice were performed using a HEKA 10 amplifier (HEKA Instrument) and an ITC Analog Digital Converter (HEKA) in whole-cell voltage clamp or current clamp configuration. The currents were filtered with a built-in 5 kHz 8-pole Bessel filter and digitized at 50 kHz. The currents were analysed using Clampfit 10.3 (Molecular Devices) and graphs were plotted in Prism 5 (Graphpad). Experiments were performed 1–3 days after plating. The Trpv1$^{Cre}$ and the GCaMP3 reporter mice were acquired from Jackson Laboratory. DRG neurons were prepared for imaging from 25-week-old $Trpv1^{Cre/GCaMP3}$ mice.

The extracellular solution contained in mM: 150 NaCl, 5 KCl, 10 HEPES, 10 glucose, 2 $CaCl_2$, 1 $MgCl_2$. The intracellular solution contained (values in mM): 130 KCl, 10 HEPES, 10 EGTA, 1 $CaCl_2$, 1 $MgCl_2$, 2 MgATP, 1 NaGTP, 4 NaCl and 4 PhosphoCreatine.

Thick-walled electrodes (Harvard Apparatus, $1.17 \times 0.87\,mm$, external and internal diameter, respectively) were pulled with a Sutter P-197 puller to a final resistance of 3–5 MΩ. After the breakthrough, the intracellular solution was allowed to dialyse the intracellular medium for at least 1 min before the beginning of the recordings. Series resistance compensation reached values between 70 and 90%. Neurons were selected to have a leak current $< -100\,pA$ on breakin at $-60\,mV$. All experimental procedures were carried out in accordance with the State of Berlin Animal Welfare requirements and were approved by this authority.

**Intracellular calcium imaging.** DRG neurons plated on a 5-mm glass coverslip were placed in a recording chamber of $300\,\mu l$ volume (Harvard Apparatus) and were continuously perfused with extracellular solution at a rate of $2\,ml\,min^{-1}$. The wt neuron cells were loaded with Cal-520 ($5\,\mu M$, AAT-Bioquest) calcium dye for 1 h at $37\,^{\circ}C$ in the presence of 0.02% pluronic acid dissolved in Ringer's solution (values in mM): 140 NaCl, 5 KCl, 2 $CaCl_2$, 2 $MgCl_2$, 10 HEPES, 10 glucose, adjusted to pH 7.4. CAP (100 nM, Tocris) was dissolved in extracellular solution from a stock concentration of 10 mM in ethanol. Fluorescent images were acquired with Metafluor Software (Molecular Devices) and analysed in Clampfit. All experiments were performed at room temperature. The dye excitation was performed at $\lambda = 480\,nm$. The results were plotted as the change in fluorescence over baseline fluorescence ($\Delta F/F$) as a function of time (s).

**Ex-vivo skin nerve preparation.** The skin-nerve preparation was used as previously described, to record from single primary afferents[52]. Mice were killed with $CO_2$ inhalation for 2–4 min, followed by cervical dislocation. The saphenous nerve and the shaved skin of the hind limb were dissected free and placed in an organ bath at $32\,^{\circ}C$. The chamber was perfused with a synthetic interstitial fluid (SIF buffer) composed of (in mM): 123 NaCl, 3.5 KCl, 0.7 $MgSO_4$, 1.7 $NaH_2PO_4$, 2.0 $CaCl_2$, 9.5 Na-gluconate, 5.5 glucose, 7.5 sucrose and 10 HEPES at a pH of 7.4. The skin was placed with the corium side up in the organ bath for pharmacological application to the receptive fields of single sensory units. The saphenous nerve was placed in an adjacent chamber on a mirror, and under microscopy fine filaments were teased from the nerve and placed on the recording electrode. Electrical isolation was achieved with mineral oil. Signals from the filaments were amplified (Neurolog System, Digitimer Ltd) and sampled using a data acquisition system (PowerLab 4.0, ADInstruments). The receptive fields of individual mechanically sensitive C-fibre units were identified by manually probing the skin with the blunted tip of a glass probe. Their conduction velocities (calculated by dividing conduction distance over electrical latency for the first spike) were determined using an electrical stimulator (1 MΩ). All C-fibres studied were identified by mechanical probing and were heat sensitive (C-MH). The conduction velocities were in the C-fibre range ($<1\,m\,s^{-1}$). The thermal sensitivity of identified C-fibres was tested with a computer-controlled peltier device ($3 \times 5\,mm$, Yale University Medical School, Medical Instruments, New Haven, USA). A heat ramp was delivered from 32 to $48\,^{\circ}C$ at a rate of $1\,^{\circ}C\,s^{-1}$ to the mechanically localized receptive fields of single C-fibres. The receptive fields of heat-sensitive C-fibres were isolated with a metal cylinder ring (5 mm inner diameter, 10 mm in height and 1.44 g weight) for the administration of the drugs (AzCA4 or CAP). The metal

ring was tested for leakage before drug application. A stock solution of AzCA4 was prepared in DMSO and diluted in SIF buffer. One hundred microlitres of a 1-μM AzCA4 solution dissolved in SIF buffer was applied into the ring. The pharmacological testing protocol with AzCA4 had four distinct successive phases: (1) 60 s recording time before the application of AzCA4; (2) 60 s recording after application of AzCA4; (3) 20 s of recording during photostimulation of AzCA4 with light-emitting diode (LED) light, $\lambda = 365\,nm$ (Mic-LED-365, Prizmatix); (4) 20 s of recording during photoinhibition of AzCA4 with LED light, $\lambda = 460\,nm$ (UHP-LED-460, Prizmatix). AzCA4 was kept in the ring for 300 s and a second photostimulation with LED light 360 nm for 20 s was applied to test the reproducibility of the initial responses. One hundred microlitres of CAP (1 μM, Sigma) were administered onto the receptive field of another set of heat-sensitive C-fibres for 20 s before washout. Recordings were obtained for 1 min before and 2 min after CAP administration. Spikes were discriminated offline, with the spike histogram extension of the software. Data were obtained from six skin-nerve preparations (2 $Trpv1^{-/-}$ and 4 wt C57Bl/6 N mice). All experimental procedures were carried out in accordance with the State of Berlin Animal Welfare requirements and were approved by this authority.

**Compound switching.** Compound switching for electrophysiology in HEK293T cells was achieved using a Polychrome V (Till Photonics) monochromator (intensity versus wavelength screen Supplementary Fig. 12) and the light beam was guided via a fibre-optic cable through the microscope objective and operated by the amplifier and PatchMaster software (HEKA Electronik).

Compound switching for electrophysiology in DRG neurons was achieved using a Prizmatix Mic-LED-365 high-power ultraviolet LED light source for illumination at $\lambda = 365\,nm$ and the Prizmatix UHP-Mic-LED-460 ultra-high-power LED light source for illumination at $\lambda = 460\,nm$. The light beam was guided by a fibre-optic cable and pointed directly towards the cells from above at an angle of about 45° from the side. The distance between the end of the cable and the cells was no greater than 2 cm. Ultraviolet illumination during intracellular calcium imaging was also performed using this light source.

**Compound synthesis and characterization.** All reagents and solvents were purchased from commercial sources (Sigma-Aldrich, TCI Europe N.V., Strem Chemicals and so on) and were used without further purification unless otherwise noted. Tetrahydrofuran was distilled under a $N_2$ atmosphere from Na/benzophenone before use. Triethylamine was distilled under a $N_2$ atmosphere from $CaH_2$ before use. Further dry solvents such as ethyl acetate, benzene, dichloromethane, toluene, ethanol and methanol were purchased from Acros Organics as 'extra dry' reagents and used as received. Solvents were degassed by sparging the freshly distilled solvent with argon gas in a Schlenk flask under ultrasonication using a Bandelin Sonorex RK510H ultrasonic bath for 20 min before use. Reactions were monitored by thin-layer chromatography on pre-coated, Merck Silica gel 60 $F_{254}$ glass-backed plates and the chromatograms were first visualized by ultraviolet irradiation at 254 nm, followed by staining with aqueous ninhydrin, anisaldehyde or ceric ammonium molybdate solution and finally gentle heating with a heat gun. Flash silica gel chromatography was performed using silica gel ($SiO_2$, particle size 40-63 μm) purchased from Merck.

Ultraviolet visible spectra were recorded using a Varian Cary 50 Bio UV-Visible Spectrophotometer with Helma SUPRASIL precision cuvettes (10 mm light path). All compounds were dissolved at a concentration of 25 μM in DMSO. Switching was achieved using a Polychrome V (Till Photonics) monochromator (intensity versus wavelength screen Supplementary Fig. 12). The illumination was controlled using PolyCon3.1 software and the light was guided through a fibre optic cable with the tip pointed directly into the top of the sample cuvette.

All NMR spectra were measured on a BRUKER Avance III HD 400 (equipped with a CryoProbe). Multiplicities in the following experimental procedures are abbreviated as follows: s, singlet; d, doublet; t, triplet; q, quartet; quint, quintet; sext, sextet; hept, heptet; br, broad; m, multiplet. Proton chemical shifts are expressed in parts per million (p.p.m., $\delta$ scale) and are referenced to the residual protium in the NMR solvent (CDCl$_3$: $\delta = 7.26$, D$_6$-DMSO: $\delta_H = 2.50$). Carbon chemical shifts are expressed in p.p.m. ($\delta$ scale) and are referenced to the carbon resonance of the NMR solvent (CDCl$_3$: $\delta = 77.16$, D$_6$-DMSO: $\delta = 39.52$). Note: owing to the trans/cis isomerization of some compounds containing an azobenzene functionality, more signals were observed in the $^1H$ and $^{13}C$ spectra than would be expected for the pure trans-isomer. Only signals for the major trans-isomer are reported; however, the identities of the remaining peaks were verified by two-dimensional correlation spectroscopy, heteronuclear single quantum coherence and heteronuclear multiple bond correlation experiments. The atom-numbering system is defined as depicted in Supplementary Fig. 13. All relevant synthetic details and NMR spectra can be found in Supplementary Figs 14–46 and the Supplementary Methods.

Infrared spectra were recorded as neat materials on a PERKIN ELMER Spectrum BX-59343 instrument. For detection, a SMITHS DETECTION DuraSam-plIR II Diamond ATR sensor was used. The measured wave numbers are reported in $cm^{-1}$.

Low- and high-resolution electron ionization mass spectra were obtained on a MAT CH7A mass spectrometer. Low- and high-resolution electrospray ionization mass spectra were obtained on a Varian MAT 711 MS instrument operating in either positive or negative ionization modes.

## References

1. Wang, X. Lipid signaling. *Curr. Opin. Plant Biol.* **7**, 329–336 (2004).
2. Shevchenko, A. & Simons, K. Lipidomics: coming to grips with lipid diversity. *Nat. Rev.* **11**, 593–598 (2010).
3. *IUPAC Gold Book—Fatty Acids* 1307 (Blackwell Scientific Publications, 1997).
4. Szallasi, A., Cortright, D. N., Blum, C. A. & Eid, S. R. The vanilloid receptor TRPV1: 10 years from channel cloning to antagonist proof-of-concept. *Nat. Rev. Drug Discov.* **6**, 357–372 (2007).
5. Chuang, H. H. et al. Bradykinin and nerve growth factor release the capsaicin receptor from PtdIns(4,5)P2-mediated inhibition. *Nature* **411**, 957–962 (2001).
6. Clapham, D. E. TRP channels as cellular sensors. *Nature* **426**, 517–524 (2003).
7. Venkatachalam, K. & Montell, C. TRP channels. *Annu. Rev. Biochem.* **76**, 387–417 (2007).
8. Gavva, N. R. Body-temperature maintenance as the predominant function of the vanilloid receptor TRPV1. *Trends Pharmacol. Sci.* **29**, 550–557 (2008).
9. Tominaga, M. et al. The cloned capsaicin receptor integrates multiple pain-producing stimuli. *Neuron* **21**, 531–543 (1998).
10. Holzer, P. Local effector functions of capsaicin-sensitive sensory nerve endings: involvement of tachykinins, calcitonin gene-related peptide and other neuropeptides. *Neuroscience* **24**, 739–768 (1988).
11. Caterina, M. J. et al. The capsaicin receptor: a heat-activated ion channel in the pain pathway. *Nature* **389**, 816–824 (1997).
12. Jordt, S. E., Tominaga, M. & Julius, D. Acid potentiation of the capsaicin receptor determined by a key extracellular site. *Proc. Natl Acad. Sci. USA* **97**, 8134–8139 (2000).
13. Cuypers, E., Yanagihara, A., Karlsson, E. & Tytgat, J. Jellyfish and other cnidarian envenomations cause pain by affecting TRPV1 channels. *FEBS Lett.* **580**, 5728–5732 (2006).
14. Siemens, J. et al. Spider toxins activate the capsaicin receptor to produce inflammatory pain. *Nature* **444**, 208–212 (2006).
15. Ross, R. Anandamide and vanilloid TRPV1 receptors. *Br. J. Pharmacol.* **140**, 790–801 (2003).
16. Huang, S. M. et al. An endogenous capsaicin-like substance with high potency at recombinant and native vanilloid VR1 receptors. *Proc. Natl Acad. Sci. USA* **99**, 8400–8405 (2002).
17. Brand, L. et al. NE-19550: a novel, orally active anti-inflammatory analgesic. *Drugs Exp. Clin. Res.* **13**, 259–265 (1987).
18. Melck, D. et al. Unsaturated long-chain N-acyl-vanillyl-amides (N-AVAMs): vanilloid receptor ligands that inhibit anandamide-facilitated transport and bind to CB1 cannabinoid receptors. *Biochem. Biophys. Res. Commun.* **262**, 275–284 (1999).
19. Davis, J. B. et al. Vanilloid receptor-1 is essential for inflammatory thermal hyperalgesia. *Nature* **405**, 183–187 (2000).
20. Hu, W. P., Guan, B. C., Ru, L. Q., Chen, J. G. & Li, Z. W. Potentiation of 5-HT3 receptor function by the activation of coexistent 5-HT2 receptors in trigeminal ganglion neurons of rats. *Neuropharmacology* **47**, 833–840 (2004).
21. Hanack, C. et al. GABA blocks pathological but not acute TRPV1 pain signals. *Cell* **160**, 759–770 (2015).
22. Velema, W. A., Szymanski, W. & Feringa, B. L. Photopharmacology: beyond proof of principle. *J. Am. Chem. Soc.* **136**, 2178–2191 (2014).
23. Mourot, A., Tochitsky, I. & Kramer, R. H. Light at the end of the channel: optical manipulation of intrinsic neuronal excitability with chemical photoswitches. *Front. Mol. Neurosci.* **6**, 5 (2013).
24. Fehrentz, T., Schönberger, M. & Trauner, D. Optochemical genetics. *Angew. Chem. Int. Ed. Engl.* **50**, 12156–12182 (2011).
25. Stein, M., Breit, A., Fehrentz, T., Gudermann, T. & Trauner, D. Optical control of TRPV1 channels. *Angew. Chem. Int. Ed. Engl.* **52**, 9845–9848 (2013).
26. Morgan, C. G., Thomas, E. W., Yianni, Y. P. & Sandhu, S. S. Incorporation of a novel photochromic phospholipid molecule into vesicles of dipalmitoyl-phosphatidylcholine. *Biochim. Biophys. Acta Biomembr.* **820**, 107–114 (1985).
27. Cao, E., Liao, M., Cheng, Y. & Julius, D. TRPV1 structures in distinct conformations reveal activation mechanisms. *Nature* **504**, 113–118 (2013).
28. Hellwig, N., Albrecht, N., Harteneck, C., Schultz, G. & Schaefer, M. Homo- and heteromeric assembly of TRPV channel subunits. *J. Cell Sci.* **118**, 917–928 (2005).
29. Bevan, S. et al. Capsazepine: a competitive antagonist of the sensory neurone excitant capsaicin. *Br. J. Pharmacol.* **107**, 544–552 (1992).
30. Walpole, C. S. et al. The discovery of capsazepine, the first competitive antagonist of the sensory neuron excitants capsaicin and resiniferatoxin. *J. Med. Chem.* **37**, 1942–1954 (1994).
31. Cavanaugh, D. J., Chesler, A. T., Bráz, J. M., Shah, N. M. & Basbaum, A. I. Restriction of TRPV1 to the peptidergic subset of primary afferent neurons follows its developmental downregulation in nonpeptidergic neurons. *J. Neurosci.* **31**, 10119–10127 (2011).
32. Bölcskei, K. et al. Investigation of the role of TRPV1 receptors in acute and chronic nociceptive processes using gene-deficient mice. *Pain* **117**, 368–376 (2005).
33. Huang, J., Zhang, X. & McNaughton, P. Inflammatory pain: the cellular basis of heat hyperalgesia. *Curr. Neuropharmacol.* **4**, 197–206 (2006).
34. Sugiuar, T., Bielefeldt, K. & Gebhart, G. F. TRPV1 function in mouse colon sensory neurons is enhanced by metabotropic 5-hydroxytryptamine receptor activation. *J. Neurosci.* **24**, 9521–9530 (2004).
35. Moriyama, T. et al. Sensitization of TRPV1 by EP1 and IP reveals peripheral nociceptive mechanism of prostaglandins. *Mol. Pain* **1**, 3 (2005).
36. Ohta, T. et al. Potentiation of transient receptor potential V1 functions by the activation of metabotropic 5-HT receptors in rat primary sensory neurons. *J. Physiol.* **576**, 809–822 (2006).
37. Zariwala, H. A. et al. A Cre-dependent GCaMP3 reporter mouse for neuronal imaging in vivo. *J. Neurosci.* **32**, 3131–3141 (2012).
38. Cavanaugh, D. J. et al. Trpv1 reporter mice reveal highly restricted brain distribution and functional expression in arteriolar smooth muscle cells. *J. Neurosci.* **31**, 5067–5077 (2011).
39. Yeh, J. Z. Sodium inactivation mechanism modulates QX-314 block of sodium channels in squid axons. *Biophys. J.* **24**, 569–574 (1978).
40. Mourot, A. et al. Rapid optical control of nociception with an ion-channel photoswitch. *Nat. Methods* **9**, 396–402 (2012).
41. Binshtok, A. M., Bean, B. P. & Woolf, C. J. Inhibition of nociceptors by TRPV1-mediated entry of impermeant sodium channel blockers. *Nature* **449**, 607–610 (2007).
42. Iida, T. et al. TRPV1 activation and induction of nociceptive response by a non-pungent capsaicin-like compound, capsiate. *Neuropharmacology* **44**, 958–967 (2003).
43. Wang, Y. et al. Kinetics of penetration influence the apparent potency of vanilloids on TRPV1. *Mol. Pharmacol.* **69**, 1166–1173 (2006).
44. Ursu, D., Knopp, K., Beattie, R. E., Liu, B. & Sher, E. Pungency of TRPV1 agonists is directly correlated with kinetics of receptor activation and lipophilicity. *Eur. J. Pharmacol.* **641**, 114–122 (2010).
45. Liu, L., Lo, Y., Chen, I. & Simon, S. a. The responses of rat trigeminal ganglion neurons to capsaicin and two nonpungent vanilloid receptor agonists, olvanil and glyceryl nonamide. *J. Neurosci.* **17**, 4101–4111 (1997).
46. Chou, M. Z., Mtui, T., Gao, Y. D., Kohler, M. & Middleton, R. E. Resiniferatoxin binds to the capsaicin receptor (TRPV1) near the extracellular side of the S4 transmembrane domain. *Biochemistry* **43**, 2501–2511 (2004).
47. Jung, J. et al. Capsaicin binds to the intracellular domain of the capsaicin-activated ion channel. *J. Neurosci.* **19**, 529–538 (1999).
48. Vyklický, L., Lyfenko, A., Kuffler, D. P. & Vlachová, V. Vanilloid receptor TRPV1 is not activated by vanilloids applied intracellularly. *Neuroreport* **14**, 1061–1065 (2003).
49. Zemelman, B. V., Nesnas, N., Lee, G. A. & Miesenbo, G. Photochemical gating of heterologous ion channels: Remote control over genetically designated populations of neurons. *Proc. Natl Acad. Sci. USA* **100**, 1352–1357 (2003).
50. Höglinger, D., Nadler, A. & Schultz, C. Caged lipids as tools for investigating cellular signaling. *Biochim. Biophys. Acta Mol. Cell Biol. Lipids* **1841**, 1085–1096 (2014).
51. Voets, T. et al. The principle of temperature-dependent gating in cold- and heat-sensitive TRP channels. *Nature* **430**, 748–754 (2004).
52. Moshourab, R. A., Wetzel, C., Martinez-Salgado, C. & Lewin, G. R. Stomatin-domain protein interactions with acid-sensing ion channels modulate nociceptor mechanosensitivity. *J. Physiol.* **591**, 5555–5574 (2013).

## Acknowledgements

D.T. and J.A.F. gratefully acknowledge the Deutsche Forschungsgemeinschaft (SFB 1032) and the European Research Council (ERC Advanced Grant 268795 to D.T.) for financial support. Additional funds were obtained from grants from the Alexander von Humboldt foundation (stipend to M.M.) and a senior ERC Advanced grant (294678 to G.R.L). R.M. was supported by a fellowship from the clinical scientist programme of the MDC/Charité. We thank Dr Johannes Broichhagen, Arunas Damijonaitis, Dr David Barber, Cedric Hugelshofer and Dr Matthias Schönberger for insightful discussions leading to the preparation of the manuscript.

## Author contributions

D.T. and G.R.L. coordinated and supervised the study. J.A.F. designed and synthesized the compounds, and carried out electrophysiological and imaging experiments in collaboration with M.M. and M.S. R.M. carried out experiments with C-fibres. J.A.F., D.T., M.M. and R.M. wrote the paper with input from all other co-authors.

## Additional information

**Competing financial interests:** The authors declare no competing financial interests.

# 15

# Parts-per-million level loading organocatalysed enantioselective silylation of alcohols

Sang Yeon Park[1], Ji-Woong Lee[1] & Choong Eui Song[1]

The field of organocatalysis has blossomed over the past few decades, becoming an alternative to transition-metal catalysis or even replacing the realm of transition-metal catalysis. However, a truly powerful organocatalyst with a high turnover number (TON) and turnover frequency (TOF) while retaining high enantioselectivity is yet to be discovered. Similar to metal catalysis, extremely low catalyst loading (p.p.m. or p.p.b. levels) is the ultimate goal of the organocatalysis community. Herein we report a remarkable contribution in this context: 1 p.p.m. loading of a simple 1,1′-bi-2-naphthol-based organocatalyst was enough to achieve highly enantioselective silylation reactions of alcohols. The unprecedented TONs and excellent enantioselectivity are ascribed to the robustness of the catalyst and systematic cooperative hydrogen-bonding organocatalysis in a densely confined chiral space.

[1] Department of Chemistry, Sungkyunkwan University, 2066, Seobu-ro, Jangan-gu, Suwon-si, Gyeonggi-do 440-746, Korea. Correspondence and requests for materials should be addressed to C.E.S. (email: s1673@skku.edu).

atalysis, in principle, should offer an infinite turnover number (TON) for a desired chemical transformation by regenerating the catalyst without altering its molecular-level structure and physicochemical properties. However, limitations arise from different deactivation mechanisms and catalyst poisoning by chemical impurities and/or side products, lowering the lifetime of catalytically active species[1].

The current strong interest in asymmetric organocatalysis[2,3] has pinpointed that small organic molecules are comparable to transition-metal catalysts owing to their noble activation modes and excellent selectivity. However, further application of organocatalysts in industry is often hampered by their relatively low turnover efficiency than those of the corresponding homogeneous or heterogeneous metal-catalysed reactions[4]. Despite many efforts[4–6], the current limit of catalyst loading for asymmetric organocatalysis is usually in the range of 0.1–1 mol% (refs 6,7) for overriding of the non-selective background pathway. Thus, high turnover organocatalysis (for example, TON > 100,000) is considered as a formidable challenge[8].

Our group has developed a new type of easily accessible organocatalysts, 1,1′-bi-2-naphthol (BINOL)-based polyether catalysts 1 (Fig. 1), that bears phenols and polyether units, for asymmetric cation-binding catalysis[9–11]. Its unprecedented activation mode and remarkable enantioselectivity are explained by multiple binding and hydrogen-bonding interactions of the polyether and phenols with the metal salt to generate reactive 'chiral' anions. Strikingly, the robustness of catalyst 1 enabled the one-pot large-scale production of unnatural α-amino acids even under harsh reaction conditions (refluxing in 6 N HCl) and the subsequent quantitative recovery of the catalyst[11]. The structural simplicity and vast application potential of the catalyst stimulated us to explore other challenging catalytic asymmetric reactions.

The preparation of enantiopure secondary alcohols is of significant interest in the pharmaceutical industry; therefore, diverse methodologies have been developed for preparing these synthons with high enantiomeric excess[12–15]. The kinetic resolution of racemic alcohols is an ideal approach to access both enantiomers by one enantioselective transformation, for example, acylative kinetic resolution reactions[16]. Although silylation of alcohols is one of the most common strategies in protecting alcohol functionalities, the asymmetric silylation reactions[17–28] of functionally unbiased alcohols is still challenging[17,18,21] due to the lack of additional functional groups needed to interact with a chiral catalyst. Until now, the substrate scope of the reported silylation reactions has been mostly limited to the substrates with pendant donors (for example, additional hydroxyl or pyridyl groups), which are necessary to achieve good selectivity by increasing the binding

affinity towards catalysts. Snapper and Hoveyda demonstrated remarkable examples in this context by using amino-acid-derived bifunctional organocatalysts[25–28]. However, as a high catalyst loading (20–30 mol%) was necessary to obtain the desired products with high enantioselectivity, it may hamper the large-scale application of these organocatalysts. On the other hand, our group showed that catalysts such as 1 (X = Cl, Br or I) catalyse desilylative kinetic resolution of silyl-protected racemic alcohols in the presence of potassium fluoride. Although this approach provided a new synthetic route for chiral alcohols, a practical application was also far from satisfactory due to the high catalyst loading (20 mol%) and very low turnover frequency (TOF; 43–55% conversion after 5 days at 20 °C) (Fig. 1)[10].

Herein we report that the same catalyst 1 can catalyse the enantioselective silylation of simple aryl alkanols, which do not contain any pendant donors, exhibiting unprecedented catalytic TONs (substrate to catalyst = up to 1,000,000) and excellent enantioselectivity. Extremely high TONs and excellent enantio-selectivity were obtained by thorough catalyst structure analysis, detailed mechanistic studies and the identification of catalyst regeneration conditions.

## Results

**Reaction design.** We commenced our study by selecting 1,1,1,3,3,3-hexamethyldisilazane (HMDS) as the silylating reagent (Supplementary Table 1). HMDS is a stable, commercially available and inexpensive reagent for the trimethylsilylation of alcohols, releasing ammonia as the sole by-product. However, the weak silylating ability of HMDS is the main drawback to its application[29], particularly in asymmetric catalysis. Because HMDS is usually activated using an acidic catalyst[29], chiral Brønsted acid catalysts were chosen first for investigation of the enantioselective silylation of alcohols with HMDS; however, these catalysts resulted in very low conversions and selectivities (Supplementary Fig. 1).

Given our previous studies on catalyst 1 (refs 10,11), in spite of the lower acidity of phenolic protons than those of the common Brønsted acid catalysts[30], we presumed that catalyst 1 would act as a catalyst for the silylative kinetic resolution of racemic alcohols; the interaction between the acidic phenolic protons of the catalyst and the leaving group of the silylating reagent (Y) by hydrogen bonding would enhance the electrophilicity of Si. Moreover, the ether moieties of the catalyst would function as a Brønsted base by the association with an alcohol substrate, enhancing alcohol reactivity (Fig. 2).

**Catalyst screening.** To prove this assumption, several chiral BINOL-based bis(hydroxy) polyethers 1a–1m (Fig. 3) were screened to shed light on the relationship between the catalyst's structure and the reaction outcome. Gratifyingly, after the opti-mization of the reaction conditions (Supplementary Tables 1–3), catalysts 1c, 1d, 1e, 1f and 1g bearing electron-withdrawing groups Cl, Br, I, CF$_3$ and C$_2$F$_5$, respectively, on the 3,3′-positions showed some catalytic activity (Fig. 3). The catalytic performance of these catalysts depends significantly on the steric demand of the substituents. Thus, catalysts 1e and 1f show the best catalytic performance with s-factors of 42 and 44, respectively (Fig. 3). In contrast, other catalysts without electron-withdrawing sub-stituents at the 3,3′-positions (catalysts 1a and 1b) were inactive under identical reaction conditions. In addition, the less acidic H$_8$-BINOL-derived catalyst 1h showed negligible conversion, thus, indicating the importance of the Brønsted acidity of the phenols. To further confirm the importance of an acidic phenol moiety for catalysis, we synthesized the catalysts 1i–1l. A total loss in catalytic activity was observed when one or both phenol

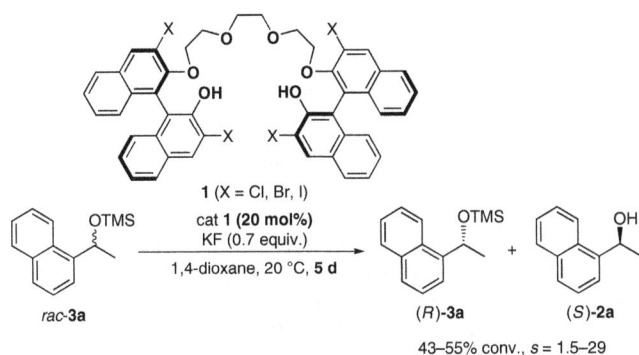

1 (X = Cl, Br, I)

rac-3a → cat 1 (20 mol%) KF (0.7 equiv.) 1,4-dioxane, 20 °C, 5 d → (R)-3a + (S)-2a

43–55% conv., s = 1.5–29

**Figure 1 | Structure of BINOL-based polyether catalyst 1 and our previous work: desilylative kinetic resolution of racemic alcohols.**
Selectivity factor (s) = (rate of fast reacting enantiomer)/(rate of slow reacting enantiomer).

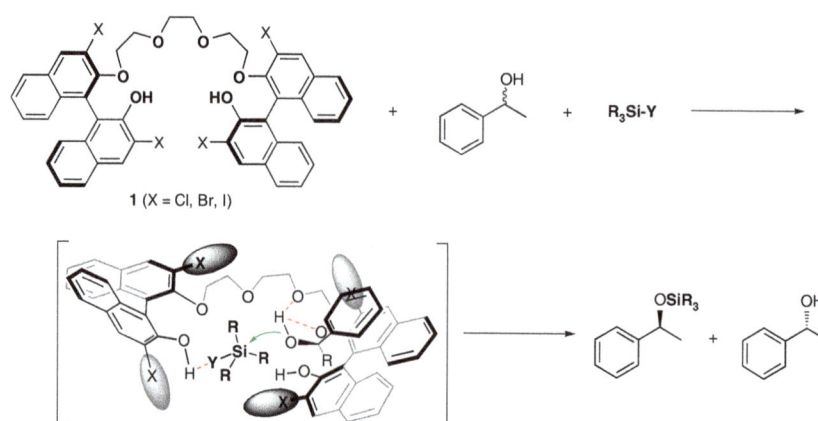

**Figure 2 | Proposed catalytic enantioselective silylation of racemic alcohols.** The proposed transition state for silylation of *rac*-alcohol is shown.

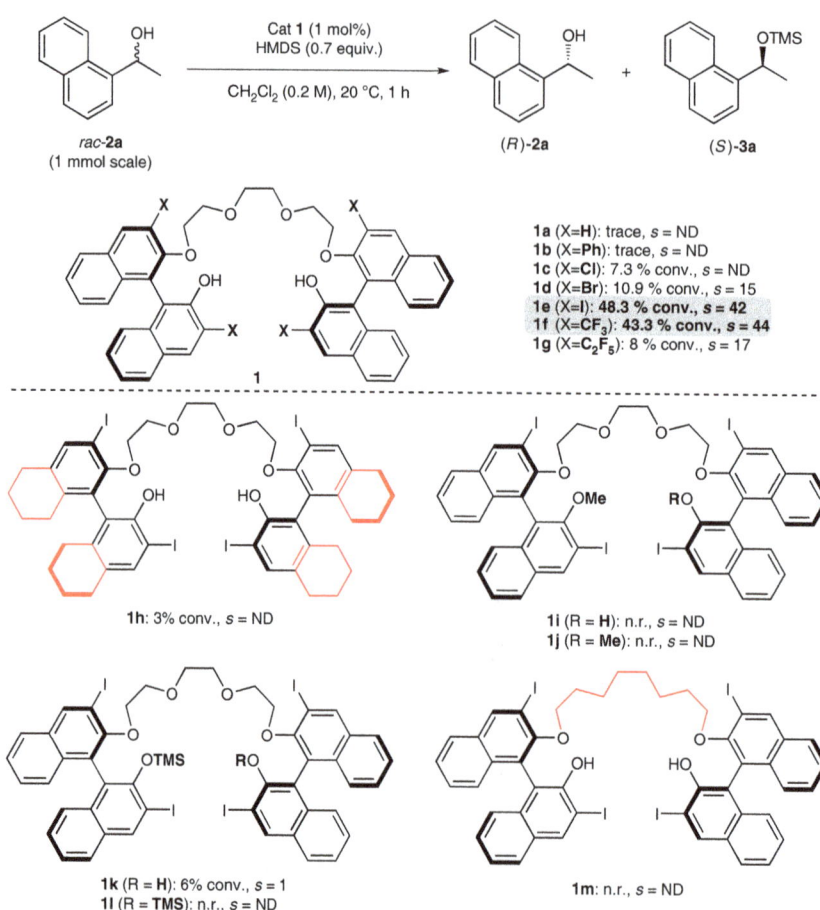

**Figure 3 | Effect of variations in catalyst structure on the silylative kinetic resolution of substrate 2a.** Calculated conversion,
$c = ee_{(R)\text{-}2a}/(ee_{(R)\text{-}2a} + ee_{(S)\text{-}3a})$. Selectivity factor $(s) = \ln((1-c)(1-ee_{(R)-2a}))/\ln((1-c)(1+ee_{(R)-2a})) = \ln(1-c(1+ee_{(S)-3a}))/\ln(1-c(1-ee_{(S)-3a}))$.

groups were replaced by OMe (**1i** and **1j**). Silylated catalysts (**1k** and **1l**) also showed no activity, excluding the possibility of a silyl-transfer reaction from the silylated catalyst to the starting alcohol to afford the desired product. The importance of the polyether chain was confirmed unambiguously by replacing the polyether with the alkyl chain (catalyst **1m**), indicating that the polyether backbone is crucial in achieving the observed catalytic activity.

The complexation of an alcohol substrate with the polyether moiety of the catalyst was clearly verified by observing the decreased $^{13}$C spin–lattice relaxation time ($T_1$) on complexation of catalyst **1e** with racemic alcohol **2a** (Supplementary Table 4).

This significant decrease in the $T_1$ value of ether moiety of **1e** indicates that the complexation of the alcohol with catalyst **1e** significantly reduces the mobility of the ether units. Moreover, the fast-reacting (S)-configured substrate ((S)-**2a**) exhibited a more decreased $T_1$ value of ether moiety than that of the slow-reacting substrate, (R)-**2a**, directly indicating a selective association of the catalyst with the fast-reacting (S)-**2a** (ref. 31). On the other hand, no change in $T_1$ values was observed when alcohol **2a** was mixed with the alkyl chain-replaced catalyst **1m**, verifying the crucial role of the polyether backbone for the interaction with alcohols (Supplementary Table 5). $^{29}$Si nuclear magnetic resonance (NMR) experiments were also used to establish the interaction between the phenolic proton of the catalyst and HMDS (Supplementary Table 6). As expected, the $^{29}$Si NMR data of HMDS in the presence of diverse phenols exhibited $^{29}$Si chemical shifts that are shifted downfield relative to the uncoordinated HMDS, dependent on phenol acidity. This result clearly indicates that the acidic phenolic proton interacts with HMDS, increasing the electrophilic character of the silicon atom. The results from NMR spectroscopy and other experimental results (Fig. 3) strongly support our proposed transition-state model shown in Fig. 2 (Y = NHSiMe$_3$).

**Optimization of the reaction conditions.** More importantly, during the optimization studies, we also realized that the major deactivation pathway of the catalyst is the silylation of the phenol moieties under the reaction conditions. In a separated set of experiments, we verified that the steric demand of the 3,3'-substituents of the catalyst is crucial to this deactivation pathway (Supplementary Table 7). As mentioned earlier, the silylated catalysts (**1k** and **1l**) are inactive towards the kinetic resolution reaction (Fig. 3).

Recognizing that the rate of desilylation of trimethylsilyl (TMS)-protected phenols using KF in oligoethylene glycol[32] is much faster than those of TMS-protected secondary alcohols, we speculated that the addition of KF and an appropriate proton

source would regenerate the catalyst *in situ* (Table 1a). As predicted, the desilylation reaction of the silylated catalysts (**1k** and **1l**) and subsequent protonation of the K salt of catalysts clearly improved TON and TOF (Table 1b). As shown in entries 1 and 7 of Table 1b, with catalyst loading of 0.01 mol% (100 p.p.m.), almost no catalytic activity was observed without the additives. However, the addition of KF and a proton source such as water, *t*-BuOH and Amberlite CG 50 (a weakly acidic cation-exchange resin)[33] was able to restore catalysis (entries 2–5). In particular, a combination of KF and Amberlite CG 50 provided the best result (entry 5). Even with 100 p.p.m. of catalyst loading, 52% conversion was obtained in 12 h with excellent selectivity (*s* = 57). This can be ascribed to the adequate acidity of Amberlite CG 50 for rapid protonation of the K salt of **1e**, releasing the insoluble polymeric K salt of Amberlite CG 50 as a by-product without interfering with the catalysis. Moreover, Amberlite CG 50 alone was completely inactive towards catalysis. In addition, we also confirmed that, under the reaction condition where KF and Amberlite CG 50 were used as additives (condition in entry 5), the unwanted desilylation process that takes place with a TMS-protected secondary alcohol such as (*R*)-**3a** did not occur (Supplementary Fig. 2) and that the relative catalytic activity of **1a**–**1f** is very similar with that shown in Fig. 3 (Supplementary Table 8).

**Table 1 | Proposed mechanism of catalyst reactivation (a) and effects of KF and proton source additives on turnover numbers (b).**

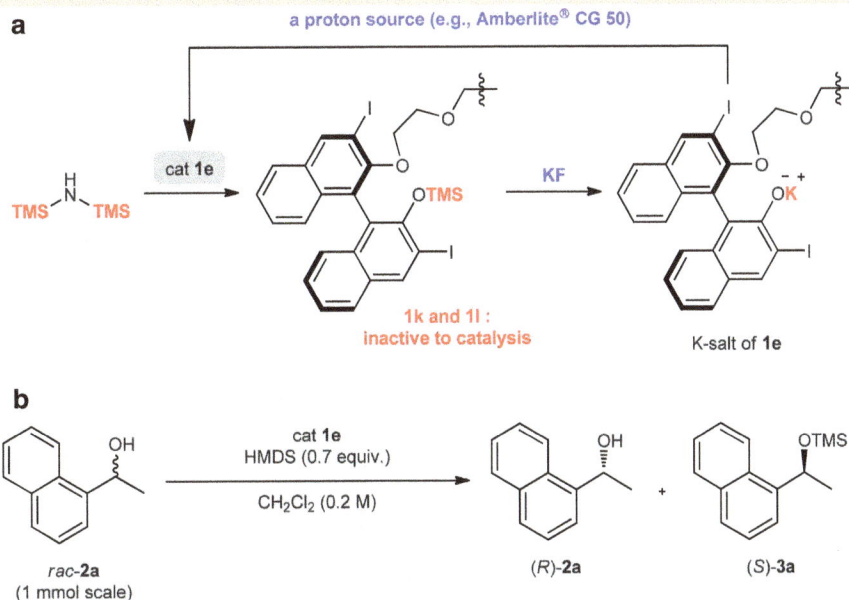

| Entry | Cat 1e (p.p.m.)* | KF† | H$^+$ source‡ | Temp. (°C) | Time | Conv. (%)§ | s-factor‖ |
|---|---|---|---|---|---|---|---|
| 1 | 100 | No | No | 20 | 12 h | 4 | ND |
| 2 | 100 | Yes | No | 20 | 12 h | 17 | 54 |
| 3 | 100 | Yes | *t*-BuOH | 20 | 12 h | 15 | 58 |
| 4 | 100 | Yes | H$_2$O | 20 | 12 h | 18 | 48 |
| 5 | 100 | Yes | CG 50 | 20 | 12 h | 52 | 57 |
| 6 | 100 | Yes | CG 50/H$_2$O¶ | 20 | 12 h | 48 | 51 |
| 7 | 100 | No | No | −30 | 24 h | 0 | ND |
| 8 | 100 | Yes | CG 50 | −30 | 24 h | 50 | 132 |
| 9 | 100 | Yes | CG 50/H$_2$O¶ | −30 | 24 h | 52 | 144 |
| 10# | 1 | Yes | CG 50 | −30 | 14 d | 33 | 99 |

*The appropriate aliquots were taken from the stock solution of **1e** in CH$_2$Cl$_2$.
†KF (1 equiv. for entries 2–6, 8 and 9; 0.5 equiv. for entry 10) was added.
‡CG 50 (0.8 equiv. for entries 5, 6, 8 and 9; 0.2 equiv. for entry 10), *t*-BuOH (0.8 equiv.) or H$_2$O (0.8 equiv.) was added.
§Calculated conversion, $c = ee_{(R)-2a}/(ee_{(R)-2a} + ee_{(S)-3a})$.
‖Selectivity factor (*s*) = (rate of fast reacting enantiomer)/(rate of slow reacting enantiomer).
¶0.8 equiv. of H$_2$O was added.
#Using 5 mmol of racemic alcohol **2a**. CG 50 = Amberlite CG 50 (10 mmol H$^+$ g$^{-1}$).

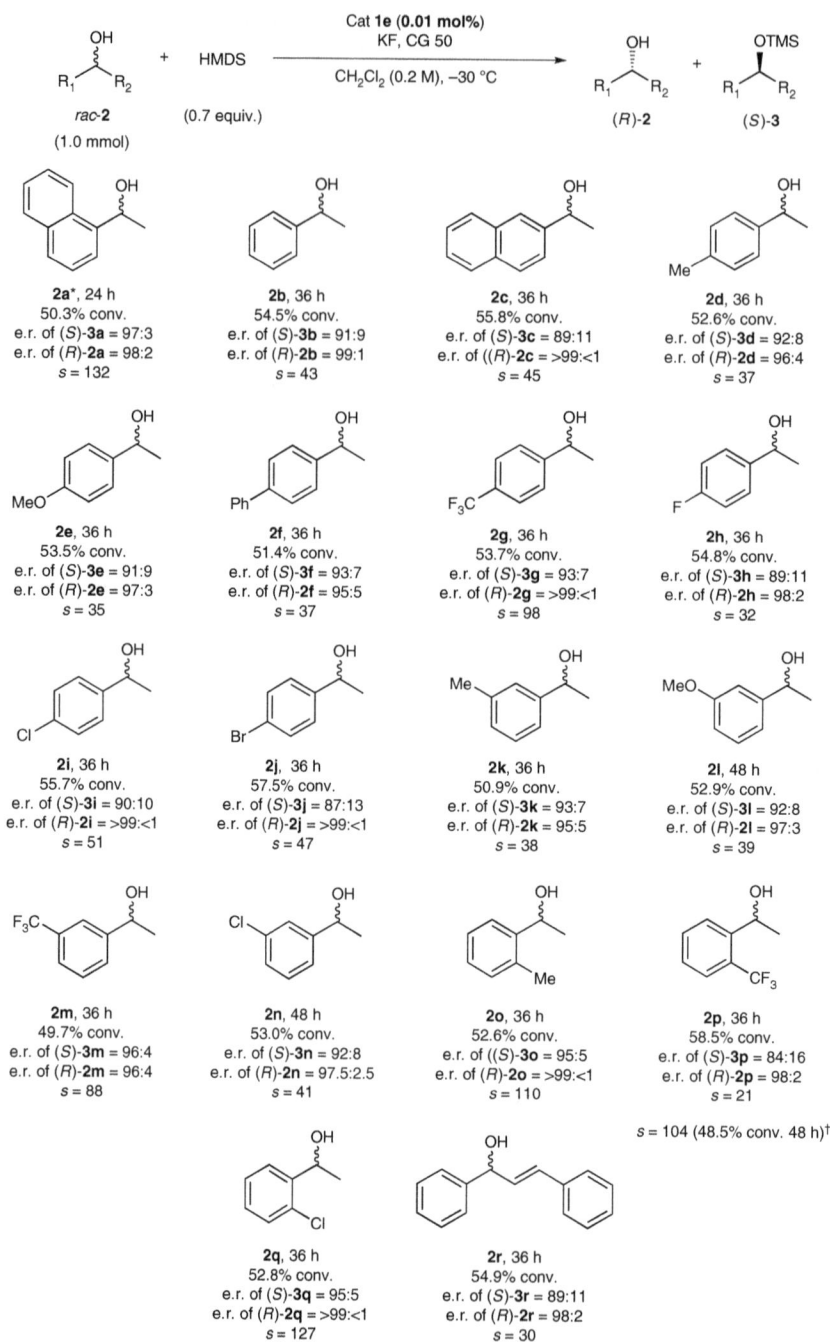

**Figure 4 | Substrate scope of the silylative kinetic resolution of racemic alcohol 2 catalysed by 1e.** The reactions were performed in the presence of 1 equiv of KF and 0.8 equiv of Amberlite CG 50 (CG 50). *Using catalyst **1c** (X = Cl, 16.8% conv., s = 20), catalyst **1d** (X = Br, 36.7% conv., s = 48), catalyst **1f** (X = CF₃, 50.8% conv., s = 50) and catalyst **1g** (X = C₂F₅, 34.1% conv., s = 30). †Using catalyst **1d** (X = Br). e.r., enantiomeric ratio.

Further improvement of selectivity was achieved by lowering the reaction temperature to − 30 °C (s = 132) (entry 8). More surprisingly, the catalyst is active even at 1 p.p.m. catalyst loading (0.0001 mol%) with excellent s-factor (s = 99 in Table 1, entry 10). The reaction proceeds with p.p.b.-level catalyst loading (100 p.p.b.); however, a significant decrease in selectivity (s = 17) was observed (Supplementary Fig. 3). To the best of our knowledge, this is the lowest catalyst loading for metal-free organocatalysis reported to date, with excellent enantio-selectivities and high TOF (TOF ∼ 1,000 h⁻¹, entry 11) also being achieved. The high turnover efficiency can be attributed to the stability of the catalyst, its insensitivity to impurities and rapid reactivation under the reaction conditions. Notably, the reactions do not require exclusion of air and moisture, and thus

solvent purification is not needed. As shown from the results of entries 6 and 9, the presence of water does not affect catalytic results, which is highly beneficial for further large-scale applications.

**Scope of the reaction.** With the optimized reaction conditions in hand, the substrate scope of our protocol was investigated, and the results are shown in Fig. 4. Various simple acyclic 1-arylalkanols **2a–2r** without any secondary binding functionality were successfully resolved with excellent selectivity factors. Both the electron-withdrawing and electron-donating substituents on the aromatic ring were tolerated under the reaction conditions. Sterically demanding *ortho*-substituted substrates were also

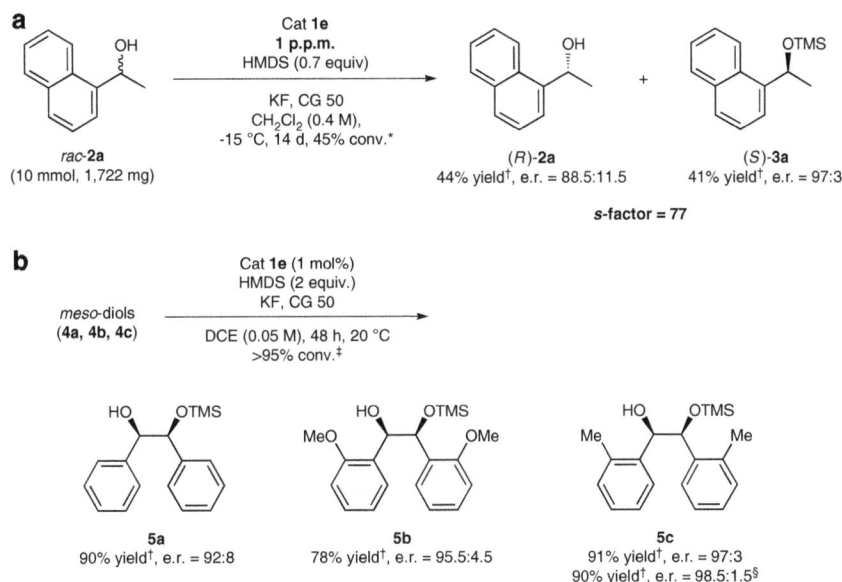

**Figure 5 | Applications of the silylation reaction protocol. (a)** Gram-scale kinetic resolution of secondary alcohol **2a** using 1 p.p.m. of catalyst: The reactions were performed in the presence of 0.5 equiv of KF and 0.2 equiv of Amberlite CG 50 (CG 50). *Calculated conversion, $c = ee_{(R)-2a}/(ee_{(R)-2a} + ee_{(S)-3a})$. †Isolated yields. **(b)** Catalytic enantioselective silylation of meso-diols **4** (0.1 mmol scale): The reactions were performed in the presence of 1 equiv of KF and 0.6 equiv of Amberlite CG 50 (CG 50). ‡The conversion was determined by ${}^1$H-NMR. †Isolated yields. §The reaction was performed at 0 °C. DCE, 1,2-dichloroethane; e.r., enantiomeric ratio.

smoothly and selectively converted to TMS-protected alcohols with excellent s-factors. Notably, the s-factors for the silylation of the sterically bulky alcohol **2p** increased from 21 to 104 by changing the catalyst **1e** (X = I) to **1d** (X = Br), a sterically less demanding catalyst, indicating that stereoselectivity can be controlled by tuning the cage size of the catalyst[34,35]. However, 1-cyclohexylethanol, which does not contain an aromatic moiety, exhibited a poor selectivity factor (s = 1.4, Supplementary Fig. 4), indicating that the π–π interaction between catalyst and substrate might play an important role in obtaining high stereoselectivity, as shown in Fig. 2.

To demonstrate the synthetic utility and potential large-scale applications of this protocol, 1.722 g (10 mmol) rac-**2a** was resolved at −15 °C by using only 0.012 mg of the catalyst (1 p.p.m. to rac-**2a**), with a high TOF (1,309 h⁻¹) and an excellent s-factor of 77 (Fig. 5a), affording TMS-ether (S)-**3a** (41% yield) and (R)-**2a** (44% yield) after simple filtration of insoluble additives (KF and CG 50) and column chromatography. Finally, to demonstrate the generality of our silylation protocol, the silylative desymmetrization reactions of meso-hydrobenzoins **4** were conducted using 1 mol% of catalyst **1e** at room temperature. As shown in Fig. 5b, monosilylated alcohols **5a**–**5c** were obtained in high yields and with excellent enantiomeric excesses (ee) (up to 97% ee). Furthermore, an essentially linear relationship between the optical purity of the catalyst and enantioselectivity of the major product **5a** was observed, indicating that the reaction involves only one catalyst in the enantio-determining step, supporting our proposed model shown in Fig. 2 (Supplementary Fig. 5).

## Discussion

In summary, a p.p.m.-level loading organocatalytic enantioselective silylation of simple alcohols was developed using the easily accessible BINOL-based polyether catalyst with the commercially available HMDS as the silylating reagent. This unprecedented catalyst turnover was achieved by adding KF and a weakly acidic cation-exchange resin, thus regenerating the active form of the catalyst. Using this protocol, diverse racemic secondary alcohols

were resolved with excellent selectivity factors under mild reaction conditions and the reaction was conducted without protecting moisture and without the use of an inert gas atmosphere. Moreover, the utility of this silylation protocol was further extended to the enantioselective silylation of meso-diols. The features of very low catalyst loading, high enantioselectivity, broad substrate scope and mild reaction conditions can make this protocol easily adaptable to the practical synthesis of diverse biologically and pharmaceutically important chiral alcohols. The extremely high turnover efficiency and selectivity achieved can be ascribed to the robustness of the catalyst and systematic cooperative hydrogen-bonding catalysis in a densely confined chiral space[34,35], which mimics the action of enzymes. In a confined space, active sites activate both the reactants simultaneously and keep them in close proximity, thus enhancing reactivity and efficiently transferring the stereochemical information. Our results may contribute to the development of an 'ideal' catalyst system based on weak interactions (hydrogen bondings, charge–charge interactions and π-interactions) such those of enzymes.

## Methods

**General procedure for catalytic asymmetric silylation of secondary alcohols.** To a solution of chiral catalyst **1e** (0.12 mg, 0.01 mol%) and secondary alcohols **2** (1.0 mmol) in distilled CH₂Cl₂ (5.0 ml), spray-dried KF (58.1 mg, 100 mol%) and Amberlite CG 50 (80 mg, 0.8 equiv.) were added in one portion. The reaction mixture was stirred at − 30 °C for 10 min, followed by addition of HMDS (0.15 ml, 0.7 equiv.). After stirring for 24–48 h at − 30 °C, the reaction mixture was filtered and the filtrate was concentrated in vacuo to afford a colourless oil. The residue thus obtained was purified by short silica gel chromatography (EA/hexanes = 1/5). Silyl ethers **3a**–**3f**, **3i**–**3j**, **3m**–**3o** and **3r** are known compounds in the literature and their spectroscopic data were consistent with previously reported values (Supplementary Methods and Supplementary References).

The enantiomeric excesses of **2** and **3** were determined by chiral high-performance liquid chromatography (HPLC) analysis. The enantiomeric excess of TMS-ether products **3** was determined after converting them to their corresponding alcohols **2** by the deprotection of its trimethylsilyl group with TBAF. Thus, the absolute configurations of **3** were determined by comparison of the retention time of HPLC with the literature data of the corresponding alcohols **2a**–**2r** (Supplementary Methods and Supplementary References).

For NMR and HPLC analysis of the compounds in this article, see Supplementary Fig. 6–20.

## References

1. Crabtree, R. H. Deactivation in homogeneous transition metal catalysis: causes, avoidance, and cure. *Chem. Rev.* **115**, 127–150 (2015).
2. List, B. & Yang, J. W. The organic approach to asymmetric catalysis. *Science* **313**, 1584–1586 (2006).
3. MacMillan, D. W. C. The advent and development of organocatalysis. *Nature* **455**, 304–308 (2008).
4. Lee, J. W. *et al.* Organotextile catalysis. *Science* **341**, 1225–1229 (2013).
5. Cozzi, F. Immobilization of organic catalysts: when, why, and how. *Adv. Synth. Catal.* **348**, 1367–1390 (2006).
6. Giacalone, F., Gruttadauria, M., Agrigento, P. & Noto, R. Low-loading asymmetric organocatalysis. *Chem. Soc. Rev.* **41**, 2406–2447 (2012).
7. Ingle, G. K., Mormino, M. G., Wojtas, L. & Antilla, J. C. Chiral phosphoric acid-catalyzed addition of thiols to *N*-acyl imines: access to chiral *N,S*-acetals. *Org. Lett.* **13**, 4822–4825 (2011).
8. Ratjen, L., van Gemmeren, M., Pesciaioli, F. & List, B. Towards high-performance Lewis acid organocatalysis. *Angew. Chem. Int. Ed.* **53**, 8765–8769 (2014).
9. Brak, K. & Jacobsen, E. N. Asymmetric ion-pairing catalysis. *Angew. Chem. Int. Ed.* **52**, 534–561 (2013).
10. Yan, H. *et al.* A chiral–anion generator: application to catalytic desilylative kinetic resolution of silyl-protected secondary alcohols. *Angew. Chem. Int. Ed.* **49**, 8915–8917 (2010).
11. Yan, H., Oh, J. S., Lee, J. W. & Song, C. E. Scalable organocatalytic asymmetric Strecker reactions catalysed by a chiral cyanide generator. *Nat. Commun.* **3**, 1212 (2012).
12. Reeves, J. T., Senanayake, C. H., Song, J. J. & Farina, V. Asymmetric synthesis of active pharmaceutical ingredients. *Chem. Rev.* **106**, 2734–2793 (2006).
13. Cobley, C. J. & Holt-Tiffin, K. E. Asymmetric routes to chiral secondary alcohols. *Pharmaceutical Technology* **34**, 6–13 (2010).
14. Huisman, G. W., Liang, J. & Krebber, A. Practical chiral alcohol manufacture using ketoreductases. *Curr. Opin. Chem. Biol.* **14**, 122–129 (2010).
15. Noyori, R. & Ohkuma, T. Asymmetric catalysis by architectural and functional molecular engineering: practical chemo- and stereoselective hydrogenation of ketones. *Angew. Chem. Int. Ed.* **40**, 40–73 (2001).
16. Müller, C. E. & Schreiner, P. R. Organocatalytic enantioselective acyl transfer onto racemic as well as meso alcohols, amines, and thiols. *Angew. Chem. Int. Ed.* **50**, 6012–6042 (2011).
17. Isobe, T., Fukuda, K., Araki, Y. & Ishikawa, T. Modified guanidines as chiral superbases: The first example of asymmetric silylation of secondary alcohols. *Chem. Commun.* **3**, 243–244 (2001).
18. Sheppard, C. I., Taylor, J. L. & Wiskur, S. L. Silylation-based kinetic resolution of monofunctional secondary alcohols. *Org. Lett.* **13**, 3794–3797 (2011).
19. Rendler, S., Auer, G. & Oestreich, M. Kinetic resolution of chiral secondary alcohols by dehydrogenative coupling with recyclable silicon-streogenic silanes. *Angew. Chem. Int. Ed.* **44**, 7620–7624 (2005).
20. Klare, H. F. T. & Oestreich, M. Chiral recognition with silicon-stereogenic silanes: remarkable selectivity factors in the kinetic resolution of donor-functionalized alcohols. *Angew. Chem. Int. Ed.* **46**, 9335–9338 (2007).
21. Karatas, B., Rendler, S., Fröhlich, R. & Oestreich, M. Kinetic resolution of donor-functionalised tertiary alcohols by Cu-H-catalysed stereoselective silylation using a strained silicon-stereogenic silane. *Org. Biomol. Chem.* **6**, 1435–1440 (2008).
22. Rendler, S. & Oestreich, M. Kinetic resolution and desymmetrization by stereoselective silylation of alcohols. *Angew. Chem. Int. Ed.* **47**, 248–250 (2008).
23. Steves, A. & Oestreich, M. Facile preparation of CF₃-substituted carbinols with an azine donor and subsequent kinetic resolution through stereoselective Si-O coupling. *Org. Biomol. Chem.* **7**, 4464–4469 (2009).
24. Weickgenannt, A., Mewald, M. & Oestreich, M. Asymmetric Si-O coupling of alcohols. *Org. Biomol. Chem.* **8**, 1497–1504 (2010).
25. Zhao, Y., Rodrigo, J., Hoveyda, A. H. & Snapper, M. L. Enantioselective silyl protection of alcohols catalysed by an amino-acid-based small molecule. *Nature* **443**, 67–70 (2006).
26. Zhao, Y., Mitra, A. W., Hoveyda, A. H. & Snapper, M. L. Kinetic resolution of 1,2-diols through highly site- and enantioselective catalytic silylation. *Angew. Chem. Int. Ed.* **46**, 8471–8474 (2007).
27. Zhao, Y., Hoveyda, A. H. & Snapper, M. L. Catalytic enantioselective silylation of acyclic and cyclic triols: application to total syntheses of cleroindicins D, F, and C. *Angew. Chem. Int. Ed.* **48**, 547–550 (2009).
28. Manville, N. *et al.* Enantioselective silyl protection of alcohols promoted by a combination of chiral and achiral Lewis basic catalysts. *Nat. Chem.* **5**, 768–774 (2013).
29. Bruynes, C. A. & Jurriens, T. K. Catalysts for silylations with 1,1,1,3,3,3-hexamethyldisilazane. *J. Org. Chem.* **47**, 3966–3969 (1982).
30. Christ, P. *et al.* pKa Values of chiral Brønsted acid catalysts: phosphoric acids/amides, sulfonyl/sulfuryl imides, and perfluorinated TADDOLs (TEFDDOLs). *Chem. Eur. J.* **17**, 8524–8528 (2011).
31. Li, Y. & Echegoyen, L. Enatiomeric recognition between chiral triazole-18-crown-6 ligands and organic ammonium cations assessed by ¹³C and ¹H NMR relaxation times. *J. Org. Chem.* **59**, 6539–6542 (1994).
32. Yan, H., Oh, J. S. & Song, C. E. A mild and efficient method for the selective deprotection of silyl ethers using KF in the presence of tetraethylene glycol. *Org. Biomol. Chem.* **9**, 8119–8121 (2011).
33. Müller, S., Webber, M. J. & List, B. The catalytic asymmetric Fischer indolization. *J. Am. Chem. Soc.* **133**, 18534–18537 (2011).
34. Coric, I. & List, B. Asymmetric spiroacetalization catalysed by confined Brønsted acids. *Nature* **483**, 315–319 (2012).
35. Shenoy, S. R., Crisóstomo, F. R. P., Iwasawa, T. & Rebek, Jr. J. Organocatalysis in a synthetic receptor with an inwardly directed carboxylic acid. *J. Am. Chem. Soc.* **130**, 5658–5659 (2008).

## Acknowledgements

The authors are grateful for the financial support provided by the Ministry of Science, ICT and Future Planning (NRF-2014R1A2A1A01005794). In addition, we thank Prof. S.-G. Lee and Dr. M.T. Oliveira for their helpful dicsussions.

## Author contributions

C.E.S. designed and developed the strategy and the mechanistic concepts. S.Y.P. and J.W.L. performed the experiments. J.W.L. and C.E.S. wrote the manuscript. All the authors discussed the results and commented on the manuscript.

## Additional information

**Competing financial interests:** The authors declare no competing financial interests.

# 16

# New peptide architectures through C–H activation stapling between tryptophan–phenylalanine/tyrosine residues

Lorena Mendive-Tapia[1,2,3], Sara Preciado[2], Jesús García[1], Rosario Ramón[4], Nicola Kielland[4], Fernando Albericio[1,2,3,5] & Rodolfo Lavilla[4,6]

Natural peptides show high degrees of specificity in their biological action. However, their therapeutical profile is severely limited by their conformational freedom and metabolic instability. Stapled peptides constitute a solution to these problems and access to these structures lies on a limited number of reactions involving the use of non-natural amino acids. Here, we describe a synthetic strategy for the preparation of unique constrained peptides featuring a covalent bond between tryptophan and phenylalanine or tyrosine residues. The preparation of such peptides is achieved in solution and on solid phase directly from the corresponding sequences having an iodo-aryl amino acid through an intramolecular palladium-catalysed C–H activation process. Moreover, complex topologies arise from the internal stapling of cyclopeptides and double intramolecular arylations within a linear peptide. Finally, as a proof of principle, we report the application to this new stapling method to relevant biologically active compounds.

[1] Institute for Research in Biomedicine, Barcelona Science Park, Baldiri Reixac 10-12, 08028 Barcelona, Spain. [2] Department of Organic Chemistry, University of Barcelona, Martí i Franqués 1-11, 08028 Barcelona, Spain. [3] CIBER-BBN, Networking Centre on Bioengineering, Biomaterials and Nanomedicine. [4] Barcelona Science Park, Baldiri Reixac 10-12, 08028 Barcelona, Spain. [5] School of Chemistry, Yachay Tech, Yachay City of Knowledge, 100119 Urcuqui, Ecuador. [6] Laboratory of Organic Chemistry, Faculty of Pharmacy, University of Barcelona, Avda. Joan XXII s.n., 08028 Barcelona, Spain. Correspondence and requests for materials should be addressed to F.A. (email: albericio@irbbarcelona.org) or to R.L. (email: rlavilla@pcb.ub.es).

Peptides are attracting great attention as therapeutics since they combine the high selectivity, potency and low toxicity of biologics with advantages such as the conformational restrictions and the reduced costs characteristic of small molecular entities[1,2]. In addition to the large number of commercialized peptides, many others are also found in clinical phases, thereby demonstrating their validity and application as active pharmaceutical ingredients[3]. However, the general use of peptides as drugs is severely hampered by their poor pharmacokinetic features. In this regard, it is widely believed that the characterization of protein–protein interactions, a fundamental issue in deciphering biological pathways, is better tackled through structurally defined small peptides with specific sequences. Therefore, there is a need for peptides with new topological architectures; however, these are difficult to obtain even using modern synthetic procedures[4,5]. To address these problems, general strategies for peptide macrocyclization have been developed to improve the properties (cell penetration, stability, selectivity and so on) and enhance the potential of peptides as therapeutics and bioprobes, with a special focus on poorly tractable targets[6–8]. In this context, peptides constrained via a non-amide sidechain-to-sidechain linkage (stapled peptides) provide a new structural paradigm because the conformational stabilized species display remarkably stronger biological activity. In this regard, the pioneering work of Verdine et al.[9,10] on the

basis of all-hydrocarbon staples through olefin metathesis represents a breakthrough in the field.

So far, stapled peptides are generated through a variety of strategies[11], the main being the use of cysteine side chains to form disulfide bridges[12] and thioether formation (crosslinking with $\alpha,\alpha'$-dibromo-$m$-xylene[13] or aromatic nucleophilic substitutions with perfluoroaromatic reactants[14]); or with functionalized non-natural amino acids by means of biaryl linkages involving the borylated phenylalanine derivatives[15–18], ring-closing metathesis[19] and azide–alkyne cycloadditions (click chemistry)[20], and so on, usually in expensive and long stepwise syntheses. A comparative table summarizes the major strengths and weaknesses of the main conventional stapling strategies (Fig. 1a). Owing to the increasing interest in stapled peptides and other conformationally restricted structures, many efforts have been made to develop new practical and general preparative methods for their generation.

There is a considerable concern regarding the efficiency of synthetic aspects with respect to the preparation of complex bioactive compounds. In this context, the function-oriented synthesis approach to target simplified, although biologically meaningful, fragments of complex chemical entities is relevant, as it leads to practical syntheses of new specific drugs and probes, also in the field of peptides[21].

Tryptophan (Trp) has a low relative abundance in peptide and protein sequences ($\approx 1\%$ of the amino acids); however, its presence is critical for the activity of these biomolecules. Therefore, the development of new synthetic methods for the selective and straightforward chemical modification of Trp is highly significant[22].

Recently, metal-catalysed C–C coupling through direct C–H activation[23–29] has become a fundamental process in modern organic syntheses, allowing the straightforward preparation of a plethora of new structural types. In this respect, the functionalization of indoles using this approach has been extensively examined[30,31], including studies described in (refs 28, 32–35) among others. Particularly relevant to our research was the methodology of Larrosa for indole arylation using $Pd(OAc)_2$ in acidic media[36]. In particular, although the Pd-mediated C-2 arylation of Trp has been reported[37–39], it has not yet been applied to staple true peptides. We recently disclosed the direct C-2 arylation of indoles in Trp-containing peptidic sequences in an intermolecular manner, without additional requierements[37]. Although the alternative $N$-arylation processes are conceivable, such Ullmann-type reactions take place normally with Cu and Pd catalysts, in the latter case usually requiring strong bases[40]. No experimental evidence of these transformations have been recorded in our systems. The main restriction is that lower conversions are obtained for sequences that comprise methionine, cysteine or histidine residues, presumably because of selective hydrolysis of the peptide bond catalysed by bidentate palladium coordination; on-going experiments along this line show that working in nonaqueous solvents allows reasonable arylations (unpublished results). Later we reported on the arylation of Trp-diketopiperazines[41] and related transformations of Trp derivatives, leading to the generation of Fmoc-protected arylated Trps that are amenable to direct incorporation in solid-phase peptide synthesis (SPPS)[42]. These processes work well with $N$-protected peptides in dimethylformamide (DMF) or in aqueous environments.

In this work, we present a new stapling methodology involving Trp and Phe(Tyr) through a Pd-catalysed C–H activation process. The method is versatile, allowing the formation of constrained peptides of different ring sizes, amenable to solution- and solid-phase synthesis and can lead to the direct preparation of biologically meaningful peptide derivatives.

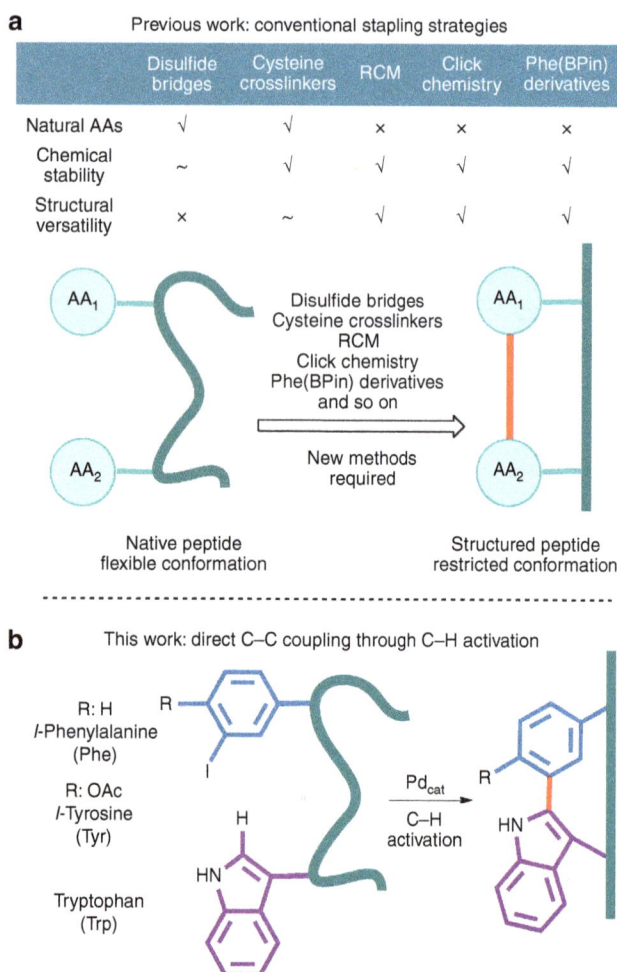

**a** Previous work: conventional stapling strategies

| | Disulfide bridges | Cysteine crosslinkers | RCM | Click chemistry | Phe(BPin) derivatives |
|---|---|---|---|---|---|
| Natural AAs | √ | √ | × | × | × |
| Chemical stability | ~ | √ | √ | √ | √ |
| Structural versatility | × | ~ | √ | √ | √ |

Disulfide bridges
Cysteine crosslinkers
RCM
Click chemistry
Phe(BPin) derivatives
and so on

New methods required

Native peptide flexible conformation

Structured peptide restricted conformation

**b** This work: direct C–C coupling through C–H activation

R: H
*l*-Phenylalanine (Phe)

R: OAc
*l*-Tyrosine (Tyr)

Tryptophan (Trp)

Pd$_{cat}$
C–H activation

**Figure 1 | Formation of stapled peptides.** (**a**) Conventional stapling methods. (**b**) Phe/Tyr-Trp stapling via a selective Pd-catalysed C–H arylation process.

## Table 1 | Results for the stapled bond formation of peptides 1 under microwave irradiation.

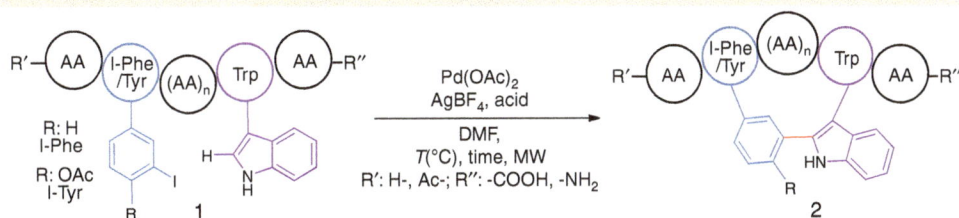

| Entry | n | Linear peptide (1) | Coupling conditions* | Stapled peptide | Conv (%)† |
|---|---|---|---|---|---|
| 1 | 1 | Ac-Ala-*m*-I-**Phe**-Ala-**Trp**-Ala-OH (**1a**) | A | 2a | 38 |
| 2 | 2 | Ac-Ala-*m*-I-**Phe**-Ala-Ala-**Trp**-Ala-OH (**1b**) | A | 2b | 100 |
| 3 | 3 | Ac-Ala-*m*-I-**Phe**-Ala-Ala-Ala-**Trp**-Ala-OH (**1c**) | A | 2c | 100 |
| 4 | 1 | Ac-Ala-*m*-I-**Tyr**-Ala-**Trp**-Ala-OH (**1d**) | A | 2d | 100 |
| 5 | 2 | Ac-Ala-*m*-I-**Tyr**-Ala-Ala-**Trp**-Ala-OH (**1e**) | A | 2e | 100 |
| 6 | 3 | Ac-Ala-*m*-I-**Tyr**-Ala-Ala-Ala-**Trp**-Ala-OH (**1f**) | A | 2f | 100 |
| 7 | 3 | Ac-*m*-I-**Phe**-Asn-Gly-Arg-**Trp**-NH₂ (**1g**) | B | 2g | 77 |
| 8 | 3 | Ac-*m*-I-**Phe**-Arg-Gly-Asp-**Trp**-NH₂ (**1h**) | B | 2h | 70 |
| 9 | 2 | H-Ala-*m*-I-**Phe**-Ser-**Trp**-Ala-OH (**1i**) | B | 2i | 39‡ |
| 10 | 1 | Ac-Ala-*m*-I-**Phe**-Val-**Trp**-Ala-OH (**1j**) | B | 2j | 71 |
| 11 | 0 | Ac-Ala-*p*-I-**Phe**-**Trp**-Ala-OH (**1k**) | B | 2k§ | 60 |

Conv, conversion; HPLC-MS, high-performance liquid chromatography-mass spectrometry; MW, microwave.
*Coupling conditions (A): 5 mol % Pd(OAc)₂, 1.0 eq. of AgBF₄, 1.5 eq. of 2-NO₂BzOH in DMF:PBS (1:1), MW 80 °C, 15 min. (B) 5 mol % Pd(OAc)₂, 2.0 eq. of AgBF₄, 1.0 eq. of TFA in DMF, MW 90 °C, 20 min.
†Conversion: estimated yield (HPLC-MS).
‡Additional MW irradiation cycles were necessary to obtain the desired product as the main compound (HPLC-MS).
§Cyclodimer **2k** was obtained in place of the putative monomeric structure **2k′**.

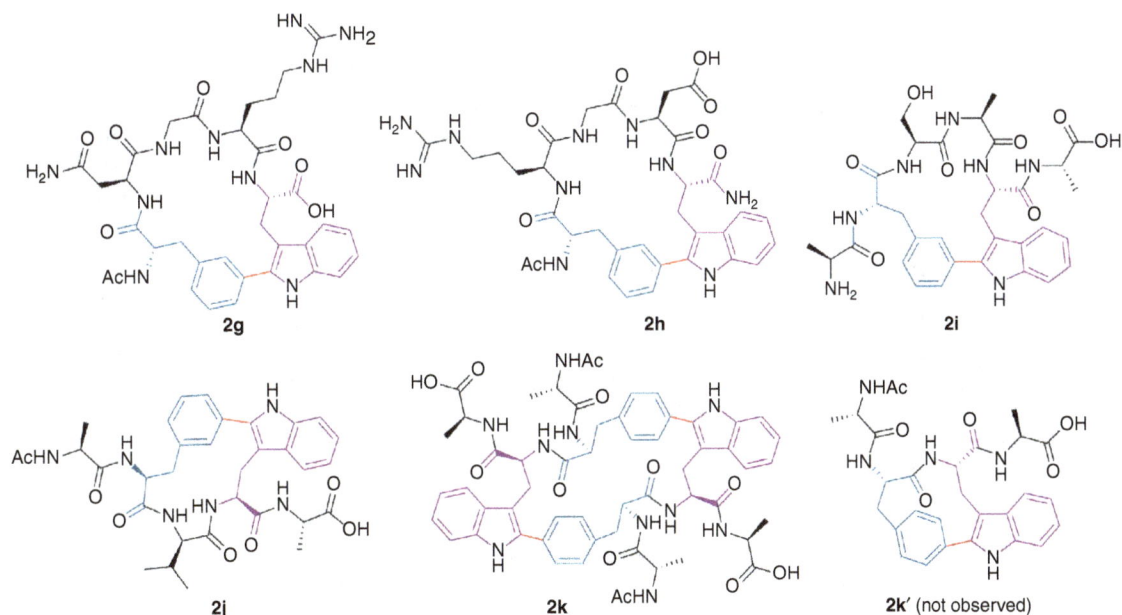

**Figure 2 | Structure of isolated locked peptides 2g–2k.** Schematic representation: Phe (blue), Trp (purple) and staple bond (red).

## Results

**Preliminary studies.** Here we present a one-step process for the synthesis of Trp–Phe(Tyr)-stapled peptides directly from commercially available precursors. The method is based on an intramolecular Pd-catalysed C–H activation reaction between a Trp residue and an iodo-phenylalanine (or tyrosine) unit (Fig. 1b).

After a preliminary modelling study, we established the structural features for the intramolecular C–H arylation, placing the iodo-substituent for Phe or Tyr amino acids in the *meta* position. The preferential distances between these latter residues and Trp ranged from one to three amino acids in a series of linear N-terminal-acetylated sequences.

The initial experiments were carried out using the conditions previously applied in the arylation of the Trp diketopiperazine:[41] Overall, 5 mol% Pd(OAc)₂, AgBF₄ (1.0 eq.) and *o*-nitrobenzoic acid (2-NO₂BzOH) (1.5 eq.) in PBS:DMF (1:1) under microwave (MW) irradiation at 80 °C for 15 min (Table 1, entries 1–6). These preliminary observations suggested that these arrangements are suitable for stapled peptide formation, affording good to excellent conversions for Trp–Phe(Tyr) staples located at i − i + 2, i − i + 3 and i − i + 4 positions (Table 1, entries 1–6). Furthermore, a series of representative sequences displaying from one to three amino acids between Trp and *m*-iodinated Phe were constrained in useful yields (Table 1, entries 7–10; Fig. 2; see Supplementary Fig. 1), in an

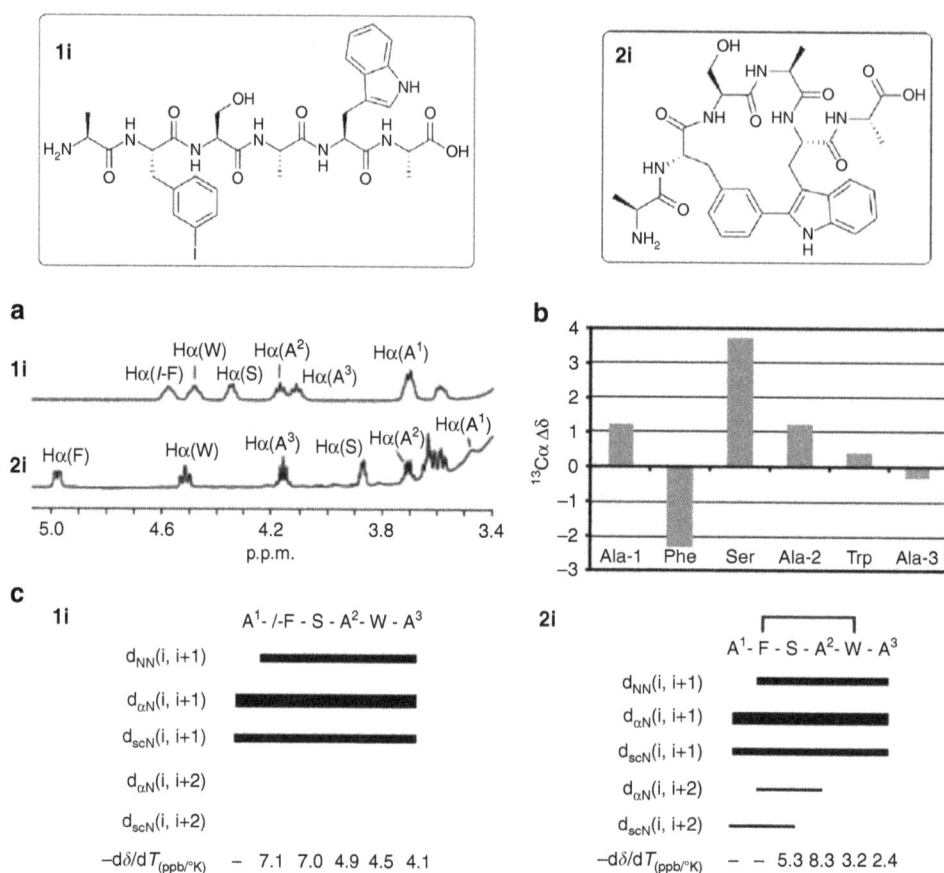

**Figure 3 | Peptide NMR spectra comparison between stapled peptide 2i and its linear counterpart 1i.** (a) NMR $H_\alpha$ region of peptide **2i** and its linear precursor **1i**. (**b**) Plot of the $^{13}C_\alpha$ chemical shift differences ($^{13}C_\alpha \Delta\delta_{cyclic-linear}$) between stapled peptide **2i** and its linear counterpart **1i**. (**c**) Summary of NOE connectivity and temperature coefficients of the NH amide protons ($\Delta\delta/\Delta T$) of peptide **1i** (bottom left) and **2i** (bottom right). The thickness of the bars reflects the intensity of the NOEs, that is, weak (➖), medium (▬) and strong (■). *I*-F: *m*-iodophenylalanine.

anhydrous medium using trifluoroacetic acid (TFA) as the acid, applying the optimized conditions previously reported for the preparation of Fmoc-protected arylated Trps[42]. Although these transformations may lead to atropoisomeric diastereomers, only one stereochemically defined structure was observed in each case. Of note is the successful application of this technique to stapled peptides containing the Asn-Gly-Arg (-NGR-) and Arg-Gly-Asp (-RGD-) tumour-homing signalling sequences (**2g** and **2h**, Table 1, entries 7 and 8, respectively).

Preliminary studies on the capacity of the RGD-containing compound **2h** and its linear precursor **1h** to inhibit cellular adhesion showed that these substances act selectively as antagonists for the $\alpha v\beta 3$ in front of $\alpha v\beta 5$ integrin receptors; compound **2h** showing a moderate $EC_{50}$ (6 μM) is more active than its linear precursor (**1h**, 26 μM, see Biochemical and Cellular Studies in Supplementary Methods). Compound **2h** is considerably less potent than cilengitide, not surprisingly as this drug has been thoroughly optimized. However, it has to be considered that even linear analogues display a much lower potency (10–10³ times, depending on the targeted integrin) than the parent cilengitide[43,44]. These preliminary results clearly show that this stapling technique preserves the activity of the natural sequence, while improving its potency, in line with pioneering experiments in macrocyclization techniques for medchem peptide development[45,46].

We next tackled the preparation of the more challenging locked peptide $i - i + 1$. Three sequences with adjacent Trp—(*ortho*-, *meta*- or *para*)-I-Phe were synthesized. A constrained C–C linked structure was obtained for the *meta*-derivative (60%),

although it could not be properly characterized. With respect to the *ortho* analogue, only the reduced peptide was detected. Interestingly, the *para*-I-Phe peptide **1k** was successfully reacted under the usual conditions to yield cyclodimer **2k** (60%, Table 1, entry 11; see Supplementary Figs 2 and 3). Presumably, the putative monomeric structure **2k′** would be highly strained (molecular models display a nonplanar phenyl ring, see Supplementary Fig. 4) and the process evolves through a ditopic pathway, in a remarkable demonstration of the process versatility.

A detailed nuclear magnetic resonance (NMR) study was performed on dimethylsulphoxide to analyse the conformational behaviour of the peptides synthesized. The NMR spectra of linear sequences **1g–1j** were indicative of flexible, unstructured peptides. $H_\alpha$ and $^{13}C_\alpha$ chemical shifts displayed values typical for random coil, and the NH temperature coefficients and $^3J(H_\alpha NH)$ couplings were within the range expected for unstructured peptides[47].

Staple bond formation caused a substantial modification of the peptide NMR spectra. Compared with their linear precursors, peptides **2g–2j** showed larger $H_\alpha$ chemical shift dispersion (Fig. 3a and Supplementary Figs 5–7), indicating less conformational flexibility. However, broad resonances were observed at 25 °C for some backbone NH protons, suggesting that peptides **2g–2j** show some flexibility. Significant differences in $^{13}C_\alpha$ chemical shifts were observed between stapled peptides **2g–2j** and their linear counterparts **1g–1j** (Fig. 3b and Supplementary Figs 5–7). The temperature dependence of the amide NH groups was also significantly affected by staple formation (Fig. 3c and Supplementary Figs 5–7). Stapled peptides showed a wider distribution of $\Delta\delta/\Delta T$ values within the amino-acid sequence and

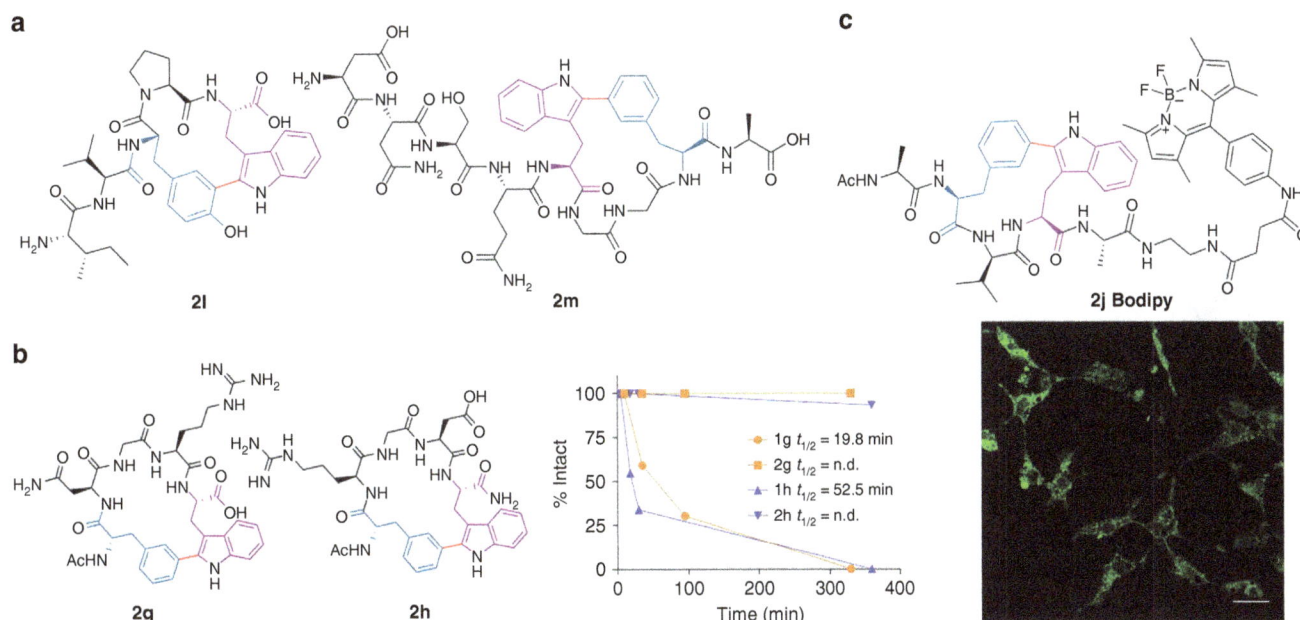

**Figure 4 | Bioactive stapled peptides, biochemical and cellular studies.** (**a**) Structure of stapled peptides **2l** (valorphin analogue) and **2m** (baratin analogue). (**b**) Proteolytic degradation assay of stapled peptides **2g** and **2h** and their linear precursors **1g** and **1h**. (**c**) Structure of labelled-stapled peptide **2j Bodipy** (left) and the corresponding confocal microscopy image of SH-SY5Y cells treated with compound **2j Bodipy** (750 nM). Scale bar, 25 μm (right).

also featured values characteristic of solvent-shielded NH groups (lower than $-3$ ppb/K). NOE connectivity was also affected by staple formation. In the case of peptides **2h**, **2i** and **2j**, several nonsequential NOEs indicated that the peptides were constrained by intramolecular cyclization (Fig. 3c and Supplementary Figs 6 and 7). Incidentally, we detect all $N_{indole}$–H atoms, and also we find the corresponding indole Cα positions consequently substituted, therefore ruling out any interference by N-arylation processes. The aryl–aryl moieties show NOE correlations consistent with a defined single configuration in each case, which matches with the arrangement predicted from molecular modelling.

Moreover, circular dichroism measurements of stapled peptides **2g** and **2h** were performed in order to detect evidence of secondary structure and were then compared with their linear precursors **1g** and **1h**, respectively (See Supplementary Figs 8–11). Stapled peptides **2** show a positive maximum at $\sim 190$ nm and minimum peaks at 206 nm, which indicate some levels of structuring. In contrast, linear peptides exhibit a more flattened profile typical of a flexible unfolded structure.

**Extension to biologically relevant peptides.** To further explore the potential of the methodology, we planned to make the stapled analogues of known linear bioactive peptides, and at the same time test the influence of new features, such as the presence of proline (Pro) in the sequence, the overall length of the chain (up to nine amino acids) and the possibility of performing the C–H arylation on the solid phase, which enables the transformation of unprotected N-terminal peptides. In this way, we prepared the stapled version of an active valorphin analogue[48], which is a potent dipeptidyl peptidase III inhibitor, displaying Pro in a five amino-acid sequence containing also Trp and Tyr (see Supplementary Fig. 12). Interestingly, this peptide is closely related to spinorphin, an endogenous antinociceptive peptide, a potent and noncompetitive antagonist at the ATP-activated human P2X3 receptor[49]. The SPPS method was used to synthesize the Fmoc-protected sequence, which was intramolecularly arylated on resin. Subsequent N-terminal

deprotection and cleavage, successfully afforded the stapled peptide **2l** (Fig. 4a). This solid-phase protocol is fully compatible with the Pd chemistry involved and, interestingly, allows the preparation of arylated sequences having unprotected terminal amino groups. Futhermore, baratin[50], a neurostimulating peptide, was stapled following a related procedure. In this way, the standard SPPS protocol was interrupted to perform the on-resin C–H activation step, to be continued with the incorporation of the final four amino acids. Afterwards, deprotection and cleavage afforded the desired derivative **2m** (Fig. 4a, see Supplementary Methods and Supplementary Fig. 12).

Next, we determined the proteolytic stability of a couple of meaningful stapled peptides, in comparison with their respective linear counterparts. We followed a chymotrypsin-based protocol, previously used to evaluate stapled peptides[51]. Plotting the HPLC-MS profiles of the couples **2g/1g** and **2h/1h** (Fig. 4b, also Supplementary Figs 13–15) clearly shows an almost total protection of the stapled peptides towards enzymatic degradation, after 5–6 h, whereas the linear precursors suffered a rapid hydrolytic cleavage to remove the N-terminal amino acid, and completely disappeared after this time lapse.

Finally, we studied the controlled labelling of the stapled peptides. The goal was to selectively attach a fluorophore to the C-terminal amino acid to trace cellular permeabilization and localization. Thus, we linked a Bodipy residue to cyclopeptide **2j** through an amino spacer (the novel Bodipy construct was designed and prepared for this purpose, see Supplementary Fig. 16 and Supplementary Methods), to get the desired labelled-stapled peptide **2j Bodipy** (Fig. 4c) in a convenient manner. This compound exhibits low cytotoxicity at 750 nM after 24 h of incubation in SH-SY5Y cells (MTT assay; see Supplementary Fig. 17), and this concentration was used for the cell penetration studies. In a preliminary flow cytometry experiment (fluorescence-activated cell sorting assay), SH-SY5Y cells resulted in brightly stained sections on incubation for 30 min (see Supplementary Fig. 18). The cells treated in this manner were analysed using confocal microscopy. The stapled peptide **2j Bodipy** was localized both in the membrane and in the cytoplasmatic region (Fig. 4c). Incidentally, the Bodipy-labelled

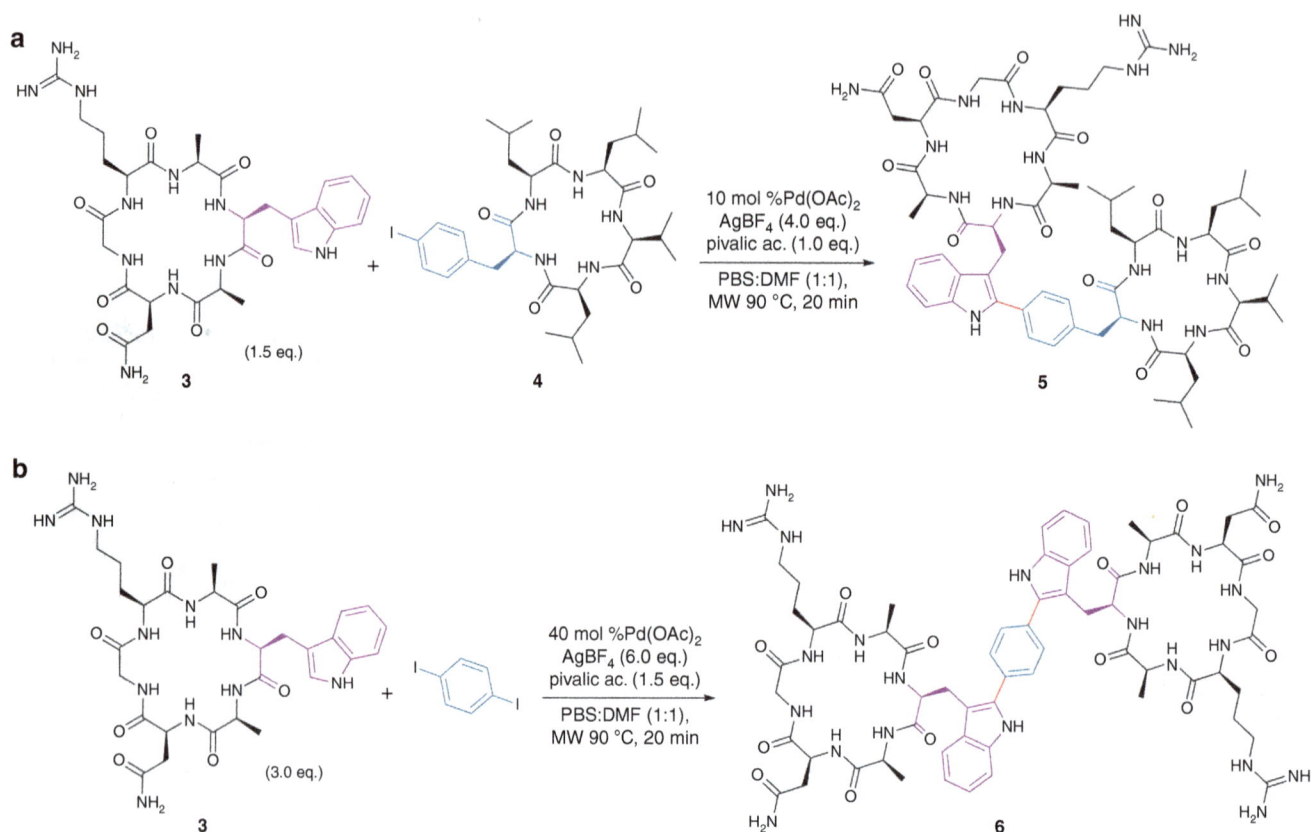

**Figure 5 | Macrocyclic conjugation via C–H activation. (a)** Intermolecular conjugation of NGR cyclopeptide **3** with the sansalvamide derivative **4** via C–H activation. **(b)** Double conjugation of the NGR cyclopeptide **3** with 1,4-diodobenzene.

linear precursor **1j** **Bodipy** also penetrates into cells in a comparable extent; however, there are appreciable differences in toxicity and cell permeability with respect to the stapled analogue (see Supplementary Figs 19 and 20). Overall, these results enable the performance of systematic bioimaging studies on these compounds.

**Intermolecular peptide conjugation through C–H activation.** Chemical conjugation can effectively link drugs to carriers, and this technique is routinely used to modify biologics, especially antibodies with therapeutic indications[52]. We next explored peptide–peptide conjugation via C–H activation to achieve bismacrocyclic peptide constructs linked through a nonhydrolysable bond.

Peptides that inhibit the new blood vessel growth (angiogenesis) have become a promising tool for treating cancer. In this context, it has been reported that cyclic forms of NGR-containing sequences (tumour-homing peptides) lead to an improvement of the anticancer activity of an associated drug[53]. We selected the previously studied Asn-Gly-Arg (NGR) array and synthesized the corresponding Trp-containing macrocycle **3** to be conjugated to a synthetic p-I-Phe-cyclopeptide derivative of the cyclic depsipeptide sansalvamide A (**4**, Fig. 5a), a natural product whose synthetic analogues have demonstrated significant anticancer activity[54]. Using a standard Pd-catalysed reaction in an aqueous medium of PBS–DMF, the two partners were successfully coupled to yield the C–C conjugate **5** (Fig. 5a).

In a second example, the conjugation of two units of the NGR Trp-containing cyclopeptide **3** with a 1,4-diiodobenzene connector via a double C–H arylation was achieved. Again, the process yielded the expected bis-NGR-adduct **6** in a single step (Fig. 5b).

To evaluate the effects of macrocyclic conjugation, the NMR spectra of conjugated peptide **5** and its cyclic precursors **3** and **4** were compared (see Supplementary Fig. 21). After conjugation, small shifts were observed in the $H_\alpha$ resonances of the sansalvamide A analogue **4**. In contrast, the $H_\alpha$ chemical shifts of the NGR-containing cycle were almost identical in peptides **5** and **3**. Comparison of $^{13}C_\alpha$ chemical shifts showed that in both cycles the larger $^{13}C_\alpha$ chemical shift change corresponded to the residue involved in the intermolecular bond formation: Trp (1.2 p.p.m.) in the NGR-containing cycle and Phe (0.6 p.p.m.) in the sansalvamide A derivative. More modest changes were observed in the $^{13}C_\alpha$ of the remaining amino acids. The resemblance of the $H_\alpha$ and $^{13}C_\alpha$ chemical shifts, which are highly sensitive to conformational changes, suggests that conjugation provided a minimal perturbation of the overall structures of peptides **3** and **4**. The NMR analysis was extended to peptide **6** and its precursor **3**. The effects of conjugation in the $H_\alpha$ and $C_\alpha$ resonances are shown in Supplementary Fig. 22.

Unfortunately, in preliminary biological studies evaluating the conjugated peptide **5** and its macrocycle precursors **3** and **4** against several cancer cell lines, only the cyclopeptide **4** showed activity (IC$_{50}$ < 10 µM). Nevertheless, the protocol seems general and offers new possibilities for peptide-based conjugation.

**Stapled cyclopeptides.** We then explored access to complex topologies, focusing on the synthesis of bicyclic peptide chemotypes. In a first approach, the synthesis of a stapled cyclopeptide was attempted by promoting the usual C–H arylation through an intramolecular Trp-I–Phe interaction. Preliminary studies showed that alanine (Ala) hexapeptides containing a m-I-Phe and Trp units, respectively, placed at positions i − i + 2 and i − i + 3 afforded the corresponding macrocycles after routine amide

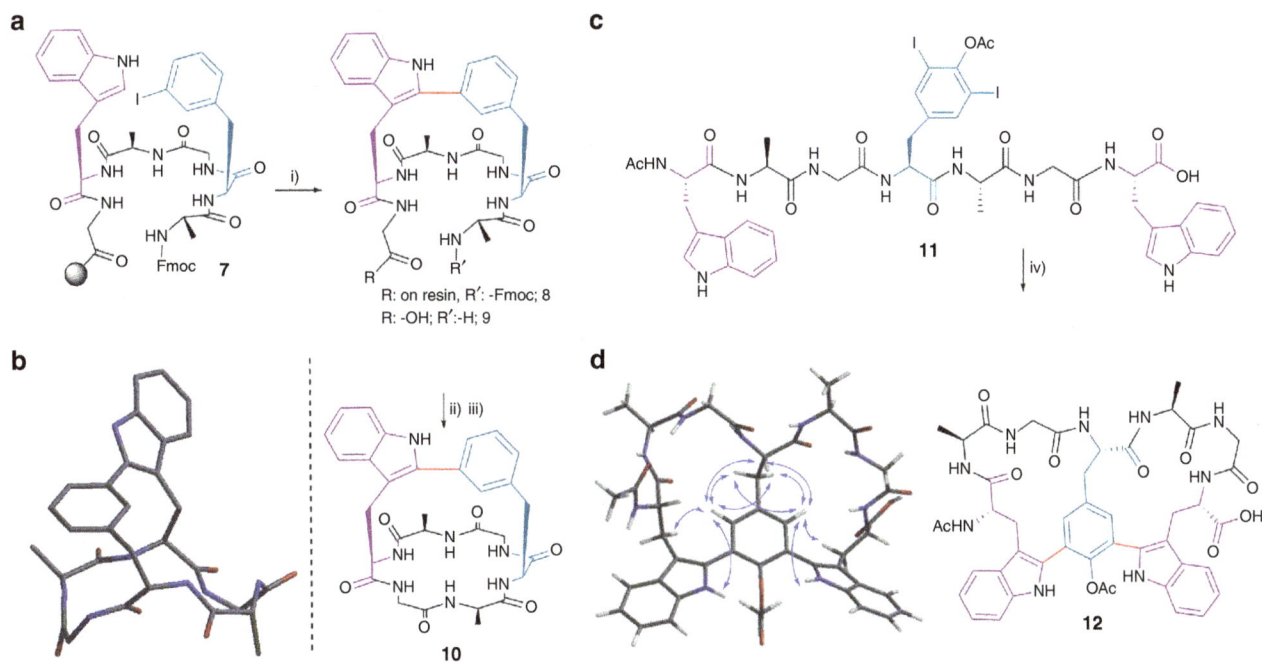

**Figure 6 | Synthesis of macrobicyclic peptides 10 and 12. (a)** Synthesis of macrobicyclic peptide **10** through solid-phase stapling. Reaction conditions: (i) Pd(OAc)$_2$ (0.05 eq.), AgBF$_4$ (1.0 eq.), 2-NO$_2$BzOH (1.5 eq.), DMF, MW 90 °C, 20 min; (ii) (1) 1% sodium diethyldithiocarbamate (DDC) in DMF. (2) Piperidine-DMF (1:4; 1 × 1 min, 2 × 5 min). (3) TFA-TIS-H$_2$O (95:2.5:2.5), r.t, 1 h; (iii) PyAOP (2.0 eq.), DIEA (6.0 eq.), DMF, r.t, 1.5 h. (**b**) Minimized geometry of compound **10** generated by the Spartan '14 suite. Hydrogens omitted for clarity. (**c**) Double C–H arylation to cyclic biaryl **12** (25% conversion, estimated by HPLC). Reaction conditions: (iv) 40 mol %Pd(OAc)$_2$, AgBF$_4$ (6.0 eq.), pivalic ac. (1.5 eq.), DMF, MW 90 °C, 20 min. (**d**) Minimized geometry of compound **12** generated by Spartan '14 suite showing the diagnostic NOE correlations (blue arrows).

coupling in quantitative yields; however, the subsequent intramolecular C–H arylations did not take place under the standard conditions, probably because of a highly restricted conformation. Remarkably, when the stapling on the linear sequence anchored to the resin **7** was carried out first on the solid phase, we obtained the corresponding NH-free amino terminal-stapled peptide **9** after cleavage from resin. Finally, we isolated the desired bicyclic compound **10** by amide cyclization (Fig. 6a,b). Furthermore, in this way, we overcame the relative limitation of irreversible protection of the N-terminal amino group, as previously reported[37]. Incidentally, preliminary results showed that increasing the Pd(OAc)$_2$ amount up to 0.2 eq. gave acceptable intermolecular arylations for NH-free amino terminal sequences (data not shown). The stapled cyclopeptide **10** was prepared in a suitable manner, involving a solid-phase arylation, thus facilitating the purification step, followed by a routine peptide coupling. This protocol may enable the synthesis of further derivatives of this attractive and unexplored structural class.

The $^1$H NMR spectrum of the linear stapled peptide **9** was characterized by a large chemical shift dispersion of the NH and H$_\alpha$ resonances. Bicyclopeptide **10** showed a slightly larger chemical shift range for the H$_\alpha$ protons, as expected for a more rigid and structured peptide. Backbone cyclization was also reflected in a large splitting of the methylene H$_\alpha$ atoms of Gly-1 and H$_\beta$ atoms of Phe. In the case of peptide **9**, the chemical shift difference between the two geminal protons was $<0.2$ (H$_\alpha$, Gly-1) and 0.1 (H$_\beta$, Phe) p.p.m., whereas for peptide **10** this difference increased to $>0.9$ (H$_\alpha$, Gly-1) and $>0.7$ (H$_\beta$, Phe) p.p.m. Backbone cyclization was further evidenced by significant changes in $^{13}$C$_\alpha$ chemical shifts and in the temperature coefficients of amide NH (Supplementary Fig. 23).

**Double stapling of linear peptides.** Biaryl bismacrocyclic peptide-derived natural products such as vancomycin[55] and complestatin[56] display very interesting bioactivity profiles and

as synthetic targets they are extremely difficult to prepare[57]. Thus, the development of new methodologies to access simplified scaffolds is crucial to enable structural diversification, thereby allowing practical medicinal chemistry and biological studies. Hence, it was envisioned that intramolecular double C–H arylation of a sequence containing a diiodinated Tyr (commercially available) flanked by two Trp units would give raise to bicyclic peptide topologies with adjacent biaryl moieties in a straightforward manner. In order to establish the conditions for this transformation, we performed some preliminary experiments where the intermolecular C–H activation of a Ac-$m,m'$-I,I-Tyr(OAc)-OH unit with 2.0 eq. of Ac-Trp-OH was tested. An increase in the amount of the Pd catalyst (40%) and AgBF$_4$ (6.0 eq.)[58] and use of a mild excess of pivalic acid resulted in a productive reaction (see Supplementary Methods). Next, we designed the linear peptide sequence **11** (Fig. 6c), which was synthesized in a straightforward manner on the solid phase. Using the previous C–H activation conditions for the diodoTyr, we successfully obtained the double stapled peptide **12** directly from the corresponding linear precursor in one step in a 25% HPLC conversion, together with monoarylated cycles and dehalogenated derivatives, which were probably produced in competitive processes (Fig. 6c). This remarkable result is the proof of principle that even these complex peptidic topologies are accessible through the present methodology. The $^1$H NMR spectrum of the double stapled peptide **12** is characterized by a wide chemical shift range for the NH and H$_\alpha$ protons (see Supplementary Fig. 134). The observed pattern of NOEs, summarized in Fig. 6d, indicates that each side of the Tyr aromatic ring faces a distinct Trp-Ala-Gly motif.

As an optimization of this methodology, we have developed the exclusive use of standard proteinogenic amino acids (Tyr), which, once coupled, are modified in situ, thus simplifying the protocol and rendering it considerably more affordable. In this way, the corresponding linear precursor of peptide **11** can be obtained through routine amide couplings followed by on-resin iodination of the Tyr residue to yield derivative **11** (41% conversion, unoptimized, see Supplementary Fig. 24)[59]. This remarkable result enables the direct synthesis of bicyclopeptide **12** to be performed on solid-phase from commercially available reagents in a single sequence.

## Discussion

Metal-catalysed arylations through C–H activation are suitable processes to gain access to minimalistic staples (two-electron) in relevant Trp-containing peptide sequences directly from Trp and iodo-phenylalanine (-tyrosine) precursors. The validation of the methodology includes the analysis of the scope of the transformation, compatibility with other amino acids and the applicability to SPPS. This approach has also been applied to constrain biologically active signalling sequences. In an intermolecular mode, the process efficiently links two Trp peptides to a benzene connector and is also useful to conjugate peptides through a C–C bond. All compounds showed a structured nature, as revealed by spectroscopic characterization. Finally, we have developed a simple protocol for the straightforward access to novel peptide topologies such as dimeric macrocycles, stapled bicyclopeptides and biaryl–biaryl species (see Supplementary Figs 4 and 25) from the corresponding linear precursors in only one step. These findings open up general access to a variety of novel constrained peptidic chemotypes, and we believe that this breakthrough will make a significant contribution to the development of a broad range of applications for peptides in biological and medicinal chemistry.

## Methods

**General.** For abbreviations and detailed experimental procedures see Supplementary Methods. For NMR analysis of the compounds, see Supplementary Figs 26–148 and Supplementary Tables 1–40.

**Peptide synthesis.** All peptides were manually synthesized on a 2-Chlorotrityl, H-Rink-Amide Chemmatrix or TentaGel S $NH_2$ resin using standard Fmoc chemistry for SPPS.

**Linear peptide cyclization.** The free-amine free-acid linear peptide (1.0 eq.) was dissolved in ACN/DMF or DMF (0.001–0.003 M) and N,N-diisopropylethylamine (DIEA) (6.0 eq.) and the corresponding coupling agents (1.5–3.0 eq., benzotriazol-1-yl-oxytripyrrolidinophosphonium hexafluorophosphate (PyBOP) with hydroxybenzotriazole (HOBt) or 1-[Bis(dimethylamino)methylene]-1H-1,2,3-triazolo[4,5-b]pyridinium 3-oxid hexafluorophosphate (HATU) and O-(benzotriazol-1-yl)-N,N,N′,N′-tetramethyluronium tetrafluoroborate (TBTU) or (7-azabenzotriazol-1-yloxy)tripyrrolidinophosphonium hexafluorophosphate (PyAOP)) were added. The solution was stirred at r.t until the cyclization was complete (1–3 h). Workup was performed by extraction with aqueous solutions of $NH_4Cl_{sat}$ and $NaHCO_{3sat}$. Organic layers were combined, dried over anhydrous sodium sulfate, filtered and concentrated under vacuum. When remaining protecting groups were present, the macrocycle was treated with a 95% TFA, 2.5% triisopropylsilane (TIS) and 2.5% $H_2O$ cocktail (3 h), washed with $Et_2O$, dissolved in $ACN:H_2O$ and lyophilized to furnish the corresponding deprotected peptide. When necessary, the macrocycle was purified by flash column chromatography on silica gel or semipreparative RP-HPLC.

**General protocol for the stapled bond formation in solution.** The linear peptide (50 mg), $AgBF_4$ (2.0 eq.), trifluoroacetic acid (1.0 eq.) and $Pd(OAc)_2$ (0.05 eq.) were placed in a MW reactor vessel in DMF (1.2 ml). The mixture was heated under MW irradiation (250 W) at 90 °C for 20 min. The residue was filtered and purified using semipreparative RP-HPLC. This process was scaled up to 0.907 mmol of peptide, affording the stapled/locked peptides **2g**–**2k** in isolated yields ranging from 1 to 32%.

**Procedure for the single macrocycle conjugation.** Cyclopeptide **4** (40.0 mg, 0.056 mmol), cyclopeptide **3** (55.3 mg, 0.084 mmol, 1.5 eq.), $AgBF_4$ (43.8 mg, 0.225 mmol, 4.0 eq.), pivalic acid (5.7 mg, 0.056 mmol, 1.0 eq.) and $Pd(OAc)_2$ (1.4 mg, 0.077 mmol, 0.1 eq.) were placed in a MW reactor vessel in 2 ml of PBS:DMF (1:1). The mixture was heated under MW irradiation (250 W) at 90 °C for 20 min. The irradiation cycle was repeated after adding a new portion of $Pd(OAc)_2$ and $AgBF_4$. The residue was filtered and partially purified in a PoraPak Rxn reverse phase column (1.63 mg, 2% estimated using HPLC-MS). A pure fraction was obtained using analytic RP-HPLC to yield pure conjugate **5**.

**Procedure for the double macrocycle conjugation.** 1,4-diiodobenzene (35 mg, 0.106 mmol), macrocycle **3** (209 mg, 0.318 mmol, 3.0 eq.), $AgBF_4$ (124 mg, 0.637 mmol, 6.0 eq.), pivalic acid (16.3 mg, 0.159 mmol, 1.5 eq.) and $Pd(OAc)_2$ (9.5 mg, 0.042 mmol, 0.4 eq.) were placed in a MW reactor vessel in 2 ml of PBS:DMF (1:1). The mixture was heated under MW irradiation (250 W) at 90 °C for 20 min. The crude product was filtered, and the workup was carried out by washing with AcOEt and then precipitating by adding ACN to the aqueous phase. The resulting precipitate was washed with ACN, decanted and dried, obtaining 159 mg of crude product (pale solid, 42% estimated using HPLC-MS). A pure fraction was obtained with semipreparative RP-HPLC to yield pure conjugate **6**.

**Typical procedure for the stapled cyclopeptide formation on the solid phase.** Once sequence **7** was synthesized on a TentaGel S $NH_2$ resin via an AB linker, the peptide anchored to the resin (139 mg, 0.145 mmol), $AgBF_4$ (28 mg, 0.144 mmol, 1.0 eq.), 2-nitrobenzoic acid (36 mg, 0.215 mmol, 1.5 eq.) and $Pd(OAc)_2$ (1.6 mg, 7.1 μmol, 0.05 eq.) in DMF (2 ml) was placed in a MW reactor vessel. The mixture was heated under MW irradiation (250 W) at 90 °C for 20 min. Eight more batches were carried out following the same procedure and were then combined. The resin was treated with 1% DDC in DMF and the Fmoc group was removed. The peptide was cleaved from the resin with a 95% TFA, 2.5% TIS and 2.5% $H_2O$ cocktail (1 h), yielding the stapled sequence **9** (85% purity, estimated using HPLC-MS). Finally, the stapled peptide **9** was cyclized (see above for standard cyclization procedure) and the crude product was purified using semipreparative RP-HPLC to yield bicyclopeptide **10** (pale solid, 18% unoptimized).

**One-step double arylation to biaryl–biaryl stapled peptide.** Linear peptide **11** (50 mg, 0.044 mmol), $AgBF_4$ (51 mg, 0.262 mmol, 6.0 eq.), pivalic acid (6.7 mg, 0.066 mmol, 1.5 eq.) and $Pd(OAc)_2$ (3.9 mg, 0.018 mmol, 0.4 eq.) were placed in a MW reactor vessel in DMF (500 μl). The mixture was heated under MW irradiation (250 W) at 90 °C for 20 min. Three more batches were carried out following the same procedure and then filtered and combined (25% conversion, estimated using HPLC-MS). A pure fraction of bicyclopeptide **12** was isolated using semipreparative RP-HPLC.

## References

1. Ramakers, B. E. I., van Hest, J. C. M. & Löwik, D. W. P. M. Molecular tools for the construction of peptide-based materials. *Chem. Soc. Rev.* **43**, 2743–2756 (2014).
2. Craik, D. J., Fairlie, D. P., Liras, S. & Price, D. The future of peptide-based drugs. *Chem. Biol. Drug Des.* **81**, 136–147 (2013).
3. Albericio, F. & Kruger, H. G. Therapeutic peptides. *Future Med. Chem.* **4**, 1527–1531 (2012).
4. Lawson, K. V., Rose, T. E. & Harran, P. G. Template-constrained macrocyclic peptides prepared from native, unprotected precursors. *Proc. Natl Acad. Sci. USA* **110**, E3753–E3760 (2013).
5. Marsault, E. *et al.* Efficient parallel synthesis of macrocyclic peptidomimetics. *Bioorg. Med. Chem. Lett.* **18**, 4731–4735 (2008).
6. Royo-Gracia, S., Gaus, K. & Sewald, N. Synthesis of chemically modified bioactive peptides: recent advances, challenges and developments for medicinal chemistry. *Future Med. Chem.* **1**, 1289–1310 (2009).
7. White, C. J. & Yudin, A. K. Contemporary strategies for peptide macrocyclization. *Nat. Chem.* **3**, 509–524 (2011).
8. Heinis, C. Tools and rules for macrocycles. *Nat. Chem. Biol.* **10**, 696–698 (2014).
9. Verdine, G. L. & Hilinski, G. J. Stapled peptides for intracellular drug targets. *Methods Enzymol.* **503**, 3–33, 2012).
10. Walensky, L. D. & Bird, G. H. Hydrocarbon-stapled peptides: principles, practice, and progress. *J. Med. Chem.* **57**, 6275–6288 (2014).
11. Lau, Y. H., de Andrade, P., Wu, Y. & Spring, D. R. Peptide stapling techniques based on different macrocyclisation chemistries. *Chem. Soc. Rev.* **44**, 91–102 (2015).
12. Góngora-Benítez, M., Tulla-Puche, J. & Albericio, F. Multifaceted roles of disulfide bonds. Peptides as therapeutics. *Chem. Rev.* **114**, 901–926 (2014).
13. Jo, H. *et al.* Development of α-helical calpain probes by mimicking a natural protein-protein interaction. *J. Am. Chem. Soc.* **134**, 17704–17713 (2012).
14. Spokoyny, A. M. *et al.* A perfluoroaryl-cysteine SNAr chemistry approach to unprotected peptide stapling. *J. Am. Chem. Soc.* **135**, 5946–5949 (2013).
15. Bois-Choussy, M., Cristau, P. & Zhu, J. Total synthesis of an atropdiastereomer of RP-66453 and determination of its absolute configuration. *Angew. Chem. Int. Ed.* **42**, 4238–4241 (2003).
16. Carbonnelle, A.-C. & Zhu, J. A novel synthesis of biaryl-containing macrocycles by a domino miyaura arylboronate formation: intramolecular suzuki reaction. *Org. Lett.* **2**, 3477–3480 (2000).
17. Afonso, A., Feliu, L. & Planas, M. Solid-phase synthesis of biaryl cyclic peptides by borylation and microwave-assisted intramolecular Suzuki–Miyaura reaction. *Tetrahedron* **67**, 2238–2245 (2011).

18. Meyer, F.-M. *et al.* Biaryl-bridged macrocyclic peptides: conformational constraint via carbogenic fusion of natural amino acid side chains. *J. Org. Chem.* **77**, 3099–3114 (2012).

19. Blackwell, H. E. & Grubbs, R. H. Highly efficient synthesis of covalently cross-linked peptide helices by ring-closing metathesis. *Angew. Chem. Int. Ed.* **37**, 3281–3284 (1998).

20. Dharanipragada, R. New modalities in conformationally constrained peptides for potency, selectivity and cell permeation. *Future Med. Chem.* **5**, 831–849 (2013).

21. Wender, P. A. Toward the ideal synthesis and molecular function through synthesis-informed design. *Nat. Prod. Rep.* **31**, 433–440 (2014).

22. Sletten, E. M. & Bertozzi, C. R. Bioorthogonal chemistry: fishing for selectivity in a sea of functionality. *Angew. Chem. Int. Ed.* **48**, 6974–6998 (2009).

23. Yu, J.-Q. & Shi, Z. *Topics in Current Chemistry* 384 (Springer, 2010).

24. Seechurn, C. C. C. J., Kitching, M. O., Colacot, T. J. & Snieckus, V. Palladium-catalyzed cross-coupling: a historical contextual perspective to the 2010 Nobel Prize. *Angew. Chem. Int. Ed.* **51**, 5062–5086 (2012).

25. Ackermann, L. Carboxylate-assisted transition-metal-catalyzed C-H bond functionalizations: mechanism and scope. *Chem. Rev.* **111**, 1315–1345 (2011).

26. McMurray, L., O'Hara, F. & Gaunt, M. J. Recent developments in natural product synthesis using metal-catalysed C-H bond functionalisation. *Chem. Soc. Rev.* **40**, 1885–1898 (2011).

27. Wencel-Delord, J., Dröge, T., Liu, F. & Glorius, F. Towards mild metal-catalyzed C-H bond activation. *Chem. Soc. Rev.* **40**, 4740–4761 (2011).

28. Daugulis, O., Do, H. & Shabashov, D. Palladium- and copper-catalyzed arylation of carbon-hydrogen bonds. *Acc. Chem. Res.* **42**, 1074–1086 (2009).

29. Noisier, F. M. & Brimble, M. A. C – H functionalization in the synthesis of amino acids and peptides. *Chem. Rev.* **114**, 8775–8806 (2014).

30. Rossi, R., Bellina, F., Lessi, M. & Manzini, C. Cross-Coupling of heteroarenes by C-H functionalization: recent progress towards direct arylation and heteroarylation reactions involving heteroarenes containing one heteroatom. *Adv. Synth. Catal.* **356**, 17–117 (2014).

31. Lebrasseur, N. & Larrosa, I. Recent advances in the C2 and C3 regioselective direct arylation of indoles. *Adv. Heterocycl. Chem.* **105**, 309–351 (2012).

32. Liégault, B., Petrov, I., Gorelsky, S. I. & Fagnou, K. Modulating reactivity and diverting selectivity in palladium-catalyzed heteroaromatic direct arylation through the use of a chloride activating/blocking group. *J. Org. Chem.* **75**, 1047–1060 (2010).

33. Wang, X., Gribkov, D. V. & Sames, D. Phosphine-free palladium-catalyzed C-H bond arylation of free (N-H)-indoles and pyrroles. *J. Org. Chem.* **72**, 1476–1479 (2007).

34. Phipps, R. J., Grimster, N. P. & Gaunt, M. J. Cu(II)-catalyzed direct and site-selective arylation of indoles under mild conditions. *J. Am. Chem. Soc.* **130**, 8172–8174 (2008).

35. Islam, S. & Larrosa, I. "On water", phosphine-free palladium-catalyzed room temperature C-H arylation of indoles. *Chem. Eur. J.* **19**, 15093–15096 (2013).

36. Lebrasseur, N. & Larrosa, I. Room temperature and phosphine free palladium catalyzed direct C-2 arylation of indoles. *J. Am. Chem. Soc.* **130**, 2926–2927 (2008).

37. Ruiz-Rodríguez, J., Albericio, F. & Lavilla, R. Postsynthetic modification of peptides: chemoselective C-arylation of tryptophan residues. *Chem. Eur. J.* **16**, 1124–1127 (2010).

38. Williams, T. J., Reay, A. J., Whitwood, A. C. & Fairlamb, I. J. S. A mild and selective Pd-mediated methodology for the synthesis of highly fluorescent 2-arylated tryptophans and tryptophan-containing peptides: a catalytic role for Pd(0) nanoparticles? *Chem. Commun.* **50**, 3052–3054 (2014).

39. Dong, H., Limberakis, C., Liras, S., Price, D. & James, K. Peptidic macrocyclization via palladium-catalyzed chemoselective indole C-2 arylation. *Chem. Commun.* **48**, 11644–11646 (2012).

40. Monguchi, Y., Marumoto, T., Takamatsu, H., Sawama, Y. & Sajiki, H. Palladium on carbon-catalyzed one-pot N-arylindole synthesis: intramolecular aromatic amination, aromatization, and intermolecular aromatic amination. *Adv. Synth. Catal.* **356**, 1866–1872 (2014).

41. Preciado, S. *et al.* Synthesis and biological evaluation of a post-synthetically modified Trp-based diketopiperazine. *Med. Chem. Comm.* **4**, 1171–1174 (2013).

42. Preciado, S., Mendive-Tapia, L., Albericio, F. & Lavilla, R. Synthesis of C-2 arylated tryptophan amino acids and related compounds through palladium-catalyzed C-H activation. *J. Org. Chem.* **78**, 8129–8135 (2013).

43. Mas-moruno, C., Rechenmacher, F. & Kessler, H. Cilengitide: the first anti-angiogenic small molecule drug candidate. Design, synthesis and clinical evaluation. *Anticancer Agents Med. Chem.* **10**, 753–768 (2010).

44. Weide, T., Modlinger, A. & Kessler, H. Spatial screening for the identification of the bioactive conformation of integrin ligands. *Top. Curr. Chem.* **272**, 1–50 (2007).

45. Boger, D. L. & Myers, J. B. Design and synthesis of a conformational analogue of deoxybouvardin. *J. Org. Chem.* **56**, 5385–5390 (1991).

46. Jackson, S. *et al.* Template-constrained cyclic peptides: design of high-affinity ligands for GPIIb/IIIa. *J. Am. Chem. Soc.* **116**, 3220–3230 (1994).

47. Dyson, H. J. & Wright, P. E. Nuclear Magnetic Resonance methods for elucidation of structure and dynamics in disordered states. *Methods Enzymol.* **339**, 258–270 (2001).

48. Chiba, T. *et al.* Inhibition of recombinant dipeptidyl peptidase III by synthetic hemorphin-like peptides. *Peptides* **24**, 773–778 (2003).

49. Jung, K.-Y. *et al.* Structure - activity relationship studies of spinorphin as a potent and selective human P2X3 receptor antagonist. *J. Med. Chem.* **50**, 4543–4547 (2007).

50. Nässel, D. R., Persson, M. G. S. & Muren, J. E. Baratin, a nonamidated neurostimulating neuropeptide, isolated from cockroach brain: distribution and actions in the cockroach and locust nervous systems. *J. Comp. Neurol.* **286**, 267–286 (2000).

51. Bird, G. H. *et al.* Hydrocarbon double-stapling remedies the proteolytic instability of a lengthy peptide therapeutic. *Proc. Natl Acad. Sci. USA* **107**, 14093–14098 (2010).

52. Du, A. W. & Stenzel, M. H. Drug carriers for the delivery of therapeutic peptides. *Biomacromolecules* **15**, 1097–1114 (2014).

53. Colombo, G. *et al.* Structure-activity relationships of linear and cyclic peptides containing the NGR tumor-homing motif. *J. Biol. Chem.* **277**, 47891–47897 (2002).

54. Pan, P.-S. *et al.* A comprehensive study of Sansalvamide A derivatives: the structure-activity relationships of 78 derivatives in two pancreatic cancer cell lines. *Bioorg. Med. Chem.* **17**, 5806–5825 (2009).

55. Boger, D. L. *et al.* Total synthesis of the vancomycin aglycon. *J. Am. Chem. Soc.* **121**, 10004–10011 (1999).

56. Wang, Z., Bois-Choussy, M., Jia, Y. & Zhu, J. Total synthesis of complestatin (chloropeptin II). *Angew. Chem. Int. Ed.* **49**, 2018–2022 (2010).

57. Feliu, L. & Planas, M. Cyclic peptides containing biaryl and biaryl ether linkages. *Int. J. Pept. Res. Ther.* **11**, 53–97 (2005).

58. Arroniz, C., Denis, J. G., Ironmonger, A., Rassias, G. & Larrosa, I. An organic cation as a silver(i) analogue for the arylation of sp2 and sp3 C–H bonds with iodoarenes. *Chem. Sci.* **5**, 3509–3514 (2014).

59. Arsequell, G. *et al.* First aromatic electrophilic iodination reaction on the solid-phase: iodination of bioactive peptides. *Tetrahedron Lett.* **39**, 7393–7396 (1998).

## Acknowledgements

This work was supported by DGICYT—Spain (project BQU-CTQ2012-30930), Generalitat de Cataluña (2014 SGR 137) and Institute for Research in Biomedicine Barcelona (Spain). We acknowledge an FPU fellowship for L.M.-T. from the Ministerio de Educación, Cultura y Deporte–Spain (MECD). NMR instruments were made available by the Scientific and Technological Centre of the University of Barcelona (CCiT UB). Dr M. J. Macias, Dr M. Vilaseca, Dr M. Teixidó and J. Garcia (IRB Barcelona), Dr F. Mitjans (BioLeitat), Dr M. Royo (Barcelona Science Park) and Professors R. Pérez-Tomás and V. Soto-Cerrato (U. Barcelona) are gratefully acknowledged for useful suggestions and for the biological assays.

## Author contributions

L.M.-T., F.A. and R.L. wrote the manuscript. L.M.-T., S.P., N.K. and R.R. performed the experiments, compound characterization and data analysis. J.G. performed the NMR studies. R.L. and F.A. supervised the research and evaluated all the data.

## Additional information

**Competing financial interests:** The authors declare no competing financial interests.

# 17

# Direct hydrodeoxygenation of raw woody biomass into liquid alkanes

Qineng Xia[1,*], Zongjia Chen[1,*], Yi Shao[1], Xueqing Gong[1], Haifeng Wang[1], Xiaohui Liu[1], Stewart F. Parker[2], Xue Han[3,4], Sihai Yang[4] & Yanqin Wang[1]

Being the only sustainable source of organic carbon, biomass is playing an ever-increasingly important role in our energy landscape. The conversion of renewable lignocellulosic biomass into liquid fuels is particularly attractive but extremely challenging due to the inertness and complexity of lignocellulose. Here we describe the direct hydrodeoxygenation of raw woods into liquid alkanes with mass yields up to 28.1 wt% over a multifunctional Pt/NbOPO$_4$ catalyst in cyclohexane. The superior performance of this catalyst allows simultaneous conversion of cellulose, hemicellulose and, more significantly, lignin fractions in the wood sawdust into hexane, pentane and alkylcyclohexanes, respectively. Investigation on the molecular mechanism reveals that a synergistic effect between Pt, NbO$_x$ species and acidic sites promotes this highly efficient hydrodeoxygenation of bulk lignocellulose. No chemical pretreatment of the raw woody biomass or separation is required for this one-pot process, which opens a general and energy-efficient route for converting raw lignocellulose into valuable alkanes.

[1] Key Laboratory for Advanced Materials, Research Institute of Industrial Catalysis, East China University of Science and Technology, Shanghai 200237, China. [2] ISIS Facility, STFC Rutherford Appleton Laboratory, Chilton, Oxfordshire OX11 0QX, UK. [3] School of Chemistry, University of Nottingham, Nottingham, NG7 2RD, UK. [4] School of Chemistry, University of Manchester, Manchester M13 9PL, UK. * These authors contributed equally to this work. Correspondence and requests for materials should be addressed to H.F.W. (email: hfwang@ecust.edu.cn) or to S.Y. (email: Sihai.Yang@manchester.ac.uk) or to Y.Q.W. (email: wangyanqin@ecust.edu.cn).

Fossil fuel consumption is projected to increase significantly in coming decades, with potentially catastrophic consequences for the environment[1]. Sustainable alternatives to crude oil are imperatively needed to bridge gaps in the supply of chemical fuels and feedstocks[2]. For the production of liquid fuels in particular, the replacement of oil-based routes by renewable biomass has received increasing attention[3–6]. Lignocellulose, as the main component of woody biomass, is composed of cellulose (40–50 wt%; a linear polymer of D-glucopyranose connected by $\beta$-1,4-glycosidic linkages), hemicellulose (16–33 wt%; a heteropolymer consisting of many different sugar monomers) and lignin (15–30 wt%; a heavily cross-linked, complex polymer with coumaryl, coniferyl and sinapyl alcohols as monomers)[7,8]. Owing to the complexity of lignocellulosic biomass and its notorious resistance to chemical transformation, energy-efficient and cost-effective production of liquid fuels from lignocellulose remains a mammoth challenge. So far, two strategies have been reported to address this challenge: (i) separation of lignocellulose into isolated sugars and lignin followed by biological or chemical (hydrolysis) processing[9–11]; (ii) thermochemical treatment of lignocellulose to produce upgradeable intermediates, such as bio-oils by pyrolysis or syngas by gasification, coupled with subsequent catalytic upgrading[12,13]. Thermochemical processes offer the total conversion of lignocellulose, but are often non-selective and intractable, and the resultant bio-oils or syngas need to be upgraded for further utilisation. Although hydrolysis-based approaches offer selective production of liquid fuels, they are generally multistep and thus very energy-intensive[14]. Moreover, the lignin by-products generated from the hydrolysis of lignocellulose are usually burned as a low-value fuel[15]. Powerful drivers therefore exist to develop alternative efficient and selective strategies to directly convert raw lignocellulose into liquid fuels.

Direct conversion of raw lignocellulose into alcohols and phenols was realised recently in exceptional cases[16–17]. However, successful direct production of hydrocarbon fuels (that is, total removal of oxygen) is mostly achieved so far from separated components of lignin or cellulose[18–20]. For example, the conversion of lignin into alkanes and methanol has been reported through a two-step process (chemical pretreatment and sequential hydrogenolysis and hydrogenation)[18]. Recently, the one-pot conversion of cellulosic feedstock into liquid alkanes in biphasic reaction systems (organic + water) were also reported over Ir-ReO$_x$/SiO$_2$–H-ZSM-5 or tungstosilicic acid–Ru/C catalysts[19,20]. The industrial Shell/GTI hydropyrolysis and Virent Energy System's approaches are also known to directly convert sugars or raw biomass into liquid fuels[21,22]. The former is based on a catalytic thermal–chemical technique, which reacts at very high temperature (350–540 °C) (ref. 21). Virent's approach converts water-soluble oxygenated hydrocarbons into $C_{4+}$ hydrocarbons, alcohols and/or ketones in aqueous phase or vapour phase. This is achieved by aqueous phase reforming of water-soluble oxygenates, followed by condensation and deoxygenation[22]. More recently, a three-catalyst system was reported to convert raw biomass into liquid alkanes and other mono-functional hydrocarbons over layered LiTaMoO$_6$ combined with Ru/C in aqueous phosphoric acid medium[23]. Gaining in-depth understanding on the reaction mechanism is of fundamental importance for the development of improved catalytic systems.

Here we report that, by using a multifunctional Pt/NbOPO$_4$ catalyst, raw woody biomass can be directly converted into liquid alkanes in high yields in a single-phase medium (cyclohexane) with cellulose, hemicellulose and lignin fractions in solid woods being converted into hexane, pentane and alkylcyclohexanes, respectively (Fig. 1), representing direct conversion of raw lignocellulose into liquid alkanes under mild conditions over a single catalyst. Importantly, no chemical pretreatment (for example, hydrolysis and separation) to the raw wood is required for this process, and thus, tremendous energy savings can be potentially gained in comparison with the existing thermochemical- and hydrolysis-based approaches. More significantly, the pathway for this novel catalytic reaction was systematically investigated by control experiments, and the molecular mechanism for the rate-determining step in this conversion studied by *in situ* inelastic neutron scattering and computational studies. These complementary investigations reveal that the NbO$_x$ species promotes the crucial C–O bond cleavage over hydrodeoxygenation of tetrahydrofuran (THF) and phenol (model units of cellulose and lignin, respectively) to hydrocarbons under mild reaction conditions.

## Results

**Direct hydrodeoxygenation of raw woody biomass.** To verify the applicability of this one-pot approach, seven different types of wood sawdusts ( < 75 µm), including both softwoods and hardwoods, were employed as feedstocks for direct hydrodeoxygenation over the Pt/NbOPO$_4$ catalyst in a cyclohexane medium (Table 1). The reactions were conducted at 190 °C and 5 MPa H$_2$ for 20 h and over 20 wt% total mass yield of liquid alkanes was achieved for all woods, among which birch wood gave the highest mass yield of 28.1 wt%. Considering that the theoretical mass yield of alkanes from raw woody biomass is limited to ∼ 50 wt% as the removed oxygen accounts for almost half of the mass loss, the yields obtained here are excellent. In addition to C$_1$–C$_6$ alkane products, surprisingly appreciable amounts of alkylcyclohexanes (for example, propylcyclohexane and ethylcyclohexane) were also detected (Supplementary Fig. 1), indicating that not only the cellulose and hemicellulose but also the lignin fraction in sawdusts were converted into alkanes. Obviously, the source/texture of lignocellulose had a significant influence on both mass and carbon yields of the alkane products. In general, higher yields of hexanes and pentanes were achieved from softwoods: the carbon yields of hexanes and pentanes on the basis of cellulose and hemicellulose fractions reached 72.8 and 69.3% on average, respectively. These yields are surprisingly high and even comparable to those using isolated cellulose as feedstock[19,20]. Indeed, pure cellulose was tested as model material for carbohydrate fractions in raw woody biomass to confirm the performance of the catalyst. A total of 71.5% yield of hexanes and 8.7% yield of pentanes (by C–C cleavage) were achieved from cellulose conversion with excellent stability (Supplementary Table 1 and Supplementary Fig. 2). On the other hand, the yield of alkylcyclohexanes produced from hardwoods is much higher than that from softwoods, with an average carbon yield of 34.0% from hardwood (here only monomer alkylcyclohexanes were determined). It is worth noting that this yield is very high because there is a large proportion of C–C linkages in lignin structure (30–34% for hardwoods and 43–51% for softwoods on average)[15], which are hardly cleaved under such mild reaction conditions, thus resulting in a maximum theoretical carbon yield of monomer alkylcyclohexanes at 44–49% from hardwoods and 24–32% from softwoods (Supplementary Note 1). This result indicates that the catalyst has excellent performance for the direct hydrogenolysis of C–O–C linkages of lignin and total hydrodeoxygenation of the resultant lignin monomers. To further confirm this, diphenyl ether and phenol, which possess aromatic ether and hydroxyl functionalities, respectively, were tested as model compounds over Pt/NbOPO$_4$. A total of 99.9% yield of cyclohexane was achieved from both substrates, demonstrating the efficient cleavage of ether bond in lignin by this catalyst (Supplementary Table 1).

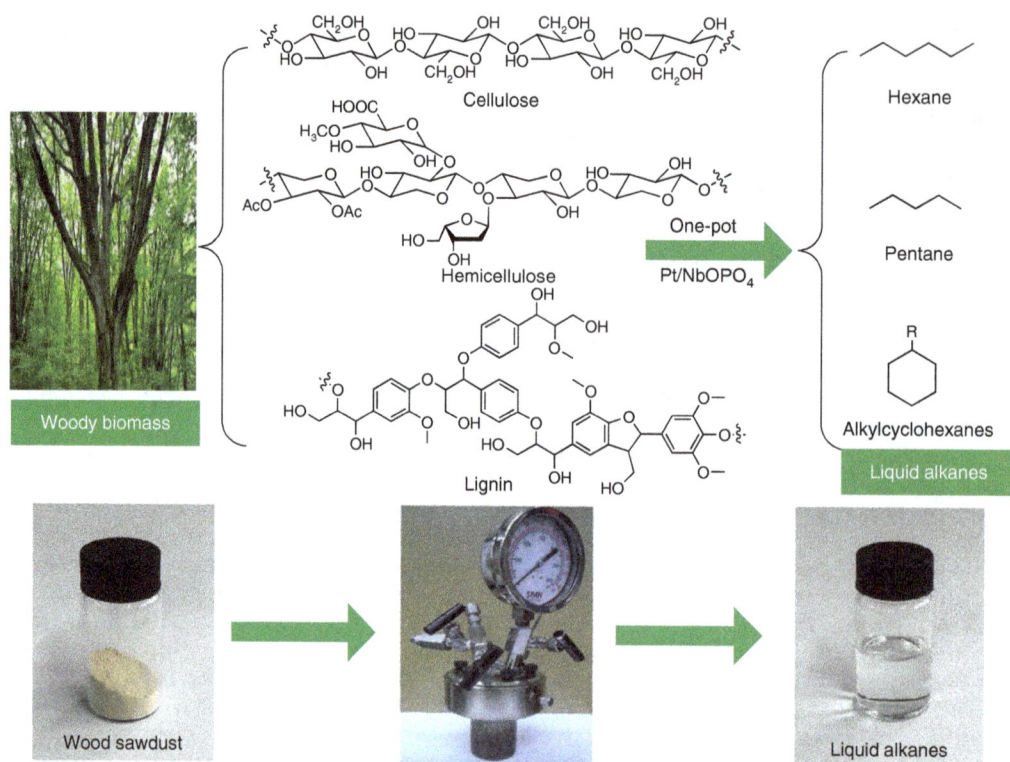

**Figure 1 | Schematic representation of direct hydrodeoxygenation of raw woody biomass into liquid alkanes.** Raw woody biomass can be directly converted into liquid alkanes over Pt/NbOPO₄ catalyst in cyclohexane medium, with cellulose, hemicellulose and lignin fractions in solid woods being converted into hexanes, pentanes and alkylcyclohexanes, respectively.

**Table 1 | Summary of direct hydrodeoxygenation of various woody biomass over Pt/NbOPO₄.***

| Woody biomass | Softwood | | | | Hardwood | | |
|---|---|---|---|---|---|---|---|
| | Pine | White pine | Larch | Fir | Camphor | Birch | Poplar |
| *Lignocellulose content* | | | | | | | |
| Hemicellulose (wt%) | 9.3 | 10.4 | 13.1 | 10.3 | 24.8 | 25.7 | 28.2 |
| Cellulose (wt%) | 47.2 | 52.3 | 49.4 | 43.5 | 45.1 | 48.7 | 45.3 |
| Lignin (wt%) | 32.6 | 29.5 | 28.3 | 33.9 | 22.6 | 20.4 | 22.8 |
| *Mass yield/carbon yield (wt%/mol%)*[†] | | | | | | | |
| Pentanes[‡] | 4.1/80.8 | 3.6/63.4 | 4.8/67.2 | 3.7/65.9 | 7.3/54.0 | 10.2/73.1 | 7.1/46.2 |
| Hexanes[§] | 18.7/74.8 | 19.3/69.4 | 20.4/77.7 | 16.0/69.2 | 15.2/63.4 | 13.1/50.7 | 9.7/40.3 |
| Alkylcyclohexanes[‖] | 2.1/9.6 | 2.2/11.2 | 2.6/13.7 | 2.3/10.1 | 5.1/33.7 | 4.8/35.1 | 5.1/33.3 |
| Total liquid alkanes (wt%) | 24.9 | 25.1 | 27.8 | 22.0 | 27.6 | 28.1 | 21.9 |
| Others (wt%)[¶] | 1.7 | 1.7 | 2.0 | 1.4 | 1.9 | 2.3 | 1.6 |
| Residue (wt%) | 31.5 | 27.9 | 27.1 | 33.4 | 17.4 | 14.8 | 16.7 |

*The reactions were conducted at 190 °C and 5 MPa H₂ for 20 h. Feedstock (0.20 g), Pt/NbOPO₄ (0.20 g) and cyclohexane (6.46 g) were put into a 50 ml stainless-steel autoclave.
[†]Mass yields were calculated by the equation: mass yield of pentanes (hexanes, alkylcyclohexanes) = [mass of pentanes (hexanes, alkylcyclohexanes)]/[mass of feedstock input]. Carbon yields were calculated by the equation: carbon yield of pentanes (hexanes, alkylcyclohexanes) = [mass of carbon in pentanes (Hexanes, alkylcyclohexanes)]/[mass of carbon in hemicellulose (cellulose, lignin)].
[‡]Pentanes include *n*-pentane and iso-pentane.
[§]Hexanes include *n*-hexane and iso-hexane.
[‖]Only monomer alkylcyclohexanes were determined here, including methylcyclohexane, ethylcyclohexane, isopropylcyclohexane and propylcyclohexane.
[¶]Others were mainly CO₂ and C₁–C₄ alkanes.

**Clarification of the unique activity of Pt/NbOPO₄.** A variety of catalysts with different combinations of support and metal (that is, Pt/NbOPO₄, Pt/H–ZSM-5, Pt–ReO$_x$/SiO₂, Pt–ReO$_x$/C and Pd, Ru, Rh loaded NbOPO₄) were tested with birch sawdust as feedstock for this reaction to clarify the unique activity of Pt/NbOPO₄ (Table 2). With Pt supported on H–ZSM-5, which possesses similar acidity to NbOPO₄ but does not contain transition metal oxide (Supplementary Fig. 3), only 8.7 wt% yields of liquid alkanes was obtained (versus 28.1 wt% for Pt/NbOPO₄). This suggests that the NbO$_x$ species of NbOPO₄ support has a significant promotion effect in this reaction. Such promotion effect on C–O cleavage was investigated recently in transition metal oxides of NbO$_x$ and ReO$_x$ (refs 24–26). To provide more insight, two other ReO$_x$ support with reduced acidity, ReO$_x$/SiO₂ and ReO$_x$/C, were tested for comparison, and alkane yields of 11.4 and 9.8 wt%, respectively, were obtained with a small amount of mono-functional hydrocarbons detected (for example, tetrahydropyran). This result suggests that the promotion effect should be accompanied by sufficient acidity to achieve efficient C–O cleavage in this reaction. The NbOPO₄ support fulfills these two requirements (that is, surface NbO$_x$ species and sufficient acidity), and thus posesses the best catalytic activity for this

**Table 2 | The results of direct hydrodeoxygenation of birch wood over different catalysts.\***

| Entry | Catalyst | Mass yield of liquid alkanes (wt%) | | | | |
|---|---|---|---|---|---|---|
| | | Pentanes[†] | Hexanes[‡] | Alkylcyclohexanes[§] | Total | Residue |
| 1 | Pt/NbOPO$_4$ | 10.2 | 13.1 | 4.8 | 28.1 | 14.8 |
| 2 | Pt–H–ZSM-5 | 4.9 | 1.3 | 2.5 | 8.7 | 35.8 |
| 3 | Pt–ReO$_x$/SiO$_2$[||] | 5.2 | 3.6 | 2.6 | 11.4 | 26.9 |
| 4 | Pt–ReO$_x$/C[||] | 6.4 | 1.5 | 1.9 | 9.8 | ND |
| 5 | Pd/NbOPO$_4$ | 8.1 | 6.2 | 3.5 | 17.8 | 29.4 |
| 6 | Ru/NbOPO$_4$ | 7.3 | 8.0 | 3.9 | 19.2 | 36.3 |
| 7 | Rh/NbOPO$_4$ | 8.2 | 5.0 | 4.0 | 17.2 | 31.5 |

\*The reactions were conducted at 190 °C and 5 MPa H$_2$ for 20 h. Feedstock (0.2 g), catalyst (0.2 g), and cyclohexane (6.46 g) were put into a 50 ml stainless-steel autoclave. All metal loading was 5 wt%.
†Pentanes include n-pentane and iso-pentane.
‡Hexanes include n-hexane and iso-hexane.
§Only monomer alkylcyclohexanes were determined, including methylcyclohexane, ethylcyclohexane, isopropylcyclohexane and propylcyclohexane.
||The loading amount of Re was 1 by the molar ratio of Re to Pt.

**Table 3 | Summary of direct hydrodeoxygenation of birch sawdust and cellulose under various reaction conditions over Pt/NbOPO$_4$.\***

| Entry | Feedstock | Carbon yield of liquid alkanes (%) | | |
|---|---|---|---|---|
| | | Hexanes[†] | Pentanes[‡] | alkylcyclohexanes |
| 1[§] | Birch sawdust | 72.4 | 52.3 | 34.6 |
| 2[||] | Birch sawdust | 69.1 | 49.7 | 35.9 |
| 3 | Cellulose | 71.5 | 8.7 | ND[¶] |
| 4[#] | Cellulose | 40.3 | 4.4 | ND |
| 5[**] | Celloluse | 65.2 | 8.2 | ND |
| 6[††] | Cellulose | 41.3 | 4.6 | ND |
| 7[††,‡‡] | Celloluse | 37.5 | 4 | ND |

\*Unless otherwise specified, the reactions were conducted at 190 °C and 5 MPa H$_2$ for 20 h. Feedstock (0.2 g), catalyst (0.2 g), and cyclohexane (6.46 g) were put into a 50 ml stainless-steel autoclave.
†Hexanes include n-hexane and iso-hexane.
‡Pentanes include n-pentane and iso-pentane.
§Tridecane was employed as the reaction solvent.
||0.4 g of birch sawdust, 0.4 g of catalyst, and 6.46 g of cyclohexane were put into a 50 ml stainless-steel autoclave.
¶"ND" is the abbreviation of "not determined".
#Ashes (minerals) obtained by calcination of 1 g birch sawdust at 500 °C for 3 h were added into the reaction system and reaction for 8 h.
\*\*The cellulose was unmilled and used directly.
††The reaction time was 8 h.
‡‡Data shown are for the fourth run of the stability test.

reaction. On the metal side, Pd, Ru and Rh were tested by loading them onto the NbOPO$_4$ support, and moderate yields (17.2–19.2 wt%) of liquid alkanes were achieved from the hydrodeoxygenation of birch wood. The reason that Pt gave the highest yield among the studied metals is due to its supreme activity for H$_2$ activation and hydrogenation (Supplementary Table 2). From these control experiments, we rationalise the superior performance of Pt/NbOPO$_4$ by a synergistic effect between Pt, NbO$_x$ species and the acidic sites on the support (including both Brönsted acid sites on PO$_4$ and Lewis acid sites on NbO$_x$, Supplementary Fig. 4).

**Study of the activity and stability of the catalyst.** To investigate the applicability and recyclability of the catalyst, Pt/NbOPO$_4$ was tested under various reaction conditions and the results are summarised in Table 3. As the products of this conversion were only alkanes, they act as an additional solvent to drive the sequential reactions without the need of solvent separation or recycling after each run. Cyclohexane was used as a solvent to facilitate the analysis because it does not overlap with any product. Alternatively, other alkane (for example, tridecane) can be used as a solvent for this reaction to afford similar results (Table 3, entry 1). Moreover, the reaction with double solid

loadings was attempted, and there was no obvious decline of the alkane yields (Table 3, entry 2), indicating that this process is capable of dealing with higher solid loadings. Mineral poisoning is widely known to reduce the catalyst activity in biomass conversion. To probe more insight, extra ashes obtained by calcination of 1 g birch sawdust at 500 °C for 3 h were added into the system to test the cellulose conversion. The result suggests that a fivefold amount of ash has no influence on the catalytic activity of this catalyst (Table 3, entries 4 and 6). Moreover, the stability of the catalyst at lower alkane yields was tested by shortening the reaction time to 8 h. Small decreases on the yields of hexanes and pentanes were observed after four successive runs (hexanes: from 41.3 to 37.5%; pentanes: from 4.6 to 4.0%; Table 3, entries 6 and 7, Supplementary Fig. 2). Characterisations of the catalyst before and after the reaction showed small reductions on the BET surface area and the Pt dispersion (Supplementary Note 2), consistent with the observed small decreases on the yields of hexanes and pentanes. On the whole, the catalyst showed good and consistent catalytic performance in repeated runs in this one-pot process. This could be due to two reasons: first, the reaction was carried out under mild conditions (190 °C), which retards the significant aggregation of Pt particles (Supplementary Fig. 5); second, the use of non-aqueous single-phase medium (cyclohexane) hinders leaching and structural change of the catalyst (Supplementary Fig. 6). Indeed, the ICP analysis of the reaction solution suggested that the concentration of Pt, P or Nb was all below the detection limit, confirming the absence of catalyst leaching during the reaction.

**Studies of the representative reaction pathway.** In attempting to reveal the reaction pathway, model compounds were used to simplify the original reaction system for easier detection of intermediates and products. Diphenyl ether and phenol were chosen as the model compounds of the lignin fraction to investigate the cleavage of ether bond (as mentioned above), and cellobiose was used to represent the carbohydrate fraction to elucidate the conversion of cellulose and hemicellulose to alkanes. To better monitor the possible intermediates, the reaction was carried out at a lower temperature of 170 °C, and water was added into the reaction mixture after reaction to extract water-soluable intermediates. After 1 h reaction, small amounts of glucose (2), sorbitol (3), sorbitan (4) and isosorbide (5) were detected by HPLC (Supplementary Fig. 7), and a mixture of 1-dehydroxyl-glucose (6), 1,6 anhydro glucose (7), 2-hydroxymethyl-tetrahydropyran (8), 5-methyl-THF-2-methanol (9) as well as many other undefined intermediates in the aqueous phase were detected by GC–mass spectrometry (MS) (Supplementary Fig. 8). As no water existed at the beginning of this reaction, this result

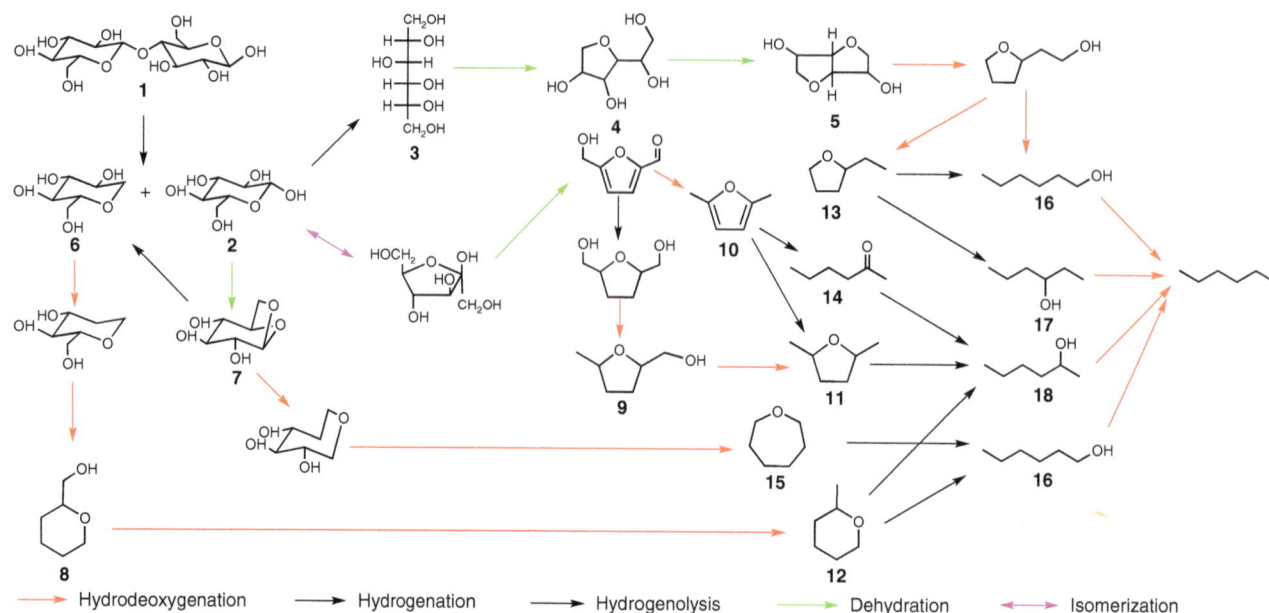

**Figure 2 | Representative reaction pathways for the direct hydrodeoxygenation of cellobiose.** According to all the intermediates detected, it is possible to deduce the main reaction pathways of the direct conversion of cellobiose to hexane. The reaction occurred by the direct hydrogenolysis of the $\beta$-1,4 linkage to D-glucose (**2**) and 1-deoxy-D-glucose (**6**) first, and then **6** was converted to hexane by sequential hydrogenolysis via 2-hydroxymethyl-tetrahydropyran (**8**), 2-methyltetrahydropyran (**12**) and hexanols (**16**) and (**18**), whereas the conversion of **2** has three main reaction pathways: (i) hydrogenated to sorbitol (**3**) and then dehydrated to sorbitan (**4**) and isosorbide (**5**), followed by sequential hydrogenolysis via 2-ethyl-THF (**13**) and hexanols (**16**) and (**17**). (ii) Isomerised to fructose and then dehydrated to 5-hydroxymethylfurfural followed by hydrogenation and sequential hydrogenolysis via 2,5-dimethylfuran (**10**), 2-hexanone (**14**), 5-methyl-THF-2-methanol (**9**), 2,5-dimethyl-THF (**11**) and 2-hexanol (**18**). (iii) Dehydrated to 1,6-anhydroglucose (**7**) followed by sequential hydrogenolysis via oxepane (**15**) and n-hexanol (**16**).

indicates that the $\beta$-1,4 linkage in cellobiose was cleaved by direct hydrogenolysis rather than hydrolysis, as a result of the excellent performance of Pt/NbOPO$_4$ in hydrogenolysis. When further reacted for 6 h, appreciable amounts of hexane, 2,5-dimethylfuran (**10**), 2,5-dimethyl-THF (**11**), 2-methyltetrahydropyran (**12**), 2-ethyl-THF (**13**), hexanone (**14**), oxepane (**15**) and hexanols (**16–18**) were observed in the organic layer (Supplementary Fig. 9). In the aqueous phase, isosorbide (**5**) and 2-hydroxymethyl-tetrahydropyran (**8**) were the main residuals (Supplementary Fig. 10), indicating that **5** and **8** are the two major intermediates in cellulose conversion. When further reacted for 24 h, no residual was detected in the aqueous phase, and a large amount of **11–13** and **16–18** were observed in the organic phase, suggesting that the ring-opening of THF derivatives and the subsequent hydrodeoxygenation are the rate-determining steps (Supplementary Fig. 11). Therefore, the direct conversion of cellobiose to hexane occurs via the hydrogenolysis of the $\beta$-1,4 linkage to **2** and **6** first, which then undergo a combination of hydrogenolysis, dehydration, hydrogenation and isomerisation reactions. The main reaction pathways of cellobiose conversion were proposed in Fig. 2, and was further confirmed by the isolated reactions of the four main intermediates (Supplementary Fig. 12).

**Inelastic neutron scattering studies.** As discussed above and previously reported[24], the ring-opening and the sequential hydrodeoxygenation of THF derivatives (for example, 5-hydroxymethylfurfural) are problematic and often represent the major barrier for the conversion of bio-derived furans into alkanes. Indeed, it is the rate-determining step for this conversion. Direct visualisation of the interaction between adsorbed THF (as a model compund for various THF derivitives generated from this reaction; the use of THF instead of THF-derivitives gives a clearer interpretation of the results) and the catalyst surface

is crucial to understand the molecular details of binding, ring-opening and hydrodeoxygenation of THF into alkanes (butane in this case). INS is a powerful neutron spectroscopy technique to investigate the dynamics of hydrogenous compounds by exploiting the high-neutron scattering cross-section of hydrogen (82.02 barns)[27]. In addition, INS is not subjected to any optical selection rules, and the calculation of INS spectra from DFT calculations is straightforward (Supplementary Note 3). Here, we have successfully combined *in situ* INS and DFT to investigate the vibrational properties of the THF–Pt/Nb$_2$O$_5$ system to reveal the mechanism of the challenging hydrodeoxygenation of THF. NbOPO$_4$ has a large amount of surface P–OH groups, the vibrational peaks of which will overlap with signals of adsorbed THF. Therefore, we here used Pt/Nb$_2$O$_5$ instead for a clearer interpretation of the experimental observation. It is worth noting that Pt/Nb$_2$O$_5$ has a similar catalytic reactivity to Pt/NbOPO$_4$ for this reaction under the same conditions, as evidenced by the direct comparison of the yield and selectivity data for THF conversion in Supplementary Table 3. To the best of our knowledge, this is the first example of using INS/DFT to study the mechanism of catalytic biomass conversion.

The INS spectrum of the bare catalyst gives a clean background with no prominent features (details on the discussion of background spectra are given in Supplementary Note 4 and Supplementary Figs 13 and 14). In comparison, the INS spectrum of the catalyst on THF adsorption at 130 °C shows a significant increase in total intensity, demonstrating the binding of THF to the catalyst surface (Fig. 3a). Comparison of the difference spectrum before and after THF adsorption on the catalyst (that is, signals for adsorbed THF) and that of the solid THF shows a few changes (Fig. 3d). Peaks at low energy (below 200 cm$^{-1}$), assigned to the translational and rotational modes of THF, shift to lower energy with a continuum profile, suggesting that the

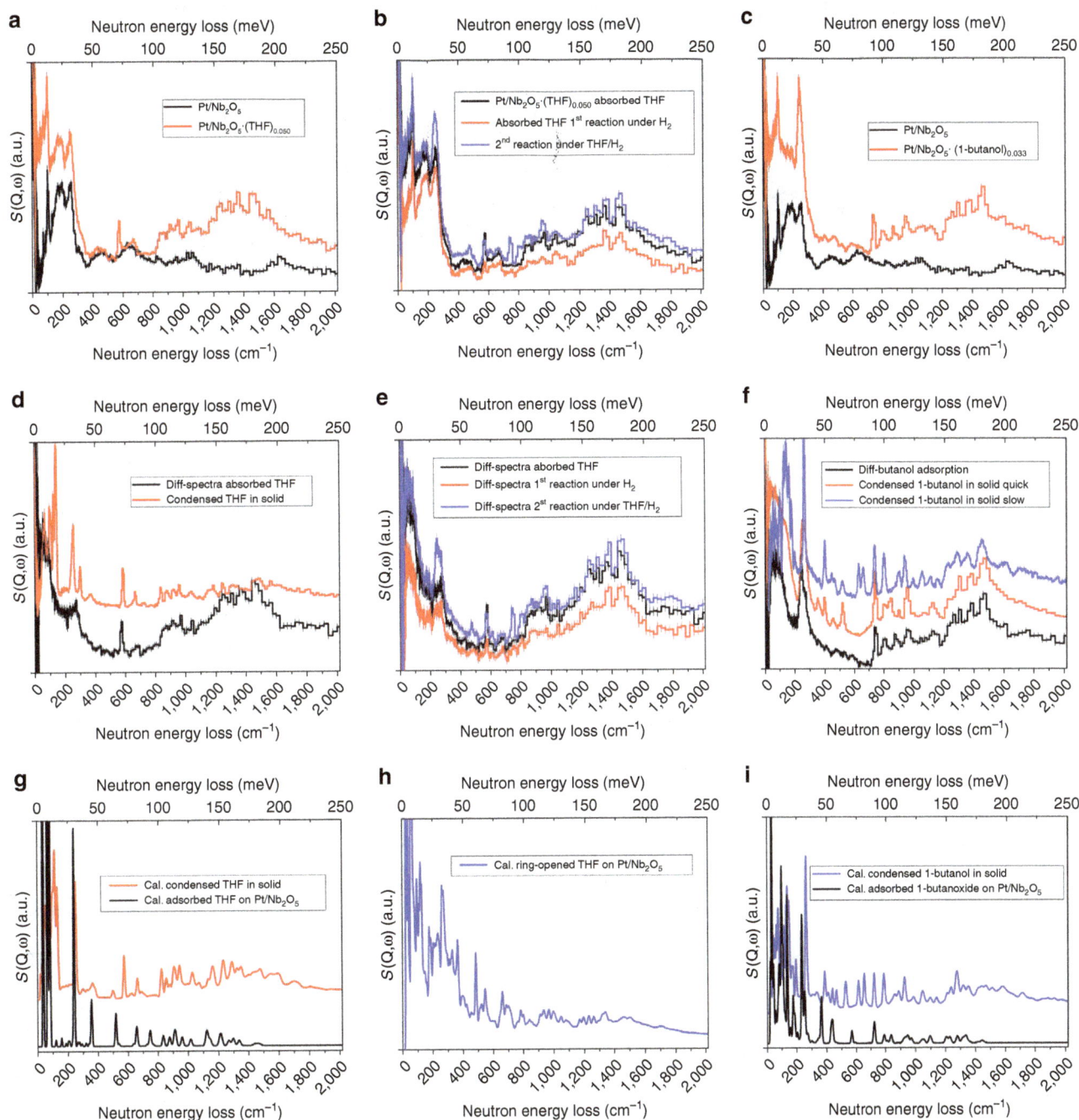

**Figure 3 | Inelastic neutron scattering spectra for Pt/Nb$_2$O$_5$ during the catalytic hydrodeoxygenation of THF.** INS spectrum of the reduced catalyst was used throughout for calculations of the difference spectra. No abscissa scale factor was used throughout this report for INS calculations. NbOPO$_4$ has a large amount of surface –OH groups, which have no specific activity in THF conversion, but will generate additional INS peaks overlapping with signals from substrates. Pt/Nb$_2$O$_5$ has a similar catalytic reactivity as Pt/NbOPO$_4$ for the hydrodeoxygenation of THF and was therefore used in INS study instead. Comparison of the experimental INS spectra for bare catalyst and the THF (**a**) and 1-butanol (**c**) adsorbed catalyst. Comparison of the difference plots for experimental INS spectra of bare and THF (**b**) and 1-butanol (**f**) adsorbed catalyst, and the experimental INS spectra of condensed THF (**d**) and 1-butanol (**f**) in the solid state. (**b**) Comparison of the experimental INS spectra for the two hydrodeoxygenation reactions of adsorbed THF on the catalyst, and difference plots (with reduced catalyst as background) are shown in (**e**). The first reaction was carried out at 130 °C in pure H$_2$ flow for 10 min and the second reaction was carried out at 130 °C in THF/H$_2$ flow for 5 h, that is fresh THF vapour was fed into the reaction cell continuously with H$_2$ as carrier gas during the reaction. (**g**) Comparison of the calculated INS spectra for condensed THF in solid and adsorbed THF on the catalyst. (**h**) View of the calculated INS spectrum for ring-opened THF on the catalyst. (**i**) Comparison of the calculated INS spectra for condensed 1-butanol in solid and adsorbed 1-butanoxide on the catalyst.

adsorbed THF molecules are disordered over the catalyst surface, and have restricted translational motion owing to the strong binding to the catalyst. The very strong peak at 251 cm$^{-1}$, assigned to the torsional mode of the C2–C3 bond of THF ring, reduced significantly in intensity, indicating the loss of this

motion on adsorption. The peak at 300 cm$^{-1}$, likely due to the combination of the lattice mode at $\sim 60$ cm$^{-1}$ and the strong mode at 251 cm$^{-1}$, concurrently reduced in intensity. In addition, the ring deformation mode at 587 cm$^{-1}$ in solid THF shifts to 571 cm$^{-1}$ when adsorbed on the catalyst. A structural

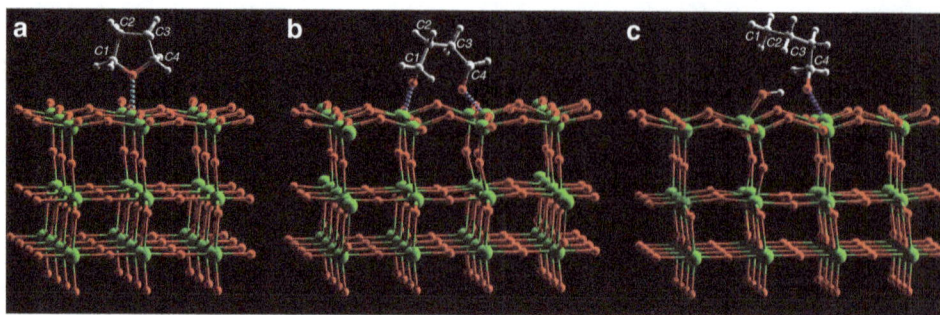

**Figure 4 | View of the optimised structural models.** (**a**) adsorbed THF, (**b**) reacted (ring-opened) THF and (**c**) 1-butanol after adsorption to generate 1-butanoxide and surface hydroxyl on the (001) plane of the catalyst Pt/Nb$_2$O$_5$ (Nb: green; O: red; C; grey; H: white).

model of solid THF and THF adsorbed on Nb$_2$O$_5$ were optimised by DFT, respectively, and calculated INS spectra were produced (Fig. 3g; Supplementary Figs 15 and 16). Comparison of the INS spectra suggests that THF is likely adsorbed intact via interaction between its O($\delta-$) centre to open Nb($\delta+$) site (O$\cdots$Nb = 2.33 Å) on the surface (Fig. 4a) and that ring-opening of THF does not happen immediately on adsorption and/or in the absence of H$_2$. The calculation also confirms the downshift of the ring deformation mode but slightly overestimates its magnitude.

The adsorbed THF underwent a first catalytic conversion in H$_2$ flow for 10 min at 130 °C. The INS spectrum of the first reacted catalyst shows a large decrease in intensity (Fig. 3b,e), suggesting that the adsorbed THF underwent fast catalytic conversion to butane, which was sequentially swept out of the cell as confirmed by MS. In particular, the peaks at 1,244 and 1,308 cm$^{-1}$ (assigned to –CH$_2$– twisting and internal ring deformation of THF, respectively) disappeared completely, confirming the cleavage of the THF ring. An optimised structural model of ring-opened THF on Nb$_2$O$_5$ suggests that the adsorbed THF interacts with the very strong Lewis acid sites (Nb$^{5+}$), and the ring opens via binding to two adjacent Nb$^{5+}$ centres simultaneously (O$\cdots$Nb = 1.98 Å) (Figs 3h and 4b). Comparison of the calculated INS spectrum for ring-opened THF and experimental difference spectrum, however, does not conclusively suggest the presence of this intermediate bound on the catalyst. This could be due to two possible reasons: (i) the ring-opened intermediate is highly active and was hydrogenated instantly, and thus cannot be captured effectively; (ii) the amount of this intermediate is too low to be detected as no THF feedstock was provided during the reaction. To enrich the intermediate on the catalyst, a second catalytic reaction was conducted in THF/H$_2$ flow for 5 h, and the production of butane observed continuously by MS. The INS spectrum of the second reacted catalyst indeed shows an increase in intensity even compared with that of THF-adsorbed catalyst, confirming the presence of additional substrates on the catalyst surface (Fig. 3b). The corresponding difference spectra confirm the presence of adsorbed THF, and more importantly, four new peaks at 245, 477, 744 and 805 cm$^{-1}$ were observed (Fig. 3e), indicating the presence of additional hydrogenous species bound on the catalyst.

To reveal the identity of this intermediate/s, the adsorption of 1-butanol on the catalyst was studied because 1-butanoxide bound to Nb$^{5+}$ is predicted to be relatively stable (Fig. 3i). The INS spectrum of 1-butanol adsorbed on the catalyst shows a large increase in intensity, and comparison of the difference spectrum (that is, signals for adsorbed 1-butanol) and that of solid 1-butanol shows significant changes (Fig. 3c,f). The most important one is the disappearance of the peak at 839 cm$^{-1}$ (assigned to the out-of-plane C–O–H bending mode, detailed discussions are given in Supplementary Note 4 and Supplementary Fig. 17), clearly suggesting 1-butanol underwent

deprotonation on adsorption to give 1-butanoxide bound to the surface Nb$^{5+}$ sites (Fig. 4c and Supplementary Fig. 18). The conformation-dependent bands between 300 and 600 cm$^{-1}$ in the INS spectrum of solid 1-butanol all disappeared on adsorption, indicating that the alkyl chain of adsorbed 1-butanoxide is not in the all-*trans* conformation as found in crystalline 1-butanol (Supplementary Note 4). In addition, the peak at 257 cm$^{-1}$ (assigned to the methyl torsion) shifts to lower energy at 243 cm$^{-1}$ on binding to the catalyst. Moreover, an additional peak at 1,265 cm$^{-1}$ (assigned to C4 chain deformation) is present for bound 1-butanoxide on Nb$^{5+}$.

Comparison of the INS spectra of adsorbed 1-butanoxide and the second reacted catalyst gives a clear message: three out of the four new peaks at 245, 744, and 805 cm$^{-1}$ (assigned as methyl torsion, –CH$_2$CH$_2$– rocking and –CH$_2$CH$_2$CH$_3$ rocking of 1-butanoxide, respectively) of the second reacted catalyst are consistent with the presence of 1-butanoxide bound on the surface (Fig. 3e,f). The remaining new peak at 477 cm$^{-1}$ is conformation-dependent and consistent with the presence of a *gauche* conformer of C4 chain of bound 1-butanoxide[28]. A final INS spectrum for the reactivated catalyst shows no prominent feature (Supplementary Fig. 19), confirming the absence of formation of residual hydrocarbonaceous species and thus demonstrating the high efficiency of catalyst regeneration in cycling experiments.

The calculated and experimental INS spectra for solid THF and 1-butanol show excellent agreement; however, those for the guest-bound-catalysts exhibit a number of discrepancies (Fig. 3d,i). It is worth noting that the experimental data were collected on a disordered system (poorly crystalline mesoporous metal oxide and disordered substrates), wheras the calculations assume fully periodic structures. A poorly crystalline system generates multiple sites with different binding energies, inducing the broadening of INS peaks. The presence of different conformers of the butane chain on the catalyst also induces discrepancy as the calculation was done with the *gauche* conformer with the lowest energy only and the low energy modes are very sensitive to the chain conformation. Indeed, the INS spectra for disordered and crystalline 1-butanol have suggested that the crystallinity of the system can induce significant changes to both peak intensity and positions (Supplementary Fig. 17). Moreover, the calculation used a model with "flat" surface, in which there is no interaction with the surface, other than through the THF/1-butanol oxygen atom. On the real, rough surface, for example, with steps (Supplementary Fig. 5), the hydrogen atoms on C1 and C4 of THF (or C1 and C2 of 1-butanol) will also be enabled to interact with the surface, further enhancing the binding and broadening of the INS bands.

In this study, we concentrate primarily on the experimental observation, which has confirmed that (i) adsorbed THF molecules on the catalyst have, in principle, an intact structure

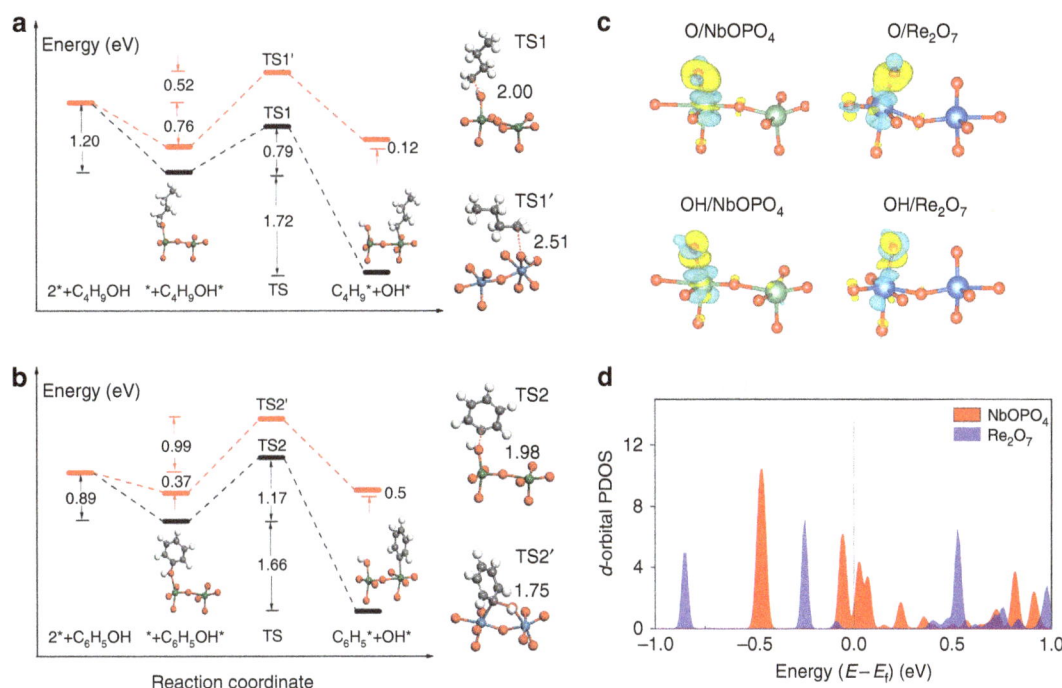

**Figure 5 | Calculated results of computational studies.** (**a,b**) Calculated energy profiles of C–O bond cleavage of $C_4H_9OH$ and $C_6H_5OH$, respectively. Black and red lines indicate $NbOPO_4(100)$ and $Re_2O_7(010)$, respectively. The structures of the initial and final states on $NbOPO_4(100)$ are shown, whereas the transition states involving C–O bond cleavage of $C_4H_9OH$ and $C_6H_5OH$ are depicted in TS1, TS1',TS2 and TS2', with the elongated C–O bond lengths shown. Emerald balls represent Nb atoms, dark blue for Re, white for H, grey for C and red for O. (**c**) The isosurfaces of charge density difference for O and OH adsorption on $NbOPO_4(100)$ and $Re_2O_7(010)$ surfaces. For clarity, only the local active sites are shown. The regions depicted in yellow indicate charge accumulation and light blue for charge depletion. Emerald balls represent Nb atoms, dark blue for Re, white for H and red for O. (**d**) The d-orbital projected density of states for the surface $Nb_{5c}$ and $Re_{5c}$ atoms, demonstrating their relative energies, in which the energy is aligned to the Fermi level ($E_F$).

with reduced motion (esp. for the C2–C3 torsion and ring deformation modes) owing to the strong binding to the $Nb^{5+}$ sites; (ii) ring-opening of adsorbed THF (cleavage of C–O bond) occurs rapidly in the presence of $H_2$ as shown by the loss of the internal THF ring deformation mode; (iii) 1-butanoxide bound to the surface $Nb^{5+}$ site is a relatively stable reaction intermediate, consistent with calculations (see below). Therefore, this catalytic conversion of THF follows adsorption, binding, ring-opening, partial hydrogenation and complete hydrodeoxygenation, and the surface $Nb^{5+}$ sites played an important role in this reaction, particularly for the binding and activation of THF substrates. It is worth noting that here the role of Pt is believed to dissociate $H_2$ and provide [H], and such role is not exclusive in the rate-determining C–O–C bond cleavage step, considering that similar performance can be obtained by substituting Pt with Pd loaded on $NbOPO_4$ for the hydrodeoxygenation of cellulose into alkanes (Supplementary Table 3). However, other possible roles of Pt in the whole reaction such as strengthening acidity through interface interactions[29] may not be fully ruled out, which is beyond the core point of this work and will be studied further in future works.

**Computational studies of the catalytic origin of $NbOPO_4$.** To probe more insight, first-principle calculations were conducted to examine the crucial C–O bond breaking process of phenol and 1-butanol, which are the model compound of lignin and important intermediate in THF conversion, respectively. Notably, the widely used $ReO_x$ catalyst in biomass conversion was also tested for comparison. The well-ordered flat $NbOPO_4(100)$ and $Re_2O_7(010)$ surfaces, which have exposed five-coordinated $Nb_{5c}$ and $Re_{5c}$ centres as the main binding sites, were chosen as the substrates (Supplementary Methods, Supplementary Figs 20,21 and Supplementary Table 4).

For 1-butanol conversion on $NbOPO_4$, the calculation indicates that 1-butanol efficiently adsorbs on $Nb_{5c}$ with a corresponding adsorption energy of $-1.20$ eV, which is evidently stronger than that on $Re_2O_7$ ($-0.76$ eV), indicating that $NbOPO_4(100)$ possesses a stronger binding ability. Subsequently, with the aid of the surface $Nb_{5c}$, the C–O bond can break with each OH and butyl occupying a $Nb_{5c}$ site as the product; this process is strongly exothermic by 1.72 eV and gives a barrier of only 0.79 eV (Fig. 5a), implying its feasibility in both thermodynamics and kinetics (Supplementary Note 5 and Supplementary Table 5). In contrast, on $Re_2O_7$, 1-butanol dissociation is an endothermic process and has to overcome a larger barrier of 1.28 eV (Supplementary Note 5). By comparing the energy profiles (Fig. 5a), it is conclusive that $NbOPO_4$ demonstrates an inherently better performance for 1-butanol deoxygenation than $Re_2O_7$. Likewise, with respect to C–O bond cleavage of phenol, we performed the same calculation, which yields a similar conclusion that $NbOPO_4$ catalyses phenol dissociation more efficiently with a lower barrier than $Re_2O_7$ (Fig. 5b and Supplementary Note 5). Therefore, the high catalytic activity of $NbOPO_4$ can be ascribed to better adsorption capability of surface $Nb_{5c}$ and a lower activation barrier in comparison to $Re_2O_7$. Correlation between the C–O bond dissociation barriers and their corresponding adsorption energies (Supplementary Fig. 22) shows that stronger the bond strength of $M_{5c}$–O (M = Nb, Re) leads to easier the C–O bond cleavage, which is in line with the principle of Brønsted–Evans–Polanyi relationship[30–32]. In other words, the outstanding binding ability of $NbOPO_4$ is one determining factor in the efficient deoxygenation of cellulose and lignin. We also calculated and compared the adsorption energies of $NbOPO_4$ with two other typical catalysts ($Re_2O_7$, $ZrO_2$) toward various intermediate species, such as OH, O, THF, butyl and butoxy (Supplementary

Table 6). Indeed, $NbOPO_4$ demonstrates the strongest binding ability seamlessly, whereas $ZrO_2$, with the weakest adsorption, is rarely applied for cellulose conversion to alkane in practice.

We are now in a position to elucidate the inherent mechanism of $NbOPO_4$ exhibiting such an exceptional binding ability. An electronic structure analysis was made to understand the bond properties of $M_{5c}$–$O(OH)$ and the energy level of surface $M_{5c}$. The isosurface of charge density difference shows electron accumulation between O (or OH) and $Nb_{5c}$ (or $Re_{5c}$) (Fig. 5c), indicating a typical covalent bond character. Therefore, the binding strength with $O(OH)$ is mainly decided by the size and energy level of the $d$ orbital of $Nb_{5c}$ and $Re_{5c}$ cations. The projected density of states on the $d$-orbitals of surface $Nb_{5c}$ and $Re_{5c}$ indicates that the $d$ band of $Nb_{5c}$ near the Fermi level ($E_F$) is more delocalised comparing to $Re_{5c}$, and more importantly, the energy level of the highest occupied $d$ bands were evidently higher for $Nb_{5c}$ with some metallic character across $E_F$ (Fig. 5d). According to the frontier orbital theory, from these $d$ bands features one can rationalise such a strong binding ability of $NbOPO_4$ in facilitating deoxygenation reaction.

## Discussion

We have presented a one-pot catalytic process for the direct production of liquid alkanes from a wide variety of raw woody biomass over $Pt/NbOPO_4$ catalyst with excellent mass and carbon yields in a cyclohexane medium. This one-pot approach avoids the separation of raw biomass into isolated components and the use of an alkane solvent further simplifies downstream separation, as the alkane products can be used as solvents for the next run. The exceptional activity of the $Pt/NbOPO_4$ catalyst enabled direct conversion of raw woody biomass into liquid alkanes under mild conditions (190 °C) over a single multifunctional catalyst. The superior efficiency of this catalyst for direct hydrodeoxygenation of lignocellulose is found to originate from the synergistic effect between Pt, $NbO_x$ species and acidic sites. This brand new one-pot route requires no chemical pretreatment or separation of the raw woody biomass and thus tremendous energy savings can be potentially gained in comparison to the existing thermochemical and hydrolysis-based approaches for production of liquid fuels and chemical feedstocks from lignocellulose.

## Methods

**Catalyst preparation.** $NbOPO_4$ used here was synthesised by a hydrothermal method at pH = 2 according to literature[33]. $M/NbOPO_4$ (M = Pt, Pd, Rh, Ru) catalysts were prepared by incipient wetness impregnation of $NbOPO_4$ with aqueous solutions of $Pt(NO_3)_2$, $Pd(NO_3)_2$, $Rh(NO_3)_3$ and $RuCl_3$, respectively. After impregnation, the catalysts were dried at 100 °C for 12 h, followed by calcination in air at 500 °C for 3 h. $Pt/H$-ZSM-5 was prepared by the same procedure. $Pt$-$ReO_x/SiO_2$ and $Pt$-$ReO_x/C$ were prepared by sequential incipient wetness impregnation. Pt was loaded onto the support first, dried and calcined before the process was repeated for Re. The loading of Pt, Pd, Rh, Ru was 5 wt% in all cases, for $Pt$-$ReO_x/SiO_2$ and $Pt$-$ReO_x/C$, the loading amount of Re was 1 by the molar ratio of Re to Pt. The dispersion of $Pt/NbOPO_4$ catalysts with different Pt loading amount was presented in Supplementary Fig. 23.

**Reaction system and product analysis.** The direct hydrodeoxygenation of wood sawdusts was conducted in a 50 ml Teflon-lined stainless-steel autoclave. In a typical run, feedstock (0.20 g), catalyst (0.20 g) and cyclohexane (6.46 g) were put into the reactor, which was then sealed, purged three times with $H_2$ and charged to an initial pressure of 5.0 MPa with $H_2$. The reactor was then slowly heated to 190 °C under vigorous stirring and held at this temperature for 20 h. After the reaction finished, the reactor was quenched in an ice/water bath. The gas phase was carefully collected in a gas bag and analysed by GC equipped with a packed column, a methaniser (for $CO_2$ detection) and a flame ionisation detector (FID). The products in the liquid phase were qualitatively analysed by GC-MS and quantitatively analysed by GC–flame ionisation detector equipped with a HP-5 column. The yields of liquid alkanes were determined by adding dodecane as an internal standard after reaction. Mass yields were calculated by the equation: mass yield of pentanes (hexanes, alkylcyclohexanes) = [mass of pentanes (hexanes, alkylcyclohexanes)]/[mass of feedstock input]. Carbon yields were calculated

by the equation: carbon yield of pentanes (hexanes, alkylcyclohexanes) = [mass of carbon in pentanes (hexanes, alkylcyclohexanes)]/[mass of carbon in hemicellulose (cellulose, lignin)].

**Reaction pathways study.** The direct hydrodeoxygenation of cellobiose was carried out in a similar way to that of wood sawdusts. After the reaction was quenched in an ice/water bath, 6.5 g of water was added into the autoclave under vigorous stirring for a few minutes to extract the hydrophilic intermediates and to dissolve cellobiose for total analysis. The aqueous phase was analyzed by HPLC (Agilent 1200 series) equipped with a Shodex SUGAR SC-1011 column and a differential refractive index detector and by GC-MS to monitor the intermediates and unreacted cellobiose. The organic phase was analysed by GC-MS to observe the lipophilic intermediates.

**Neutron scattering experiments.** INS spectra were recorded on the TOSCA spectrometer at the ISIS Facility at the STFC Rutherford Appleton Laboratory (UK). TOSCA is an indirect geometry crystal analyser instrument that provides a wide dynamic range (16–4,000 cm$^{-1}$) with resolution optimised in the 50–2,000 cm$^{-1}$ range[34]. In this region TOSCA has a resolution of 1.25% of the energy transfer. The $Pt/Nb_2O_5$ (34.7 g) catalyst was loaded into an *in situ* catalysis cell with a copper vacuum seal and connected to a gas handling system. The sample was heated at 300 °C (5 °C/min ramping) under He for 3 h to remove any remaining trace water before the experiment. The sample was cooled to room temperature and a weight loss of 0.1 g was noted, assigned to loss of adsorbed water. The samples were cooled to < 15 K during data collection by a closed cycle refrigerator cryostat. The procedure of the *in situ* catalysis experiment with the INS measurements is summarised in Supplementary Fig. 24. INS spectra for condensed THF (2.31 g) and 1-butanol (2.36 g) in the solid state were measured in a flat-plate sample container below 12 K. INS spectra for condensed THF and 1-butanol were used to (i) calculate the amount of adsorbed THF and 1-butanol onto the catalyst in each case by the integration of the peak areas; (ii) identify and compare the vibrational modes for the adsorbed and the free molecules.

Estimation of the amount of adsorbed THF on the catalyst: on the basis of the relative intensities of the INS peak at ~600 cm$^{-1}$ and the known mass (2.31 g) of THF in the condensed sample, there is 0.42 g of adsorbed THF present on the catalyst in the neutron beam.

Estimation of the amount of adsorbed 1-butanol on the catalyst: on the basis of the relative intensities of the INS peak at ~740 cm$^{-1}$ and the known mass (2.36 g) of 1-butanol in the condensed sample, there is 0.30 g of adsorbed 1-butanol present on the catalyst in the neutron beam.

Adsorption of THF was carried out by flowing THF vapour (~200 mbar) in He (1.1 bar, 0.21 min$^{-1}$; this flow condition was used throughout the study) over the catalyst at 130 °C for 3 h. The cell was then flushed briefly with pure He flow for 2 min to remove the free and weakly bound THF on the catalyst, sealed and cooled to below 15 K for INS data collection. The adsorbed THF underwent the first catalytic conversion in pure $H_2$ flow for 10 min at 130 °C, and production of butane was observed instantly by mass spectrometry. The cell was then flushed with He to remove free butane and $H_2$, sealed and cooled again for INS collection to detect the presence of possible reaction intermediates.

## References

1. Armaroli, N. & Balzani, V. The future of energy supply: challenges and opportunities. *Angew. Chem. Int. Ed.* **46**, 52–66 (2007).
2. Murray, J. & King, D. Climate policy: oil's tipping point has passed. *Nature* **481**, 433–435 (2012).
3. Huber, G. W., Iborra, S. & Corma, A. Synthesis of transportation fuels from biomass: chemistry, catalysts, and engineering. *Chem. Rev.* **106**, 4044–4098 (2006).
4. Sutton, A. D. et al. The hydrodeoxygenation of bioderived furans into alkanes. *Nat. Chem* **5**, 428–432 (2013).
5. Roman-Leshkov, Y., Barrett, C. J., Liu, Z. Y. & Dumesic, J. A. Production of dimethylfuran for liquid fuels from biomass-derived carbohydrates. *Nature* **447**, 982–985 (2007).
6. Rubin, E. M. Genomics of cellulosic biofuels. *Nature* **454**, 841–845 (2008).
7. McKendry, P. Energy production from biomass (part 1): overview of biomass. *Bioresour. Technol.* **83**, 37–46 (2002).
8. Mäki-Arvela, P., Holmbom, B., Salmi, T. & Murzin, D. Y. Recent progress in synthesis of fine and specialty chemicals from wood and other biomass by heterogeneous catalytic processes. *Catal. Rev. -Sci. Eng.* **49**, 197–340 (2007).
9. Sun, Y. & Cheng, J. Y. Hydrolysis of lignocellulosic materials for ethanol production: a review. *Bioresour. Technol* **83**, 1–11 (2002).
10. Alonso, D. M., Bond, J. Q. & Dumesic, J. A. Catalytic conversion of biomass to biofuels. *Green Chem.* **12**, 1493–1513 (2010).
11. Dugar, D. & Stephanopoulos, G. Relative potential of biosynthetic pathways for biofuels and bio-based products. *Nat. Biotechnol.* **29**, 1074–1078 (2011).
12. Mohan, D., Pittman, C. U. & Steele, P. H. Pyrolysis of wood/biomass for bio-oil: a critical review. *Energy Fuels* **20**, 848–889 (2006).

13. Bridgwater, A. V. Review of fast pyrolysis of biomass and product upgrading. *Biomass Bioenergy* **38**, 68–94 (2012).
14. Huber, G. W., Cortright, R. D. & Dumesic, J. A. Renewable alkanes by aqueous-phase reforming of biomass-derived oxygenates. *Angew. Chem. Int. Ed.* **43**, 1549–1551 (2004).
15. Zakzeski, J., Bruijnincx, P. C. A., Jongerius, A. L. & Weckhuysen, B. M. The catalytic valorization of lignin for the production of renewable chemicals. *Chem. Rev.* **110**, 3552–3599 (2010).
16. Matson, T. D., Barta, K., Iretskii, A. V. & Ford, P. C. One-pot catalytic conversion of cellulose and of woody biomass solids to liquid fuels. *J. Am. Chem. Soc.* **133**, 14090–14097 (2011).
17. Li, C., Zheng, M., Wang, A. & Zhang, T. One-pot catalytic hydrocracking of raw woody biomass into chemicals over supported carbide catalysts: simultaneous conversion of cellulose, hemicellulose and lignin. *Energy Environ. Sci* **5**, 6383–6390 (2012).
18. Yan, N. *et al.* Selective degradation of wood lignin over noble-metal catalysts in a two-step process. *ChemSusChem* **1**, 626–629 (2008).
19. Liu, S., Tamura, M., Nakagawa, Y. & Tomishige, K. One-pot conversion of cellulose into n-hexane over the Ir-ReO$_x$/SiO$_2$ catalyst combined with HZSM-5. *ACS Sustain. Chem. Eng.* **2**, 1819–1827 (2014).
20. de Beeck, B. O. *et al.* Direct catalytic conversion of cellulose to liquid straight-chain alkanes. *Energy Environ. Sci.* **8**, 230–240 (2015).
21. Del Paggio, A. A. *et al.* A process for producing hydrocarbons. WO patent WO 2013/074434 A1 (Shell Oil Company, USA 2013).
22. Cortright, R. D. & Blommel, P. G. Synthesis of liquid fuels and chemicals from oxygenated hydrocarbons. US patent US 2008/0300434 A1 (Virent, Inc., USA 2008).
23. Liu, Y. *et al.* One-pot catalytic conversion of raw lignocellulosic biomass into gasoline alkanes and chemicals over LiTaMoO$_6$ and Ru/C in aqueous phosphoric acid. *ACS Sustain. Chem. Eng.* **3**, 1745–1755 (2015).
24. Xia, Q. *et al.* Pd/NbOPO$_4$ multifunctional catalyst for the direct production of liquid alkanes from aldol adducts of furans. *Angew. Chem. Int. Ed.* **53**, 9755–9760 (2014).
25. Burch, R. *et al.* Catalytic hydrogenation of tertiary amides at low temperatures and pressures using bimetallic Pt/Re-based catalysts. *J. Catal.* **283**, 89–97 (2011).
26. Chia, M. *et al.* Selective hydrogenolysis of polyols and cyclic ethers over bifunctional surface sites on rhodium-rhenium catalysts. *J. Am. Chem. Soc.* **133**, 12675–12689 (2011).
27. Mitchell, P. C. H., Parker, S. F., Ramirez-Cuesta, A. J. & Tomkinson, J. *Vibrational Spectroscopy with Neutrons, with Applications in Chemistry, Biology, Materials Science, and Catalysis* (World Scientific, Singapore, 2005).
28. Ohno, K., Yoshida, H., Watanabe, H., Fujita, T. & Matsuura, H. Conformational study of 1-butanol by the combined use of vibrational spectroscopy and ab initio molecular orbitol calculations. *J. Phys. Chem.* **98**, 6924–6930 (1994).
29. Hibbitts, D., Tan, Q. & Neurock, M. Acid strength and bifunctional catalytic behavior of alloys comprised of noble metals and oxophilic metal promoters. *J. Catal.* **315**, 48–58 (2014).
30. Michaelides, A. *et al.* Identification of general linear relationships between activation energies and enthalpy changes for dissociation reactions at surfaces. *J. Am. Chem. Soc.* **125**, 3704–3705 (2003).
31. Wang, S. *et al.* Universal transition state scaling relations for (de)hydrogenation over transition metals. *Phys. Chem. Chem. Phys.* **13**, 20760–20765 (2011).
32. van Santen, R. A. *et al.* Reactivity theory of transition-metal surfaces: a Brønsted – Evans – Polanyi linear activation energy—free-energy analysis. *Chem. Rev.* **110**, 2005–2048 (2010).
33. Zhang, Y. *et al.* Mesoporous niobium phosphate: an excellent solid acid for the dehydration of fructose to 5-hydroxymethylfurfural in water. *Catal. Sci. Technol* **2**, 2485–2491 (2012).
34. Parker, S. F. *et al.* Recent and future developments on TOSCA at ISIS. *J. Phys. Conf. Ser.* **554**, 012003 (2014).

## Acknowledgements

This project was supported financially by the National Natural Science Foundation of China (No. 21273071, 21101063, 91545103 and 21073060), the Shanghai Rising-Star Program (14QA1401100), the Fundamental Research Funds for the Central Universities, University of Manchester and the Science and Technology Facilities Council (STFC) in UK. The Rutherford Appleton Laboratory is thanked for access to neutron beam facilities (TOSCA). Computing resources (time on the SCARF cluster for the CASTEP calculations) was provided by STFC's e-Science facility.

## Author contributions

Q.N.X., Y.S. and X.H.L. helped in preparation and characterisation of catalysts, and performing the catalytic reactions. Z.J.C., H.F.W. and X.Q.G. used DFT calculations. Q.N.X., S.F.P., X.H. and S.Y. helped in the collection, analysis and DFT modelling of neutron scattering data. Y.Q.W. conceived the overall direction of the project. All the authors discussed the results and co-wrote the manuscript.

## Additional information

**Competing financial interests:** The authors declare no competing financial interests.

# 18

# Fast, greener and scalable direct coupling of organolithium compounds with no additional solvents

Erik B. Pinxterhuis[1], Massimo Giannerini[1], Valentín Hornillos[1] & Ben L. Feringa[1]

Although the use of catalytic rather than stoichiometric amounts of metal mediator in cross-coupling reactions between organic halides and organometallic counterparts improves significantly the atom economy and waste production, the use of solvents and stoichiometric generation of main-group byproducts (B, Sn and Zn) hamper the 'greenness' and industrial efficiency of these processes. Here we present a highly selective and green Pd-catalysed cross-coupling between organic halides and organolithium reagents proceeding without additional solvents and with short reaction times (10 min). This method bypasses a number of challenges previously encountered in Pd-catalysed cross-coupling with organolithium compounds such as strict exclusion of moisture, dilution and slow addition. Operational ease of this protocol combines the use of industrially viable catalysts loadings (down to 0.1 mol%), scalability of the process (tested up to 120 mmol) and exceptionally favourable environmental impact (E factors in several cases as low as 1).

---

[1] Faculty of Mathematics and Natural Sciences, Stratingh Institute for Chemistry, University of Groningen, Nijenborgh 4, 9747AG Groningen, The Netherlands. Correspondence and requests for materials should be addressed to V.H. (email: v.hornillos.gomez.recuero@rug.nl) or to B.L.F. (email: b.l.feringa@rug.nl).

The development of greener, more efficient and simple reaction methodologies sets a priority for the synthetic chemistry community both in industry and in academia as well[1]. Solvents are mainly responsible for the environmental impact of synthetic procedures, being, in general, the largest contributors to the magnitude of the E factor (E = organic waste (kg)/product (kg)), a value introduced by Sheldon et al.[2] to measure the 'greenness' of a chemical process[3]. Thus, reduction or elimination of solvents from organic reactions is of major concern in chemical process development[4–7]. Higher energy needs, toxicity, safety hazards and massive waste treatment are direct implications of the use of large volumes of solvents that negatively affect both costs and environment. Inspired by the 12 principles of Green Chemistry[8], the development of sustainable products is committed to reduce or, possibly, prevent the use of traditional solvents that still, as today, represent the major share of chemical waste production (up to 80%).

An ideal solution to the above-mentioned issues is to completely exclude the solvent from the reaction medium. These so-called solvent-free conditions often lead to additional improvements in other critical parameters as well, such as the catalytic loading (generally lower), the speed of the reaction (generally higher) and the volume/output ratio[9].

A particular challenging class of transformations in this respect are the widely used transition metal-catalysed reactions. Despite the central role played by Pd-catalysed cross-coupling reactions of organometallic compounds with organo-(pseudo)halides, both in industrial[10,11] and in academic laboratories[12–15], the corresponding solvent-free variants have been scarcely reported. Although boron compounds have been engaged in solvent-free cross-coupling reactions, thus far the use of microwaves[16,17], ball mill[18,19] and/or high temperatures are required (Fig. 1a).

Our group has recently described methods for the palladium-catalysed direct cross-coupling of highly reactive organolithium reagents[20,21] (among the most versatile and widely used reagents in organic synthesis) with organic halides under mild conditions[22–29]. The extreme reactivity of organometallic reagents such as organolithium compounds commonly dictates highly delicate conditions such as low temperatures, dilution, slow addition and so on, to achieve high conversion and selectivity in their chemical transformations. In the case of Pd-catalysed cross-coupling reactions directly applying organolithium compounds, the use of toluene as a solvent and slow addition of a previously diluted solution of organolithium reagent are key factors to obtain high selectivity and good yields, while avoiding the notorious lithium–halogen exchange and homocoupling side reactions (Fig. 1b)[30–32].

Despite the formidable challenge presented by the quest to control selectivity when mixing organolithium reagents with neat organohalides, owing to the possibility for numerous competing reactions, we show here the development of the first Pd-catalysed solvent-free cross-coupling of highly polar organometallic compounds that, through a concise and simple procedure, affords the desired coupled product with excellent selectivities within 10 min and in many cases with E factors as low as 1 (Fig. 1c).

## Results

**Preliminary observations.** Inspired by the report of García-Álvarez and colleagues[33] on the use of deep eutectic solvents (DES, mostly obtained by mixing a quaternary ammonium salt as choline chloride with a hydrogen-bond donor such as glycerol or water) for the 1,2-addition of Grignard and organolithium reagents to ketones, we set out to explore the Pd-catalysed cross-coupling reaction of organolithium compounds and organic halides employing these green solvents. Despite the high reactivity of organolithium compounds towards protic solvents, we were delighted to find that the reaction between an excess of PhLi (2–10 eq) and 1-bromonaphthalene using 10 mol% of Pd catalyst in a type III DES proceeded with good selectivity, although in low yield (28–53% conversion, see Supplementary Table 1). We hypothesized that probably small droplets of substrate containing high concentration of catalyst were formed, and that

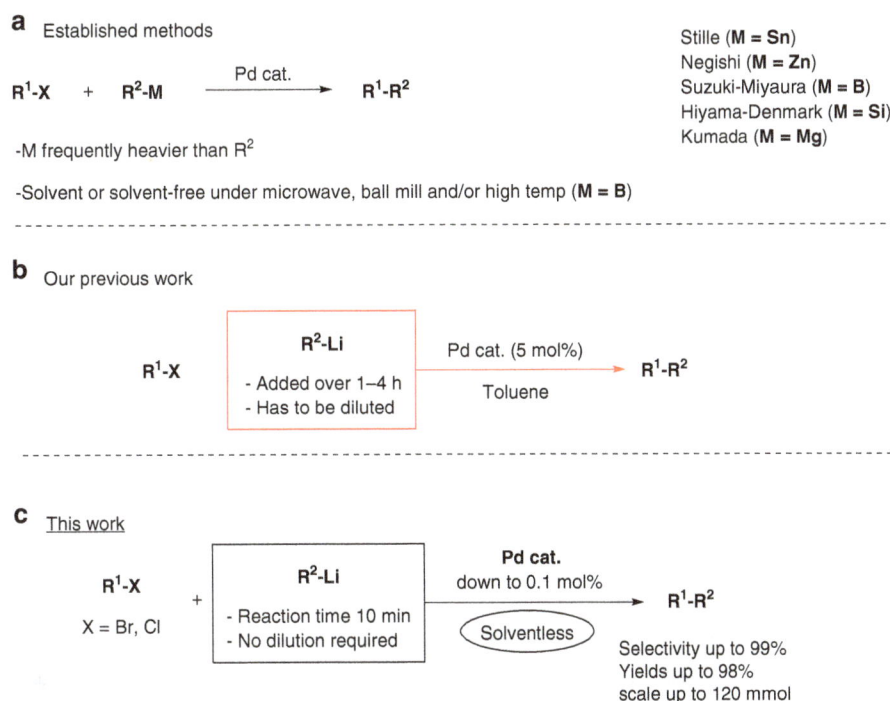

**Figure 1 | State-of-the-art overview.** (**a**) Established methods for Pd-catalysed cross-coupling reactions. (**b**) Catalytic cross-coupling with organolithium compounds. (**c**) A fast, highly scalable and solvent-free direct cross-coupling of organolithium compounds.

the reaction was taking place directly in the organic phase rather than in the DES phase. However, owing to quenching of the organolithium reagent by the solvent, the conversions obtained were low. We questioned whether the innate reactivity of organolithium compounds could be turned into an inherent advantage offering the possibility to develop a solvent-free Pd-catalysed cross-coupling protocol, which proceeds within minutes, without the support of any additional device (microwave, ball mills and so on), with low catalytic loading and at ambient temperature without the use of strictly inert conditions.

**Reaction conditions optimization.** We set out to investigate the reaction between 4-methoxybromobenzene **1a**, a reluctant aryl bromide in coupling reactions[34], and commercially available phenyllithium, as successful conditions for the coupling of these two substrates would most probably apply to a wide variety of other coupling partners as well (Table 1). All the reactions were carried out by adding the organolithium compound (without prior dilution) to a neat mixture of catalyst and organic halide over 10 min at room temperature (RT). Moreover, we employed a 1-mmol scale to illustrate the synthetic utility of the method. Reactions using the *in situ*-prepared palladium complex Pd/ XPhos (generated by mixing $Pd_2(dba)_3$ with XPhos)[35], previously reported to be effective for other Pd-catalysed cross-coupling reactions with aryllithium reagents[25,26], afforded the cross-coupling product **2a** ($R = $ Ph) within 10 min, although in the presence of significant amounts of the undesired homocoupling product (Table 1, entry 1). By employing Pd-PEPPSI-IPent

catalyst[36], the selectivity was raised to 90% at the expense of the homocoupling product (Table 1, entry 2). Nonetheless, we were delighted to find that the commercially available air- and temperature-stable Pd-PEPPSI-IPr catalyst, which is seven times cheaper than Pd-PEPPSI-IPent[37], afforded full conversion and nearly perfect selectivity ($>95\%$,) towards the coupled product **2a** ($R = $ Ph) at RT in $<10$ min, avoiding the formation of dehalogenation or homocoupling side products **3** and **4** (Table 1, entry 3). Importantly, the high selectivity was maintained while lowering the catalyst loading to 1.5 mol% (Table 1, entry 4, 84% isolated yield). With an efficient catalyst for $Csp^2$–$Csp^2$ cross-coupling in hands, we then turn our attention to the challenging $Csp^3$–$Csp^2$ solvent-free cross-coupling with alkyllithium compounds. The direct use of commercially available *n*-BuLi, one of the most reactive organometallic reagents, in combination with Pd-PEPPSI-IPr led to the desired product **2v** ($R = n$-Bu), although with slightly diminished selectivity (Table 1, entry 5). Further screening of catalysts showed that the use of commercially available $Pd[P(tBu)_3]_2$ catalyst[38] restored the selectivity ($>95\%$) towards the coupled product **2v** ($R = n$-Bu) with excellent (82%) isolated yield (Table 1, entry 6). Importantly, when this reaction was performed using an extremely low catalyst loading (0.1 mol %), product **2v** ($R = n$-Bu) was still obtained in high conversion and selectivity (Entry 7).

**Scope and applicability.** To our delight, the optimized conditions proved to be general and could be applied successfully to the solvent-free cross-coupling of a variety of aryllithium (**2a**–**2u**)

---

**Table 1 | Screening of different ligands.**

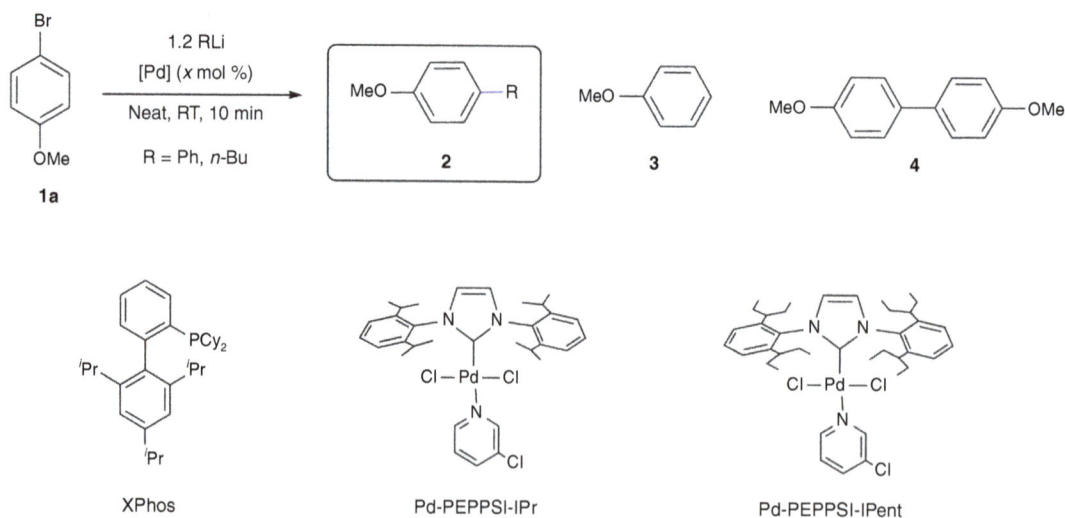

| Entry[*] | R | [Pd] | [Pd] (x mol %) | 2:3:4[†] |
|---|---|---|---|---|
| 1 | Ph | $Pd_2(dba)_3$/XPhos | 3 | **2a**, 85:3:12 |
| 2 | Ph | Pd-PEPPSI-IPent | 3 | **2a**, 90:2:8 |
| 3 | Ph | Pd-PEPPSI-IPr | 3 | **2a**, >95:-:- |
| 4 | Ph | Pd-PEPPSI-IPr | 1.5 | **2a**, >95:-:-[‡] |
| 5 | *n*-Bu | Pd-PEPPSI-IPr | 1.5 | **2v**, 88:2:10 |
| 6 | *n*-Bu | $Pd[P(t$-$Bu)_3]_2$ | 1.5 | **2v**, >95:-:-[§] |
| 7 | *n*-Bu | $Pd[P(t$-$Bu)_3]_2$ | 0.1 | **2v**, 86:9:5[‖] |

dba, dibenzylideneacetone.
*Conditions: the commercial organolithium reagent (1.2 mmol) was added to a mixture of **1** (1 mmol) and palladium catalyst over 10 min.
†**2:3:4** Ratios determined by GC analysis.
‡84% Yield.
§82% Yield.
‖91% Conversion.

**Table 2 | Pd-Catalysed cross-coupling of organolithium reagents with aryl halides under solventless conditions[*–||].**

2a, 84%  2b, 87%  2c, 87%  2d, 90%  2e, 98%  2f, 91%  2g, 91%  2h, 92%  2i, 97%

2j, 84%  2k, 84%  2l, 96%  2m, 98%  2n, 87%  2o, 88%  2p, 81%

2q, 85%  2r, 87%  2l, 87%[‡] X = Cl  2s, 84% X = Cl  2t, 86% X = Cl  2u, 84% X = Cl  2v, 82%  2w, 85%

2x, 94%  2y, 82%  2z, 84%  2aa, 86%  2ab, 97%[§]  2ac, 95%[§]  2ad, 83%  2ae, 84%  2af, 85%

[*]Conditions: RLi (1.2 mmol) was added to a mixture of organic bromide (1 mmol) and catalyst over 10 min. X, Br unless otherwise noted (1l,1s,1t,1u; X = Cl). Selectivity 2:3,4 > 95%. Compound 2 was extracted with Et₂O or AcOEt, after quenching the reaction mixture with minimum amount of sat. aq. NH₄Cl.
[†]All yields given are isolated yields.
[‡]3 Mol% of catalyst was employed for the reaction with 2-thienyllithium.
[§]GC yield: product was not isolated due to volatility issues.
[||]For limitations of the method, see Supplementary Fig. 1.

and the even more reactive alkyllithium reagents (**2v–2af**), in all cases affording the products with high selectivity within minutes (Table 2). The remarkably fast cross-coupling methodology gave excellent results in combination with non-commercially available aryllithium reagents obtained through common preparative procedure, such as lithium/halogen exchange (**2f–2h**) and *ortho*-directed lithiation. Illustrative is the case of the highly hindered bis-*ortho*-substituted 2,6-dimethoxy-phenyllithium, used in the synthesis of compounds **2i–2k**, which was prepared by direct metalation of 1,3-dimethoxybenzene. In all cases, the organolithium reagents were prepared using the minimal amount of ethereal solvent to maintain them soluble (see Supplementary Methods). Despite the higher reactivity and basicity of alkyl-lithium reagents when compared with (hetero)aryllithium compounds, we were delighted to find high selectivities and yields also for a variety of Csp³–Csp² cross-coupling products. This includes the use of different alkyllithium compounds as *n*-BuLi,

*n*-HexLi and the smallest MeLi with electron-rich and electron-poor arylbromides (Table 2, **2v–2ad**). The bifunctional C(sp³)-(trimethylsilyl)methyllithium reagent[28] also couples with excellent selectivity, providing synthetically versatile benzylsilanes **2ae** and **2af**. The lack of dehalogenated side products from these Csp³–Csp² cross-coupling reactions demonstrates that no competing β-hydride elimination/olefin dissociation[24] occurs (which is another competing pathway besides the formation of dehalogenated and isomerized products). A limitation so far for this protocol employing the Pd[P(*t*-Bu)₃]₂-based catalytic system is that secondary alkyllithium reagents such as *i*-PrLi and *s*-BuLi led to the formation of dehalogenation products.

Despite the conditions of highly concentrated reaction partners, various observations highlight how the reaction proceeds exclusively under catalyst control. Thus, the reaction of 1-bromonaphthalene **1b** resulted, with both aryl- and alkyllithium, in the corresponding coupled products (**2b**, **2f**,

**2k**, **2x** and **2aa**) without the formation of regioisomers, indicating that benzyne intermediates via 1,2-elimination are not formed. Apart from liquid substrates, even solid bromofluorene **1af** was successfully employed, despite the acidity of the benzylic protons ($pK_a = 22$). The reaction of $n$-BuLi and MeLi with $p$-chloro-bromobenzene occurs selectively with no detectable chloride displacement (Table 2, compounds **2w** and **2ab**).

Sterically hindered bromides **1c–1e**, known for being more reluctant substrates in the synthesis of biaryls[23], were also successfully coupled at RT in 10 min, indicating that the transmetallation step takes place rapidly, under these conditions, inducing a fast coupling process.

The dramatic effect of the solvent-free conditions in enhancing the reaction rate was demonstrated in the coupling of commercially available 2-thienyllithium, which, according to our previous observations, required the addition of stoichiometric amounts of tetramethylethylenediamine as the activating agent and elevated temperatures (40 °C) to react[22]. Under solvent-free conditions, 2-thienyllithium reacted smoothly at RT within 10 min, in high selectivity and yield, without the use of any additive (see compounds **2l–2r**). We have recently showed that the cross-coupling of 2-alkoxy-substituted arylbromides with organolithium is plagued by fast bromine–lithium exchange induced by the ortho-directing alkoxy unit. The use of the corresponding aryl chlorides, inherently less prone to halogen/lithium exchange, is thus mandatory, to afford selectively the product and prevent side products formation[27]. However, to our surprise, under our solvent-free protocol, aryl bromides **1n**, **1o** and 3,3′-dibromo-BINOL **1p**, could all be coupled successfully with 2-thienyllithium in high selectivity (>95%) and with excellent yield, avoiding the notorious bromine–halogen exchange (Table 2). To emphasize the versatility of the new method, it has to be noted that the only previous reported synthesis of BINOL derivate **2p** required the preparation of the corresponding bis-trifluoroboronate BINOL derivative and further reaction with 2-bromothiophene under microwave conditions[39]. The use of acetal-protected aldehyde **1q** was also tolerated without the cleavage of the protecting group. As in the case of alkyllithium compounds, 1-bromo-3-chlorobenzene **1r** reacted with 2-thienyllithium using Pd-PEPPSI-IPr catalyst selectively, leaving the chloride untouched. Nevertheless, the electron-poor aryl chloride **1l** reacted readily with 2-thienyllithium and electron-rich chlorides **1t** and **1u** were also easily coupled under the optimized conditions using more reactive PhLi at RT (Table 2).

A major issue often associated with solvent-free reactions is the homogeneity of the reaction medium, in particular with solid starting materials. However, the methodology presented here provides high selectivity and yields when solid substrates such as **1h**, **1p**, **1q**, **1ae** and **1af** were used in combination with aryllithium and TMSCH$_2$Li compounds. Unfortunately, the presence of dehalogenated product was observed for the use of $n$-HexLi and $n$-BuLi in combination with solid substrates (for preliminary results using Grignard reagents in similar conditions, see Supplementary Fig. 2).

**Scalability of the protocol**. In organic chemistry, problems in the scaling up of batch reactions have been known to arise from various issues including inefficient mixing and lack of heat transfer. To test whether this novel method is suitable to be performed on a larger scale, the cross-coupling between $n$-BuLi and 1-bromonaphthalene **1b** was tested on multigram scale with catalyst loading as low as 0.1 mol%. We were pleased to find that the scale of the reaction had little effect on the selectivity, although the presence of a small amount of dehalogenated side product was observed (Table 3). It is noteworthy that the cross-coupling was found to maintain its effectiveness even at 120 mmol scale employing 0.4 mol% of catalyst, providing exceptional E factors as low as 0.8 (Table 3, entry 3). It should be emphasized that typical E factors in the range of 5–100 are seen in transformations producing fine chemicals and pharmaceuticals[2]. Importantly, after the addition of the organolithium compound, the crude product was quenched, washed with water and dried, giving the desired product in reagent-grade quality within 60 min, including all the operations.

**Potential of the methodology in synthetic application**. To demonstrate the advantages of the new method, we have compared it with some established cross-coupling methodologies currently used in the production of two typical building blocks for pharmaceuticals and conjugated polymers for light-emitting devices. The first example deals with the preparation of a key intermediate (**2ag**) for the synthesis of a patented melanin concentrating hormone receptor ligand (Fig. 2a)[40] involved in the treatment of eating disorders, weight gain, obesity, depression and anxiety. The reaction between 1-bromo-4-chlorobenzene **1ag** and 2-thienyllithium under the optimized reaction conditions provided the cross-coupling product **2ag** in high yield and selectivity within 10 min at RT (E factor: 5.4). In sharp contrast, the reported procedure (E factor: 41) involves the corresponding thienylboronic acid, needs a mixture of DME/H$_2$O heated at reflux for 4 h, 2 eq of base and requires the corresponding highly reactive aryl iodide. As noted by Lipschutz *et al.*[3], aqueous

---

**Table 3 | Catalyst-loading effect for the cross-coupling of 1-bromonaphthalene with *n*-BuLi.**

| Entry[‡] | ArBr | [Pd] | Reaction time (min) | Conv. (%) | 2x:3:4[*] |
|---|---|---|---|---|---|
| 1 | 5 mmol (1.03 g) | 0.4 mol% (11 mg) | 20 | Full | 92:3:5 |
| 2 | 10 mmol (2.05 g) | 0.1 mol% (7 mg) | 20 | 95 | 88:7:5 |
| 3 | 120 mmol (27 g) | 0.4 mol% (250 mg) | 30 | Full | 91:7:2[†] |

[‡]Conditions: *n*-BuLi was added to a mixture of **1b** and Pd[P(t-Bu)$_3$]$_2$ over 20–30 min.
[*]**2x:3b:4b** Ratios determined by GC analysis and $^1$H NMR.
[†]**2x**: 19.9 g, 90% isolated yield.

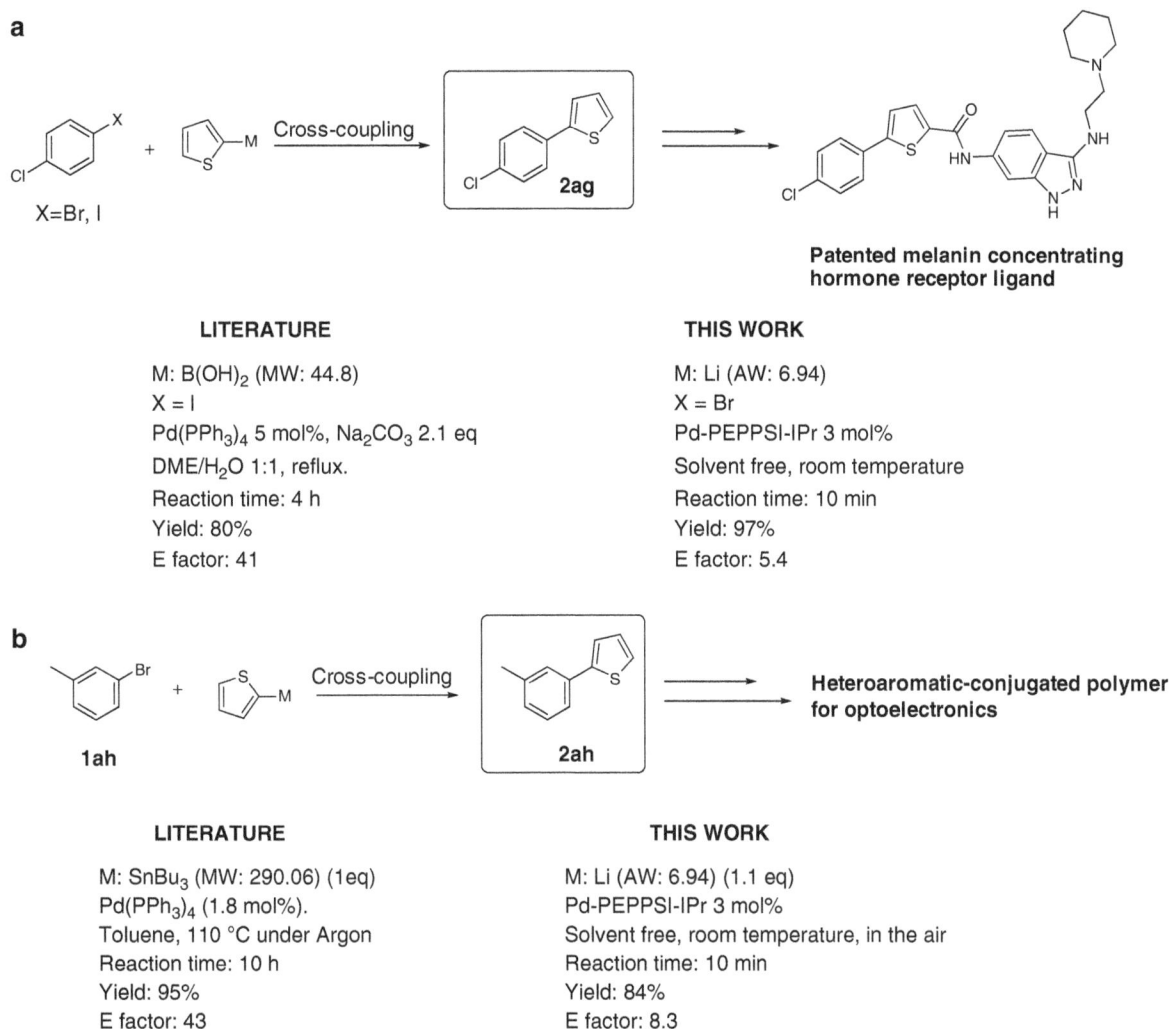

**Figure 2 | Synthetic applicability.** Comparison between established methods and the present cross-coupling protocol with organolithium reagents in the synthesis of key intermediates for a melanin-concentrating hormone receptor ligand (**a**) and conjugated polymer for optoelectronic devices (**b**).

workup may also be taken into account in the calculation of the E-factor. In this case, the advantages of using low amounts of solvents become even more evident: although the preparation of **2ag** through conventional Suzuki cross-coupling has an E-factor as high as 84, taking into account the extraction workup, thanks to the reduced amounts of extraction volumes needed, the E-factor obtained for our protocol is 15.4, marking a dramatic difference of 68.6 units between the two protocols. Notably, LiBr is the major byproduct of the reaction.

The second example illustrates the synthesis of a heteroaromatic monomer employed in the preparation of polymeric materials for optoelectronic devices (Fig. 2b) (ref. 41). Coupling between 2-thienyllithium and 3-methyl-1-bromobenzene **1ah** gives access to the desired compound **2ah**, at RT, in very good yield within 10 min and with an E factor five times lower than that reported in synthetic methodology (110 °C, 10 h in toluene using an organotin compound with a molecular weight four times higher than that of 2-thienyllithium). In the last example, in addition to the remarkable differences in reaction time (10 versus 600 min) and temperature (20 °C versus 110 °C), the use of solvent-free cross-coupling of 2-thienyllithium also avoids potentially toxic tin wastes and their often difficult removal, and prevents the use of strictly inert atmosphere required for the coupling of the corresponding tin reagent.

## Discussion

We have discovered that, in sharp contrast to all common reaction protocols using highly reactive organometallics such as organolithium compounds, the Pd-catalysed cross-coupling of arylbromides and alkyl- or aryllithium reagents under neat conditions proceeds exceptionally fast and is selective. Based on this finding we have developed a general solvent-free methodology for the Pd-catalysed direct cross-coupling of organolithium compounds with organic halides, under ambient conditions, providing high yields and excellent selectivities. Fast reaction times (10 min), lower catalyst loading (down to 0.1 mol%), high scalability, operational simplicity (see Methods) and the possibility to avoid the use of a strictly inert atmosphere, of syringe pumps and of additives such as tetramethylethylenediamine are key feature of this methodology. Compared with reported Pd-catalysed cross-coupling, this methodology is particularly attractive, owing to the strongly reduced environmental impact, that is, outstanding volume:time:output ratio, limited amount and low toxicity of the waste and five- to tenfold reduction in E-factor. The use of stable and commercially available catalysts, commercial or readily available and inexpensive organolithium reagents and the applicability to a wide variety of organic bromides are additional factors that contribute to the prospect of these C–C bond formations in the art of synthesis. For supporting videos illustrating the ease of operation, see Supplementary Movies 1–2.

## Methods

The corresponding organolithium reagent (1.2 eq) was added over a mixture of substrate (1.0 eq) and catalyst (1.5–3 mol%) at RT for 10 min. After the addition was completed, a saturated solution of aqueous $NH_4Cl$ was added and the mixture was extracted with AcOEt or $Et_2O$. The organic phases were combined and dried with anhydrous $Na_2SO_4$. Evaporation of the solvent under reduced pressure afforded the crude product that was then filtered over a silica gel plug to afford the pure product. For NMR spectra of the compounds in this article, see Supplementary Figs 3–73.

## References

1. Dach, R., Song, J. J., Roschangar, F., Samstag, W. & Senanayake, C. H. The eight criteria defining a good chemical manufacturing process. *Org. Process Res. Dev.* **16**, 1697–1706 (2012).
2. Sheldon, R. A., Arends, I. W. C. E. & Hanefeld, U. *Green Chemistry and Catalysis* (Wiley-VCH, 2007).
3. Lipshutz, B. H., Isley, N. A., Fennewald, J. C. & Slack, E. D. On the way towards greener transition-metal-catalyzed processes as quantified by E factors. *Angew. Chem. Int. Ed.* **52**, 10952–10958 (2013).
4. Tanaka, K. & Toda, F. *Solvent-Free Organic Synthesis* (Wiley-VCH, 2003).
5. Nelson, W. M. *Green Solvents for Chemistry-Perspectives and Practice* (Oxford Univ. Press, 2003).
6. Li, C.-J. Organic reactions in aqueous media with a focus on carbon–carbon bond formations: a decade update. *Chem. Rev.* **105**, 3095–3166 (2005).
7. Simon, M. O. & Li, C.-J. Green chemistry oriented organic synthesis in water. *Chem. Soc. Rev.* **41**, 1415–1427 (2012).
8. Anastas, P. T. & Warner, J. C. *Green Chemistry: Theory and Practice* (Oxford Univ. Press, 1998).
9. Singh, M. S. & Chowdhury, S. Recent developments in solvent-free multicomponent reactions: a perfect synergy for eco-compatible organic synthesis. *RSC Adv.* **2**, 4547–4592 (2012).
10. Magano, J. & Dunetz, J. R. Large-scale applications of transition metal-catalyzed couplings for the synthesis of pharmaceuticals. *Chem. Rev.* **111**, 2177–2250 (2011).
11. *Topics in Organometallic Chemistry* (eds Beller, M. & Blaser, H. -U.) Vol. 42 (Springer, 2012).
12. *Metal-Catalyzed Cross-Coupling Reactions* (eds de Meijere, A. & Diederich, F.) (Wiley-VCH, 2004).
13. *Transition Metals for Organic Synthesis* (eds Beller, M. & Bolm, C. (Wiley-VCH, 2004).
14. *Handbook of Organopalladium Chemistry for Organic Synthesis* (ed. Negishi, E.-I.) (Wiley-Interscience, 2002).
15. Wu, X.-F., Anbarasan, P., Neumann, H. & Beller, M. From noble metal to Nobel Prize: palladium-catalyzed coupling reactions as key methods in organic synthesis. *Angew. Chem. Int. Ed.* **49**, 9047–9050 (2010).
16. Brooker, M. D., Cooper, Jr S. M., Hodges, D. R., Carter, R. R. & Wyatt, J. K. Studies of microwave-enhanced Suzuki–Miyaura vinylation of electron-rich sterically hindered substrates utilizing potassium vinyltrifluoroborate. *Tetrahedron Lett.* **51**, 6748–6752 (2010).
17. El Akkaoui, A., Berteina-Raboin, S., Mouaddib, A. & Guillaumet, G. Direct arylation of imidazo[1,2-b]pyridazines: microwave-assisted one-pot suzuki coupling/Pd-catalysed arylation. *Eur. J. Org. Chem.* 862–871 (2010).
18. Bernhardt, F., Trotzki, R., Szuppa, T., Stolle, A. & Ondruschka, B. Solvent-free and time-efficient Suzuki–Miyaura reaction in a ball mill: the solid reagent system KF–Al2O3 under inspection. *Beilstein J. Org. Chem.* **6**, 7 (2010).
19. Schneider, F. & Ondruschka, B. Mechanochemical solid-state Suzuki reactions using an *in situ* generated base. *ChemSusChem* **1**, 622–625 (2008).
20. *The Chemistry of Organolithium Compounds* (ed. Rappoport, Z. & Marek, I.) (Wiley-VCH, 2004).
21. *Lithium Compounds in Organic Synthesis* (ed. Luisi, R. & Capriati, V.) (Wiley-VCH, 2014).
22. Giannerini, M., Fañanás-Mastral, M. & Feringa, B. L. Direct catalytic cross-coupling of organolithium compounds. *Nat. Chem.* **5**, 667–672 (2013).
23. Giannerini, M., Hornillos, V., Vila, C., Fañanás-Mastral, M. & Feringa, B. L. Hindered aryllithium reagents as partners in palladium-catalyzed cross-coupling: synthesis of tri- and tetra-ortho-substituted biaryls under ambient conditions. *Angew. Chem. Int. Ed.* **52**, 13329–13333 (2013).
24. Vila, C., Giannerini, M., Hornillos, V., Fañanás-Mastral, M. & Feringa, B. L. Palladium-catalysed direct cross-coupling of secondary alkyllithium reagents. *Chem. Sci.* **5**, 1361–1367 (2014).
25. Hornillos, V., Giannerini, M., Vila, C., Fañanás-Mastral, M. & Feringa, B. L. Direct catalytic cross-coupling of alkenyllithium compounds. *Chem. Sci.* **6**, 1394–1398 (2015).
26. Hornillos, V., Giannerini, M., Vila, C., Fañanás-Mastral, M. & Feringa, B. L. Catalytic direct cross-coupling of organolithium compounds with aryl chlorides. *Org. Lett.* **15**, 5114–5511 (2013).
27. Castelló, L. M. et al. Pd-catalyzed cross-coupling of aryllithium reagents with 2-alkoxy-substituted aryl chlorides: mild and efficient synthesis of 3,3′-diaryl BINOLs. *Org. Lett.* **17**, 62–65 (2015).
28. Heijnen, D., Hornillos, V., Corbet, B. P., Giannerini, M. & Feringa, B. L. Palladium-catalyzed C(sp3)–C(sp2) cross-coupling of (trimethylsilyl) methyllithium with (hetero)aryl halides. *Org. Lett.* **17**, 2262–2265 (2015).
29. Vila, C., Hornillos, V., Giannerini, M., Fañanás-Mastral, M. & Feringa, B. L. Palladium-catalysed direct cross-coupling of organolithium reagents with aryl and vinyl triflates. *Chem. Eur. J.* **20**, 13078–13083 (2014).
30. Pace, V. & Luisi, R. Expanding the synthetic portfolio of organolithiums: direct use in catalytic cross-coupling reactions. *ChemCatChem* **6**, 1516–1519 (2014).
31. Capriati, V., Perna, F. M. & Salomone, A. 'The great beauty' of organolithium chemistry: a land still worth exploring. *Dalton Trans.* **43**, 14204–14210 (2014).
32. Firth, J. D. & O'Brien, P. Cross-Coupling knows no limits: assessing the synthetic potential of the palladium-catalysed cross-coupling of organolithiums. *ChemCatChem* **7**, 395–397 (2015).
33. Vidal, C., García-Álvarez, J., Hernán-Gómez, A., Kennedy, A. R. & Hevia, E. Introducing deep eutectic solvents to polar organometallic chemistry: chemoselective addition of organolithium and grignard reagents to ketones in air. *Angew. Chem. Int. Ed.* **53**, 5969–5973 (2014).
34. Denmark, S. E., Smith, R. C., Tau, W. T. & Muhuhi, J. M. Cross-coupling reactions of aromatic and heteroaromatic silanolates with aromatic and heteroaromatic halides. *J. Am. Chem. Soc.* **131**, 3104–3118 (2009).
35. Martin, R. & Buchwald, S. Palladium-catalyzed Suzuki – Miyaura cross-coupling reactions employing dialkylbiaryl phosphine ligands. *Acc. Chem. Res.* **41**, 1461–1473 (2008).
36. Valente, C. et al. The development of bulky palladium NHC complexes for the most-challenging cross-coupling reactions. *Angew. Chem. Int. Ed.* **51**, 3314–3332 (2012).
37. Source: www.sigmaaldrich.com. (The Netherlands) (October 2015).
38. Fu, G. C. The development of versatile methods for palladium-catalyzed coupling reactions of aryl electrophiles through the use of P(t-Bu)3 and PCy3 as ligands. *Acc. Chem. Res.* **41**, 1555–1564 (2008).
39. Recsei, C. & McErlean, S. P. Synthesis of modified binol-phosphoramidites. *Tetrahedron* **68**, 464–480 (2012).
40. Ronsheim, M. & Araldi, G.-L. Indazole derivatives useful as melanin concentrating receptor ligands. PCT patent application WO 2008061109 (A2) (2008).
41. Kawamoto, K. & Tominaga, T. Light emitting element. JP2008306170 (A) (2008).

## Acknowledgements

This work was supported financially by the European Research Council (Advanced Investigator Grant Number 227897 to B.L.F.). The Netherlands Organization for Scientific Research (NWO-CW); funding from the Ministry of Education, Culture and Science (Gravitation programme 024.001.035), The Royal Netherlands Academy of Arts and Sciences (KNAW) and NRSC-Catalysis are gratefully acknowledged.

## Author contributions

E.B.P., M.G. and V.H. carried out the experimental work. V.H., M.G. and B.L.F. wrote the manuscript. B.L.F. and V.H. guided the research.

## Additional information

**Competing financial interests:** The authors declare no competing financial interests.

# 19

# Highly selective hydrogenation of arenes using nanostructured ruthenium catalysts modified with a carbon–nitrogen matrix

Xinjiang Cui[1], Annette-Enrica Surkus[1], Kathrin Junge[1], Christoph Topf[1], Jörg Radnik[1], Carsten Kreyenschulte[1] & Matthias Beller[1]

Selective hydrogenations of (hetero)arenes represent essential processes in the chemical industry, especially for the production of polymer intermediates and a multitude of fine chemicals. Herein, we describe a new type of well-dispersed Ru nanoparticles supported on a nitrogen-doped carbon material obtained from ruthenium chloride and dicyanamide in a facile and scalable method. These novel catalysts are stable and display both excellent activity and selectivity in the hydrogenation of aromatic ethers, phenols as well as other functionalized substrates to the corresponding alicyclic reaction products. Furthermore, reduction of the aromatic core is preferred over hydrogenolysis of the C–O bond in the case of ether substrates. The selective hydrogenation of biomass-derived arenes, such as lignin building blocks, plays a pivotal role in the exploitation of novel sustainable feedstocks for chemical production and represents a notoriously difficult transformation up to now.

[1] Leibniz-Institute for Catalysis, University of Rostock, Albert Einstein Street, 29a, Rostock 18059, Germany. Correspondence and requests for materials should be addressed to M.B. (email: matthias.beller@catalysis.de).

Catalytic hydrogenations constitute economic and clean transformations for many pharmaceutical and petrochemical processes. Due to the low price of the reductant and the 100% atom efficiency of the overall reaction, they adopt a privileged position in methodological tool box of the chemical industry. In the future, this type of transformation is expected to become even more interesting and essential because of the increasing importance of the valorization of oxygen-rich biomass. Till now, the vast majority of industrially produced chemicals depend on fossil raw materials namely petroleum, natural gas and coal. Hence, the conversion of renewable biomass to higher value-added platform chemicals currently attracts considerable attention[1–4]. In this respect, significant research activities are focusing on the valorization of lignin, which is one of the most abundant available feedstocks[5,6]. After lignin depolymerization, highly oxygenated aromatic monomers are obtained. For the consecutive utilization of such lignin-derived compounds, both heterogeneous materials and homogeneous catalysts based on Pd[7–9], Pt[10–12], Ru[13–15], Rh[16] and Ni[17–29] have been developed which allow for reductive C–O cleavage reactions that produce alcohols or alkene bio-building blocks (Fig. 1). Apart from this general approach, we believe it is highly desirable to develop alternative strategies based on selective arene hydrogenation to effectively utilize lignin-derived compounds and other oxygenated (hetero)arenes as feedstock for both bulk and fine chemical production.

In general, selective hydrogenation of aromatic rings plays an important role in the generation of all kinds of aliphatic derivatives, which are crucial starting materials in synthesis of polymers, resins, dyes and fine chemicals. Hence, this catalytic transformation represents also an attractive candidate for the investigation of the selective hydrogenation of lignin-derived fragments. The resulting alicyclic ethers constitute a class of very promising intermediates in the production of fine chemicals and bio-fuels[9,30,31]. Although hydrogenations of the arene rings of lignin building blocks are well-known, reports on the development of catalysts for the selective generation of the alicyclic ethers from such compounds are scarce[16].

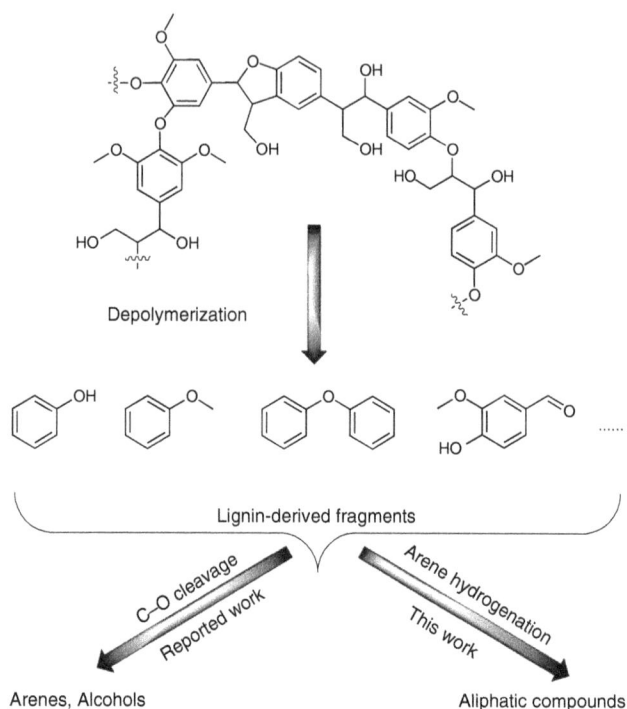

**Figure 1 | Valorization of lignin-derived building blocks.** A new route demonstrated.

Nowadays, heterogeneous ruthenium nanoparticles (Ru-NPs) represent state-of-the-art catalysts for selective hydrogenation of aromatic rings which are also lower in price ($4,200\$\,ozt^{-1}$) compared with other noble metals (for example, Pd: $60,450\$\,ozt^{-1}$). By tuning shape and size of the NPs, type of supports and even by adding functionalized ligands, the performance of these materials has been greatly improved[32–38]. Despite all these efforts, the reported systems exhibit only low selectivity for the hydrogenation of highly reactive benzylic ethers and related derivatives. On the basis of our recent work on the development of metal NPs modified with nitrogen-doped graphene layers (NGrs)[39–46], we thought that such materials might allow for more selective hydrogenations. The incorporation of nitrogen atoms into a carbon matrix has been proven to affect the catalytic activity/selectivity of the resulting materials tremendously[47,48]. Hence, many efforts have been devoted in recent years to the development of more active N-doped carbon materials including their usage as supports in catalysis[49–52]. As an instructive example in the context of lignin valorization, a Pd@CN$_{0.132}$ catalyst was prepared and successfully applied in the hydrogenation of vanillin[53].

Here, we describe the preparation, characterization and catalytic testing of novel ruthenium-based NP immobilized on a N-doped carbon support. The resulting optimal catalyst allows for unique hydrogenation of all kinds of substituted arenes including lignin-derived aromatic compounds to give the aliphatic congeners in both high activity and selectivity.

## Results

**Material preparation and characterization.** At the start of our work, we synthesized different Ru-NPs immobilized on Vulcan powder (Ru@NDCs-X; X labels the pyrolysis temperature). To incorporate nitrogen atoms mainly three different sources (dicyanamide (DCA), cyanamide and phenanthroline) were used (Supplementary Fig. 1). The preparation of these catalysts commenced with the impregnation of Vulcan powder with an ethanolic solution of a nitrogen-ligated RuCl$_3$ complex. The Ru@NDC nano-composites were obtained upon solvent evaporation and subsequent pyrolysis at 600, 800 or 900 °C under inert conditions. Preliminary screening of all the different catalysts revealed best results for the DCA-based materials (Table 1). Applying DCA as ligand elemental analysis of the resulting catalysts indicated that the nitrogen content dropped from 0.65 to 0.4 wt% as the pyrolysis temperature is increased from 600 to 900 °C (Supplementary Table 1 and Fig. 3). These findings are in agreement with the results obtained by XPS. Further evidence of N-doping is provided by X-ray photoelectron spectroscopy (XPS). As shown in Fig. 2e and Supplementary Fig. 5, three types of nitrogen were found for the samples pyrolysed at 600, 700 and 800 °C, namely: pyridinic, pyrrolic and NO$_x$ species, respectively. For the sample heat-treated at 600 °C, the amount of pyrrolic N was found to be only marginally higher than the pyridinic congener (49.7 versus 41.2%), whereas in the sample pyrolysed at 700 °C the conditions moved into reverse (37.4 versus 55.8%). For the 800 °C sample, the pyrrolic N was found to be the dominating species again (78.0 versus 12.3%). The quantitative analysis revealed a near-surface amount of 1.7 at% N after treatment at 600 °C and 700 °C, which decreased slightly to 1.4 at% at 800 °C. To better investigate the effect of doped-nitrogen, we synthesized Ru@NDCs-800 with different nitrogen content by adjusting the amount of DCA in the synthetic process. Interestingly, decreasing the ligand amount to 100 mg, the N content decreased to 0.66 at%, whereas no obvious change was observed for the catalyst prepared using 400 mg of ligand (Supplementary Table 3). The amount of N dropped significantly to 0.5 at% after treatment at 900 °C. This low nitrogen content hampered the determination of the various modifications of the nitrogen. On the basis of XPS analyses, and in agreement with other NP immobilized on N-doped carbon supports[54], we tentatively assign the corresponding signals to pyridinic and graphitic N, respectively. Due to the influence of the content and valence of metal-based species on the heterogeneous hydrogenation, Ru 3p XPS analyses of Ru@NDCs-catalysts were measured. These analyses suggest the formation of a $RuO_2$ phase and interaction of the active Ru sites with nitrogen species (Supplementary Fig. 4)[55]. To gain more insight into the morphology and structure of Ru@NDCs-800, high-resolution transmission electron microscopy analysis of the material was performed. The Ru-NPs are finely dispersed on the support and the average size of the NPs amount to $1.82 \pm 0.11$ nm (Fig. 2a). On raising the pyrolysis temperature to 900 °C, the particle size of the resulting material increased to 2.5–3.0 nm (see Supplementary Fig. 2). However, in the absence of DCA the Ru-NPs aggregated to form larger entities ranging from 10 to 20 nm particles size. These results revealed that the Ru-NPs were prevented from aggregation upon decomposition of DCA and concomitant N incorporation into the carbon matrix during the course of the pyrolysis process. In addition, the formed Ru-NPs might be stabilized by the pyridinic N and pyrrolic N because their lone-pair electrons can serve as metal coordination sites[56]. Figure 2b shows the images of the crystal plane of Ru (101), and the corresponding plane spacing was found to be 0.2 nm. This result is in good agreement with the value obtained by X-ray diffraction (Supplementary Fig. 6). As shown from the Figs 2c and 2d, the Ru-based N-doped carbon material develops a pronounced, well-textured morphology upon heat treatment at 800 °C.

**Catalysis.** After the initial screening, the performance of the prepared materials resulting from DCA was evaluated in more detail. Here, the hydrogenation of phenol to cyclohexanol with $H_2$ pressures ranging from 5 bar to 20 bar was chosen as an industrially relevant benchmark reaction, which is of interest for bulk polyester processes[57–59]. We commenced our survey by control experiments with Ru-free carbon and specifically prepared the N-doped carbon, but in neither case any product formation was observed (Table 1, entries 1 and 2). A standard Ru/C catalyst obtained by pyrolysis at 800 °C in the absence of ligand (DCA) exhibited only moderate activity at 5 bar $H_2$ and 40 °C (Table 1, entry 3). To our delight, the activity of the N-modified Ru@NDC-800 catalyst is considerably increased and an almost quantitative yield of cyclohexanol is achieved at low temperature and pressure (Table 1, entry 6). In contrast, the Ru@NDCs samples originating

### Table 1 | Benchmark reaction: variation of catalysts.

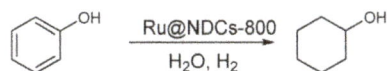

| Entry | Catalyst | Temperature (°C) | Pressure (bar) | Y (%)* | Sel (%) | TOF[†] |
|---|---|---|---|---|---|---|
| 1 | C | 40 | 20 | 0 | — | — |
| 2 | NC | 40 | 20 | 0 | — | — |
| 3 | Ru/C-800 | 40 | 5 | 52 | 98 | 22 |
| 4 | Ru@NDCs-600 | 40 | 5 | 60 | 97 | 25 |
| 5 | Ru@NDCs-700 | 40 | 5 | 49 | 96 | 21 |
| 6 | Ru@NDCs-800 | 40 | 5 | 95 | 99 | 40 |
| 7 | Ru@NDCs-900 | 40 | 5 | 10 | 99 | 4 |
| 8 | Ru@NDCs-800 | 25 | 5 | 91 | 98 | 38[‡] |
| 9 | Ru@NDCs-800 | 40 | 5 | 50 | 99 | 21[§] |
| 10 | Ru@NDCs-800 | 40 | 5 | 93 | 99 | 39[||] |
| 11 | Ru@NDCs-800-2 | 40 | 5 | 94 | 99 | 40 |
| 12 | Ru@NDCs-800-3 | 40 | 5 | 95 | 99 | 40 |
| 13 | Ru@NDCs-800-4 | 40 | 5 | 93 | 98 | 39 |
| 14 | Ru@NDCs-800-5 | 40 | 5 | 93 | 99 | 39 |
| 15 | Ru@NDCs-800-6 | 40 | 5 | 90 | 99 | 38 |

Sel, selectivity; TOF, turnover Frequency; Y, yield. Reaction conditions: 0.5 mmol phenol, 20 mg catalyst, 2 ml $H_2O$, 2 h.
*Determined by GC-FID using dodecane as internal standard.
†TOF was calculated based on 3 wt% Ru loading.
‡20 h.
§100 mg ligand used.
||400 mg ligand used.

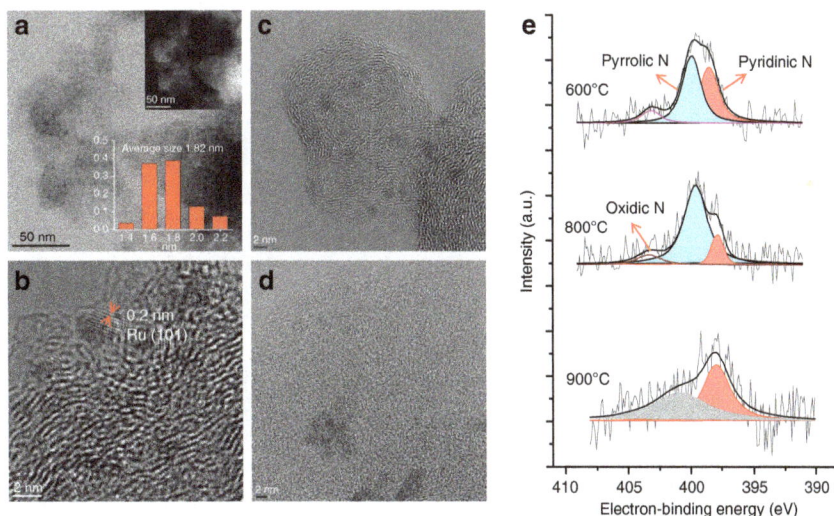

**Figure 2 | High-resolution transmission electron microscopy images.** Bright field HRSTEM images (**a,c**) and HRTEM (**b**) of the Ru@NDCs-800 (inset of **a**: HAADF image), the HRTEM image (**d**) of the Ru@C-800 and N1s XPS spectrum (**e**) of Ru@NDCs.

from lower pyrolysis temperature featured significantly lower reactivity (Table 1, entries 4, 5 and 7). When the reaction was carried out for 1 h, a significantly higher activity was also obtained in the presence of Ru@NDCs-800 (Supplementary Table 3). However, Ru@NDCs-600 or Ru@NDCs-700 might show higher TOFs based on active Ru atoms on the surface of the NPs.

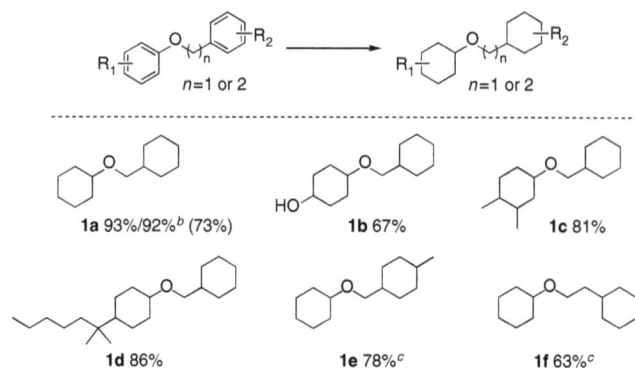

**Figure 3 | Selective hydrogenation of benzylic ethers using Ru@NDCs-800.** Reaction conditions: 0.5 mmol substrate, 20 mg catalyst, 2 ml isopropanol, 20 bar $H_2$, 60 °C, 24 h. Isolated yields, value in parentheses is obtained using commercial 5% Ru/C. [b]50 bar $H_2$. [c]GC yield.

Moreover, the Ru@NDCs sample synthesized using low amount of ligand gave considerably lower conversion, whereas similar activity was obtained with the sample prepared in the presence of 400 mg of ligand (Table 1, entries 9 and 10). This also demonstrates the importance of a critical amount of nitrogen.These experimental findings demonstrate the importance of nitrogen for the activity of the catalyst. Next, the reusability of the Ru@NDCs-800 nano-composite was examined since catalyst recyclability represents an integral part in the economic assessment of chemical transformations. The active material was separated from the reaction mixture via centrifugation and reused directly for five times. Gratifyingly, no obvious deactivation was detected and the yield of the desired product amounted to 93% at the sixth cycle (Table 1, entries 6, 11–15).

**Selective hydrogenation of arenes.** The functional group tolerance survey of various arene substrates was conducted with the Ru@NDCs-800 catalyst. As shown in Table 2, the hydrogenation of methyl benzoate was complete at room temperature within 12 h (Table 2, entry 1). Furthermore, benzamide and phthalimide underwent selective reduction retaining the amide and imide groups in the substrates (Table 2, entries 2 and 3). In case of quinoline, which is used as a feedstock in the production of

**Table 2 | Selective arene hydrogenation with Ru@NDCs-800*.**

| Entry | Substrate | Product | Y/Sel (%)[†] |
|:---:|:---:|:---:|:---|
| 1 | | | 98c[‡] |
| 2 | | | 95 |
| 3 | | | 94 |
| 4 | | | 87 |
| 5 | | | 81 |
| 6 | | | 90/94 (75/77) |
| 7 | | | 85[§,‖] |
| 8 | | | 93[‖,¶] |

Sel, selectivity; Y, yield.
*Reaction conditions: 0.5 mmol substrate, 20 mg catalyst, 2 ml isopropanol, 10 bar $H_2$, 60 °C, 24 h.
[†]Isolated yields, value in parentheses is obtained using commercial 5% Ru/C.
[‡]Room temperature.
[§]80 °C, 12 h.
[‖]Yields deteminted by $^1H$ NMR.
[¶]100 °C, 20 bar, 12 h.

**Table 3 | Selective hydrogenation of aromatic ethers*.**

| Entry | Substrate | Product | Yields[†] |
|-------|-----------|---------|--------|
| 1 | | | 94 |
| 2 | | | 93 |
| 3 | | | 89 |
| 4 | | | 88 |
| 5 | | | 82 |
| 6 | | | 84 |
| 7 | | | 93 |
| 8 | | | 88 |
| 9 | | | 90 |
| 10 | | | 87[‡] |
| 11 | | | 84[‡] |
| 12 | | | 79[‡] |
| 13 | | | 86[§] |
| 14 | | | 81[§] |

*Reaction conditions: 0.5 mmol phenol, 20 mg catalyst, 2 ml isopropanol, isolated yields. 10 bar $H_2$, 60 °C, 24 h.
[†]Isolated yields.
[‡]Determined by GC-FID using dodecane as internal standard.
[§]80 °C, 20 bar.

specialty chemicals, the heterocyclic 1,2,3,4-tetrahydroquinoline is obtained in 87% yield, whereas selective hydrogenation of the pyridine ring occurred (Table 2, entry 4). Ru@NDCs-800 also exhibited good activity in the hydrogenation of methyl benzoyl-formate, affording a cyclohexyl-tagged tertiary alcohol in 81% yield (Table 2, entry 5). The hydrogenation of benzyl alcohol proceeded smoothly with almost quantitative formation of the corresponding saturated alicyclic alcohol. It should be noted that the yield of the latter product decreased to 75% when commercial Ru/C was used as catalyst. This lower yield is explained by concomitant C–O bond cleavage which leads to the formation of a considerable amount of methyl cyclohexane (Table 2, entry 6). N-ethylcarbazole is regarded as a promising future hydrogen-storage material[60–62]. Hence, we tested our Ru@NDCs-800 catalyst in the hydrogenation of this peculiar heterocycle.

Interestingly, at 10 bar and 100 °C N-octahydroethylcarbazole is obtained in 85% yield, demonstrating the possibility of selective partial hydrogenation. Simply by increasing the temperature to 100 °C, the fully hydrogenated perhydro-derivative was isolated in 93% yield (Table 2, entries 7 and 8).

**Hydrogenation of benzylic compounds.** In general, hydro-genolysis of benzylic compounds is observed in the course of hydrogenation processes. Especially, in benzylic ethers and alco-hols the cleavage of the C–O bond easily occurrs due to the lower C–O bonding dissociation energy. For example, the bonding dissociation energy for benzylic ethers is 220 kJ mol$^{-1}$ compared with 310 kJ mol$^{-1}$ for biaryl ethers and 290 kJ mol$^{-1}$ for ary-lethyl ethers[27]. Indeed, in the hydrogenation of benzyl phenyl

**Table 4 | Selective hydrogenation of biaryl ethers\*.**

| Entry | Substrate | Product | Yields (%)[†] |
|-------|-----------|---------|---------------|
| 1 | | | $84/86^{‡}$ $(65/66^{‡})$ |
| 2 | | | 87 |
| 3 | | | 89 |
| 4 | | | 85 |
| 5 | | | 91 |
| 6 | | | 89 |
| 7 | | | 86 |
| 8 | | | 81 |

\*Reaction conditions: 0.5 mmol phenol, 20 mg catalyst, 2 ml isopropanol, 20 bar $H_2$, 24 h.
†Isolated yields, value in parentheses is obtained using commercial 5% Ru/C.
‡Selectivity.

ether using commercial Ru/C partial cleavage of the benzylic C–O linkage is observed. Hence, a mixture of products including the fully hydrogenated ether, cyclohexanol and methyl cyclohexane is obtained (Fig. 3, **1a**). However, applying the novel Ru@NDCs-800 catalyst gave excellent selectivity for the hydrogenation of the arene rings. As shown in Table 3, benzyl phenyl ether was smoothly converted to the corresponding alicyclic ether in 93% yield. Even at 50 bars of hydrogen no significant C–O cleavage is observed. Similarly, substrates with pendant functional groups on the phenyl moiety such as hydroxyl, methyl and even long alkyl chains were well-tolerated and gave high yields of the corresponding *cis/trans* ethers (ratio 1.5–4:1; Fig. 3, **1b–1d**). In addition, 1-methyl-4-(phenoxymethyl)benzene was also converted to the corresponding ether in 78% yield (Fig. 3, **1e**). All these results demonstrate the specific behaviour of *N*-doped ruthenium nano-composites for selective hydrogenation of aromatic rings in the presence of the benzylic C–O bonds.

As pointed out *vide supra* various phenethyl ethers are easily available as platform chemicals from abundant lignin. As an example for this class of compounds the hydrogenation of phenethyl ether was investigated. Gratifyingly, the Ru@NDCs-800 showed good activity for arene hydrogenation of phenethoxybenzene (Fig. 3, **1f**).

**Hydrogenation of aromatic ethers**. To expand the scope of this arene hydrogenation, next we investigated the reactivity of various alkyl aryl ethers. Anisole, ethoxybenzene, butoxybenzene and (octyloxy)benzene were readily converted to the corresponding products in yields ranging from 88–94% (Table 3, entries 1–4). Further alkyl and methoxy substituents had no negative impact on the catalytic activity and the fully hydrogenated products were obtained as *cis/trans* regioisomers (3–9:1;

Table 3, entries 5–8). Arene hydrogenation of 1-methoxynaphthalene was also realized with 90% of the desired ethers formed. In addition, 2,3-dihydrobenzofuran and benzofuran were converted to the corresponding ethers in 84–87% yield. Notably, tetrahydrofurfuryl alcohol is obtained by selective hydrogenation of furfuryl alcohol, which is manufactured industrially. Moreover, the successful hydrogenation of 4-propylguaiacol and vanillin, which are both directly accessible from lignin depolymerization, was demonstrated and good yields were achieved (Table 3, entries 13 and 14). Having verified the catalytic activity of Ru@NDCs-800 in the selective hydrogenation of alkyl aryl ethers, we tested the established catalytic protocol in the transformation of lignin-derived biphenyl ethers.

As shown in Table 4, the conversion of a broad range of substituted biaryl ethers into the corresponding aliphatic ethers proceeded smoothly. The parent compound afforded the corresponding ether in 84% yield, whereas the hydrogenation of alkyl-substituted derivatives provided slightly higher product yields ranging from 85 to 89% (Table 4, entries 1–4).

Performing the latter reaction in the presence of commercial 5% Ru/C, the yield for the desired product decreased to 65% due to C–O cleavage. Finally, dibenzo-fused five- and six-membered heterocyclic substrates displayed similar reactivity and the yield of the corresponding saturated tricyclic ethers amounted to 81–91% (Table 4, entries 5–8). To investigate the high selectivity of the catalytic system, the hydrogenolysis of aliphatic ethers such as methoxycyclohexane and oxydicyclohexane were tested and no C–O cleavage occurred. This result revealed that hydrogenation of aromatic rings is favoured compared with the cleavage of C–O bonds. The experimental findings were also substantiated by variation of the solvent (see Supplementary Table 2), whereas higher selectivity was obtained using cyclohexane as reaction medium by virtue of its property to inhibit the hydrogenolysis reaction.

## Discussion

In summary, we developed for the first time a new type of ruthenium nano-composite immobilized on a carbon support. The described material contains finely dispersed ruthenium, which is in contact with a specific carbon–nitrogen matrix. The resulting catalysts (Ru@NDCs) are easily obtained in a practical and scalable two-step method via pyrolysis of simple ruthenium trichloride and inexpensive DCA. Combination of different analytic methods including XPS revealed the formation of graphitic nitrogen, which is identified as the functional prerequisite for the development of catalytic activity. The optimal catalyst exhibited good-to-excellent activity in the selective hydrogenation of arenes, particularly in the transformation of aromatic ethers to the corresponding alicyclic compounds with preservation of the phenyl- and benzyl C–O bonds. The utility of the catalyst opens new avenues for the valorization of lignin-derived aromatic compounds to provide novel sustainable platform chemicals. In addition, industrially relevant processes such as the hydrogenation of phenol proceed under mild conditions in a green manner.

## Methods

**General.** Unless otherwise specified, reagents and solvents were purchased from Aldrich, Fluka, Acros and Strem commercially and used as received. All compounds were characterized by $^1$H NMR, $^{13}$C NMR, GC-MS spectroscopy. $^1$H and $^{13}$C NMR spectra were recorded on Bruker Avance 300 (300 MHz) or 400 (400 MHz) NMR spectrometers. The $^1$H and $^{13}$C NMR chemical shifts are reported relative to the centre of solvent resonance (CDCl$_3$: 7.26 (1H), 77.16 (13C)). EI mass spectra were recorded on an MAT 95XP spectrometer (70 eV, Thermo Electron Corporation). For GC analysis, HP 6890 chromatograph with a 29 m HP5 column was used. GC-MS analysis was conducted on an Agilent GC-MS-HP5890 instrument. The products were isolated from the reaction mixture by solvent evaporation.

**Typical preparation of Ru@NDCs-catalysts.** RuCl$_3$ (0.5 mmol) and DCA (200 mg) were dissolved in ethanol and stirred for 2 h at 60 °C. Then, carbon support (VULCAN XC72R, ordered from PT (Cabot Indonesia) was added and the suspension was stirred at 60 °C for a further period of 2 h. After that, the mixture was cooled to room temperature and dried *in vacuo* at 60 °C for 2 h, and it was grinded to a fine powder which was subsequently pyrolyzed at 600, 700, 800 or 900 °C for 2 h under an argon atmosphere.

**Catalytic hydrogenation of phenol.** In a 4 ml reaction vial equipped with a magnetic stirring bar, phenol (0.5 mmol) was mixed with 2 ml IPA. Then 20 mg of the ruthenium-based catalyst (20 mg) was added. The reaction vials were fitted with cap and needle and then placed into a 300-ml autoclave. The autoclave was purged thrice with H$_2$ (10 bar), pressurized to 5 bar H$_2$, placed into an aluminium block, heated to 40 °C and the reaction vessels were stirred for 2 h. After completion of the reaction, the autoclave was cooled to room temperature, *n*-dodecane was added to the reaction mixture as external standard and the mixture was diluted with ethyl acetate (20 ml), followed by filtration and analysis of a sample by GC and GC-MS. The crude reaction mixture was concentrated *in vacuo* and the obtained product was analysis by NMR.

For NMR analysis of the compounds in this article, see Supplementary Methods and Supplementary Figs 7–62.

## References

1. Corma, A., Iborra, S. & Velty, A. Chemical routes for the transformation of biomass into chemicals. *Chem. Rev.* **107**, 2411–2502 (2007).
2. Huber, G. W., Iborra, S. & Corma, A. Synthesis of transportation fuels from biomass: chemistry, catalysts, and engineering. *Chem. Rev.* **106**, 4044–4098 (2006).
3. Zakzeski, J., Bruijnincx, P. C. A., Jongerius, A. L. & Weckhuysen, B. M. The catalytic valorization of lignin for the production of renewable chemicals. *Chem. Rev.* **110**, 3552–3599 (2010).
4. Cornella, J., Zarate, C. & Martin, R. Metal-catalyzed activation of ethers via C–O bond cleavage: a new strategy for molecular diversity. *Chem. Soc. Rev.* **43**, 8081–8097 (2014).
5. Barta, K. & Ford, P. C. Catalytic conversion of nonfood woody biomass solids to organic liquids. *Acc. Chem. Res.* **47**, 1503–1512 (2014).
6. Hanson, S. K. & Baker, R. T. Knocking on wood: base metal complexes as catalysts for selective oxidation of lignin models and extracts. *Acc. Chem. Res.* **48**, 2037–2048 (2015).

7. Zhao, C., Kou, Y., Lemonidou, A. A., Li, X. B. & Lercher, J. A. Highly selective catalytic conversion of phenolic bio-oil to alkanes. *Angew. Chem. Int. Ed.* **48**, 3987–3990 (2009).
8. Zhao, C., He, J. Y., Lemonidou, A. A., Li, X. B. & Lercher, J. A. Aqueous-phase hydrodeoxygenation of bio-derived phenols to cycloalkanes. *J. Catal.* **280**, 8–16 (2011).
9. Li, Z., Assary, R. S., Atesin, A. C., Curtiss, L. A. & Marks, T. J. Rapid ether and alcohol C–O bond hydrogenolysis catalyzed by tandem high-valent metal triflate plus supported Pd catalysts. *J. Am. Chem. Soc.* **136**, 104–107 (2014).
10. Hong, D. Y., Miller, S. J., Agrawal, P. K. & Jones, C. W. Hydrodeoxygenation and coupling of aqueous phenolics over bifunctional zeolite-supported metal catalysts. *Chem. Commun.* **46**, 1038–1040 (2010).
11. Ohta, H., Kobayashi, H., Hara, K. & Fukuoka, A. Hydrodeoxygenation of phenols as lignin models under acid-free conditions with carbon-supported platinum catalysts. *Chem. Commun.* **47**, 12209–12211 (2011).
12. Zhu, X. L., Lobban, L. L., Mallinson, R. G. & Resasco, D. E. Bifunctional transalkylation and hydrodeoxygenation of anisole over a Pt/HBeta catalyst. *J. Catal.* **281**, 21–29 (2011).
13. Yan, N., Yuan, Y. A., Dykeman, R., Kou, Y. A. & Dyson, P. J. Hydrodeoxygenation of lignin-derived phenols into alkanes by using nanoparticle catalysts combined with bronsted acidic ionic liquids. *Angew. Chem. Int. Ed.* **49**, 5549–5553 (2010).
14. Xu, H. J. *et al.* Ionic liquid modified montmorillonite-supported Ru nanoparticles: highly efficient heterogeneous catalysts for the hydrodeoxygenation of phenolic compounds to cycloalkanes. *Catal. Sci. Technol.* **4**, 2658–2663 (2014).
15. Nichols, J. M., Bishop, L. M., Bergman, R. G. & Ellman, J. A. Catalytic C–O bond cleavage of 2-Aryloxy-1-arylethanols and its application to the depolymerization of lignin-related polymers. *J. Am. Chem. Soc.* **132**, 16725–16725 (2010).
16. Chatterjee, M., Ishizaka, T., Suzuki, A. & Kawanami, H. An efficient cleavage of the aryl ether C–O bond in supercritical carbon dioxide-water. *Chem. Commun.* **49**, 4567–4569 (2013).
17. Zhang, J. G. *et al.* A series of NiM (M = Ru, Rh, and Pd) bimetallic catalysts for effective lignin hydrogenolysis in water. *ACS Catal.* **4**, 1574–1583 (2014).
18. Ferrini, P. & Rinaldi, R. Catalytic biorefining of plant biomass to non-pyrolytic lignin bio-oil and carbohydrates through hydrogen transfer reactions. *Angew. Chem. Int. Ed.* **53**, 8634–8639 (2014).
19. Wang, X. Y. & Rinaldi, R. A route for lignin and bio-oil conversion: dehydroxylation of phenols into arenes by catalytic tandem reactions. *Angew. Chem. Int. Ed.* **52**, 11499–11503 (2013).
20. Zaheer, M., Hermannsdorfer, J., Kretschmer, W. P., Motz, G. & Kempe, R. Robust heterogeneous nickel catalysts with tailored porosity for the selective hydrogenolysis of aryl ethers. *ChemCatChem* **6**, 91–95 (2014).
21. Samant, B. S. & Kabalka, G. W. Hydrogenolysis-hydrogenation of aryl ethers: selectivity pattern. *Chem. Commun.* **48**, 8658–8660 (2012).
22. Wang, X. Y. & Rinaldi, R. Solvent effects on the hydrogenolysis of diphenyl ether with raney nickel and their implications for the conversion of lignin. *ChemSusChem* **5**, 1455–1466 (2012).
23. Wang, X. Y. & Rinaldi, R. Exploiting H-transfer reactions with RANEY (R) Ni for upgrade of phenolic and aromatic biorefinery feeds under unusual, low-severity conditions. *Energy Environ. Sci.* **5**, 8244–8260 (2012).
24. Sturgeon, M. R. *et al.* Lignin depolymerisation by nickel supported layered-double hydroxide catalysts. *Green Chem.* **16**, 824–835 (2014).
25. Zhang, J. G. *et al.* Highly efficient, NiAu-catalyzed hydrogenolysis of lignin into phenolic chemicals. *Green Chem.* **16**, 2432–2437 (2014).
26. Chieffi, G., Giordano, C., Antonietti, M. & Esposito, D. FeNi nanoparticles with carbon armor as sustainable hydrogenation catalysts: towards biorefineries. *J. Mater. Chem. A* **2**, 11591–11596 (2014).
27. He, J. Y., Zhao, C. & Lercher, J. A. Ni-catalyzed cleavage of aryl ethers in the aqueous phase. *J. Am. Chem. Soc.* **134**, 20768–20775 (2012).
28. Molinari, V., Giordano, C., Antonietti, M. & Esposito, D. Titanium nitride-nickel nanocomposite as heterogeneous catalyst for the hydrogenolysis of aryl ethers. *J. Am. Chem. Soc.* **136**, 1758–1761 (2014).
29. Sergeev, A. G., Webb, J. D. & Hartwig, J. F. A heterogeneous nickel catalyst for the hydrogenolysis of aryl ethers without arene hydrogenation. *J. Am. Chem. Soc.* **134**, 20226–20229 (2012).
30. Schneider, H. J., Ahlhelm, A. & Muller, W. An oxidative ether cleavage with para-nitroperbenzoic acid. *Chem. Ber. Recl.* **117**, 3297–3302 (1984).
31. Amati, A. *et al.* Catalytic processes of oxidation by hydrogen peroxide in the presence of Br-2 or HBr. Mechanism and synthetic applications. *Org. Process Res. Dev.* **2**, 261–269 (1998).
32. Lara, P. *et al.* Ruthenium nanoparticles stabilized by N-heterocyclic carbenes: ligand location and influence on reactivity. *Angew. Chem. Int. Ed.* **50**, 12080–12084 (2011).
33. Pieters, G. *et al.* Regioselective and stereospecific deuteration of bioactive aza compounds by the use of ruthenium nanoparticles. *Angew. Chem. Int. Ed.* **53**, 230–234 (2014).

34. Novio, F. *et al.* Surface chemistry on small ruthenium nanoparticles: evidence for site selective reactions and influence of ligands. *Chem. Eur. J* **20,** 1287–1297 (2014).

35. Machado, B. F. *et al.* Understanding the surface chemistry of carbon nanotubes: toward a rational design of Ru nanocatalysts. *J. Catal.* **309,** 185–198 (2014).

36. Gutmann, T. *et al.* Hydrido-ruthenium cluster complexes as models for reactive surface hydrogen species of ruthenium nanoparticles. solid-state H-2 NMR and quantum chemical calculations. *J. Am. Chem. Soc.* **132,** 11759–11767 (2010).

37. Cui, X. J., Shi, F. & Deng, Y. Q. Ionic liquid templated preparation of Ru/SiO2 and its activity in nitrobenzene hydrogenation. *ChemCatChem.* **4,** 333–336 (2012).

38. Maegawa, T. *et al.* Efficient and practical arene hydrogenation by heterogeneous catalysts under mild conditions. *Chem. Eur. J.* **15,** 6953–6963 (2009).

39. Jagadeesh, R. V. *et al.* Nanoscale Fe2O3-based catalysts for selective hydrogenation of nitroarenes to anilines. *Science* **342,** 1073–1076 (2013).

40. Jagadeesh, R. V., Junge, H. & Beller, M. Green synthesis of nitriles using non-noble metal oxides-based nanocatalysts. *Nat. Commun.* **5,** 4123 (2014).

41. Westerhaus, F. A. *et al.* Heterogenized cobalt oxide catalysts for nitroarene reduction by pyrolysis of molecularly defined complexes. *Nat. Chem.* **5,** 537–543 (2013).

42. Jagadeesh, R. V. *et al.* Selective oxidation of alcohols to esters using heterogeneous Co3O4-N@C catalysts under mild conditions. *J. Am. Chem. Soc.* **135,** 10776–10782 (2013).

43. Stemmler, T. *et al.* General and selective reductive amination of carbonyl compounds using a core-shell structured Co3O4/NGr@C catalyst. *Green Chem.* **16,** 4535–4540 (2014).

44. Stemmler, T., Surkus, A. E., Pohl, M. M., Junge, K. & Beller, M. Iron-catalyzed synthesis of secondary amines: on the way to green reductive aminations. *ChemSusChem.* **7,** 3012–3016 (2014).

45. Pisiewicz, S., Stemmler, T., Surkus, A. E., Junge, K. & Beller, M. Synthesis of amines by reductive amination of aldehydes and ketones using Co3O4/NGr@C catalyst. *ChemCatChem.* **7,** 62–64 (2015).

46. Banerjee, D. *et al.* Convenient and mild epoxidation of alkenes using heterogeneous cobalt oxide catalysts. *Angew. Chem. Int. Ed.* **53,** 4359–4363 (2014).

47. Yang, S. B., Feng, X. L., Wang, X. C. & Mullen, K. Graphene-based carbon nitride nanosheets as efficient metal-free electrocatalysts for oxygen reduction reactions. *Angew. Chem. Int. Ed.* **50,** 5339–5343 (2011).

48. Lee, J. S., Wang, X. Q., Luo, H. M., Baker, G. A. & Dai, S. Facile ionothermal synthesis of microporous and mesoporous carbons from task specific ionic liquids. *J. Am. Chem. Soc.* **131,** 4596-+ (2009).

49. Tan, Q., Higashibayashi, S., Karanjit, S. & Sakurai, H. Enantioselective synthesis of a chiral nitrogen-doped buckybowl. *Nat. Commun.* **3,** 891 (2012).

50. Yan, K. *et al.* Modulation-doped growth of mosaic graphene with single-crystalline p-n junctions for efficient photocurrent generation. *Nat. Commun.* **3,** 1280 (2012).

51. Zhao, Y., Nakamura, R., Kamiya, K., Nakanishi, S. & Hashimoto, K. Nitrogen-doped carbon nanomaterials as non-metal electrocatalysts for water oxidation. *Nat. Commun.* **4,** 2390 (2013).

52. Zheng, F. C., Yang, Y. & Chen, Q. W. High lithium anodic performance of highly nitrogen-doped porous carbon prepared from a metal-organic framework. *Nat. Commun.* **5,** 5261 (2014).

53. Xu, X. *et al.* Synthesis of palladium nanoparticles supported on mesoporous N-doped carbon and their catalytic ability for biofuel upgrade. *J. Am. Chem. Soc.* **134,** 16987–16990 (2012).

54. Jagadeesh, R. V. *et al.* Cobalt-based nanocatalysts for green oxidation and hydrogenation processes. *Nat. Protoc.* **10,** 916–926 (2015).

55. Hien, N. T. B. *et al.* Ru-N-C hybrid nanocomposite for ammonia dehydrogenation: influence of N-doping on catalytic activity. *Materials* **8,** 3442–3455 (2015).

56. Lin, L., Zhu, Q. & Xu, A. W. Noble-metal-free Fe-N/C catalyst for highly efficient oxygen reduction reaction under both alkaline and acidic conditions. *J. Am. Chem. Soc.* **136,** 11027–11033 (2014).

57. Liu, H. Z., Jiang, T., Han, B. X., Liang, S. G. & Zhou, Y. X. Selective phenol hydrogenation to cyclohexanone over a dual supported Pd-Lewis acid catalyst. *Science* **326,** 1250–1252 (2009).

58. Duan, H. H. *et al.* Ultrathin rhodium nanosheets. *Nat. Commun.* **5,** 3093 (2014).

59. Yoon, Y., Rousseau, R., Weber, R. S., Mei, D. H. & Lercher, J. A. First-principles study of phenol hydrogenation on Pt and Ni catalysts in aqueous phase. *J. Am. Chem. Soc.* **136,** 10287–10298 (2014).

60. Wan, C., An, Y., Xu, G. H. & Kong, W. J. Study of catalytic hydrogenation of N-ethylcarbazole over ruthenium catalyst. *Int. J. Hydrogen Energy* **37,** 13092–13096 (2012).

61. Wang, Z. H., Tonks, I., Belli, J. & Jensen, C. M. Dehydrogenation of N-ethyl perhydrocarbazole catalyzed by PCP pincer iridium complexes: evaluation of a homogenous hydrogen storage system. *J. Organomet. Chem.* **694,** 2854–2857 (2009).

62. Eblagon, K. M. *et al.* Study of catalytic sites on ruthenium for hydrogenation of N-ethylcarbazole: implications of hydrogen storage via reversible catalytic hydrogenation. *J. Phys. Chem. C.* **114,** 9720–9730 (2010).

## Acknowledgements

This work was supported by the state of Mecklenburg Vorpommern. We thank Dr Carsten Kreyenschulte, Dr Jabor Rabeah and Dr Jörg Radnik for technical assistance.

## Author contributions

X.C. planned and conducted the experiments, and analysed the results. C.T., K.J. and M.B. directed the project. X.C., C.T. and M.B. participated in writing the manuscript. A.-E.S. worked on the pyrolysis of the Ru catalysts. J.R. and C.K. are responsible for the XPS and TEM analysis.

## Additional information

**Competing financial interests:** The authors declare no competing financial interests.

# 20

# Synthesis of tetra- and octa-aurated heteroaryl complexes towards probing aromatic indoliums

Jun Yuan[1,*], Tingting Sun[2,*], Xin He[1], Ke An[2], Jun Zhu[2] & Liang Zhao[1]

Polymetalated aromatic compounds are particularly challenging synthetic goals because of the limited thermodynamic stability of polyanionic species arising from strong electrostatic repulsion between adjacent carbanionic sites. Here we describe a facile synthesis of two polyaurated complexes including a tetra-aurated indole and an octa-aurated benzodipyrrole. The imido trinuclear gold(I) moiety exhibits nucleophilicity and undergoes an intramolecular attack on a gold(I)-activated ethynyl to generate polyanionic heteroaryl species. Their computed magnetic properties reveal the aromatic character in the five-membered ring. The incorporation of the aurated substituents at the nitrogen atom can convert non-aromaticity in the parent indolium into aromaticity in the aurated one because of hyperconjugation. Thus, the concept of hyperconjugative aromaticity is extended to heterocycles with transition metal substituents. More importantly, further analysis indicates that the aurated substituents can perform better than traditional main-group substituents. This work highlights the difference in aromaticity between polymetalated aryls and their organic prototypes.

[1] Key Laboratory of Bioorganic Phosphorus Chemistry and Chemical Biology (Ministry of Education), Department of Chemistry, Tsinghua University, Beijing 100084, China. [2] State Key Laboratory of Physical Chemistry of Solid Surfaces and Collaborative Innovation Center of Chemistry for Energy Materials (iChEM), Fujian Provincial Key Laboratory of Theoretical and Computational Chemistry and Department of Chemistry, College of Chemistry and Chemical Engineering, Xiamen University, Xiamen 361005, China. * These authors contributed equally to this work. Correspondence and requests for materials should be addressed to J.Z. (email: jun.zhu@xmu.edu.cn) or to L.Z. (email: zhaolchem@mail.tsinghua.edu.cn).

ryl- and heteroaryl-Au(I) compounds have recently attracted substantial attention because of their crucial intermediate role in versatile gold-catalysed transformations for the formation of new carbon–carbon and carbon–heteroatom bonds[1–5]. Structural characterization and reactivity studies of the organogold intermediates facilitated comprehension of the reaction mechanisms and provided inspiration for developing new synthetic methodologies[6–9]. Moreover, great interest has also arisen in the exploration of rich reactivity of aryl-Au(I) compounds. They can serve as an organometallic partner for metal-catalysed (for example, Pd and Ni) cross-coupling reactions[10], as a nucleophile to react with diverse appropriate electrophiles[11], and as a key intermediate to undergo a Au(I/III)-catalysed C-H activation and cross-coupling reaction in the presence of oxidants[12,13]. In addition, due to the attractive aurophilic interaction[14,15] the $d^{10}$ gold(I) atom holds a great potential to aggregate into a polynuclear cluster. This structural characteristic perplexes the mechanism studies of gold-catalysed reactions, especially when bi- or polynuclear gold catalysts were used[16,17]. Therefore, the structure and reactivity studies of polyaurated organometallics may not only lead to new reagents for use in synthesis but also help understand the reaction mechanisms of Au-catalysed reactions. Furthermore, detailed structural characterization of polymetalated aryls may also provide a new insight into our current ideas about bonding between metal ions and organic species. However, relative to numerous reported mono-metalated aryls[18–20] derived from direct electrophilic metalation or halogen-lithium-metal exchange reactions[21], to date polymetalated aryl compounds still represent a very rare class of molecules[22,23] that are synthetically hard to reach by conventional methods. The formidable challenge of this synthesis lies in the difficult generation of multitopic anions and the limited thermodynamic stability of polyanionic species due to strong electrostatic repulsion between adjacent carbanionic sites.

On the other hand, the concept of hyperconjugative aromaticity was first proposed in 1939 by Mulliken[24], who considered that the saturated $CH_2$ group in cyclopentadiene could contribute 'pseudo' $2\pi$ electrons by hyperconjugation to the four olefinic $\pi$ electrons, thus leading to aromatic cyclopentadiene with $6\pi$ electrons. Later, Schleyer, Nyulászi and O'Ferrall extended this concept and found that electropositive substituents (for example, $SiH_3$, $GeH_3$ and $SnH_3$) led to enhanced hyperconjugative aromaticity whereas electronegative substituents (for example, F, Cl) resulted in anti-aromatic five-membered rings[25–27]. Very recently, Houk revealed that the transition state distortion energies for the Diels–Alder reactions of 5-substituted cyclopentadienes with ethylene or maleic anhydride were directly related to the hyperconjugative aromaticity[28]. We found that hyperconjugation could play an important role in triplet state aromaticity[29]. However, all the previous studies on hyperconjugative aromaticity are mainly limited to carbocycles and all the substituents considered are main-group elements only.

In this contribution, we disclose a facile synthesis of two cationic polymetalated heteroaryl compounds including a tetra-aurated indolium and an octa-aurated benzodipyrrole complex. We find that the $\mu_3$-imido aurated species exhibit good nucleophilicity and undergo an intramolecular attack on a Au(I)-activated ethynyl to generate polyanionic heteroaryls. The polyaurated heteroaryls exhibit reasonable stability upon exposure to air. The theoretical investigation reveals that the incorporation of the aurated substituents at the nitrogen atom can not only convert non-aromaticity in the parent indolium into aromaticity in aurated one due to the hyperconjugation but also lead to the most aromatic indolium. Thus, our findings not only extend the concept of hyperconjugative aromaticity to transition metal-involved heterocycles and accounts for the excellent stability of two polyaurated heteroaryls but also reveal better performance of transition metal-involved substituents over the traditional main-group ones on hyperconjugative aromaticity. In contrast to metallabenzene complexes where a metal atom is incorporated directly into a ring[30–32], the polymetalated heteroaryl compounds reported in this work represent a class of metalla-aromatics where the aromaticity is introduced by the hyperconjugation of transition metal substituents.

## Results

**Synthesis of tetra- and octa-aurated heteroaryl complexes.** It is well-known that the gold(I) centre can act as an excellent soft Lewis acid to activate carbon–carbon multiple bonds. Many mononuclear organogold(I) compounds have been synthesized by the reactions of allene/alkene/alkyne substrates and nucleophiles with stoichiometric gold(I) compounds[18–20,33]. For example, $\sigma$- and $\pi$-activation of the ethynyl group of o-alkynylanilines by a gold(I) species led to the formation of 2- and 3-aurated indole complexes, respectively[34,35]. In view of the acknowledged isolobality of the proton and the LAu$^+$ fragment (L = phosphine or carbene ligands)[36], we therefore assume that an aurated imido species in an o-alkynylaniline derivative may act like an amine group to undergo a similar annulation reaction to react with the $\pi$- and/or $\sigma$-activated ethynyl group.

Sharp and co-workers have previously reported that a primary amine group readily reacts with the $\mu_3$-oxo trinuclear gold(I) compound [(PPh$_3$Au)$_3$($\mu_3$-O)](BF$_4$) ([**Au$_3$O**]) to form a $\mu_3$-imido trinuclear gold(I) compound[37,38]. According to this synthesis, we then carried out the reaction of [**Au$_3$O**] (1 equiv.) with o-(trimethylsilylethynyl)aniline **1** (trimethylsilyl = TMS) aiming at the synthesis of the gold(I) imido complex **3** (Fig. 1). The reaction mixture experienced a colour change from light brown to yellow. Diffusion of diethyl ether to the resulting yellow solution yielded colourless crystals of complex **2**. Structural analysis by X-ray crystallography (*vide infra*) revealed that **2** has an unexpected tetra-aurated indolyl structure instead of the NAu$_3$ cluster aggregate. This unusual transformation can be rationalized in the light of the following experimental facts, and divided as two steps: the auration of the amino group in **1** and the subsequent nucleophilic attack of the aurated imido on an Au(I)-activated carbon–carbon triple bond.

The electrospray ionization mass spectroscopy (ESI-MS) analysis of the reaction mixture of **1** and [**Au$_3$O**] revealed the strongest signal at $m/z = 1564.2502$ that can be ascribed to the 1+ species of **3** ($\sim$1564.2548), substantiating that the $\mu_3$-imido trinuclear gold(I) species **3** indeed forms in the **1**-to-**2** transformation process (Supplementary Fig. 1). In order to clarify what kind of activation ($\sigma$-, $\pi$- or $\sigma$,$\pi$-mode)[39] involved in the **1**-to-**2** transformation, we employed the inner alkyne **5** to react with [**Au$_3$O**] based on the synthetic procedure for **2**. The $\mu_3$-imido trinuclear gold(I) complex **6** was obtained as a crystalline solid in almost quantitative yield (crystal structure is shown in Supplementary Fig. 2). No cyclization products were found in this transformation. The obtaining of **6** not only provides further evidence for the formation of the NAu$_3$ species but also suggests the necessity of the $\sigma$-activation in the **1**-to-**2** transformation. Since the reaction of **1** and [**Au$_3$O**] is too rapid to isolate the $\sigma$-activated complex **4**, we next conducted the reaction between a simplified analogue PhC≡CTMS and [**Au$_3$O**]. The $^1$H-NMR monitoring explored the formation of the known complex PhC≡CAuPPh$_3$ in the reaction mixture, suggesting that [**Au$_3$O**] can break the C–Si bond in the TMS-C≡C moiety and engender the formation of a $\sigma$-bonded gold acetylide. This result affords a rationale for the $\sigma$-activation of **1** in the **1**-to-**2** transformation

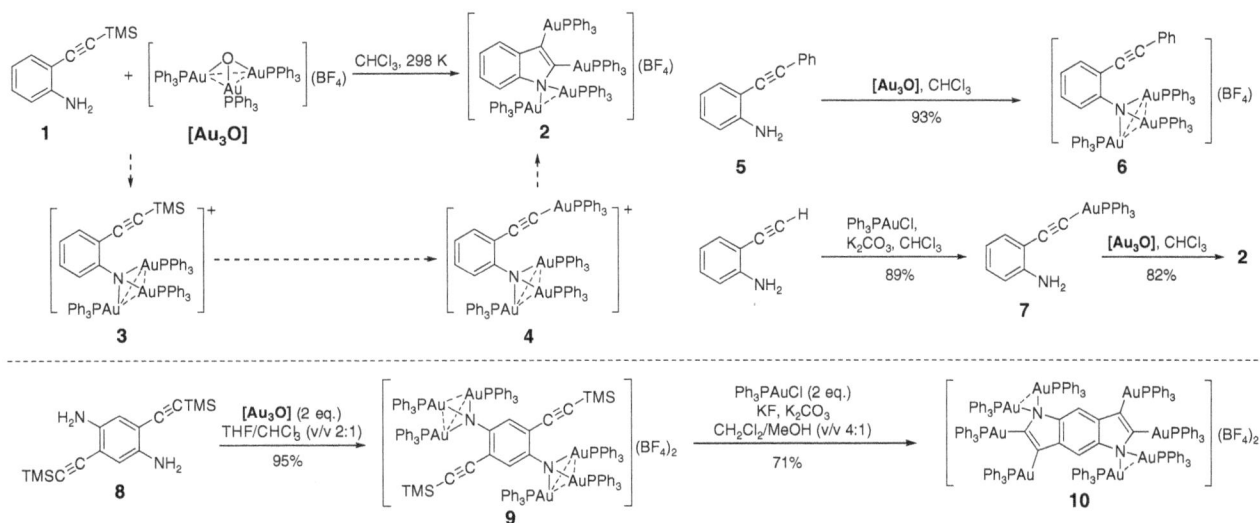

**Figure 1 | Synthesis of tetra-aurated indole 2 and octa-aurated benzodipyrrole 10.** The structures of complexes **2**, **6**, **9** and **10** have been determined by X-ray crystallography.

that should include the [**Au₃O**]-induced C–Si bond cleavage as well. In addition, we found the σ-aurated complex **7**, which derived from the reaction of 2-ethynylaniline with PPh₃AuCl, can react with [**Au₃O**] to produce **2** quantitatively within 5 min as evidenced in the ¹H-NMR monitoring (Supplementary Fig. 3). The **7**-to-**2** transformation supports the necessity of the σ-activation and meanwhile embodies the remarkable nucleophilicity of the aurated imido moiety.

We then carried out the reaction of the ditopic starting material **8** with [**Au₃O**] to purposefully synthesize the poly-aurated benzodipyrrole complex **10** based on the synthetic procedure for complex **2** (Fig. 1). However, this reaction gave a mixture of several gold-containing species which were observed in NMR spectroscopy and ESI-MS analysis. This outcome may result from several incomplete transformations including the auration of NH₂, the σ-metalation of TMS-C≡C and the cyclization between the imido and the ethynyl group as similar as the **1**-to-**2** transformation. Fortunately, in a mixed solvent of tetrahydrofuran (THF) and chloroform (v/v = 2:1) the reaction of **8** with [**Au₃O**] (2 equiv.) produced a yellow precipitate of complex **9**. This transformation is quantitative, so further separation is not necessary. X-ray crystallographic analysis disclosed two μ₃-imido trinuclear gold(I) moieties in the crystal structure of **9** (Supplementary Fig. 4). In view of the prerequisite σ-activation for the **7**-to-**2** transformation, we then conducted the reaction of **9** with Ph₃PAuCl (2 equiv.) in the presence of potassium carbonate and fluoride in order to replace the two TMS groups by two gold(I)-containing units. Amazingly, this transformation directly resulted in the formation of the octa-aurated benzodipyrrole complex **10**, which was characterized by X-ray crystallography (*vide infra*).

Molecular structures of the tetra-aurated indole complex **2** and the octa-aurated benzodipyrrole complex **10** were unambiguously determined by X-ray crystallographic analysis and confirmed by elemental analysis, multinuclear NMR spectroscopy and ESI-MS. As shown in Fig. 2a, crystal structure of **2** is composed of a tetra-aurated indolyl trianion, four coordinative AuPPh₃ units plus a BF₄⁻ counter anion. The nitrogen and carbon atoms of the indolyl moiety in **2** are coplanar. Therein, carbon atoms C1 and C2 each is coordinated by a AuPPh₃ unit with the Au–C bond lengths in the range of 2.01(3)–2.05(3) Å, comparable with the values in reported mononuclear gold(I)-indole complexes[34,35]. Both gold atoms (Au3 and Au4) maintain the typical linear

geometry with the P–Au–C bond angles of 175.4(9)° and 175.6(7)°, respectively. The indolyl nitrogen atom N1 is bonded by two geminal gold atoms (N1-Au1 = 2.17 (2) Å and N1-Au2 = 2.08 (2) Å), which together with C1 and C8 constitute a distorted tetrahedral bonding geometry for N1. The two gold atoms are held together by a strong aurophilic interaction[14,15] with the Au···Au distance of 2.862(2) Å. Although the whole structure of **2** was determined by X-ray crystallography and its composition was further substantiated by elemental analysis and ESI-MS; however, due to crystal twinning it is difficult to acquire the precise dimension of the indolyl skeleton in **2**. The mean deviation of all C–C and C–N bond lengths is as large as 0.04 Å.

Crystal structure of **10** can be described as a higher congener of **2** (Fig. 2b). To the best of our knowledge, complex **10** represents the first aurated benzodipyrrole compound. The negatively charged carbon and nitrogen atoms in the benzodipyrrole skeleton are mono- or di-aurated as similar as in **2**. An inversion centre is located in the centroid of the benzodipyrrole skeleton. Thus, two identical tetra-aurated pyrrolyl units are included in the crystal structure of **10**. The Au–C bond lengths in **10** lie in the range of 2.019(11)–2.036(8) Å, a bit shorter than the Au–N distances in the range of 2.082(11)–2.094(10) Å. The Au1···Au2 distance of 2.879(1) Å is comparable with the value in complex **2**. Although there are a number of examples containing the μ₃-imido trinuclear gold(I) cluster[37,38,40,41], the geminally diaurated fashion of a heteroaryl nitrogen atom as shown in **2** and **10** has not been reported before. In view of the isolobality of the proton and the LAu⁺ fragment[36], such diaurated indolyl structure can be considered as an analogue of indolium. However, scrutiny of the dimension of the benzodipyrrole moiety in **10** reveals that the cyclic five-membered ring exhibits very good aromaticity with the C–N and C–C bond lengths being equalized and comparable with the parent indole.

Complexes **2** and **10** have very good stability upon exposure to air and moisture. In solution, these two complexes can also keep their structures intact as evidenced in ESI-MS and NMR. In the ESI mass spectrum of **2**, an isotopically well-resolved peak at $m/z = 1950.2641$ corresponding to the $[2 - BF_4]^+$ species was observed (Supplementary Fig. 5). The ¹H-NMR spectrum of **2** clearly showed two multiplets at 7.12 and 7.82 p.p.m., which can be assigned to the four protons on the benzene ring of the indolyl skeleton (Supplementary Fig. 6). ESI mass spectrum of **10**

**a**

**b**

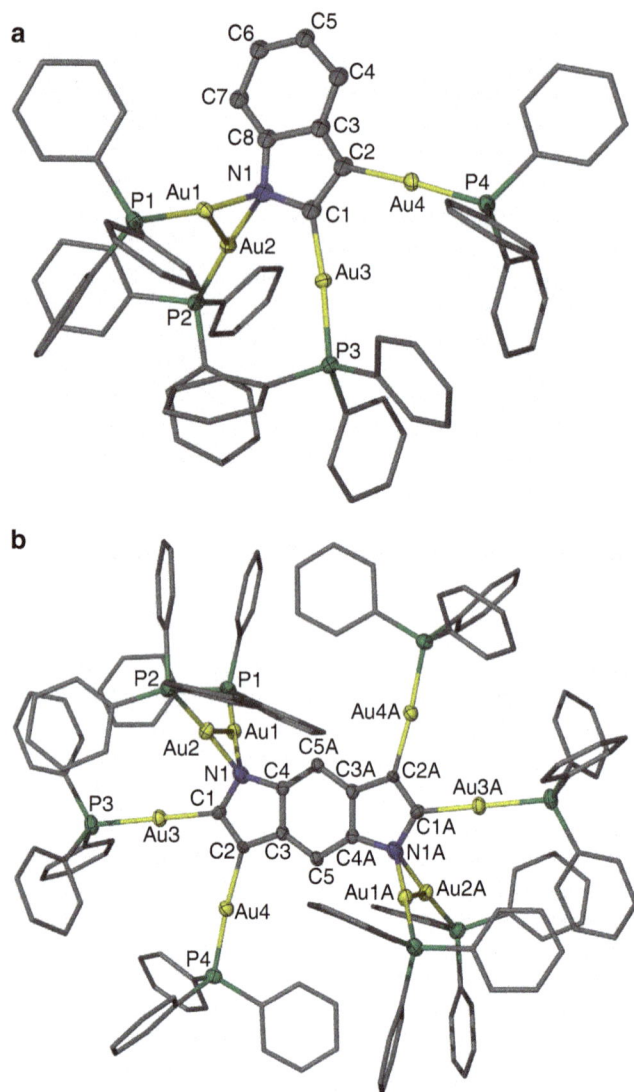

**Figure 2 | Crystal structures of heteroaryl-gold(I) complexes. (a)** tetra-aurated complex **2** and **(b)** octa-aurated complex **10**. Hydrogen atoms and tetrafluoroborate counter anions are omitted for clarity. Dimension of the benzodipyrrolium moiety in **10** (Å): N1-C1 1.414(14); N1-C4 1.470(14); C1-C2 1.344(13); C2-C3 1.417(12); C3-C4 1.387(14); C3-C5 1.393(13); and C4-C5A 1.381(13).

revealed a strong and isotopically well-resolved peak at $m/z = 1911.7428$ corresponding to the $[10 - 2BF_4]^{2+}$ species (Supplementary Fig. 7). Its $^1$H-NMR spectrum gave a sharp singlet at 8.29 p.p.m. due to two isolated proton atoms on the central benzene ring of the benzodipyrrole moiety (Supplementary Fig. 8). $^{31}$P-NMR spectra of **2** and **10** revealed two similar resonance peaks at around 33.6 and 39.4 p.p.m. (Supplementary Fig. 9), which can be assigned to the AuPPh$_3$ units attached on the nitrogen and carbon atoms.

**DFT computations on aurated heteroaryl complexes.** The equalized bond lengths and good stability of the polyaurated heteroaryl complexes **2** and **10** prompt us to examine whether the auration has a special influence on the aromaticity of the heteroaryl moieties. Density functional theory (DFT) computations[42] have then been carried out to help understand the aromaticity of the tetra- and octa-aurated heteroaryl complexes. For simplified model complexes **2'** and **10'**, the PH$_3$ ligands was

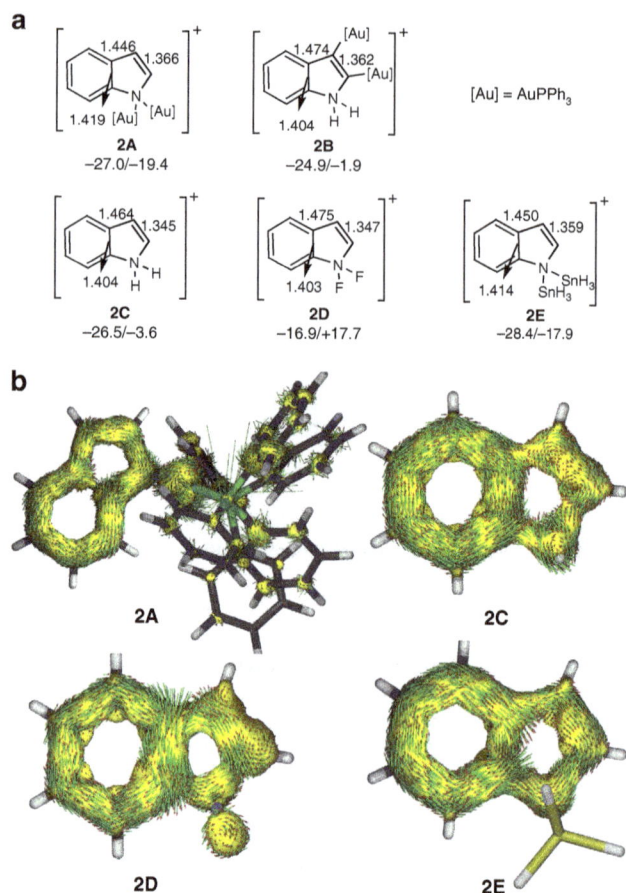

**a**

**b**

**Figure 3 | Hyperconjugative (anti)aromaticity in indolium derivatives.** **(a)** The selected C-C bond lengths (Å), NICS(1)$_{zz}$ values (p.p.m.) on the rings of **2A-2E**. The NICS(1)$_{zz}$ values given before and after the '/' are those computed at 1 Å above the geometrical centres of six- and five-membered rings, respectively. **(b)** ACID isosurfaces of **2A**, **2C**, **2D** and **2E** by the π contribution. Current density vectors are plotted onto the ACID isosurface of 0.03 to indicate dia- and para-tropic ring currents. The magnetic field vector is orthogonal with respect to the ring plane and points upward.

used to replace the PPh$_3$ ligands of the cationic moiety of **2** and **10**. The nucleus-independent chemical shift (NICS) computations[43,44] show that the NICS(1)$_{zz}$ values[29,45] for the five-membered ring (5MR) in **2'** and **10'** are −16.6 and −14.2 p.p.m., respectively (Supplementary Fig. 10). In general, negative values indicate aromaticity and positive values account for anti-aromaticity. These NICS(1)$_{zz}$ values in the 5MR are in sharp contrast to that of the parent indolium **2C** ( − 3.6 p.p.m., Fig. 3), indicating an aromaticity switch from non-aromatic to aromatic. Note that the Au···Au distance in **2'** is 3.303 Å, which is much longer than the experimental value. Thus, the PPh$_3$ ligands including dispersion corrections have been considered for the optimization of complexes **2A** and **2B**. Indeed, the Au···Au distance in **2A** becomes 2.889 Å, indicating that the large ligands and dispersion corrections play an important role in reproducing the aurophilic interaction. Further NICS calculations on complexes **2A** and **2B** suggest that it is the aurated substituents on the nitrogen atom rather than those on the $sp^2$-hybridized carbon atom that are responsible for the aromaticity change. Specifically, the NICS(1)$_{zz}$ value in the 5MR of **2A** is close to that of **2'**, whereas complexes **2B** has the similar NICS(1)$_{zz}$ value with the parent indolium **2C**.

Previous study by Schleyer and Nyulászi[25] indicated that distannylcyclopentadiene is nearly as aromatic as furan according to various aromaticity indices. To evaluate the magnitude of the aromaticity in **2A**, we performed the NICS calculation on distannylindolium **2E**. The NICS$(1)_{zz}$ value in the 5MR of **2E** is $-17.9$ p.p.m., which is less negative than that of **2A**, suggesting higher aromaticity in **2A** over **2E**. Thus complex **2A** could be considered as the most aromatic indolium reported so far. The highest aromaticity of **2A** is also supported by highest delocalized C–C bonds in the 5MR. Specifically, the C–C bond length alternation (0.080 Å) is smallest among **2A**-**2E**. The important role of hyperconjugation in aromaticity of **2A** is further supported by the anti-aromaticity in the 5MR of difluoroindolium **2D**, indicated by the significant positive NICS$(1)_{zz}$ value ($+17.7$ – p.p.m.).

The hyperconjugatvie aromaticity in **2A** is further supported by the anisotropy of the current-induced density analysis[46,47]. As shown in Fig. 3b, the current density vectors plotted on the anisotropy of the current-induced density isosurface indicate a diatropic ring current in the 5MRs of **2A** and **2E** in the $\pi$ system whereas the paratropic ring current is displayed in the 5MR of **2D**. No clear diatropic or paratropic ring current could be found in the 5MR of **2C**, suggesting its non-aromaticity. Thus, the concept of hyperconjugative aromaticity, firstly proposed by Mulliken[24] and later confirmed by Schleyer, Nyulászi, O'Ferrall and Houk[25-28], has now been first extended to transition metal substituted heterocycles[45]. More importantly, the aurated substituents can outperform the traditional main group substituents. These calculation results nicely confirm the aromatic essence of the geminally diaurated heteroaryls and account for the excellent stability of complexes **2** and **10**.

## Photoluminescent properties of aurated heteroaryl complexes.
Previous studies have explored that the heavy-atom effect of gold significantly influences the spin–orbit coupling between states of different spin multiplicity, thus enhancing intersystem crossing

from singlet to triplet excited states and facilitating radiative decay of the triplet excited state[48,49]. We therefore embarked on photophysical studies on four structurally well-defined polyaurated complexes **2**, **6**, **9** and **10**. **2** and **6** have similar absorption bands below 375 nm. In contrast, the absorption spectra of **9** and **10** extend to $\sim 475$ nm (Supplementary Fig. 11). This bathochromic shift could be rationalized by the extended $\pi$ system from **2** and **6** to **9** and **10**. In addition, the red shift is also reproduced qualitatively by time-dependent DFT calculations of **2'** and **10'** (the simplified complex of **10** with $PH_3$ in place of $PPh_3$). Specifically, the calculated intense absorption bands of complexes **2'** and **10'** at $\lambda = 321$ and 382 nm are consistent with the bands observed at $\lambda = 307$ and 355 nm of **2** and **10**, respectively. These computed absorption bands can be assigned to the electronic transitions HOMO→LUMO$+1$ and HOMO→LUMO$+2$ (HOMO is denoted as highest occupied molecular orbital whereas LUMO refers to lowest unoccupied molecular orbital, Supplementary Fig. 12), respectively, which are clearly a ligand-to-metal excitation nature.

Upon excitation at 325 nm (for **2** and **6**) and 345 nm (for **9** and **10**) at 298 K in dichloromethane, all four complexes are luminous with the emission spectra spanning the range from 350 to 650 nm (Fig. 4). The luminescence intensity of all four complexes is diminished by exposure to oxygen (Supplementary Fig. 13). In addition, the emission spectra of **2**, **9** and **10** show vibronic structures with a mean peak-to-peak separation of 1,010, 1,190 and 1,247 cm$^{-1}$, respectively. At low temperature (77 K), the emission from 450 to 550 nm becomes more intense and structured emission appears in the spectra of **6**, **9** and **10**. The emission lifetimes of the strongest emission for every complex at 77 K were collected and summarized in Table 1. The long lifetimes in microsecond-scale together with the oxygen-quenching of the emission and the large Stokes shifts indicate a phosphorescence parentage from a triplet-excited state. time-dependent DFT calculations on the model complex **2'** in the

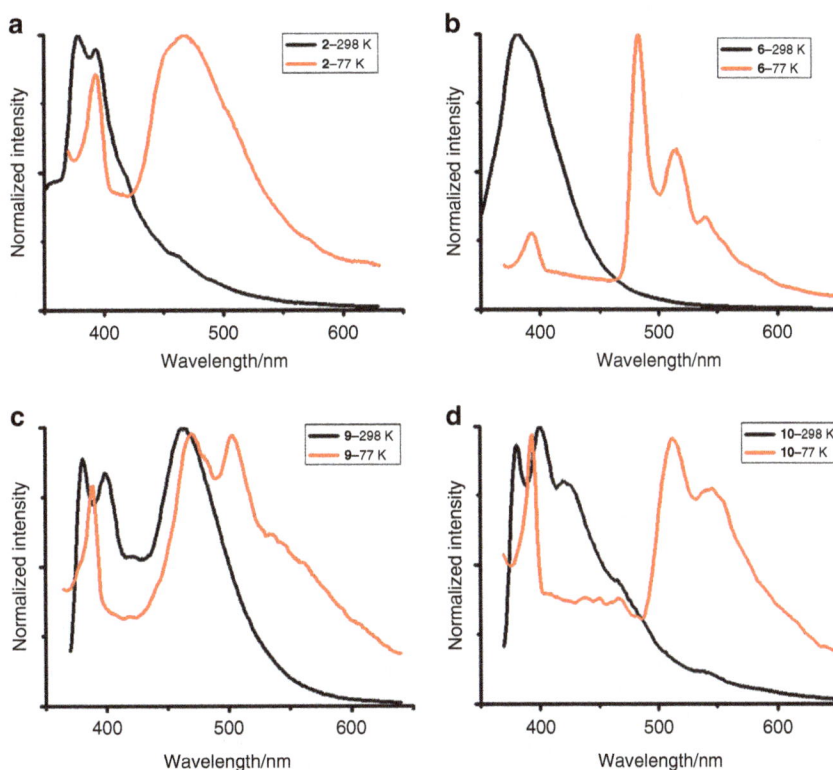

**Figure 4 | Emission spectra of polynuclear gold complexes.** Emission spectra of (**a**) **2**, (**b**) **6**, (**c**) **9** and (**d**) **10** collected at 298 and 77 K in dichloromethane ($c = 5 \times 10^{-6}$ M). Excitation: 325 nm (for **2** and **6**) and 345 nm (for **9** and **10**).

**Table 1 | Photophysical parameters.**

| Complex | Emission | | |
| --- | --- | --- | --- |
| | T/K | $\lambda_{max}$/nm ($\tau_o$/µs) | $\Phi_{em}$* |
| **2** | 298 | 378, 393, 464 | 0.04 |
| | 77 | 393, 468(219)† | |
| **6** | 298 | 381 | 0.37 |
| | 77 | 393, 482(609)†, 515, 540 | |
| **9** | 298 | 380, 398, 421, 462 | 0.14 |
| | 77 | 389, 470(62)‡, 505, 567 | |
| **10** | 298 | 381, 400, 421, 468, 541 | 0.02 |
| | 77 | 394, 512(53)‡, 546 | |

*Measured at 298 K using quinine sulfate dihydrate as a standard.
†Measured in a 2-methyltetrahydrofuran glass at 77 K.
‡Measured in a EtOH-MeOH-CH$_2$Cl$_2$ (40:20:1 v/v) glass at 77 K.

lowest triplet state (T$_1$) indicate that the calculated emission of **2'** at $\lambda_{max} = 383$, 412 and 474 nm are consistent with the bands observed at $\lambda = 378$, 393 and 464 nm of complex **2**, respectively. In addition, calculations reveal that the three emission bands are ligand-to-metal, metal-centred and ligand-centred, respectively (Supplementary Fig. 14). Gray and co-workers have once proposed a plausible excited-state hypothesis accounting for the temperature-dependent emission of aurated aryls[49]. It includes the decay of a singlet excited state to an excimer, which catalyses or pumps the formation of a triplet. High temperatures suppress excimer formation, resulting in emission difference. Furthermore, quantum yield measurements at 298 K revealed that the imido trinuclear gold(I) complexes **6** (37%) and **9** (14%) have higher yields than **2** (4%) and **10** (2%). This difference is possibly ascribed to the crowded arrangement of the AuPPh$_3$ units in **6** and **9**, which impedes non-radiative decay processes by blocking molecular rotations and vibrations of PPh$_3$ ligands.

## Discussion

The synthesis of the polyaurated heteroaryls **2** and **10** is unprecedented and unusual. Stepwise transformations from 2-ethynylaniline through **7**-to-**2** and from **8** through **9**-to-**10** highlight two alternative pathways towards the synthesis of polyaurated aryls no matter which step (the auration of the amine group and the $\sigma$-metalation of the TMS-substituted ethynyl group) takes place first. A common prerequisite for the synthesis of **2** and **10** is the $\sigma$-activation of the carbon–carbon triple bond, which accounts for the unsuccessful annulation of internal alkyne substrates such as **5**. Considering the isolobality of the proton and the LAu$^+$ fragment, the $\mu_3$-imido aurated species is analogous to an ammonium species that is lack of nucleophilicity. We hypothesize that the NAu$_3$ species in **3** and **9** exhibit nucleophilicity on condition that the trinuclear gold cluster aggregate dissociates into low nuclearity species. This speculation is partially evidenced by the transformation of **9** to **10**. In this transformation process, the stoichiometric ratio of **9** to Ph$_3$PAuCl is 1:2. The number of [AuPPh$_3$] units in **9** plus two equivalents Ph$_3$PAuCl is equal to that of **10**, while the number of the nitrogen-bonded gold unit decreases from three in **9** to two in **10**. This result suggests the release of a AuPPh$_3$ unit from the imido NAu$_3$ aggregate. This AuPPh$_3$ unit may translocate to finally attach on a carbon centre of the resulting heteroaryls. If replacing Ph$_3$PAuCl (2 equiv.) by the same amount of Cy$_3$PAuCl (Cy = cyclohexyl) in the **9**-to-**10** transformation, the ESI-MS of the reaction mixture revealed several Au$_8$-benzodipyrrole species containing different ligand combinations such as [6PPh$_3$ + 2 PCy$_3$], [5PPh$_3$ + 3PCy$_3$] and [4PPh$_3$ + 4PCy$_3$], suggesting the dissociation and intermolecular exchange of phosphine ligands (Supplementary Fig. 15).

On the other hand, hyperconjugation, a weak interaction in chemistry, has a strong effect on the aromaticity. Simply tuning the substituents can lead to an aromaticity switch from non-aromaticity to (anti)aromaticity. Previous studies on hyperconjugative aromaticity mainly focus on the carbocycles. Examples on heterocycles are particularly rare and all the substituents reported so far contain main group elements only. Our calculations first demonstrate that the substituent containing the transition metal gold is also able to work for hyperconjugative aromaticity. More importantly, the introduction of *d*-orbitals from the transition metal can have a stronger effect on the hyperconjugative aromaticity in comparison with that of *p*-orbitals from traditional main group elements, thus opening an avenue to the design of various novel metalla-aromatics.

In summary, we have demonstrated a facile synthetic method of producing polyaurated heteroaryls. Two alternative pathways by activating the ethynyl group at first and then metalating the amine group or *vice versa* both accomplished the synthesis of unprecedented polyaurated indole and benzodipyrrole complexes. Theoretical analysis revealed an unexpected aromaticity change due to the hyperconjugation caused by the introduction of aurated substituents, which can perform better than the traditional main group substituents, extending the concept of hyperconjugative aromaticity to the transition metal-involved heterocycles. This study showcases a promising approach to acquire more complicated and intricate organometallics by forming saturated polymetalated units. Structure and reactivity comprehension of these polymetalated organometallics foresees deep research in bonding nature and mechanistic studies and stimulates the advancement of more efficient and versatile synthetic methods.

## Methods

**General.** All commercially available reagents were used as received. Flash column chromatography was performed on silica gel (100–200). The solvents used in this study were dried by 4 Å molecular sieves. $^1$H-NMR, $^{13}$C-NMR and $^{31}$P-NMR spectra were recorded by using 400 MHz spectrometers. Chemical shifts are reported in ppm versus tetramethylsilane with either tetramethylsilane or the residual solvent resonance used as an internal standard. Synthetic procedures for complexes **6**, **7** and **9** are summarized in Supplementary Methods. ESI-MS spectra of complexes **6** and **9** are shown in Supplementary Figs 16 and 17, respectively. $^1$H-, $^{31}$P- and $^{13}$C-NMR spectra of complexes **7** (Supplementary Figs 18–20), **6** (Supplementary Figs 21–23), **2** (Supplementary Figs 24–26), **9** (Supplementary Figs 27–29) and **10** (Supplementary Figs 30–32) are shown in Supplementary Information.

**Synthesis of 2.** Complex **7** (10 mg, 0.017 mmol) and oxotris((triphenylphosphine)gold) tetrafluoroborate (25.7 mg, 0.017 mmol) were dissolved in CHCl$_3$ (2 ml) with stirring at room temperature over 20 min. Addition of hexane into the reaction mixture afforded a white solid of **2**. Yield: 82% (30 mg, 0.014 mmol). Single crystals of **2** were obtained by vapour diffusion of diethyl ether into a CHCl$_3$ solution of **2**. $^1$H-NMR (400 MHz, CDCl$_3$): δ 7.87–7.80 (m, 2H); 7.48 (dd, $J = 12.3$, 7.8 Hz, 10H); 7.38 (dd, $J = 12.4$, 7.5 Hz, 20H); 7.30 (d, $J = 7.7$ Hz, 12H); 7.15–7.11 (m, 2H); and 7.05 (td, $J = 7.8$, 1.8 Hz, 18H). $^{13}$C-NMR (100 MHz, CDCl$_3$): δ 134.2, 134.1, 132.1, 131.6, 130.5, 129.9, 129.5, 129.4, 129.3, 129.2, 128.6, 122.7, 121.9, 120.1 and 114.7. $^{31}$P-NMR (162 MHz, CDCl$_3$): δ 39.43, 33.61. IR (KBr, cm$^{-1}$): 3052, 1964, 1890, 1812, 1667, 1615, 1587, 745, 710 and 692; HR-MS (ESI): calcd. for [M-BF$_4$]$^+$ (C$_{80}$H$_{64}$Au$_4$NP$_4$) 1950.2646, found 1950.2641. Elemental Analysis: Calcd. for (C$_{80}$H$_{64}$Au$_4$BF$_4$NP$_4$ + 1/2(CHCl$_3$)): C, 46.85; H, 3.37; and N, 0.65. Found: C, 46.88; H, 3.31; and N, 0.78.

**Synthesis of 10.** To a mixture of **9** (32.3 mg, 0.010 mmol) and 2 equiv. of PPh$_3$AuCl (9.9 mg, 0.020 mmol) in dichloromethane (2 ml) were added a methanolic solution (0.5 ml) of potassium fluoride (2.3 mg, 0.030 mmol) dropwise and 20 equiv. of K$_2$CO$_3$ (27.6 mg, 0.20 mmol). The mixture was stirred overnight in the dark. The solvent was removed by vacuum and the residue was redissolved in CH$_2$Cl$_2$. The resulting mixture was filtered through celite. Addition of Et$_2$O to the filtrate afforded a yellow solid of **10**. Yield: 71% (28.4 mg, 0.0071 mmol). Single crystals of **10** were obtained by vapour diffusion of diethyl ether into a CHCl$_3$ solution of **10** after 2 days. $^1$H-NMR (400 MHz, CDCl$_3$): δ 8.29 (s, 2H); 7.48-7.36 (m, 64H); and 7.09-6.97 (m, 56H). $^{13}$C-NMR (100 MHz, CDCl$_3$): δ 134.1, 131.9, 131.4, 130.6, 130.0, 129.2 and 105.8. $^{31}$P-NMR (162 MHz, CDCl$_3$): δ 39.48, 33.56.

IR (KBr, cm$^{-1}$): 3049, 1964, 1890, 1812, 1585, 746, 709 and 692; HR-MS (ESI): calcd. for [M–2BF$_4$]$^{2+}$ (C$_{154}$H$_{122}$Au$_8$N$_2$P$_8$) 1911.7428; found 1911.7428. Elemental analysis: Calcd. for (C$_{154}$H$_{122}$Au$_8$B$_2$F$_8$N$_2$P$_8$ + CH$_2$Cl$_2$): C, 45.98; H, 3.22; and N, 0.68. Found: C, 45.67; H, 3.12; and N, 0.85.

**X-ray crystallographic analysis.** Data for complexes **2**, **6**, **9** and **10** (CCDC-1435045) were collected at 173 K with Mo-Kα radiation ($\lambda = 0.71073$ Å) on a Rigaku Saturn 724+ CCD diffractometer with frames of oscillation range 0.5°. All structures were solved by direct methods, and non-hydrogen atoms were located from difference Fourier maps. All non-hydrogen atoms were subjected to anisotropic refinement by full-matrix least-squares on $F^2$ by using the SHELXTL program unless otherwise noticed[50]. All crystal structure figures were drawn by using X-seed program[51]. Structural refinement details are summarized in Supplementary Methods. See also Supplementary Data 1.

**Photophysical measurements.** Photophysical measurement of low-temperature glass samples was carried out with the solid sample loaded in a quartz tube inside a quartz-walled Dewar flask. Liquid nitrogen was placed into the Dewar flask for low temperature (77 K). Luminescence quantum yields for solution samples were measured by the optical dilute method. All values are relative to quinine sulfate in 1.0 M H$_2$SO$_4$ as being 0.55 as the reference. Luminescent decay experiments were measured on an Edinburgh FLS920 spectrometer with μF920 microsecond flash lamp (pulse width < 2 μs) as light source.

# References

1. Hashmi, A. S. K. Gold-catalyzed organic reactions. *Chem. Rev.* **107**, 3180–3211 (2007).
2. Gorin, D. J., Sherry, B. D. & Toste, F. D. Ligand effects in homogeneous Au catalysis. *Chem. Rev.* **108**, 3351–3378 (2008).
3. Li, Z., Brouwer, C. & He, C. Gold-catalyzed organic transformations. *Chem. Rev.* **108**, 3239–3265 (2008).
4. Corma, A., Leyva-Pérez, A. & Sabater, M. J. Gold-catalyzed carbon-heteroatom bond-forming reactions. *Chem. Rev.* **111**, 1657–1712 (2011).
5. Dorel, R. & Echavarren, A. M. Gold(I)-catalyzed activation of alkynes for the construction of molecular complexity. *Chem. Rev.* **115**, 9028–9072 (2015).
6. Cheong, P. H.-Y., Morganelli, P., Luzung, M. R., Houk, K. N. & Toste, F. D. Gold-catalyzed cycloisomerization of 1,5-allenynes via dual activation of an ene reaction. *J. Am. Chem. Soc.* **130**, 4517–4526 (2008).
7. Hashmi, A. S. K. et al. Simple gold-catalyzed synthesis of benzofulvenes-gem-diaurated species as 'instant dual-activation' precatalysts. *Angew. Chem. Int. Ed.* **51**, 4456–4460 (2012).
8. Gómez-Suárez, A. & Nolan, S. P. Dinuclear gold catalysis: are two gold centers better than one? *Angew. Chem. Int. Ed.* **51**, 8156–8159 (2012).
9. Ye, L., Wang, Y., Aue, D. H. & Zhang, L. Experimental and computational evidence for gold vinylidenes: generation from terminal alkynes via a bifurcation pathway and facile C-H insertions. *J. Am. Chem. Soc.* **134**, 31–34 (2012).
10. Hirner, J. J., Shi, Y.-L. & Blum, S. A. Organogold reactivity with palladium, nickel, and rhodium: transmetalation, cross-coupling, and dual catalysis. *Acc. Chem. Res.* **44**, 603–613 (2011).
11. Boogaerts, I. I. F. & Nolan, S. P. Carboxylation of C-H bonds using N-heterocyclic carbene gold(I) complexes. *J. Am. Chem. Soc.* **132**, 8858–8859 (2010).
12. Cambeiro, X. C., Boorman, T. C., Lu, P. & Larrosa, I. Redox-controlled selectivity of C-H activation in the oxidative cross-coupling of arenes. *Angew. Chem. Int. Ed.* **52**, 1781–1784 (2013).
13. Ball, L. T., Lloyd-Jones, G. C. & Russell, C. A. Gold-catalyzed oxidative coupling of arylsilanes and arenes: origin of selectivity and improved precatalyst. *J. Am. Chem. Soc.* **136**, 254–264 (2014).
14. Pyykkö, P. Theoretical chemistry of gold. *Angew. Chem. Int. Ed.* **43**, 4412–4456 (2004).
15. Schmidbaur, H. & Schier, A. Aurophilic interactions as a subject of current research: an up-date. *Chem. Soc. Rev.* **41**, 370–412 (2012).
16. Sherry, B. D., Maus, L., Laforteza, B. N. & Toste, F. D. Gold(I)-catalyzed synthesis of dihydropyrans. *J. Am. Chem. Soc.* **128**, 8132–8133 (2006).
17. Jin, L. et al. Trinuclear gold clusters supported by cyclic(alkyl)(amino)carbene ligands: mimics for gold heterogeneous catalysts. *Angew. Chem. Int. Ed.* **53**, 9059–9063 (2014).
18. Johnson, A., Laguna, A. & Gimeno, M. C. Axially chiral allenyl gold complexes. *J. Am. Chem. Soc.* **136**, 12812–12815 (2014).
19. Heng, W. Y., Hu, J. & Yip, J. H. K. Attaching gold and platinum to the rim of pyrene: a synthetic and spectroscopic study. *Organometallics* **26**, 6760–6768 (2007).
20. Hashmi, A. S. K., Braun, I., Rudolph, M. & Rominger, F. The role of gold acetylides as a selectivity trigger and the importance of gem-diaurated species in the gold-catalyzed hydroarylating-aromatization of arene-diynes. *Organometallics* **31**, 644–661 (2012).
21. Winter, C. H., Seneviratne, K. N. & Bretschneider-hurley, A. Permetalated aromatic compounds. *Comment Inorg. Chem.* **19**, 1–23 (1996).
22. Wei, Q.-H., Zhang, L.-Y., Yin, G.-Q., Shi, L.-X. & Chen, Z.-N. Luminescent heteronuclear Au$^I$$_5$Ag$^I$$_8$ complexes of {1,2,3-C$_6$(C$_6$H$_4$-4)$_3$}$^{3-}$ (R = H, CH$_3$, Bu$^t$) by cyclotrimerization of arylacetylides. *J. Am. Chem. Soc.* **126**, 9940–9941 (2004).
23. Heckler, J. E., Zeller, M., Hunter, A. D. & Gray, T. G. Geminally diaurated gold(I) aryls from boronic acids. *Angew. Chem. Int. Ed.* **51**, 5924–5928 (2012).
24. Mulliken, R. S. Intensities of electronic transitions in molecular spectra IV. Cyclic dienes and hyperconjugation. *J. Chem. Phys.* **7**, 339–352 (1939).
25. Nyulászi, L. & Schleyer, P. v. R. Hyperconjugative π-aromaticity: how to make cyclopentadiene aromatic. *J. Am. Chem. Soc.* **121**, 6872–6875 (1999).
26. Lawlor, D. A. et al. Hyperaromatic stabilization of arenium ions: cyclohexa- and cycloheptadienyl cations—experimental and calculated stabilities and ring currents. *J. Am. Chem. Soc.* **133**, 19729–19742 (2011).
27. Fernández, I., Wu, J. I. & Schleyer, P. v. R. Substituent effects on 'hyperconjugative' aromaticity and antiaromaticity in planar cyclopolyenes. *Org. Lett.* **15**, 2990–2993 (2013).
28. Levandowski, B. J., Zou, L. & Houk, K. N. Schleyer hyperconjugative aromaticity and Diels–Alder reactivity of 5-substituted cyclopentadienes. *J. Comput. Chem.* **37**, 117–123 (2016).
29. Sun, H., An, K. & Zhu, J. Triplet state aromaticity: NICS criterion, hyperconjugation, and charge Effects. *Chem. Asian J.* **11**, 234–240 (2016).
30. Landorf, C. W. & Haley, M. M. Recent advances in metallabenzene chemistry. *Angew. Chem. Int. Ed.* **45**, 3914–3936 (2006).
31. Dalebrook, A. F. & Wright, L. J. Metallabenzenes and metallabenzenoids. *Adv. Organomet. Chem.* **60**, 93–177 (2012).
32. Fernández, I., Frenking, G. & Merino, G. Aromaticity of metallabenzenes and related compounds. *Chem. Soc. Rev.* **44**, 6452–6463 (2015).
33. Alcaide, B. & Almendros, P. Progress in allene chemistry. *Chem. Soc. Rev.* **43**, 2886–2887 (2014).
34. Gimeno, A. et al. Competitive gold-activation modes in terminal alkynes: an experimental and mechanistic study. *Chem. Eur. J.* **20**, 683–688 (2014).
35. Zeng, X., Kinjo, R., Donnadieu, B. & Bertrand, G. Serendipitous discovery of the catalytic hydroammoniumation and methylamination of alkynes. *Angew. Chem. Int. Ed.* **49**, 942–945 (2010).
36. Raubenheimer, H. & Schmidbaur, H. Gold chemistry guided by the isolobal concept. *Organometallics* **31**, 2507–2522 (2012).
37. Ramamoorthy, V. & Sharp, P. R. Late-transition-metal μ-oxo and μ-imido complexes. 6. Gold( I) imido complexes. *Inorg. Chem.* **29**, 3336–3339 (1990).
38. Ramamoorthy, V., Wu, Z., Yi, Y. & Sharp, P. R. Preparation and decomposition of gold(I) hydrazido complexes: gold cluster formation. *J. Am. Chem. Soc.* **114**, 1526–1527 (1992).
39. Hashmi, A. S. K. Isolable vinylgold intermediates–first access to phantoms of homogeneous gold catalysis. *Gold Bull.* **42**, 275–279 (2009).
40. Shan, H., Yang, Y., James, A. J. & Sharp, P. R. Dinitrogen bridged gold clusters. *Science* **275**, 1460–1462 (1997).
41. Sharp, P. R. Oxo and imido ligands in late transition metal chemistry. *J. Chem. Soc. Dalton Trans.* 2647–2657 (2000).
42. Frisch, M. J. et al. *Gaussian 09, Revision D.01* (Gaussian, Inc., 2013).
43. Schleyer, P. v. R., Maerker, C., Dransfeld, A., Jiao, H. & Hommes, N. J. R. v. E. Nucleus-independent chemical shifts: a simple and efficient aromaticity probe. *J. Am. Chem. Soc.* **118**, 6317–6318 (1996).
44. Chen, Z., Wannere, C. S., Corminboeuf, C., Puchta, R. & Schleyer, P. v. R. Nucleus-independent chemical shifts (NICS) as an aromaticity criterion. *Chem. Rev.* **105**, 3842–3888 (2005).
45. Fallah-Bagher-Shaidaei, H., Wannere, C. S., Corminboeuf, C., Puchta, R. & Schleyer, P. v. R. Which NICS aromaticity index for planar π rings is best? *Org. Lett.* **8**, 863–866 (2006).
46. Herges, R. & Geuenich, D. Delocalization of electrons in molecules. *J. Phys. Chem. A.* **105**, 3214–3220 (2001).
47. Geuenich, D., Hess, K., Köhler, F. & Herges, R. Anisotropy of the induced current density (ACID), a general method To quantify and visualize electronic delocalization. *Chem. Rev.* **105**, 3758–3772 (2005).
48. Partyka, D. V., Esswein, A. J., Zeller, M., Hunter, A. D. & Gray, T. G. Gold(I) pyrenyls: excited-state consequences of carbon-gold bond formation. *Organometallics* **26**, 3279–3282 (2007).
49. Gao, L. et al. Mono- and di-gold(I) naphthalenes and pyrenes: syntheses, crystal structures, and photophysics. *Organometallics* **28**, 5669–5681 (2009).
50. Sheldrick, G. M. *SHELXL-97* (Univ. Göttingen (1997).
51. Atwood, J. L. & Barbour, L. J. Molecular graphics: from science to art. *Cryst. Growth Des.* **3**, 3–8 (2003).

## Acknowledgements

Financial support by NNSFC (21132005, 21421064, 21522206, and 21573179), MOST (2013CB834501) and Top-Notch Young Talents Program of China is gratefully acknowledged. We are grateful to Profs. Mei-Xiang Wang, Qing-Zheng Yang, Lian Duan and Ming-Tian Zhang for data measurements and helpful discussions.

## Author contributions

L.Z. and J.Z. conceived and supervised the project. The synthetic experiments and photophysical measurements were carried out by J.Y. and X.H. T.S. and K.A. performed the theoretical calculation. L.Z. and J.Z. co-wrote the manuscript. All authors discussed the results and commented on the manuscript.

## Additional information

**Competing financial interests:** The authors declare no competing financial interests.

# 21

# A method for controlling the synthesis of stable twisted two-dimensional conjugated molecules

Yongjun Li[1], Zhiyu Jia[1], Shengqiang Xiao[2], Huibiao Liu[1] & Yuliang Li[1]

Thermodynamic stabilization ($\pi$-electron delocalization through effective conjugation) and kinetic stabilization (blocking the most-reactive sites) are important considerations when designing stable polycyclic aromatic hydrocarbons displaying tunable optoelectronic properties. Here, we demonstrate an efficient method for preparing a series of stable two-dimensional (2D) twisted dibenzoterrylene-acenes. We investigated their electronic structures and geometries in the ground state through various experiments assisted by calculations using density functional theory. We find that the length of the acene has a clear effect on the photophysical, electrochemical, and magnetic properties. These molecules exhibit tunable ground-state structures, in which a stable open-shell quintet tetraradical can be transferred to triplet diradicals. Such compounds are promising candidates for use in nonlinear optics, field effect transistors and organic spintronics; furthermore, they may enable broader applications of 2D small organic molecules in high-performance electronic and optical devices.

[1] CAS Key Laboratory of Organic Solids, Beijing National Laboratory for Molecular Science (BNLMS), Institute of Chemistry, Chinese Academy of Sciences, Beijing 100190, China. [2] State Key Laboratory of Advanced Technology for Materials Synthesis and Processing, Wuhan University of Technology, Wuhan 430070, China. Correspondence and requests for materials should be addressed to Yongjun Li (email: liyj@iccas.ac.cn) or to Yuliang Li (email: ylli@iccas.ac.cn).

In the last two decades, methodologies for the controlled synthesis of large π-extended polycyclic aromatic hydrocarbons (PAHs) have been established using bottom-up principles[1-11]. The properties of PAHs depend heavily on their degrees of π-extension, shapes, widths and edge topologies[1,3,4,12-14]. Many linear acenes, including tetracene, pentacene and related derivatives, have been synthesized to produce highly desirable electronic properties, including remarkable charge-carrier mobilities[15-19]. Open-shell PAHs, which feature one or more π-electrons not tightly paired into the bonding molecular orbital in the ground state, are particularly interestingly structures that have properties very different from those of typical PAHs having closed-shell electronic structures[20-23]. The unique electronic structures of two-dimensional (2D) open-shell PAHs can impart attractive electronic, optical and magnetic properties for applications in materials science[24-26].

Acenes become increasingly reactive as the number of rings increases, with the central ring being the most reactive[27,28]. As a result, the central rings in these molecules are susceptible to oxidation, photodegeneration and Diels–Alder reactions[29,30]. Open-shell molecules are particularly vulnerable to degradation reactions[22,23]; therefore, instability remains a key obstacle affecting their practical applications. In most cases, thermodynamic stabilization (π-electron delocalization through effective conjugation) and kinetic stabilization (blocking of the most-reactive sites) are both necessary to obtain stable materials[18,31]. One strategy is to annulate the aromatic rings onto two neighbouring rings, creating 2D acene analogues[15]. Changing from the linear character of a condensed array (peri-condensed) to an angular geometry (cata-condensed) results in at least two sextets of π-electrons, increasing the stability relative to that of a linear analogue having only one sextet[32]. Tuning a planar structure to a nonplanar structure through twisting or saddling can also enhance the kinetic stability as a result of steric blocking[21,33].

2D acene analogues that represent various fragments of graphene have received significant attention because of their electronic properties[6,9,10,15]. Molecules of this class have large, planar π-surfaces allowing high-intermolecular surface overlap[34-36]. The packing of such molecules is dictated by multiple interactions that can effectively increase the dimensionality of the electronic structure, leading to enhanced transport properties[37,38]. Many 2D planar PAHs[11,32,34,36,39-41] and contorted polycyclic aromatics[5,42-44] have been synthesized and characterized recently.

In this paper, we report the transformation of a one-dimensional (1D) single conjugated dibenzoterrylene through the attachment of conjugated acene 'arms' to generate a 2D molecular structure. High dimensionality within the single molecule was assured by hybridizing dibenzoterrylene with acenes of different conjugation lengths; stability was provided by benzoannulating the active rings of acenes and inducing contortions from the cove or fjord regions. This protocol is applicable to construct various 2D twisted nanographene-like structures in a rapid manner.

## Results

### Design of target compounds

Figure 1a showed the conceptual approaches toward twisted 2D acenes. We positioned electron-donating/withdrawing groups (for example, $CH_3$, OMe, CN, F) to alter the optical absorbance and the energy levels of the highest occupied molecular orbital (HOMO) and lowest unoccupied molecular orbital (LUMO) of the tetracene-dibenzoterrylene hybrid, in addition to tuning the molecular packing in the solid state (Fig. 1b). We introduced phenanthrenyl groups to further

increase the steric constraint, and naphthyl and perylenyl groups to elongate the acenes and allow annulation of benzene rings. Using this approach, 2D twisted acenes varying from closed-shell to open-shell structures were available through extension of the acenes vertical to the benzoterrylene and benzoannulation at the zigzag edge. Compound **4** can be considered (Fig. 1c) as a hybrid of two pentacene-annulated-tribenzo-octacenes, in which two phenalenyl radicals (5,10 positions of octacene) are annulated with a central tetracene unit, or having evolved from a dibenzo-pentacene with the reactive $5',7'$-positions of the pentacene connected and protected together. In both cases, one more Clar sextet was formed on resonance to tetraradical state, which also predicted the stability of the tetraradical.

**Synthesis of target compounds.** Figure 2 displays the synthesis of **1**, starting from the bromination of the amine **5** (ref. 45) with $Br_2$ followed by conversion of the amino group to the iodide through diazotization. Compound **7** was coupled with 2,7-bis(4,4,5,5-tetramethyl-1,3,2-dioxaborolan-2-yl)pyrene through Suzuki–Miyaura cross-coupling to give the key intermediate **8**. This compounds was coupled with a series of $RB(OH)_2$ derivatives **9** to provide the precursors **10a–g** for Scholl-type oxidative cyclodehydrogenation, in which $FeCl_3$ was used as both a Lewis acid and an oxidant. *tert*-Butyl groups were introduced to increase the solubility of the final product. A phenanthrenyl group was used to further increase the steric constraint, leading to precursor **11**; naphthyl and perylenyl groups were introduced to elongate the acenes and introduce the annulation benzene rings, providing the precursors **12** and **13**, respectively.

For **10e**, C–C bonds were formed *ortho* and *para* to the *meta* fluorine atom, leading to a mixture where the fluorine atoms were located at the $R_1$ and $R_3$ positions, as supported in the X-ray structure (see below). For **11**, the Scholl reaction at the K-region (convex armchair edge) of the phenanthrenyl group led to **2** in 57%, which is too crowded at the fjord regions to allow two benzene rings to get sufficiently close to form a further C–C bond[46]. π-Conjugation extended compound **3** was obtained from **12** in 78% yield. The oxidative cyclohydrogenation of **13** is time-dependent. When the reaction time was <2 h, we obtained some incompletely cyclized products, which were difficult to separate from the target **4**; when the reaction time was longer than 12 h we obtained some byproducts with similar molecular weight as **4** but with different absorption spectra. The black compound **4** is well-soluble in $C_2H_2Cl_4$ and o-dichlorobenzene (dark-green solution), partially soluble in $CH_2Cl_2$ and $CHCl_3$, we could purify it (in 51% yield) through column chromatography on deactivated silica gel with $CH_2Cl_2$/MeOH (5/1). We characterized all of the intermediates and final products, except for **4**, using $^1H$ and $^{13}C$ nuclear magnetic resonance (NMR) spectroscopy and high-resolution mass spectrometry (MS); we characterized compound **4** only through high-performance liquid chromatography, high-resolution MS and elemental analysis.

**Conformations characterization of compounds 1–3.** At first, we performed density functional theory (DFT) optimization to gain insight into the molecular conformations of these 2D acenes. The nonplanar conformations of these compounds arose from severe steric strain between the two C–H bonds at the cove or fjord regions, as can be seen clearly in the model of **3** in Fig. 3a. DFT calculations revealed that **3** has five possible conformations (Fig. 3a); the hexacene units can take on conformations that are twisted or anti-folded with the central naphthalene unit, while the outer naphthalene rings in the hexacenes can be contorted out of the plane defined by the dibenzoterrylene core in the *cis* or *trans* direction: cAA-**3**, tAA-**3**, AT-**3**, cTT-**3** and tTT-**3**. Compound **1**

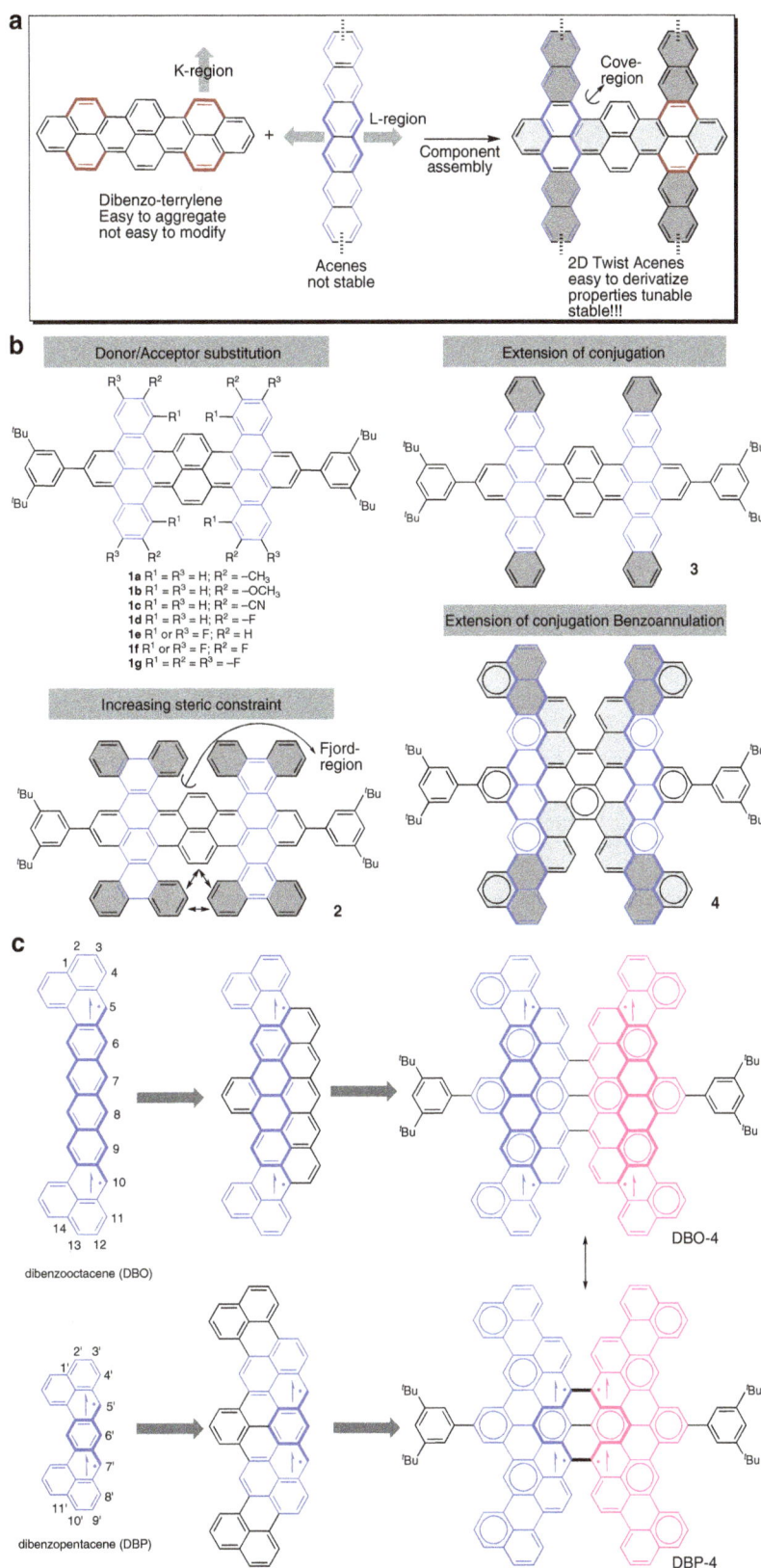

**Figure 1 | Conceptual approaches toward twisted 2D acenes.** (**a**) π-Components of 2D acenes. (**b**) Molecular structures of 2D acenes with increasing steric constraint (**2**), extension of conjugation (**3**) and with both extension of conjugation and benzoannulation at the L-region (zigzag edge; **4**). (**c**) Chemical structural evolution of **4**: (top) Benzoannulation of octacene to generate phenalenyl radicals annulated with a central tetracene and further pentacene-annulation on the other side to provide more stable diradicals, two of which connect in the central area; (bottom) evolution from a dibenzopentacene with benzo- and naphtheno-annulation, with the reactive 5',7'-positions of the pentacene connected and protected together.

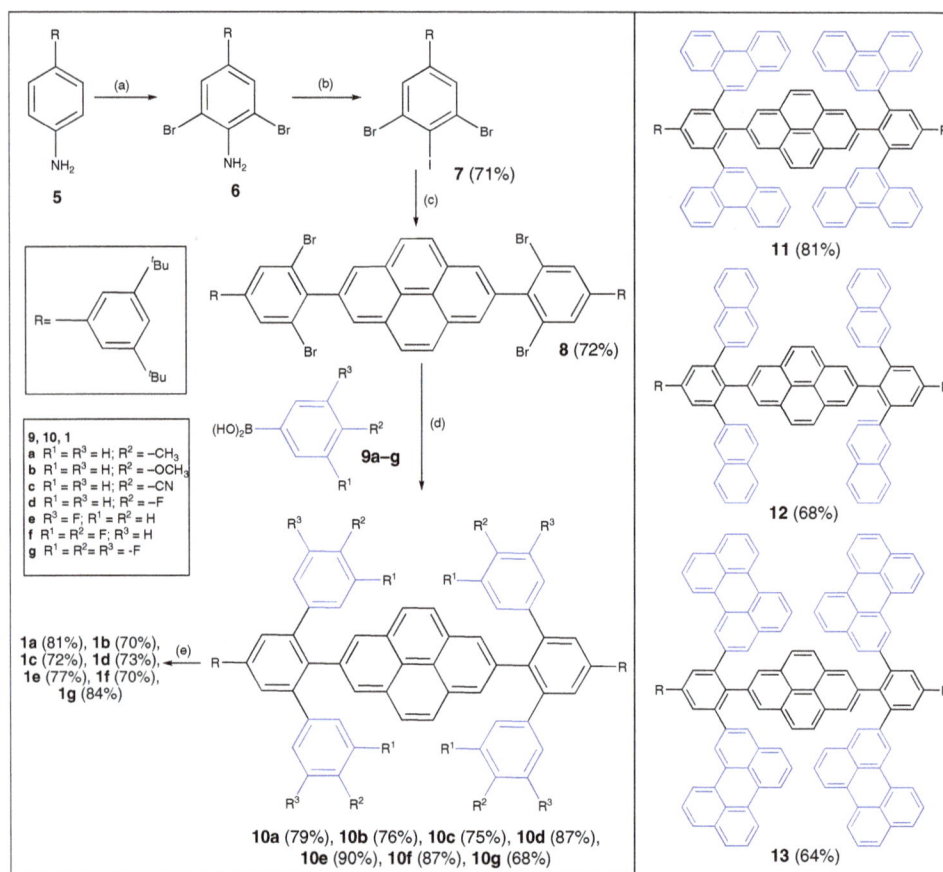

**Figure 2 | Synthesis of the tetracene-dibenzoterrylene hybrids 1a–f and the precursors 11–13 for generation of 2–4.** (a) Br$_2$, DCM/MeOH; (b) NaNO$_2$, AcOH/H$_2$SO$_4$, KI/I$_2$; (c) 2,7-bis(4,4,5,5-tetramethyl-1,3,2-dioxaborolan-2-yl)pyrene, Pd(PPh$_3$)$_4$, Na$_2$CO$_3$, THF/H$_2$O; (d) Pd(PPh$_3$)$_4$, Na$_2$CO$_3$, toluene/ EtOH/H$_2$O; (e) FeCl$_3$, CH$_3$NO$_2$, CH$_2$Cl$_2$, 4 h.

existed in conformations analogous to those of **3** (Supplementary Fig. 69; Fig. 3b displays the most stable conformation of **1a**). In all of these cases, the $C_2$-symmetric *c*AA structures were calculated to be more stable than the other conformers by 3.03–24.65 kJ mol$^{-1}$ (Supplementary Table 1).

We grew single crystals through diffusion of acetone into a solution of **1e** in CHCl$_3$ (Supplementary Table 2). Figure 3d displays the crystal structure of **1e**. The fluorinated outer aromatic rings are contorted, in the same direction, out of the plane defined by the dibenzoterrylene core, resulting in a saddle-like conformation, consistent with the calculated gas-phase structure (*c*AA-**1e**). The degree of twisting ($\theta$), measured as the dihedral angle between the central naphthalene and the fluorinated outer benzene/terminal benzo group, were in the range of 32.2–37.3° (Fig. 3e). Molecules of **1e** were stacked in columns along the *c*-axis (Fig. 3f). Within a columnar stack, two saddle molecules with curvature in the same direction formed a π-dimer with effective π–π surface overlap; this π-dimer then stacked with another π-dimer of opposite molecular orientation.

For compound **2**, we predicted a similar set of five possible conformations (Supplementary Fig. 70), with *c*TT-**2** calculated as the most stable because of additional steric strain between the phenanthrenyl moieties (Fig. 3c). Fortunately, this zero-dipole-moment conformer *c*TT-**2** could be separated through column chromatography. Through slow diffusion of acetone into a solution of *c*TT-**2** in CHCl$_3$, were obtained crystals for X-ray analysis, although they diffracted only weakly (Supplementary Fig. 71). A partial solution revealed that the phenanthrenyl units were twisted with respect to the central naphthalene unit in the *cis* direction. Although frozen in a single conformation as a result of

crystal packing constraints in the solid state, these 2D acenes are highly flexible in solution and change their shape constantly on heating (Supplementary Fig. 72).

**Optical and electrochemical properties of compounds 1–3.** We measured the ultraviolet–visible absorption spectra of compounds **1–3** in CH$_2$Cl$_2$ solution (Fig. 4a–d, Supplementary Fig. 73, Supplementary Table 3). Absorption bands appeared in the ranges 300–400 nm and 400–600 nm with molar absorptivities of >50,000 M$^{-1}$. The substituents and lengths of the acenes appeared to influence the π-stacking of these 2D acenes. For **1**, three peaks in the range 400–600 nm, attributed to 0→0, 0→1 and 0→2 vibrational absorptions, were good handles for observation of its aggregation. With the increasing of concentration, the intensity of the 0→0 absorption decreased; in particular, the 0→0 absorption became a shoulder for the CN-substituted **1c**, even disappearing for the trifluorine-substituted **1g**. These results suggested the permanent formation of H-aggregated π-dimers, as in the crystal structure of **1e** (Fig. 3d) and the molecular modelling of **1g** (Fig. 3g), in solution. On increasing the concentration, the intensity of the 0→0 vibrational absorption decreased further (Supplementary Fig. 74), with the 0→2 vibrational absorptions becoming the main absorption bands in the spectra of spin-coated thin film of **1c–f**, consistent with further H-aggregation of their π-dimers. In the thin films of **1a** and **1b**, in addition to H-aggregation we also observed some J-aggregates, as indicated by the red-shifting of new bands. Such J-aggregates were also found for compounds **2** and **3** with bulky acenes, with red-shifting of the absorption maxima ($\lambda_{max}$).

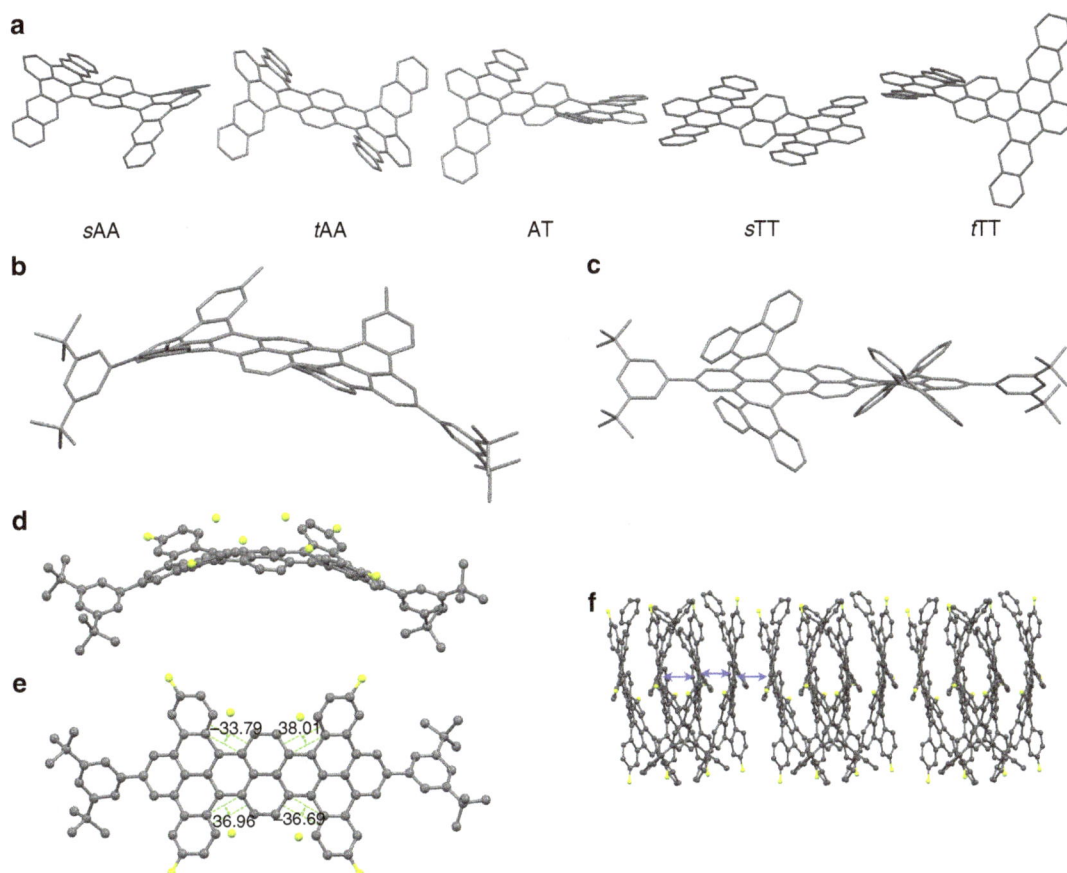

**Figure 3 | Conformational flexibility of the 2D acenes.** (a) Possible molecular conformations of **3** and (**b,c**) the most stable conformations of (**b**) **1a** and (**c**) **2**, calculated at the B3LYP/6-31G (**d**) level of theory. cAA-: anti-folded and anti-folded in the *cis* direction; tAA-: anti-folded and anti-folded in the *trans* direction; AT: anti-folded and twisted; cTT-: twisted and twisted in the *cis* direction; tTT-: twisted and twisted in the *trans* direction. (**d**) Side and (**e**) top views of the π-backbone of **1e** from crystal structure analysis; (**f**) molecular packing of **1e**, viewed along the c-axis of the unit cell; the hydrogen atoms and *tert*-butyl groups have been omitted for clarity; carbon and fluorine atoms are coloured grey and green, respectively.

The photoluminescence features of these compounds in dilute solution ($10^{-3}$ mmol l$^{-1}$) were complementary of their absorption spectra; quantum yields were in the range 0.52–0.89 (Supplementary Table 3). We observed large bathochromic emission features in the emission spectra of these compounds on concentrating or going from solution to the solid state, consistent with strong aggregation. For the tetracene-dibenzoterrylene **1**, we observed a bathochromic emission shift from 108 to 190 nm; the more bulky compounds **2** and **3** featured shifts of only 24 and 71 nm, respectively.

We used cyclic voltammetry (CV) and differential pulse voltammetry to examine the electrochemical behaviour of **1–3** in CH$_2$Cl$_2$ containing 0.1 M TBAPF$_6$ as the supporting electrolyte (Supplementary Fig. 75, Supplementary Table 3). The CV traces of **1a–c**, **1e**, **1f** and **2** each featured two well-defined, reversible oxidation potentials. The cyclic voltammogram of **3** was more remarkable, displaying three oxidation potentials for four electrons at $+0.25$ V (**1e**), $+0.51$ V (**1e**) and $+0.72$ V (**2e**), all as fully reversible waves (Fig. 4e). Compound **1d** exhibited a similar oxidation behaviour, with signals at $+0.42$ V (**1e**), $+0.63$ V (**1e**) and $+0.87$ V (**2e**) (versus Fc/Fc$^+$), implying multi-charge storage ability[47] for both **1d** and **3**. Compared with **1a**, the electron-donating MeO groups decreased the oxidation potentials by 70 mV, while the electron-withdrawing groups increase the oxidation potentials more evidently; for example, CN groups shifted the oxidation potentials by $\sim$350 mV. The oxidation peaks shifted to higher potentials on increasing the number of fluorine atoms (50–70, 250 and 500 mV for one, two and three fluorine substituents, respectively). Extension of the conjugation of the acene units also decreased the oxidation potentials; for example, to 0.25 V for **3** with its six conjugated benzene rings and to 0.21 V for **2** with its eight conjugated benzene rings. Consistent with these findings, **2** and **3** appear to be susceptible to slow oxidative degradation in solution; no such degradation appeared in the solutions of the other compounds. All of these compounds exhibited two reduction potentials, with the effects of the substituents on the reduction potentials occurring in the same direction as those on the oxidation potentials, but to a lesser extent. Accordingly, the band gaps of these species were influenced by the substituents and the conjugation length, with electron acceptors decreasing the HOMO energy level and, thereby, increasing the band gap (see Fig. 4f, especially for the trifluorinated derivative).

We performed the DFT calculations at the B3LYP/6-31G(d) level to examine the electronic structures of the dibenzoterrylene-acene hybrids. The calculated frontier orbitals indicated that they maintained the characteristics of their individual components. The HOMO and LUMO orbitals of **1a** are mainly localized in the central pyrene unit (Supplementary Table 4). We observed nodal planes perpendicular to the molecular skeleton in the HOMOs and LUMOs of these compounds, with the electron densities distributed symmetrically on the two sides of the nodal plane. The HOMO-1 and LUMO+1 orbitals of **1a** are comparable with the HOMO-2 and LUMO+2 orbitals, respectively, of

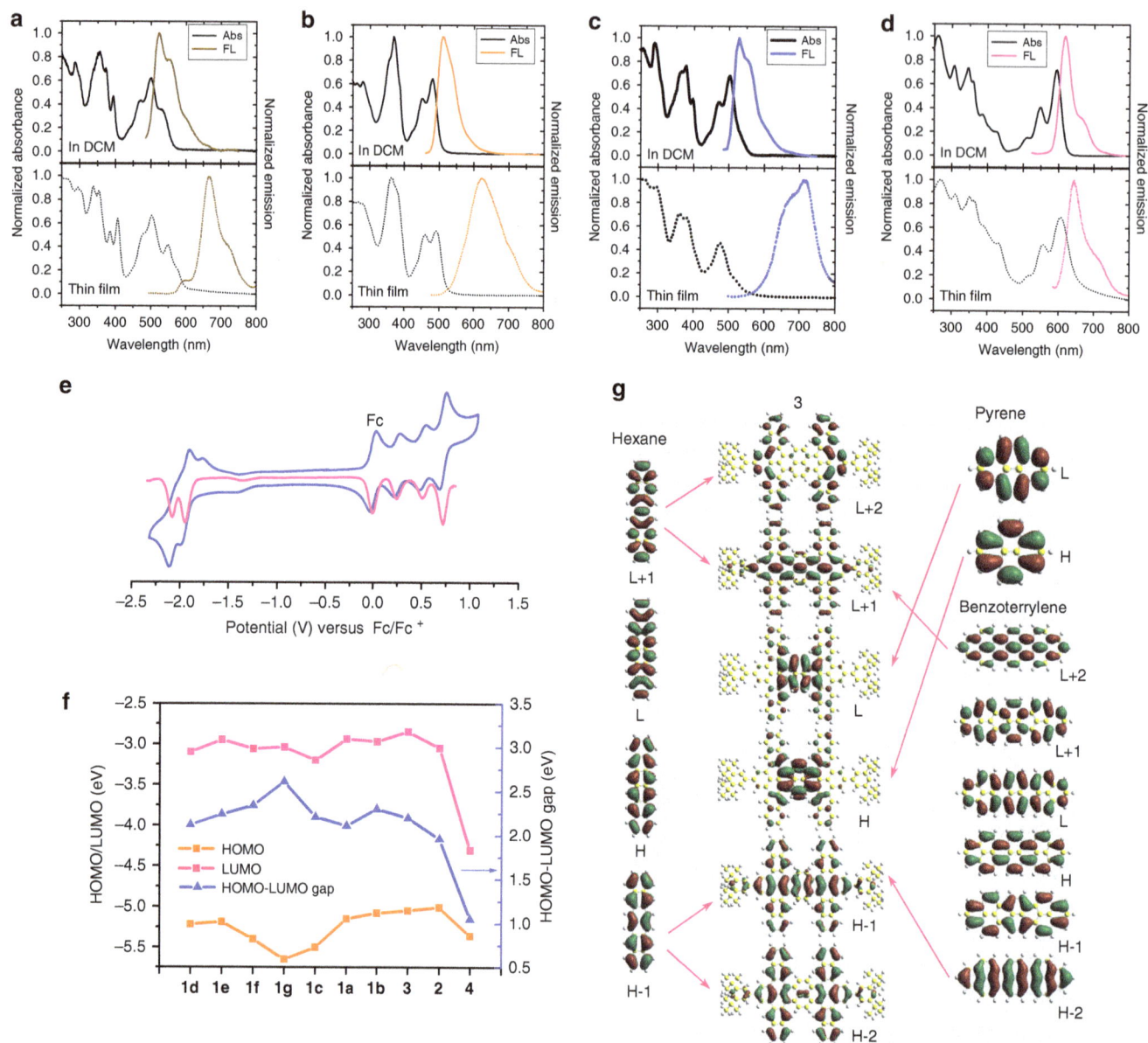

**Figure 4 | Optical, electrochemical behaviour and orbital analysis of compounds 1–3.** Absorbance (black) and emission (coloured) spectra of (**a**) **1a**, (**b**) **1g**, (**c**) **1c** and (**d**) **2** recorded in $CH_2Cl_2$ solution (solid lines) and in the solid state (spin-cast from $CH_2Cl_2$ solutions; dotted lines). (**e**) CV (blue) and DPV(red) traces of **3** in $CH_2Cl_2$ (0.1 mM) at room temperature. Scan rate: 100 mV s$^{-1}$; working electrode: glassy carbon; reference electrode: Ag wire; electrolyte: TBAPF$_6$. Fc/Fc$^+$: ferrocene/ferrocenium. (**f**) HOMO and LUMO energies and band gaps of **1–4**. (**g**) Frontier orbitals analysis of **3** calculated at the B3LYP/6-31G (**d**) level.

dibenzoterrylene. The other compounds had similarly appearing orbitals, except for **1f**, where the strongly electron-withdrawing CN group affected the polarization of its orbitals (Supplementary Table 4). Interestingly, extension of the acenes vertical to the benzoterrylene, as in **3**, led to more independent characteristics for the pyrene, benzoterrylene and acene subcomponents (Fig. 4g). Increasing the steric constraint of these acenes led to compounds exhibiting the properties of the central benzoterrylene (for example, for **2** in Supplementary Fig. 76). Time-dependent DFT calculations revealed that the absorption behaviour of these compounds was the result of a mixture of pyrene-, dibenzoterrylne- and acene-like orbitals (Supplementary Fig. 77, Supplementary Tables 5–11). For **1**, the main band in the range 400–600 nm, corresponding to the HOMO–LUMO transition, according to the calculation, was caused by a transition from a pyrene-like HOMO to a pyrene-like LUMO. The second band, near 350 nm, due to the HOMO − 1 →LUMO + 1 transition, was

related to the transition between the dibenzoterrylene-like HOMO-2 and LUMO + 2. The small sharp peak near 400 nm arose from the transition from a pyrene-like HOMO to a dibenzoterrylene-like LUMO + 2. We observed similar transition behaviour for **3**, although the band in the 300–400 nm region was red-shifted by ∼ 50 nm. Consistent with the molecular orbital profiles, **2** displayed a dibenzoterrylene-like HOMO→LUMO transition for the main band near 600 nm, with a transition between the dibenzoterrylene-like HOMO-2 and LUMO + 2 for the band near 350 nm. We also observed dibenzoterrylene-like HOMO-2→LUMO and HOMO→LUMO + 2 transitions for the weak band at 344 nm.

**Optical and magnetic properties of open-shell compound 4.** Dark-green compound **4** provides an absorption spectrum (Fig. 5b) displaying bands with maxima at 834, 675 and 585 nm

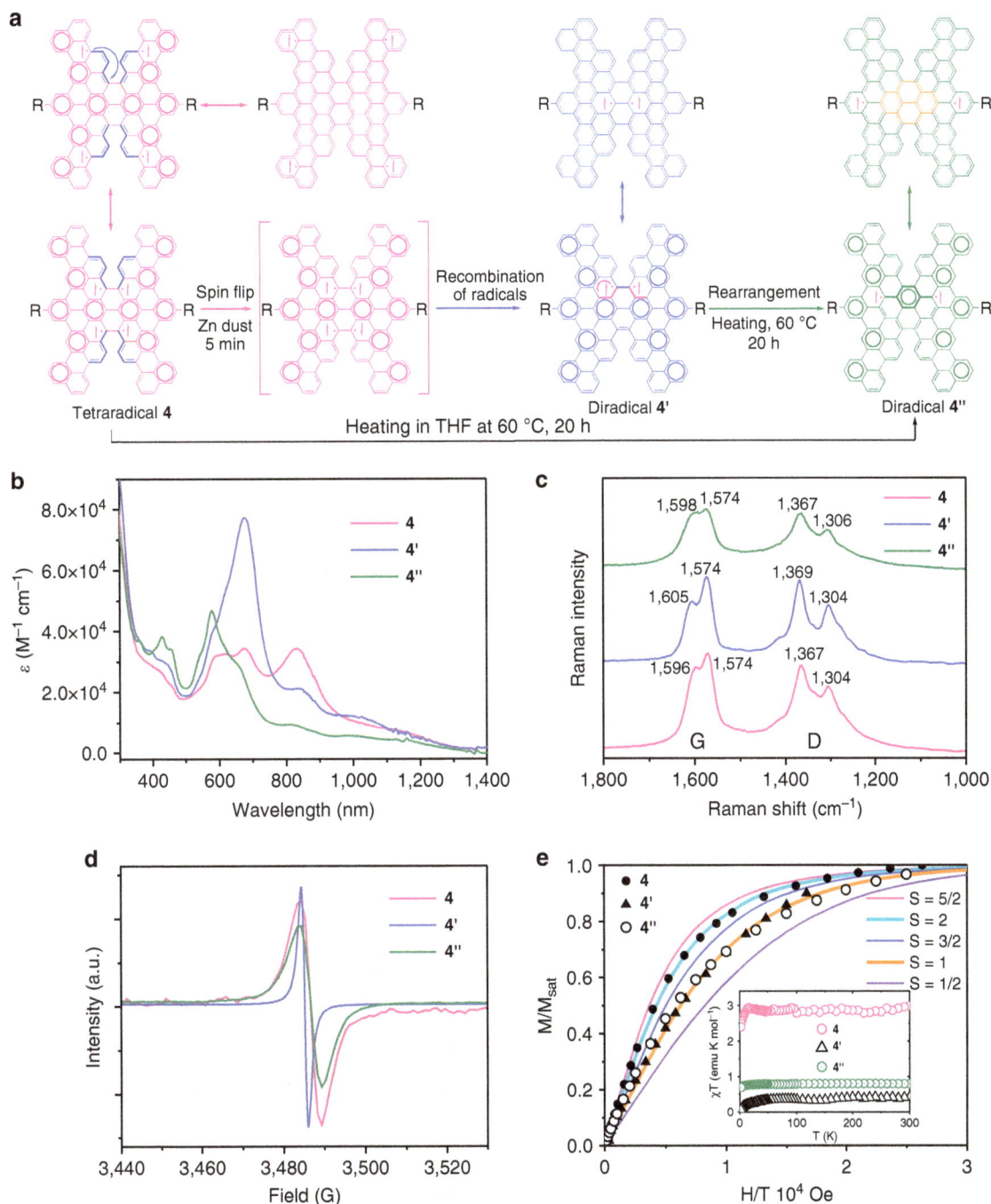

**Figure 5 | Spectroscopic characterization of compound 4.** (**a**) Transformation of tetraradical **4** to diradicals **4'** or **4''** mediated with spin catalysis by Zinc or trace radicals in THF, and the resonance structures of **4**, **4'** and **4''**. (**b**) UV–vis–near-infrared spectra of **4**, **4'** and **4''** in $CH_2Cl_2$. (**c**) Raman spectra of solid **4**, **4'** and **4''**. (**d**) ESR spectra of chloroform solutions of **4**, **4'** and **4''** recorded at room temperature. (**e**) The magnetization of **4**, **4'** and **4''** are plotted as M/$M_{sat}$ versus H/T, with Brillouin functions for S = 5/2, 2, 3/2,1 and 1/2. The fitting parameters for **4**, **4'** and **4''** are S = 2.08 (1.9 K), 1.02 (3 K), 1.14 (1.9 K) and the corresponding $M_{sat}$ = 0.8, 0.37, 0.88 μB, respectively. Inset plots: the magnetic susceptibility of **4** (2,000 Oe), **4'** (10,000 Oe) and **4''** (2,000 Oe) are plotted as χT versus T.

with a tail into the near-infrared region (ca. 1,200 nm). We calculated an optically measured band gap of ~1.03 eV. The CV trace of **4** (Supplementary Fig. 75) in 1,2-dichlorobenzene (0.1 M tetrabutylammonium hexafluorophosphate) displayed a one-step irreversible oxidation and a one-step reduction. The HOMO and LUMO energies, calculated through CV, were −5.31 and −4.26 eV, respectively, corresponding to a band gap of 1.05 eV, very close to that determined *in silico* (1.1 eV). In addition, we observed almost no fluorescence from **4** in solution and in the solid state.

Raman spectroscopy is a very useful tool for characterizing benzene-type aromatic radicals[48], because characteristic benzene vibrational Raman bands exist near 1,600 $cm^{-1}$ that are very sensitive to the electronic configuration within the six-membered benzene ring, either 'benzoquinoidal' or 'benzoaromatic.' The Raman spectrum of **4** featured two sets of peaks, much like the D and G bands of graphene (Fig. 5c). The peak near 1,367 $cm^{-1}$ corresponds to the breathing vibrations of $sp^2$-hybridized carbon domains in aromatic rings (that is, the D band); we assign the peak near 1,596 $cm^{-1}$ to the first-order scattering of the $E_{2g}$

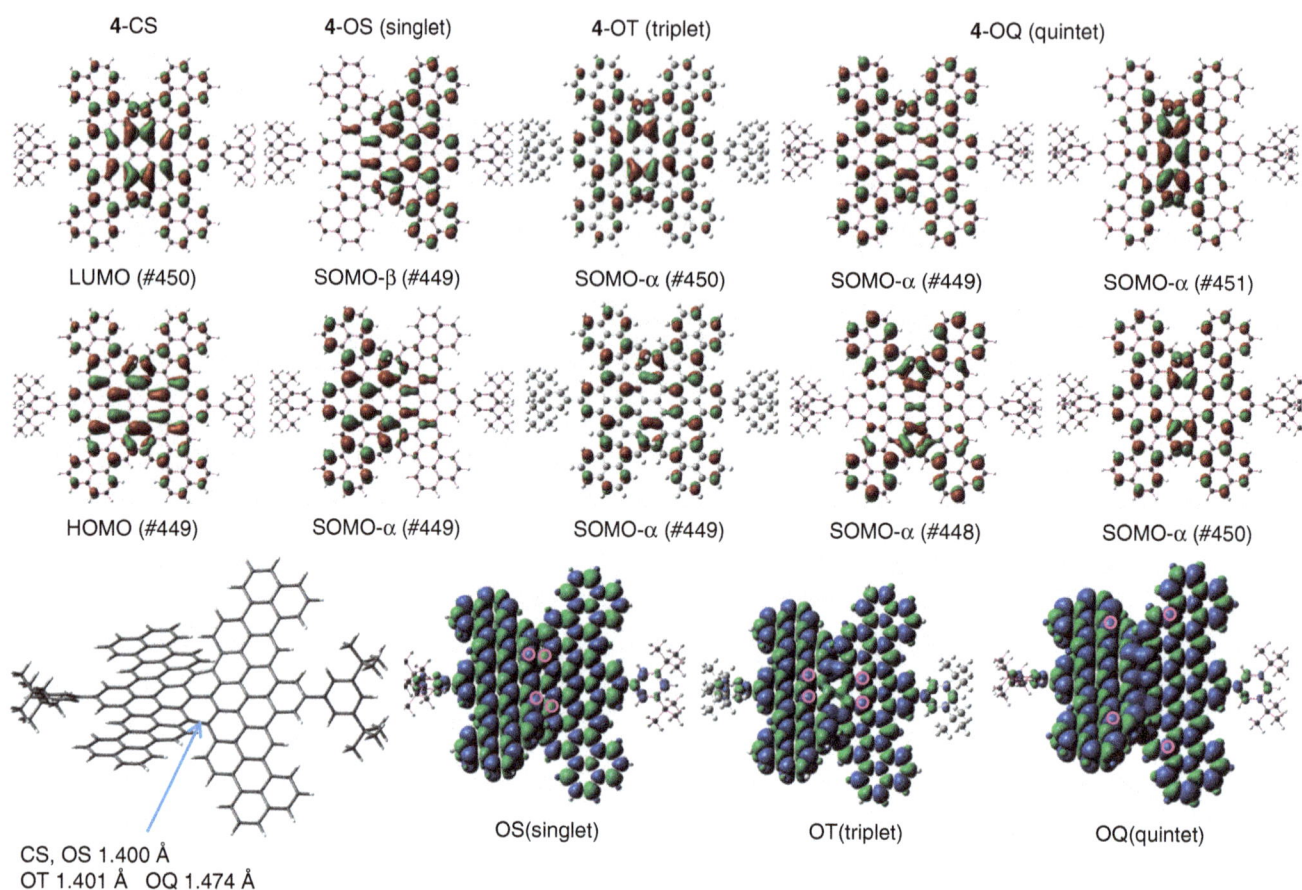

**Figure 6 | Orbitals and spin-density distributions for closed-shell and open-shell compound 4.** Calculated (UCAM-B3LYP) HOMOs and LUMOs for **4**-CS, SOMOs and spin-density distributions for **4** in the singlet (**4**-OS, S = 0), triplet (**4**-OT, S = 1) and quintet state (**4**-OQ, S = 2). Blue and green surfaces represent α- and β-spin densities, respectively. The red marks indicate the atoms exhibiting highest spin-density. The central bond connecting the two pentacene-annulated-tribenzo-octacene units in quintet-state pocesses the single-bond charateristic, which indicates the less conjugation between the two units than other states.

mode observed for in-phase stretching vibration of $sp^2$-hybridized carbon domains in aromatic rings (that is, the G band)[49]. The downfield-shifting of the D and G bands (from 1,367 to 1,304 cm$^{-1}$ and from 1,596 to 1,574 cm$^{-1}$, respectively) indicated the coexistence of 'benzoquinoidal' and 'benzoaromatic' benzene rings.

The NMR spectrum of **4** in $C_2D_2Cl_4$ did not feature any signals at room temperature, nor after cooling to $-50\,^\circ$C, suggesting the presence of a considerable paramagnetic species. In addition, the solution of **4** provided a featureless broad ESR signal (Fig. 5d), resulting from the long-distance spin – spin dipole interaction within the molecules and the extended spin-delocalization[48,50,51]. These data indicate that **4** exists as an open-shell multi-radical in the ground state.

We performed SQUID measurements for **4** in the powder form as a function of magnetic field ($H = 0 - 5 \times 10^4$ Oe and $T = 1.9$ K) and temperature ($T = 2 - 300$ K at $H = 2,000$ Oe). The value of S = 2.08, determined from the curvature of the Brillouin plots, indicate the quintet (S = 2) ground state for tetraradical **4** (Fig. 5d). The inset of Fig. 5d presents a plot of $\chi_M T(T)$ for **4**. The constant value of $\chi T$ (2.84 emu K mol$^{-1}$), as evidenced by the flatness of the $\chi T$ versus $T$ plots in the $T = 30 - 300$ K range, indicates that there is no significant change in the thermal population of spin states up to 300 K. After removing the temperature independent contribution, the numerical fit of the data by the Currie-Weiss model give Currie constant of C = 2.84 emu K mol$^{-1}$ and Weiss constant $\theta_W$ of $-0.38$ K (Supplementary Fig. 78), which indicated weak intermolecular antiferromagnetic interaction below 20 K.

Interestingly, solid **4** displayed high stability, and there was no obvious decomposition when the solid was stored under ambient air and light conditions for months. However, mixing the tetraradical **4** with Zinc dust can generate another radical **4'**, and heating the solution of **4** or **4'** in THF at 60 °C for about 20 h will transform both of them to another species **4''** completely. **4'** and **4''** showed no molecular weight change, while the strongest absorption bands of **4'** and **4''** were shifted to 678 and 575 nm, respectively, with the absorption band at 834 nm became weak (Fig. 5b and Supplementary Fig. 81). Similar Raman spectra (Fig. 5c) and magnetic field dependent magnetic signal changes (Fig. 5e) indicate unequivocally that both of **4'** and **4''** are triplet biradicals at ground state (Supplementary Figs 79 and 80). The values of $\chi T \approx 1$ emu K mol$^{-1}$ were found for **4'** and **4''**, after values of $\chi T$ (0.77, 0.36 emu K mol$^{-1}$) are corrected by spin concentration ($M_{sat} = 0.88$, 0.37 μB). The transformation from tetraradical to diradicals is probably due to the fact that one pair of electrons in the tetraradical formed one covalent bond by Zinc surface catalysed spin flip[52,53] (Fig. 5a), the left two electrons delocalized on the two pentacene-annulated-tribenzo-octacene units, respectively, with parallel spin electron patterns at ground state. The obtained diradical **4'** can be rearranged to a more stable diradical **4''** on heating. Transformation directly from **4** to **4''** in hot THF processed through spin catalysis[52,53] by the trace radicals in THF, followed by the thermal rearrangement, without the observation of the intermediates. Sharpening ESR signal for **4'** (Fig. 5d), indicates that the spin – spin dipole interaction distance within the molecules of **4'** is shorter than that of **4** and **4''**. The transformation processes and the possible resonance structures of **4**, **4'** and **4''** are summarized in Fig. 5a, with the supports from the following DFT optimization of the tetraradical and diradicals. Tetraradical **4** can be considered as four phenalenyl radicals annulated by twisted H-shaped acene; intermediate diradical **4'** featured two close-contacted radical pairs like two over-lapped phenalenyl radicals; **4''** can be viewed as two phenalenyl radicals fused by central pyrene, with benzene rings surrounded. The detailed spin catalysis mediated transformation mechanism needs further investigation.

Although 1D PAHs (n-acenes) featured the antiferromagnetic ground state due to the zigzag-shaped boundaries, which cause π-electrons to localize and form spin orders at the edges, the longer acene ground states are polyradical in nature[54,55]. 2D PAHs (periacenes and circumacenes) develops a strong multiradical character with increasing zigzag chain length[56,57]. Our experimental results indicated tetraradical character for the twisted 2D molecule **4**. We performed DFT calculations at the UCAM-B3LYP level of theory to investigate the ground state of the tetraradical **4** (Fig. 6). Because of steric effects, the molecule adopted a twisted planar structure. The structure **4**-CS features extended π-electron delocalization from the pyrene core to the four annulated perylene units, with the central pyrene unit possessing the largest HOMO and LUMO coefficients (Fig. 6), suggesting that it would be the most reactive site. For the broken-symmetry singlet state **4**-OS, we observed disjointed singly occupied molecular orbitals (SOMOs), SOMO-α and SOMO-β, with orbital coefficients mainly localized at the terminal pentacene-annulated-tribenzo-octacene units. The spin-density distribution reveals delocalization with the central pyrene units having the largest density, especially the K-region carbon atoms (also the $5'$, $7'$ positions of dibenzopentacene) (Supplementary Fig. 82, Supplementary Tables 12–14); this situation is different from that found in bisphenalenyls, where the spin-density is delocalized mainly at the terminal phenalenyl units[25], with antiparallel spin electron patterns typical for this broken-symmetry singlet state. This S = 0 state can be considered as the **DBP-4** form in Fig. 1c, which also corresponds to an S = 1 state when the spin electrons parallel to each other (diradical **4'** in Fig. 5a). For the high-spin quintet state (**4**-OQ, S = 2), we observed a large spin-density at the 5,10 positions of the octacene core—positions that are blocked by the annulated pentacene, which is the **DBO-4** form (Fig. 1c, also the tetraradical **4**). However, for the S = 1 state (**4**-OT, also the diradical **4''** in Fig. 5a), the spins are delocalized on the pentacene-annulated-tribenzo-octacene units, with more distribution on the 7, 8 positions of the octacene core, well-protected by surrounding benzene rings. From the SOMO-α and SOMO-β profiles and the spin-distributions of the singlet **4**-OS and high-spin triplet **4**-OT and quintet **4**-OQ, we ascribe the good chemical stability of the tetraradical toward oxidation and dimerization/oligomerization to its thermodynamic stabilization[33], resulting from both delocalization and kinetic blocking through benzoannulation.

## Discussion

We have developed an efficient method for the preparation of a series of stable 2D twisted dibenzoterrylene-acenes. We have investigated their electronic structures and geometries in the ground state using various experimental techniques, assisted by DFT calculations. The photophysical, electrochemical, and magnetic properties of these compounds were dependent on the length of the acene. The ground-state structures in this series of molecules were tunable, with **1–3** being closed-shell hydrocarbons and **4** being an open-shell singlet tetraradical. Compound **4** is the first stable twisted 2D hydrocarbon tetraradical obtained

by hybridizing dibenzoterrylene with acene, which can be transferred to more thermodynamic stable triplet diradical. The high stability of the tetraradical and diradical arose from thermodynamic stabilization, due to delocalization and kinetic blocking through benzoanullation. We believe that 2D twisted conjugated molecules with various width and edge structures could also be synthesized in a similar approach. With the installation of different functional groups such as heterocycles on the periphery of such 2D twisted conjugated molecules, specific functionalities on them could be obtained. Combination this approach with polymerization could give great promise for the synthesis of chemically precise, multiple-dimensional polymers. Moreover, the unique optical, electronic, and magnetic properties of these extended 2D acenes suggest that they might be promising candidates for use in nonlinear optics, field effect transistors, and organic spintronics.

## Methods

**Materials.** Most of the chemical reagents were purchased from Alfa Aesar or Aldrich Chemicals and were utilized as received unless indicated otherwise. All solvents were purified using standard procedures. Column chromatography was performed on silica gel (size 200–300 mesh).

**Synthesis procedures of compounds 1–4.** For details of the synthetic procedures, see Supplementary Methods. For NMR and high-resolution mass spectra of compounds in this manuscript, see Supplementary Figs 1–68.

**Sample characterization.** $^1H$ and $^{13}C$ NMR spectra were recorded on a Bruker AVANCE 400 or Bruker AVANCE III 500WB instrument, at a constant temperature of 25 °C. Chemical shifts are reported in parts per million from low-to-high field and referenced to TMS. Matrix-assisted laser desorption/ionization Fourier transform ion cyclotron resonance (MALDI-FT-ICR-MS) MS were performed on a Bruker Solarix 9.4T FT-ICR-MS mass spectrometer. EI mass spectrometric measurements were performed on a SHIMADZU GCMS-QP2010 puls Spectrometer. Elemental analyses were recorded on a Carlo-Erba-1106 instrument. Electronic absorption spectra were measured on a JASCO V-579 spectrophotometer. Raman spectra were taken on a NT-MDT NTEGRA spectra Raman spectroscopy SPM system. CV experiments were performed using an electrochemical analyser with a conventional three-electrode configuration, glassy carbon electrode as a working electrode and Ag/AgCl as a reference electrode. All experiments were performed in $CH_2Cl_2$ with 0.1 M of $nBu_4NPF_6$ as a supporting electrolyte, the ferrocene/ferrocenium ion (Fc/Fc$^+$) as inter-reference. The $E_{1/2}$ values were determined as $1/2(E_{pa} + E_{pc})$, where $E_{pa}$ and $E_{pc}$ are the anodic and cathodic peak potentials, respectively.

**X-ray diffraction data analysis.** Crystals of **1e** and **2** were grown by slow diffusing acetone to the solutions in CHCl$_3$ solution. Single crystal X-ray diffraction data were collected on a Agilent SuperNova (Dual, Cu at zero, Atlas) diffractometer with Cu Kα radiation ($\lambda = 1.54184$ Å) from micro-focus sealed X-ray tube. Intensities were corrected for absorption effects using the multi-scan technique[58]. The structures were solved by direct methods and refined by a full matrix least squares technique based on $F^2$ using SHELXT program[59]. The refinement details: Refinement of $F^2$ against ALL reflections. The weighted R-factor wR and goodness of fit S are based on $F^2$, conventional R-factors R are based on F, with F set to zero for negative $F^2$. The threshold expression of $F^2 > 2sigma$ ($F^2$) is used only for calculating R-factors (gt) and so on, and is not relevant to the choice of reflections for refinement. R-factors based on $F^2$ are statistically about twice as large as those based on F, and R- factors based on ALL data will be even larger. For X-ray data see Supplementary Data 1 and 2.

Attempts to refine peaks of residual electron density as solvents led to 12 CHCl$_3$ and 6 acetone and some additional peaks of residual electron density which can't be refined successfully. The data were corrected for disordered electron density through use of the SQUEEZE procedure[60] as implemented in PLATON[61]. A total solvent-accessible void volume of 2,179.1 A$^3$ with a total electron count of 584.3 was found in the unit cell in four voids.

The structure solution of crystal **2** (Cell: 19.2306(4) Å, 19.4633(5) Å, 58.6850(14) Å, 90°, orthorhombic P c c n, Z = 8) shows two independent half molecules, one of which is disordered with its symmetry equivalent. It is possible that the cell chosen for the current refinement in represents a subcell and that reflections of a larger supercell were too weak to be observed. The current result represents a preliminary connectivity only. We have grown the crystals of this structure several times, the crystals are thin plate, and we tried to collect better data, however, the data reported here is the best we can obtain. The six-membered rings in the phenanthrenyl groups and terminal benzenes with t-butyl groups were flattened with AFIX 66 and DFIX. As agreed by the reviewer, although the data is poor, it is perfectly fine for the assumptions that there is a twist associated with the molecule as predicted by the DFT calculation. We present with caution some geometrical parameters: phenanthrenyl units twisted with the central naphthene unit in the *cis* direction.

ESR spectra were obtained with a Bruker E500-10/12 spectrometer. No additional hyperfine coupling was detected at 0.1 G, thus experiments were performed at 1.0 or 2 G for improved signal intensities. The sweep width was 83.89 s, time constant was 40.96 ms and the microwave power was 10.1 mW (corresponding to attenuation = 13 dB, which was sufficiently high to avoid power saturation).

Magnetic measurements were performed on a Quantum Design MPMS-XL5 SQUID magnetometer. Small amounts of sample material (between 30 and 40 mg) were put into KLF containers and brought into measuring position using a straw. The variation of the magnetization as a function of temperature at fixed external magnetic field ($T = 2$–$300$ K at $H = 2,000$ Oe) and as a function of the external field at fixed temperature ($H = 0$–$5 \times 10^4$ Oe and $T = 1.9$ K) was studied. The powder of **4** was sealed in a plastic tube. The signal of sample holder and plastic tube was deducted by measuring the sample holder and plastic tube under same conditions. A correction for the diamagnetism of the sample and the sample container was applied before calculating the susceptibilities from the magnetization data. For all other samples of solid di- and tetraradicals, the correction for diamagnetism was based on high-temperature extrapolation of the $\chi$ versus $1/T$ plots, that is, a suitable numerical factor ($M_{\mathrm{dia}}$) was added to the magnetization ($M$), until the $\chi T$ versus $T$ plot becomes flat in the high-temperature range.

**Theoretical calculations of target compounds 1-4.** For compound **1**, **2** and **3**, DFT calculations were performed using the Gaussian 03 program[62], Geometries were optimized in the gas-phase using the B3LYP functional and 6–31 g (d) basis set on all atoms. Theoretical calculations of compound **4** were carried out by using the Gaussian 09 suite of programs[63]. The initial geometry optimization of **4** was performed with the UCAM-B3LYP level of theory on the singlet state and the Handy and co-workers'[66] long-range corrected version of B3LYP 6–31G*, and all electron basis sets were used for all atoms[64–66]. The resulting DFT solution (singlet 'closed-shell': zero-spin-density on all atoms) was further tested for its stability with the STABLE = OPT keyword[67]. A spin symmetry broken DFT solution was found with lower energy. Then the Guess = Read keyword was used to perform the optimization at the UCAM-B3LYP level (singlet open-shell). For DFT data see Supplementary Data 3–6.

# References

1. Fujii, S. & Enoki, T. Nanographene and graphene edges: electronic structure and nanofabrication. *Acc. Chem. Res.* **46**, 2202–2210 (2013).
2. Yan, X., Cui, X. & Li, L.-S. Synthesis of large, stable colloidal graphene quantum dots with tunable size. *J. Am. Chem. Soc.* **132**, 5944–5945 (2010).
3. Chen, L., Hernandez, Y., Feng, X. & Muellen, K. From nanographene and graphene nanoribbons to graphene sheets: chemical synthesis. *Angew. Chem. Int. Ed.* **51**, 7640–7654 (2012).
4. Kawasumi, K. *et al.* A grossly warped nanographene and the consequences of multiple odd-membered-ring defects. *Nat. Chem.* **5**, 739–744 (2013).
5. Zhang, Q. *et al.* Facile bottom-up synthesis of coronene-based 3-fold symmetrical and highly substituted nanographenes from simple aromatics. *J. Am. Chem. Soc.* **136**, 5057–5064 (2014).
6. Narita, A. *et al.* Synthesis of structurally well-defined and liquid-phase-processable graphene nanoribbons. *Nat. Chem.* **6**, 126–132 (2014).
7. Cai, J. *et al.* Atomically precise bottom-up fabrication of graphene nanoribbons. *Nature* **466**, 470–473 (2010).
8. Treier, M. *et al.* Surface-assisted cyclodehydrogenation provides a synthetic route towards easily processable and chemically tailored nanographenes. *Nat. Chem.* **3**, 61–67 (2011).
9. Vo, T. H. *et al.* Large-scale solution synthesis of narrow graphene nanoribbons. *Nat. Commun.* **5**, 3189 (2014).
10. Ozaki, K. *et al.* One-shot K-region-selective annulative pi-extension for nanographene synthesis and functionalization. *Nat. Commun.* **6**, 6251 (2015).
11. Schuler, B. *et al.* From perylene to a 22-ring aromatic hydrocarbon in one-pot. *Angew. Chem. Int. Ed.* **53**, 9004–9006 (2014).
12. Kastler, M. *et al.* From armchair to zigzag peripheries in nanographenes. *J. Am. Chem. Soc.* **128**, 9526–9534 (2006).
13. Wang, Z. H. *et al.* Graphitic molecules with partial 'Zig/Zag' periphery. *J. Am. Chem. Soc.* **126**, 7794–7795 (2004).
14. Ritter, K. A. & Lyding, J. W. The influence of edge structure on the electronic properties of graphene quantum dots and nanoribbons. *Nat. Mater.* **8**, 235–242 (2009).
15. Zhang, L. *et al.* Unconventional, chemically stable, and soluble two-dimensional angular polycyclic aromatic hydrocarbons: from molecular design to device applications. *Acc. Chem. Res.* **48**, 500–509 (2015).
16. Anthony, J. E. The larger acenes: versatile organic semiconductors. *Angew. Chem. Int. Ed.* **47**, 452–483 (2008).
17. Chun, D., Cheng, Y. & Wudl, F. The most stable and fully characterized functionalized heptacene. *Angew. Chem. Int. Ed.* **47**, 8380–8385 (2008).
18. Purushothaman, B. *et al.* Synthesis and structural characterization of crystalline nonacenes. *Angew. Chem. Int. Ed.* **50**, 7013–7017 (2011).
19. Watanabe, M. *et al.* The synthesis, crystal structure and charge-transport properties of hexacene. *Nat. Chem.* **4**, 574–578 (2012).
20. Lambert, C. Towards polycyclic aromatic hydrocarbons with a singlet open-shell ground state. *Angew. Chem. Int. Ed.* **50**, 1756–1758 (2011).
21. Morita, Y., Suzuki, S., Sato, K. & Takui, T. Synthetic organic spin chemistry for structurally well-defined open-shell graphene fragments. *Nat. Chem.* **3**, 197–204 (2011).
22. Sun, Z., Ye, Q., Chi, C. & Wu, J. Low band gap polycyclic hydrocarbons: from closed-shell near infrared dyes and semiconductors to open-shell radicals. *Chem. Soc. Rev.* **41**, 7857–7889 (2012).
23. Abe, M. Diradicals. *Chem. Rev.* **113**, 7011–7088 (2013).
24. Kamada, K. *et al.* Strong two-photon absorption of singlet diradical hydrocarbons. *Angew. Chem. Int. Ed.* **46**, 3544–3546 (2007).
25. Kubo, T. *et al.* Synthesis, intermolecular interaction, and semiconductive behavior of a delocalized singlet biradical hydrocarbon. *Angew. Chem. Int. Ed.* **44**, 6564–6568 (2005).
26. Morita, Y. *et al.* Organic tailored batteries materials using stable open-shell molecules with degenerate frontier orbitals. *Nat. Mater.* **10**, 947–951 (2011).
27. Anthony, J. E. Functionalized acenes and heteroacenes for organic electronics. *Chem. Rev.* **106**, 5028–5048 (2006).
28. Zade, S. S. & Bendikov, M. Heptacene and beyond: the longest characterized acenes. *Angew. Chem. Int. Ed.* **49**, 4012–4015 (2010).
29. Einholz, R. & Bettinger, H. F. Heptacene: increased persistence of a $4n + 2$ $\pi$-electron polycyclic aromatic hydrocarbon by oxidation to the $4n$ $\pi$-electron dication. *Angew. Chem. Int. Ed.* **52**, 9818–9820 (2013).
30. Zade, S. S. *et al.* Products and Mechanism of acene dimerization. a computational study. *J. Am. Chem. Soc.* **133**, 10803–10816 (2011).
31. Kaur, I. *et al.* Design, synthesis, and characterization of a persistent nonacene derivative. *J. Am. Chem. Soc.* **132**, 1261–1263 (2010).
32. Pérez, D., Peña, D. & Guitián, E. Aryne cycloaddition reactions in the synthesis of large polycyclic aromatic compounds. *Eur. J. Org. Chem.* **2013**, 5981–6013 (2013).
33. Rath, H. *et al.* A stable organic radical delocalized on a highly twisted pi system formed upon palladium metalation of a mobius aromatic hexaphyrin. *Angew. Chem. Int. Ed.* **49**, 1489–1491 (2010).
34. Wu, J. S. *et al.* Controlled self-assembly of hexa-peri-hexabenzocoronenes in solution. *J. Am. Chem. Soc.* **126**, 11311–11321 (2004).
35. Pisula, W., Feng, X. & Muellen, K. Tuning the columnar organization of discotic polycyclic aromatic hydrocarbons. *Adv. Mater.* **22**, 3634–3649 (2010).
36. Chen, L. *et al.* Hexathienocoronenes: synthesis and self-organization. *J. Am. Chem. Soc.* **134**, 17869–17872 (2012).
37. Diez-Perez, I. *et al.* Gate-controlled electron transport in coronenes as a bottom-up approach towards graphene transistors. *Nat. Commun.* **1**, 31 (2010).
38. Yamamoto, Y. *et al.* Photoconductive coaxial nanotubes of molecularly connected electron donor and acceptor layers. *Science* **314**, 1761–1764 (2006).
39. Feng, X. *et al.* Controlled self-assembly of C-3-symmetric hexa-peri-hexabenzocoronenes with alternating hydrophilic and hydrophobic substituents in solution, in the bulk, and on a surface. *J. Am. Chem. Soc.* **131**, 4439–4448 (2009).
40. Hill, J. P. *et al.* Self-assembled hexa-peri-hexabenzocoronene graphitic nanotube. *Science* **304**, 1481–1483 (2004).
41. Kastler, M. *et al.* Influence of alkyl substituents on the solution- and surface-organization of hexa-peri-hexabenzocoronenes. *J. Am. Chem. Soc.* **127**, 4286–4296 (2005).
42. Xiao, S. *et al.* Controlled doping in thin-film transistors of large contorted aromatic compounds. *Angew. Chem. Int. Ed.* **52**, 4558–4562 (2013).
43. Cohen, Y. S. *et al.* Enforced one-dimensional photoconductivity in core-cladding hexabenzocoronenes. *Nano Lett.* **6**, 2838–2841 (2006).
44. Kang, S. J. *et al.* A supramolecular complex in small-molecule solar cells based on contorted aromatic molecules. *Angew. Chem. Int. Ed.* **51**, 8594–8597 (2012).
45. Marsden, J. A. & Haley, M. M. Carbon networks based on dehydrobenzoannulenes. 5. extension of two-dimensional conjugation in graphdiyne nanoarchitectures. *J. Org. Chem.* **70**, 10213–10226 (2005).
46. Luo, J., Xu, X., Mao, R. & Miao, Q. Curved polycyclic aromatic molecules that are pi-isoelectronic to hexabenzocoronene. *J. Am. Chem. Soc.* **134**, 13796–13803 (2012).
47. Echegoyen, L. & Herranz, M. A. Fullerene Electrochemistry. *Fullerenes: From Synthesis To Optoelectronic Properties* **4**, 267–293 (2002).
48. Li, Y. *et al.* Kinetically blocked stable heptazethrene and octazethrene: closed-shell or open-shell in the ground state? *J. Am. Chem. Soc.* **134**, 14913–14922 (2012).

49. Tuinstra, F. & Koenig, J. L. Raman spectrum of graphite. *J. Chem. Phys.* **53**, 1126 (1970).

50. Shimizu, A. *et al.* Alternating covalent bonding interactions in a one-dimensional chain of a phenalenyl-based singlet biradical molecule having Kekulé structures. *J. Am. Chem. Soc.* **132**, 14421–14428 (2010).

51. Sun, Z., Huang, K. & Wu, J. Soluble and stable heptazethrenebis(dicarboximide) with a singlet open-shell ground state. *J. Am. Chem. Soc.* **133**, 11896–11899 (2011).

52. Buchachenko, A. L. & Berdinsky, V. L. Electron spin catalysis. *Chem. Rev.* **102**, 603–612 (2002).

53. Khavryuchenko, O. V., Khavryuchenko, V. D. & Su, D. S. Spin catalysts: a quantum trigger for chemical reactions. *Chinese J. Catal.* **36**, 1656–1661 (2015).

54. Jiang, D.-E., Sumpter, B. G. & Dai, S. First principles study of magnetism in nanographenes. *J. Chem. Phys.* **127**, 124703 (2007).

55. Hachmann, J., Dorando, J. J., Avilés, M. & Chan, G. K.-L. The radical character of the acenes: a density matrix renormalization group study. *J. Chem. Phys.* **127**, 134309 (2007).

56. Jiang, D. E. & Dai, S. Electronic ground state of higher acenes. *J. Phys. Chem. A* **112**, 332–335 (2008).

57. Plasser, F. *et al.* The Multiradical character of one- and two-dimensional graphene nanoribbons. *Angew. Chem. Int. Ed.* **52**, 2581–2584 (2013).

58. CrysAlisPro 1.171.38.41; Agilent Technologies Ltd.: Yarton, Oxfordshire, UK, 2015.

59. Sheldrick, G. M. SHELXT- integrated space-group and crystal-structure determination. *Acta Cryst.* **A71**, 3–8 (2015).

60. van der Sluis, P. & Spek, A. L. BYPASS: an effective method for the refinement of crystal structures containing disordered solvent regions. *Acta Cryst A* **46**, 194–201 (1990).

61. Spek, A. L. *PLATON—a multipurpose crystallographic tool. Acta Cryst.* Vol. A 46, C34 (Utrecht University 1990).

62. Frisch, M. J. *et al. Gaussian 03, Revision B* (Gaussian, Inc., 2003).

63. Frisch, M. J. *et al. Gaussian09, Revision E.01* (Gaussian, Inc., 2009).

64. Lee, C., Yang, W. & Parr, R. G. Development of the Colle-Salvetti correlation-energy formula into a functional of the electron density. *Phys. Rev. B: Condens. Matter Mater. Phys.* **37**, 785–789 (1988).

65. Becke, A. D. Density-functional thermochemistry. III. The role of exact exchange. *J. Chem. Phys.* **98**, 5648–5652 (1993).

66. Yanai, T., Tew, D. & Handy, N. A new hybrid exchange − correlation functional using the coulomb-attenuating method (CAM-B3LYP). *Chem. Phys. Lett.* **393**, 51–57 (2004).

67. Seeger, R. Pople, self-consistent molecular orbital methods. XVIII. constraints and stability in Hartree-Fock theory. *J. Chem. Phys.* **66**, 3045–3050 (1977).

## Acknowledgements

This work was supported by the National Natural Science Foundation of China (21290190, 21322301, 91227113), NSFC-DFG joint fund (21261130581), the National Basic Research 973 Program of China (2012CB932901), and the 'Strategic Priority Research Program' of the Chinese Academy of Sciences (XDA09020302, XDB12010300).

## Author contributions

Yongjun Li and Yuliang Li conceived and designed the project and co-wrote the manuscript, with assistance from the other authors. Yongjun Li and Z.J. prepared the materials, Yongjun Li performed the computation, the ESR and SQUID data were analysed by S.X. and H.L., respectively. All authors discussed the results and contributed to manuscript preparation.

## Additional information

**Competing financial interests:** The authors declare no competing financial interests.

# 22

# Catalytic N-radical cascade reaction of hydrazones by oxidative deprotonation electron transfer and TEMPO mediation

Xiao-Qiang Hu[1], Xiaotian Qi[2], Jia-Rong Chen[1], Quan-Qing Zhao[1], Qiang Wei[1], Yu Lan[2] & Wen-Jing Xiao[1]

Compared with the popularity of various C-centred radicals, the N-centred radicals remain largely unexplored in catalytic radical cascade reactions because of a lack of convenient methods for their generation. Known methods for their generation typically require the use of N-functionalized precursors or various toxic, potentially explosive or unstable radical initiators. Recently, visible-light photocatalysis has emerged as an attractive tool for the catalytic formation of N-centred radicals, but the pre-incorporation of a photolabile groups at the nitrogen atom largely limited the reaction scope. Here, we present a visible-light photocatalytic oxidative deprotonation electron transfer/2,2,6,6-tetramethylpiperidine-1-oxyl (TEMPO)-mediation strategy for catalytic N-radical cascade reaction of unsaturated hydrazones. This mild protocol provides a broadly applicable synthesis of 1,6-dihydropyradazines with complete regioselectivity and good yields. The 1,6-dihydropyradazines can be easily transformed into diazinium salts that showed promising *in vitro* antifungal activities against fungal pathogens. DFT calculations are conducted to explain the mechanism.

[1] CCNU-uOttawa Joint Research Centre, Key Laboratory of Pesticide and Chemical Biology, Ministry of Education, College of Chemistry, Central China Normal University, 152 Luoyu Road, Wuhan 430079, China. [2] School of Chemistry and Chemical Engineering, Chongqing University, Chongqing 400030, China. Correspondence and requests for materials should be addressed to J.-R.C. (email: chenjiarong@mail.ccnu.edu.cn).

ynthetic chemists continuously strive for fast, selective and high yielding reactions under mild conditions. Radical reactions, especially the radical cascades, provide a potential access to such ideal transformations and have attracted considerable attention of synthetic community because of their typically mild conditions, short reaction times and high efficiency[1,2]. Although various carbon radicals have been widely used in catalytic radical-based cascade reactions[3–5], however, the chemistry of N-centred radicals in this regard remains largely unexplored because of a lack of convenient and general methods for their generation[6,7]. Known methods for their generation typically require the use of N-functionalized precursors or various toxic, potentially explosive or unstable radical initiators. Pioneered by Nicolaou's discovery of the o-iodoxybenzoic acid-mediated conversion of N-aryl amides and carbamates into the corresponding nitrogen radicals[8], the groups of Chiba[9] and Lei[10], respectively, developed two efficient methods for the generation of 1,3-diazaallyl and amidyl radicals by Cu- and Ni-catalyzed oxidative cleavage of N–H bonds of amidines and N-alkoxyamides using $O_2$ and di-tertiary butyl peroxide as terminal oxidants at high temperatures. Recently, Li[11] and Chemler[12] also independently reported the Cu- and Ag-catalyzed oxidative formation of amidyl radicals in the presence of stoichiometric $MnO_2$ and Selectfluor reagents as oxidants. Despite these impressive advancements, the search for new efficient protocols for direct catalytic conversion of the N–H bonds into the corresponding N-centred radicals under mild conditions has become an increasingly significant, yet challenging priority in the development of new N-radical cascade reactions.

In recent years, the visible-light photocatalysis has been established as a powerful technique that facilitates selectively activating organic molecules and chemical bonds to indentify new chemical reactions under mild conditions[13–16]. As the notable early studies by MacMillan[17] and Sanford[18] on neutral N-centred radical-mediated photocatalytic C–H amination of aldehydes and (hetero)arenes, several promising visible-light photocatalytic protocols have been developed by other groups for generating N-centred radicals and C–N bond formation (Fig. 1a)[19–23]. Despite their advantages, these methods require the introduction of a photolabile substituent at the nitrogen atom as a handle for

photo-activation. The use of the visible-light photocatalysis in initiating strong N–H bond activation and application in neutral N-centred radical-mediated catalytic cascade reactions have been, until recently, very limited. The Knowles' group recently reported an elegant combination of iridium photocatalyst and phosphate base for a direct homolytic cleavage of strong N–H bonds of N-arylamides to access amidyl radicals by a concerted proton-coupled electron transfer, which allowed an efficient radical cascade reaction towards N-heterocycle synthesis[24]. Exploring new reactivity of N-containing compounds in the field of visible-light photocatalysis is an integral part of our recent ongoing research endeavours[25–28]. For example, our group has recently developed a direct catalytic conversion of the N–H bonds of β,γ-unsaturated hydrazones into N-centred hydrazonyl radicals by visible-light-induced photoredox catalysis, which enables an efficient and mild approach to intramolecular alkene hydroamination and oxyamination for synthesis of 4,5-dihydropyrazole derivatives[28]. In this reaction, a highly regioselective 5-exo radical cyclization of an N-centred radical was observed. It should be noted that the groups of Han[29,30] and Chiba[31] have also independently reported stoichiometric amounts of tetramethylpiperidine-1-oxyl (TEMPO)-mediated intramolecular cyclization of hydrazonyl radicals for pyrazoline synthesis. Inspired by these studies, we considered exploration of the reactivity of hydrazones in catalytic N-radical cascade reactions to assemble biologically and synthetically important dihydropyradazine scaffolds[32], inaccessible using other thermal methods[29–31,33] or our own previous protocols.

To this end, herein, we report an oxidative deprotonation electron transfer (ODET)/TEMPO-mediation strategy for direct N–H bond activation and catalytic N-radical cascade reactions of unsaturated hydrazones (Fig. 1b). This mild protocol represents the first, to our knowledge, broadly applicable synthesis of 1,6-dihydropyradazines with good regioselectivity and yield, achieved by merge of visible-light photocatalysis and TEMPO mediation.

## Results

**Reaction design.** To realize the target catalytic N-radical cascade reaction of unsaturated hydrazones as shown in Fig. 1b, several

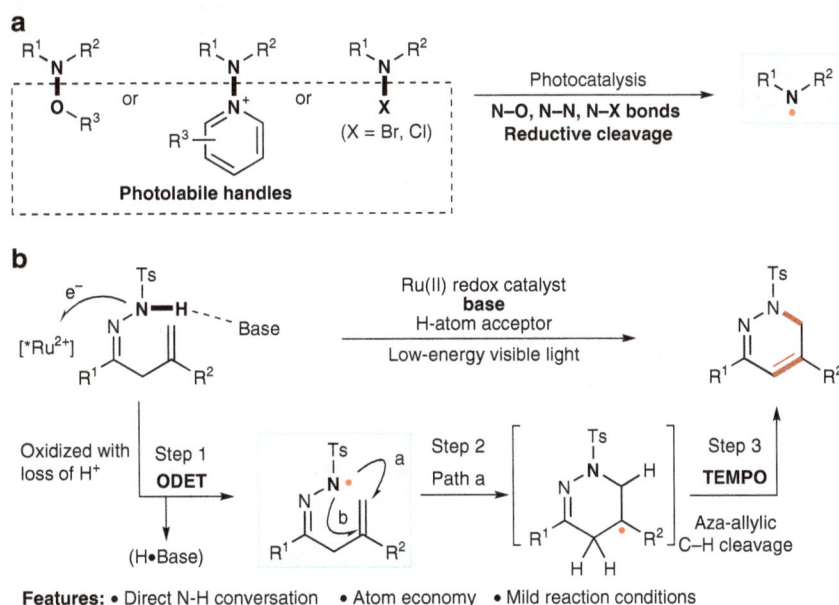

**Figure 1 | Reaction design. (a)** Visible-light-induced photocatalytic generation of N-centred radicals from N-functionalized precursors. **(b)** Our blueprint for catalytic N-radical cascade reaction of hydrazones: merge of oxidative deprotonation electron transfer (ODET) activation of N-H bond with TEMPO mediation.

major challenges would probably be encountered, such as the controlled formal homolysis of the recalcitrant N–H bond for the formation of the neutral N-centred hydrazonyl radical, regioselectivity of the N-radical cyclization step (for example, 6-*endo* and 5-*exo*, path a versus path b)[34,35] and selective homolytic activation of aza-allylic C–H bond in C-centred radical intermediate. Notably, it has been recently documented by MacMillan[36,37], Knowles[24,38,39] and our group[28,40] that the addition of a suitable Brønsted acid, Lewis acid or base could facilitate some otherwise inaccessible photocatalytic event by weakening chemical bonds of reactants and co-catalysts or modulating their redox potential. It has also been demonstrated by López and Gómez that complete 6-*endo*-selectivity over 5-*exo* ring closure in radical cyclization of C-centred radicals can be controlled by the radical property, substitution pattern at C-5 or ring strain of substrate[34,35]. Quite recently, the MacMillan group also first integrated elementary hydrogen atom transfer (HAT) process into H-bond catalysis, and achieved a highly selective photoredox α-alkylation/lactonization cascade of alcohols[41]. Based on these inspiring studies, we hypothesized that the aforementioned regioselective N-radical cascade reaction could possibly be achieved by merging visible-light photoredox with TEMPO-mediated HAT process, wherein the N–H bond might be directly converted into the corresponding N-centred hydrazonyl radical through an ODET and the aza-allylic C–H bond can probably be homolytically cleavaged by a suitable H-atom acceptor such as TEMPO[42].

To test the feasibility of this strategy, we initially conducted density functional theory (DFT) calculations on the cyclization step of N-centred radical intermediates **1a-A**, **1b-A** and **1c-A** with sterically and electronically diverse substituents at the 2-position of the alkene (Fig. 2a; see Supplementary Notes 1–3 for details). The energies given in this work are N-12//6-311 + G(d, p)//B3 LYP/6-31G(d) calculated Gibbs free energies in chloroform. See the Supplementary Information for more computational details.)[43]. As expected, both 5-*exo* and 6-*endo* N-radical-mediated radical cyclizations are possible pathways. For example, the study showed that the 5-*exo*-trig radical cyclization of **1a-A** with an activation free energy of only 8.8 kcal mol$^{-1}$ via **1a-TS1** is much more favoured than its 6-*endo*-trig variant (activation free energy of 13.5 kcal mol$^{-1}$; Fig. 1b). It was also found that **1b-A** would undergo 5-*exo* cyclization through **1b-TS1** more feasibly than its 6-*endo* cyclization via **1b-TS2**, as shown by their activation free energy (Fig. 1c, 7.5 versus 11.2 kcal mol$^{-1}$). Interestingly, the 6-*endo* cyclization of **1c-A** with a phenyl group at the 2-position of the alkene moiety proved to be easier to accomplish through **1c-TS2**, with a relatively low activation free energy of 8.7 kcal mol$^{-1}$, to give the C-centred radical intermediate **1c-C** (Fig. 1d). Encouraged by these computational results, we proceeded to perform experimental studies with these substrates to explore the feasibility of the desired 6-*endo* radical cyclization.

Under our previously developed visible-light photocatalytic conditions for hydroamination of β,γ-unsaturated hydrazones[28], substrates **1a** and **1b** indeed underwent 5-*exo* radical cyclization reactions smoothly to give the corresponding products **2a** and **2b** in 68% and 81% yields, respectively (Fig. 3a). These results also provided a solid support for the above computational investigations into these substrates. Interestingly, the reaction of **1c** resulted in the formation of a complex mixture with a complete conversion (Fig. 3b). Careful analysis of the reaction mixture revealed that an inseparable mixture of products **2c** and **3** can be obtained in 21% yield with a ratio of 1:0.9. Meanwhile, product **4** was also isolated in 16% yield, which might be formed through another radical cascade reaction between **1c** and the reaction solvent CHCl$_3$ via radical intermediate **1c-B**. The

structures of **2a-2c**, **3** and **4** were fully characterized by their $^1$H and $^{13}$C NMR spectra and mass data, and compound **4** was further characterized by single-crystal X-ray analysis (see Supplementary Fig. 79 for details). Note that the biologically significant 1,6-dihydropyridazines of type **2c** cannot be easily prepared using traditional methods[33]. These observations suggested that further optimization of reaction parameters might result in the exclusive formation of the desired 1,6-dihydropyridazines.

**Optimization of reaction conditions.** Encouraged by these initial results, we continued to optimize the reaction conditions with **1c** as a model substrate to further improve the selectivity and yield (Table 1). Inspired by the recently demonstrated wide applicability of nitroxides in organic synthesis and their unique properties[44,45], we initially focused on nitroxides as potential additives. Surprisingly, it was found that the addition of TEMPO (1.0 equiv.) did not quench the reaction; instead, it resulted in a clean reaction and gave the desired 1,6-dihydropyridazine **2c** in 89% yield (entry 1). Based on our blueprint of the reaction, we postulated that TEMPO might serve as a H-atom acceptor to abstract aza-allylic H-atom by an HAT process[41]. Then, we simply screened inorganic bases such as NaOH, Na$_2$CO$_3$ and Na$_2$HPO$_4$, and established that the base also played an important role in the reaction, with K$_2$CO$_3$ identified as the best choice (entries 2–4). With K$_2$CO$_3$ as the base, we also briefly examined several other common solvents and CHCl$_3$ proved to be the best reaction media with tetrahydrofuran, MeOH, dimethylformamide and CH$_3$CN giving relatively low yields (entries 5–8). Then, we evaluated the effect of photocatalysts on the reaction under otherwise identical conditions. Interestingly, the use of Ir(ppy)$_2$(dtbbpy)PF$_6$ as a photocatalyst provided comparable results, whereas organic photocatalyst Eosin Y was ineffective for the reaction (entries 9–10). It has been well documented that TEMPO can serve not only as a radical scavenger but also as an oxidant in transition-metal catalysis[44,45]. Thus, we continued to test several other oxidants, such as K$_2$S$_2$O$_8$ and 2,3-dichloro-5, 6-dicyano-1,4-benzoquinone (see Supplementary Table 1 for details). However, all the reactions with these oxidants resulted in a complex mixture without formation of any desired product, suggesting that TEMPO might act as a radical trap to abstract the α-hydrogen atom from intermediate **1c-C** to facilitate the target N-radical cascade reaction pathway (entries 11–12). In the control experiments with CHCl$_3$ or CH$_3$CN as the solvent, only very little or no desired products were detected in the absence of photocatalyst, base, TEMPO or light; large amounts of starting materials remained intact, highlighting the critical role of all the parameters in the reaction (entries 13–16; see Supplementary Table 2 for details).

**Substrate scope.** Under the optimized conditions, we then evaluated the substrate scope of this transformation with a variety of β,γ-unsaturated hydrazones (Fig. 4). First, we examined the effects of arene substitution using a wide range of β,γ-unsaturated hydrazones **1c-1i**. It was found that the reaction with various β,γ-unsaturated hydrazones bearing electron-neutral, electron-poor (for example, Cl, Br, CF$_3$) or electron-rich (for example, Me) substituents at the 2-, 3- or 4-position of the aromatic ring proceeded well to deliver the corresponding products **2c-2i** with yields ranging from 61 to 85%. Notably, those aryl bromides are amenable to further synthetic elaborations through transition-metal-catalyzed C–C coupling reactions. Product **2f** was also characterized by single-crystal X-ray analysis (see Supplementary Fig. 79 for details). Moreover, 2-naphthyl substituted hydrazone **1j** reacted well to afford product **2j** in 86% yield. Considering the

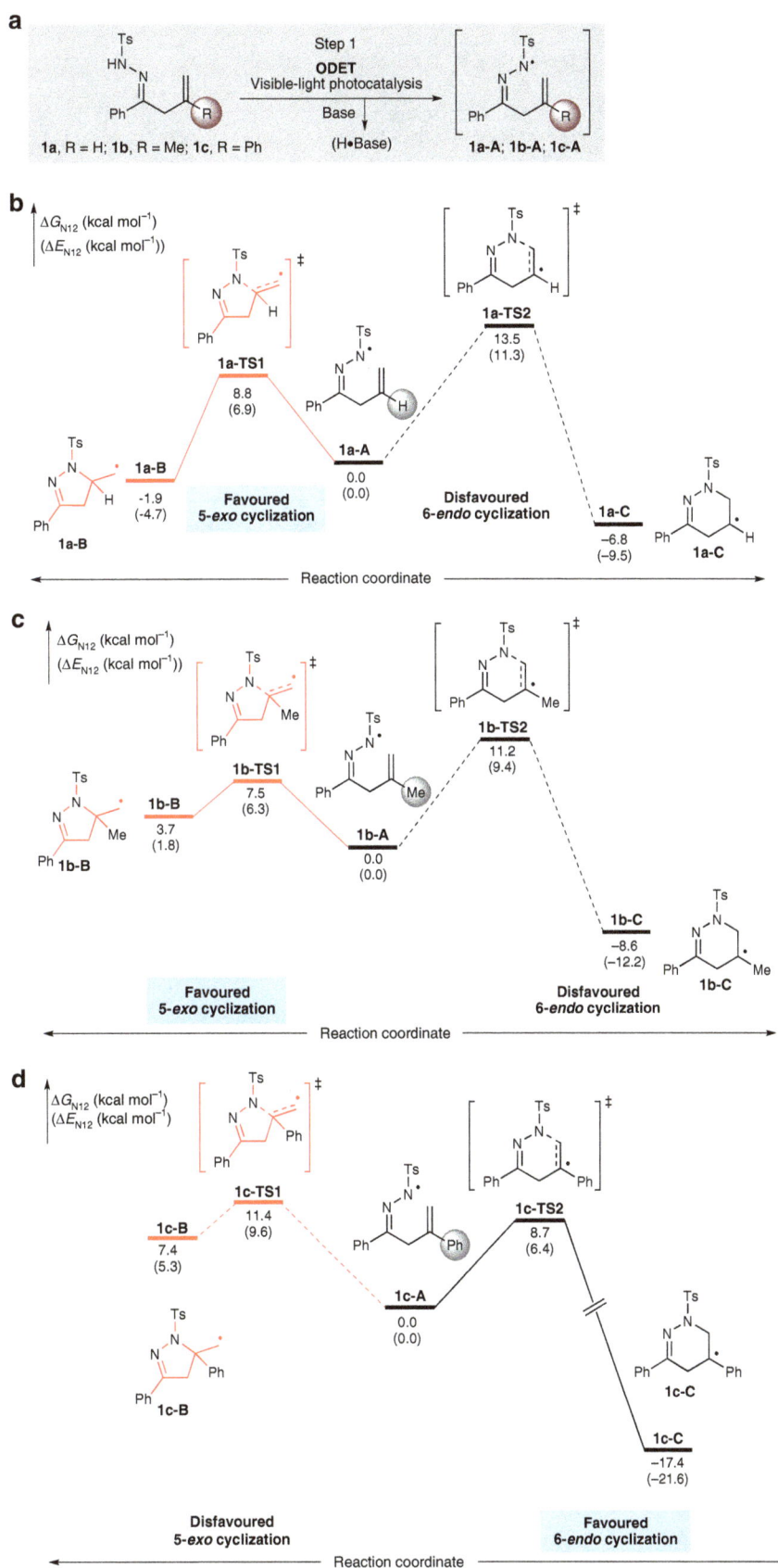

**Figure 2 | Reaction development.** (**a**) Generation of N-radicals by visible-light photocatalysis. (**b**) Free energy profiles for 5-*exo* and 6-*endo* radical cyclizations of **1a-A**. (**c**) Free energy profiles for 5-*exo* and 6-*endo* radical cyclizations of **1b-A**. (**d**) Free energy profiles for 5-*exo* and 6-*endo* radical cyclizations of **1c-A**.

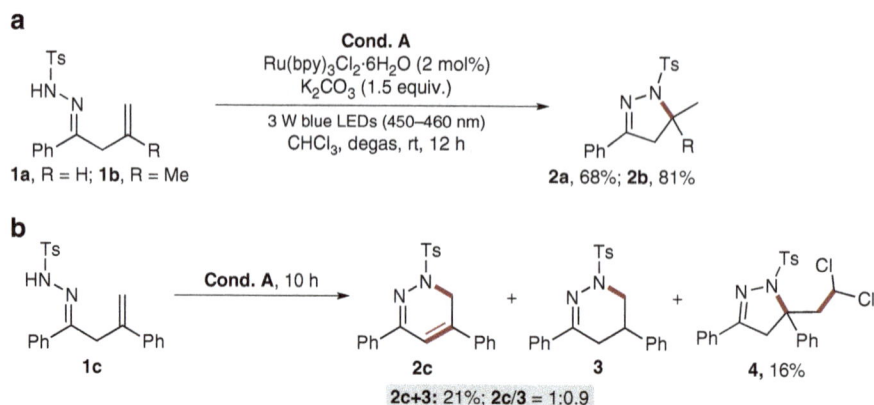

**Figure 3 | Initial results.** (**a**) Reaction of substrate **1a** and **1b** under condition **A**. (**b**) Reaction of substrate **1c** under condition **A**. Unless otherwise noted, condition **A**: reaction were run with **1** (0.2 mmol), Ru(bpy)$_3$Cl$_2 \cdot$6H$_2$O (2 mol%), K$_2$CO$_3$ (0.3 mmol), 3 W blue light-emitting diodes (450–460 nm) irradiation and CHCl$_3$ (4.0 mL) at rt for 10–12 h.

**Table 1 | Optimization of conditions for catalytic N-radical cascade reaction of unsaturated hydrazone 1c.**

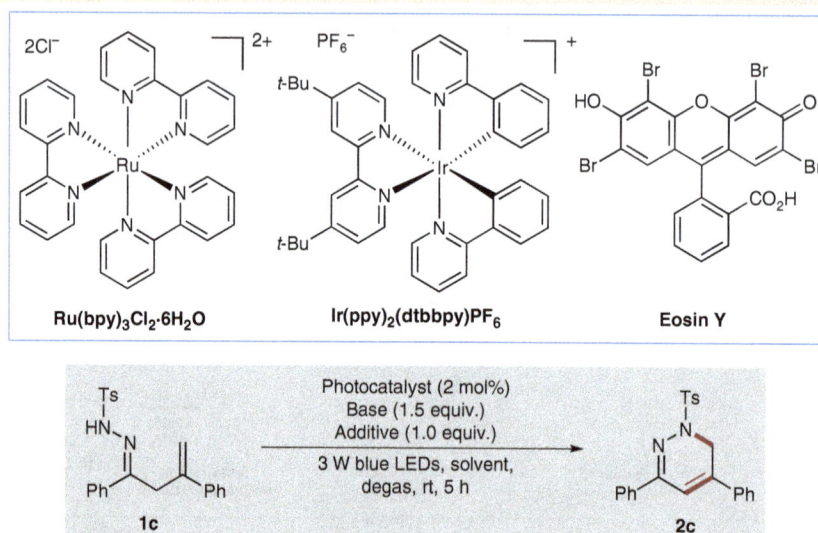

| Entry | Photocatalyst | Base | Solvent | Additive | Yield (%)* |
|---|---|---|---|---|---|
| 1 | Ru(bpy)$_3$Cl$_2 \bullet$6H$_2$O | K$_2$CO$_3$ | CHCl$_3$ | TEMPO | 89 |
| 2 | Ru(bpy)$_3$Cl$_2 \bullet$6H$_2$O | NaOH | CHCl$_3$ | TEMPO | 81 |
| 3 | Ru(bpy)$_3$Cl$_2 \bullet$6H$_2$O | Na$_2$CO$_3$ | CHCl$_3$ | TEMPO | 72 |
| 4 | Ru(bpy)$_3$Cl$_2 \bullet$6H$_2$O | Na$_2$HPO$_4$ | CHCl$_3$ | TEMPO | 8 |
| 5 | Ru(bpy)$_3$Cl$_2 \bullet$6H$_2$O | K$_2$CO$_3$ | THF | TEMPO | 51 |
| 6 | Ru(bpy)$_3$Cl$_2 \bullet$6H$_2$O | K$_2$CO$_3$ | MeOH | TEMPO | 23 |
| 7 | Ru(bpy)$_3$Cl$_2 \bullet$6H$_2$O | K$_2$CO$_3$ | DMF | TEMPO | 26 |
| 8 | Ru(bpy)$_3$Cl$_2 \bullet$6H$_2$O | K$_2$CO$_3$ | CH$_3$CN | TEMPO | 48 |
| 9 | Ir(ppy)$_2$(dtbbpy)PF$_6$ | K$_2$CO$_3$ | CHCl$_3$ | TEMPO | 83 |
| 10 | Eosin Y | K$_2$CO$_3$ | CHCl$_3$ | TEMPO | Trace |
| 11 | Ru(bpy)$_3$Cl$_2 \bullet$6H$_2$O | K$_2$CO$_3$ | CHCl$_3$ | K$_2$S$_2$O$_8$ | Trace |
| 12 | Ru(bpy)$_3$Cl$_2 \bullet$6H$_2$O | K$_2$CO$_3$ | CHCl$_3$ | DDQ | Trace |
| 13 | — | K$_2$CO$_3$ | CHCl$_3$ | TEMPO | 0 |
| 14 | Ru(bpy)$_3$Cl$_2 \bullet$6H$_2$O | — | CHCl$_3$ | TEMPO | 0 |
| 15 | Ru(bpy)$_3$Cl$_2 \bullet$6H$_2$O | K$_2$CO$_3$ | CHCl$_3$ | — | 11 |
| 16[†] | Ru(bpy)$_3$Cl$_2 \bullet$6H$_2$O | K$_2$CO$_3$ | CHCl$_3$ | TEMPO | 0 |

DDQ, 2,3-dichloro-5,6-dicyano-1,4-benzoquinone; DMF, dimethylformamide; Eosin Y, tetrabromofluorescein; TEMPO, 2,2,6,6-tetramethylpiperidine-1-oxyl; THF, tetrahydrofuran.
Reaction conditions: **1c** (0.2 mmol), photocatalyst (0.004 mmol, 2.0 mol %), TEMPO (0.2 mmol), K$_2$CO$_3$ (0.3 mmol) and solvent (4.0 ml) at room temperature for 5 h under irradiation from a 3-W blue light-emitting diodes (450–460 nm).
*Isolated yields based on **1c**.
†Without visible-light irradiation.

known medicinal chemistry, it is noteworthy that various heterocycles can be incorporated into the hydrazone substrates with no apparent deleterious effect on the reaction efficiency. For example, 2-thiophenyl and 3-indolyl substituted hydrazones were tolerated well to give the desired products **2k** and **2l** in 59% and 53% yields, respectively. More importantly, the substrate scope of

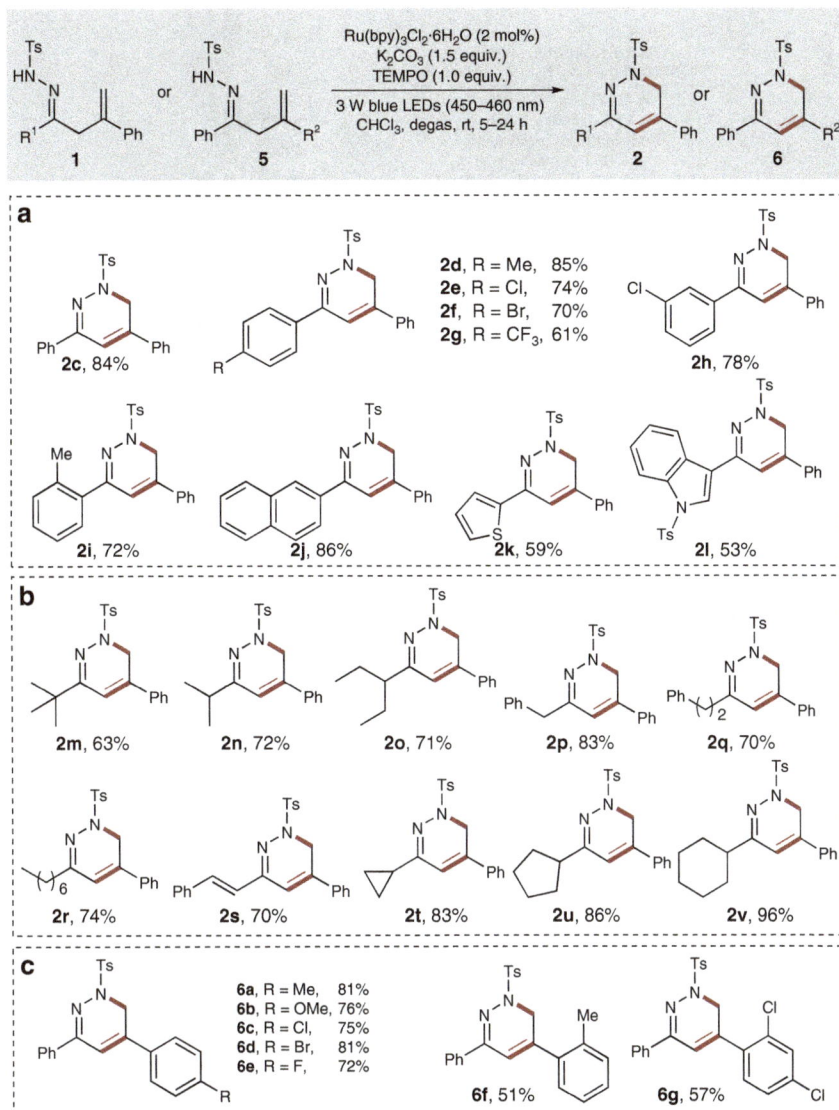

**Figure 4 | Reaction scope of unsaturated hydrazones.** (**a**) Investigation of the effects of arene substitution of hydrazones. (**b**) Substrate scope of aliphatic unsaturated hydrazones. (**c**) Substrate scope of alkene moieties. Unless otherwise noted, reactions were run with **1** or **5** (0.3 mmol), Ru(bpy)$_3$Cl$_2$·6H$_2$O (0.006 mmol, 2.0 mol%), K$_2$CO$_3$ (0.45 mmol), TEMPO (0.3 mmol) and CHCl$_3$ (6.0 ml) at rt for 5–24 h under irradiation with 3 W blue light-emitting diodes (450–460 nm).

the current protocol can be successfully extended to aliphatic β,γ-unsaturated hydrazones. Thus, the reaction with a series of linear and branched aliphatic β,γ-unsaturated hydrazones **1m-1r** can undergo the radical cascade reaction smoothly under standard conditions, although with prolonged reaction times, to afford the products **2m-2r** in 63–83% yield. The β,γ-unsaturated hydrazone **1s** bearing a styryl group also appeared to be viable for the reaction, producing a 70% yield of **2s**. Remarkably, cyclic substituents, such as cyclopropyl, cyclopentyl and cyclohexyl groups, could also be easily incorporated into the 1,6-dihydropyridazine scaffold with high yields (**2t-2v**, 83–96%).

Encouraged by these results, we proceeded to examine the scope of alkene moieties by incorporation of various substituents into the phenyl ring. As highlighted in Fig. 4c, the substitution patterns and electronic properties of the aromatic ring showed no apparent effect on the reaction efficiency either. For example, all the electron-releasing (for example, 4-Me, 2-Me and 4-MeO) and electron-withdrawing (for example, 4-F, 4-Cl, 4-Br, 2,4-2Cl) groups were well tolerated under the standard conditions, furnishing the expected products **6a-6g** in 51–81% yield.

Interestingly, during our subsequent biological studies with 1,6-dihydropyridazines **2**- and **6**-derived diazinium salts, it was found that such aromatic substituents at the 2-position of the alkene are critical to their antifungal *in vitro* activities. It should be noted that we did not detect any 5-*exo* cyclization products in all cases[29,30].

**Mechanistic investigations.** To gain additional insights into the reaction mechanism, several control experiments were conducted with model substrate **1c** (Fig. 5; see Supplementary Discussion for details). To further confirm the formation of C-centred intermediate of type **1c-C** during the reaction, common radical trapping agents (PhSeSePh or 2,6-di-tert-butyl-4-methylphenol, BHT) were added to the reaction system to capture the radical intermediate (Fig. 5a). However, no trapping products were observed; instead, only the 1,6-dihydropyridazine **2c** was produced and isolated in 85% and 83% yields, respectively. In contrast, without addition of TEMPO, the reaction with PhSeSePh as a radical trapping agent furnished a mixture of desired **2c** and selenide-adduct **7** (61% yield, 1:4 ratio; see Supplementary Fig. 78

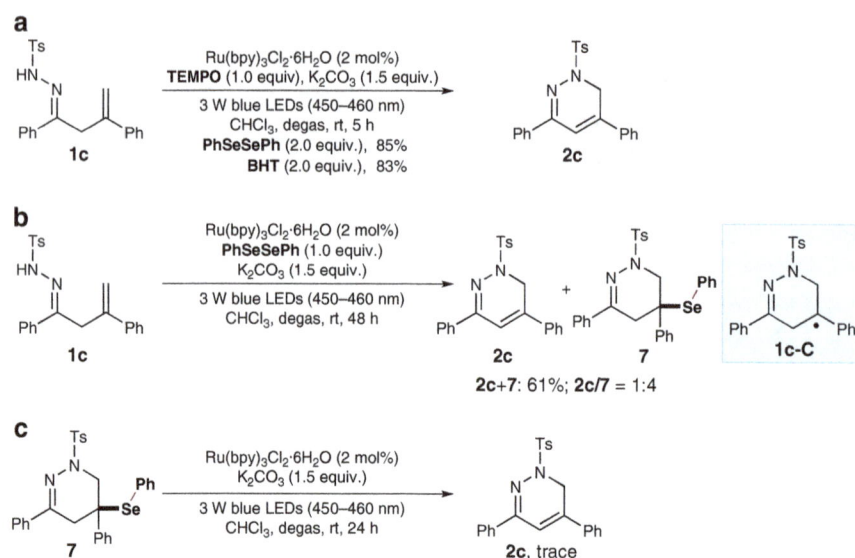

**Figure 5 | Mechanistic investigations. (a)** Trapping the C-centred intermediate by addition of PhSeSePh or BHT under the standard conditions. **(b)** Trapping the C-centred intermediate by addition of PhSeSePh under the standard conditions in the absence of TEMPO. **(c)** Control experiment with selenide-adduct **7** under the standard conditions.

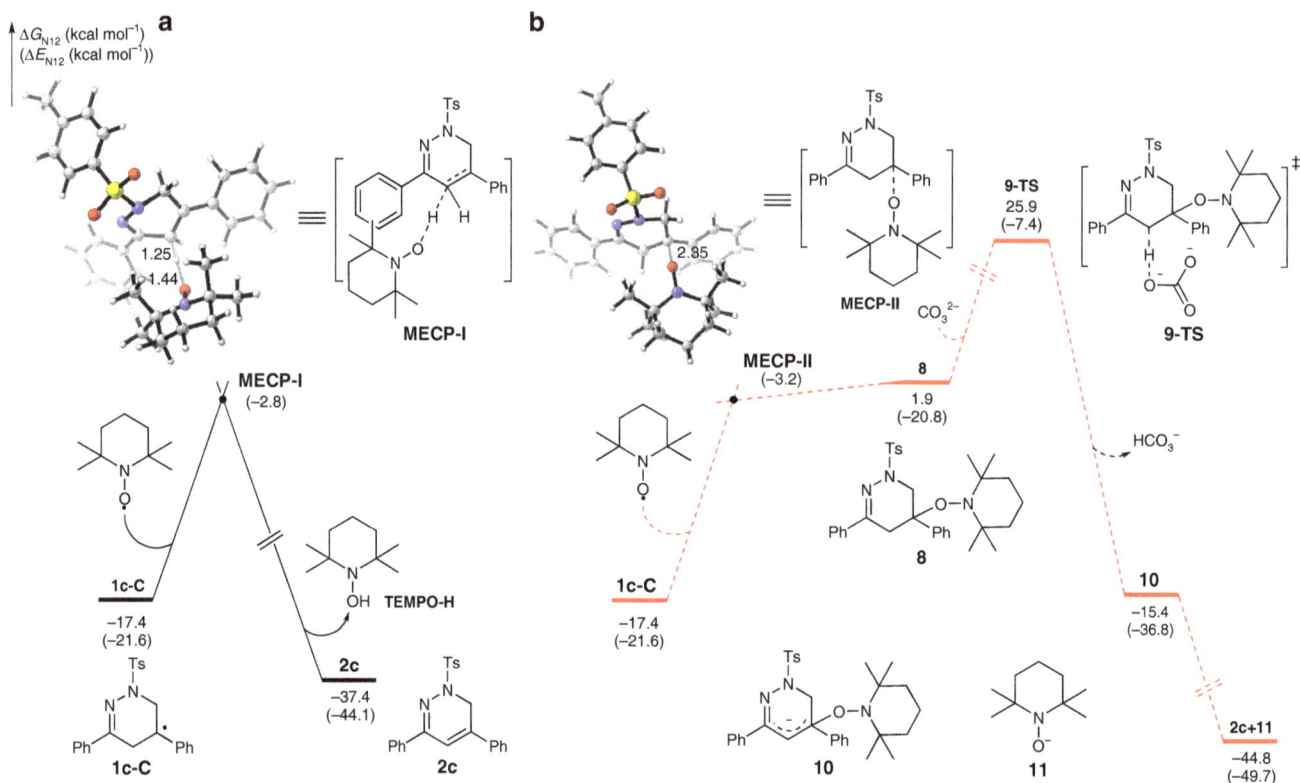

**Figure 6 | Calculation studies. (a)** Free energy profile for the transformation of C-centred radical **1c-C** into product **2c** through a TEMPO-mediated HAT process. **(b)** Free energy profile for the transformation of C-centred radical **1c-C** into product **2c** through carbon radical trapping/elimination process.

for details), and compound **7** should be formed from radical intermediate **1c-C** and PhSeSePh (Fig. 5b). Then, we obtained the pure selenide-adduct **7** by semi-preparative high-performance liquid chromatography purification and re-subjected it to the standard reaction conditions without TEMPO (Fig. 5c). However, we did not detect any desired product **2c** even after 24 h and compound **7** remained intact, suggesting that selenide-adduct **7** should not be the possible intermediate for the formation of 1,6-dihydropyridazine **2c**.

To further determine the role of TEMPO, we also calculated the free energy of the subsequent transformation of C-centred radical intermediate **1c-C** into the final product **2c** via the minimum energy crossing point (MECP; Fig. 6)[43]. As shown in Fig. 6a, the computational results showed that the TEMPO might facilitate the conversion of the intermediate **1c-C** into the final product **2c** through a TEMPO-mediated HAT process, because the calculated energy barrier ($\Delta E$) for the aza-allylic hydrogen atom abstraction via **MECP-I** is only 18.8 kcal mol$^{-1}$. Moreover,

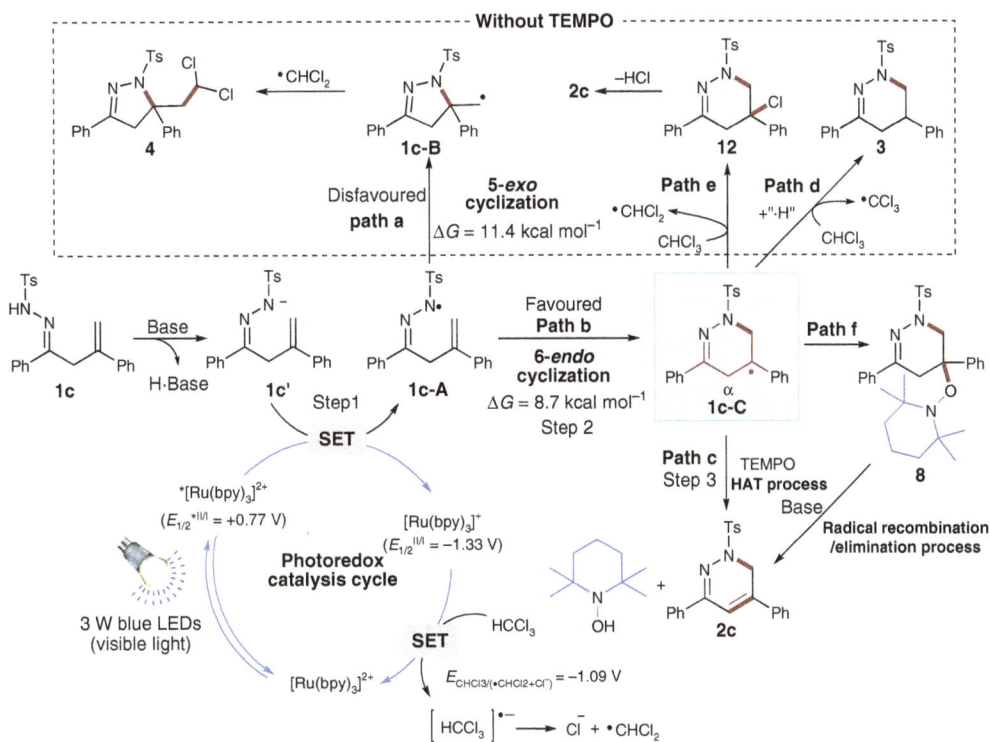

**Figure 7 | Proposed catalytic cycle.** The plausible mechanism involves oxidative deprotonation electron transfer (ODET) activation of N–H bond into N-centred radical by visible light photoredox catalysis and TEMPO-mediated N-radical cyclization.

the generation of product **2c** is exergonic by 20.0 kcal mol$^{-1}$ compared with the intermediate **1c-C**. Recently, a similar trapping of carbon radical and elimination of TEMPO-H process in the presence of base has been identified by Chiba's group as the possible pathway in TEMPO-mediated C–H bond oxygenation of oximes and hydrazones[46]. Inspired by this work, another possible pathway involving carbon radical trapping/elimination sequence of **1c-C** in the presence of base was also considered in calculation. As shown in Fig. 6b, the combination of radical **1c-C** with TEMPO occurs through **MECP-II**, and the energy barrier ($\Delta E$) of which is 18.4 kcal mol$^{-1}$. Although this energy barrier is close to that of **MECP-I** formation (Fig. 6a), the formation of TEMPO-adduct **8** is endergonic by 19.3 kcal mol$^{-1}$ compared with **1c-C**. Moreover, the activation free energy of subsequent deprotonation, which occurs via transition state **9-TS**, reaches as high as 43.3 kcal mol$^{-1}$. According to these results, the sequential combination of carbon radical **1c-C** with TEMPO and elimination process appears to be thermodynamically unfavourable. Moreover, we also intended to isolate the possible intermediate **8** upon ~50% conversion of model substrate **1c**. Unfortunately, all the attempts met failure, although a trace amount of intermediate **8** was detected by the high-resolution mass spectrometry analysis of the the reaction mixture (see Supplementary Information). Another possible pathway for base-free elimination of TEMPO-H from **8** by direct radical elimination with C–O bond homolysis is not considered as the stoichiometric base is necessary in our reaction system[47–49]. Taken together, although the calculation studies support the TEMPO-mediated HAT process as the likely mechanism for the transformation of C-centred radical intermediate **1c-C** into the final product **2c**, at present we cannot rule out the carbon radical recombination/elimination pathway (see Supplementary Figs 80 and 81 and Supplementary Notes 1-3 for details). More detailed mechanistic studies are currently underway in our laboratory.

According to our blueprint for ODET activation of N–H bond, the addition of K$_2$CO$_3$ proved be critical for the reaction as a base

and this phenomenon was indeed observed during the optimization study (Table 1, entry 14). To further evaluate the role of base in these reactions, we continue to study the mechanism of N-centred hydrazonyl radical formation by luminescence quenching experiments, NMR and electrochemical analysis with **1c** as a model substrate (see Supplementary Figs 82–86 for details). Stern–Volmer analysis demonstrated that hydrazone **1c** alone is unable to quench the excited state of *[Ru(bpy)$_3$]$^{2+}$ in dimethylformamide at 25 °C, implying that the excited state ruthenium complex does not oxidize the hydrazone **1c** directly. However, upon addition of K$_2$CO$_3$ as a base, a significant decrease of luminescence emission intensity was observed. In addition, the $^1$H NMR analysis of a solution containing both **1c** and K$_2$CO$_3$ exhibited that the addition of K$_2$CO$_3$ resulted in complete disappearance of the signal of N–H, suggesting that K$_2$CO$_3$ serve to abstract the proton of N–H bond to generate nitrogen anion intermediate **1c'** (Fig. 7 and Supplementary Information). Moreover, cyclic voltammetry data confirmed that the excited photocatalyst *Ru(bpy)$_3^{2+}$ ($E_{1/2}^{*II/I} = +0.77$ V versus SCE in CH$_3$CN) is likely to be sufficiently oxidizing to oxidize the nitrogen anion **1c'** ($E^{red} = 0.56$ V versus SCE) to generate the corresponding N-centred radical intermediate **1c-A** (Fig. 7). Taken together, although we could not completely exclude the concerted proton-coupled electron transfer mechanism at the current stage[24,38,39], the above results are more consisted with an ODET activation mechanism involving sequential deprotonation of hydrazone substrates by the K$_2$CO$_3$ and visible-light photocatalytic single-electron transfer (SET) oxidation.

Ultimately, a plausible mechanism is outlined in Fig. 7 using **1c** as an example. Initially, the β,γ-unsaturated hydrazone **1c** is transformed into anionic intermediate **1c'** upon deprotonation, which is then oxidized to the N-centred radical **1c-A** by the excited photocatalyst *[Ru(bpy)$_3$]$^{2+}$ through a SET process. Then, the key intermediate **1c-A** undergoes a 6-endo radical cyclization to afford the C-centred benzylic radical intermediate **1c-C**, which can be conveniently transformed into the final

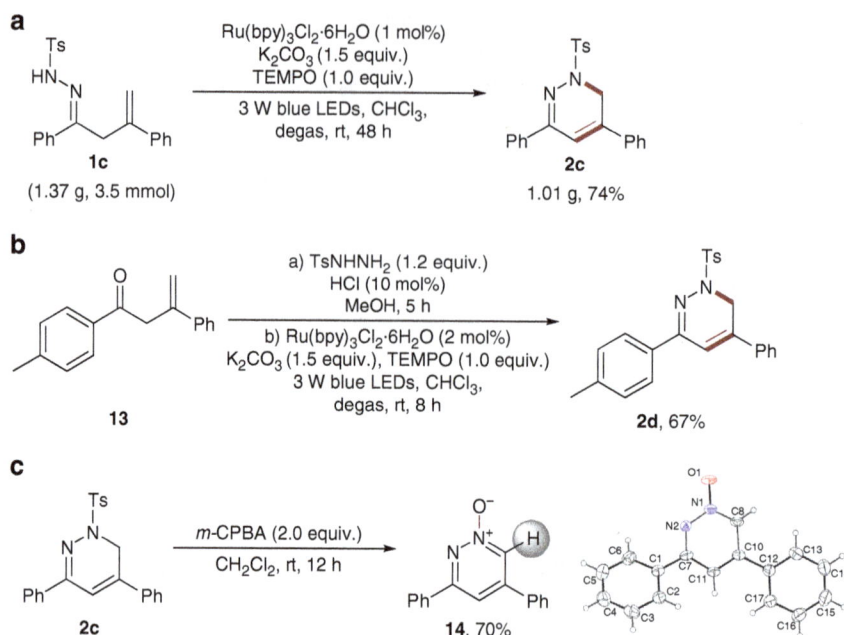

**Figure 8 | Synthetic application.** (**a**) Gram-scale reaction. (**b**) One-pot process for synthesis of 1,6-dihydropyridazine **2d**. (**c**) Synthesis of pyridazine N-oxide **14**.

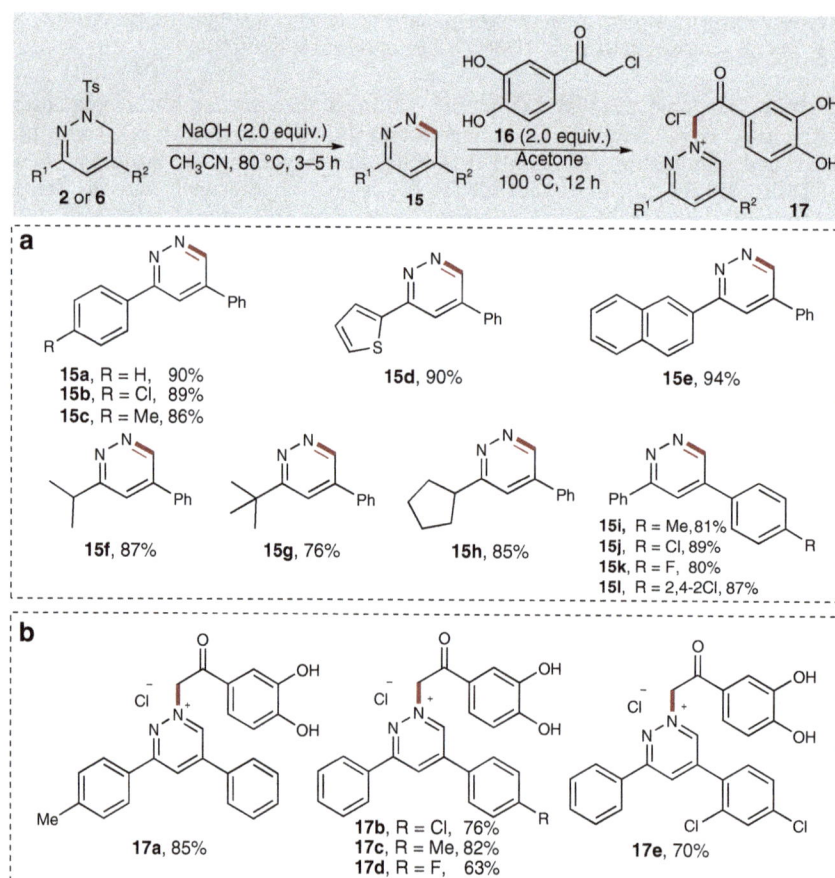

**Figure 9 | Application to the synthesis of pyridazines and diazinium salts.** (**a**) Reactions were run with **2** or **6** (0.2 mmol), NaOH (0.6 mmol) and CH$_3$CN (4.0 ml) at 80 °C for 3–5 h. (**b**) Reactions were run with **15** (0.3 mmol), **16** (0.6 mmol) and acetone (3.0 ml) at 100 °C for 12 h.

product **2c** by an HAT process in the presence of TEMPO (path c). However, as for the transformation of C-centred radical intermediate **1c-C** into the final product **2c**, at the current stage, we cannot rule out the carbon radical recombination/elimination

pathway that involves TEMPO-adduct **8** as the key intermediate (path f, see Supplementary Information). In the absence of TEMPO, the intermediate **1c-C** can abstract a hydrogen atom directly from CHCl$_3$ to give 1,4,5,6-tetrahydropyridazine **3** (path d). Meanwhile, the intermediate **1c-C** can also abstract a chlorine radical from chloroform to give rise to dichloromethyl radical and labile tertiary chloride adduct **12** intermediate[50], which can undergo facile elimination to give the product **2c**. Moreover, without addition of TEMPO, the intermediate N-centred radical **1c-A** could also undergo a 5-*exo* radical cyclization (path a) to furnish **1c-B**, partly because of the relatively small activation free energy difference between **1c-B** and **1c-C** (Fig. 2d). In the photocatalytic cycle, chloroform can regenerate the photocatalyst $[Ru(bpy)_3]^{2+}$ by an SET oxidation process with the concomitant release of the chloroform radical anion, which rapidly dechlorinated to give chloride ion and the dichloromethyl radical[51–54]. The formation of a dichloromethyl radical in the reaction was also confirmed by the isolation of side product **4**, resulting from the radical cross coupling between the dichloromethyl radical and **1c-B** intermediate.

**Synthetic application.** To further demonstrate the synthetic potential of this method, a gram-scale reaction of β,γ-unsaturated hydrazone **1c** was conducted in the presence of 1 mol% of photocatalyst under standard reaction conditions, and the desired product **2c** was still successfully obtained in 74% yield after 48 h (Fig. 8a). A key benefit of this photocatalytic radical cyclization strategy is that the β,γ-unsaturated hydrazone starting materials are easily accessed from the corresponding β,γ-unsaturated ketones and tosyl hydrazine. Thus, we examined the photocatalytic radical cyclization with β,γ-unsaturated ketone **13** and tosyl hydrazine in a two-step one-pot process (Fig. 8b). Pleasingly, the desired 1,6-dihydropyridazine **2d** was obtained in 67% overall yield. Recently, heteroaromatic N-oxides have been widely employed in transition-metal-catalyzed aromatic C–H activation/functionalization reactions to access various valuable heterocyclic molecules[55]. We found that the present method could provide a new approach to the synthesis of pyridazine N-oxides. For example, treatment of **2c** with *m*-CPBA as the oxidant resulted in the facile formation of pyridazine N-oxide **14** in a 70% yield that was also clearly determined by X-ray analysis (Fig. 8c; see Supplementary Fig. 79 for details).

Moreover, it was then established that the 1,6-dihydropyridazine products can also be easily transformed into the corresponding biologically important pyridazines under mild conditions (2.0 equiv. NaOH in $CH_3CN$ at 80 °C). As highlighted in Fig. 9a, the electronic and steric properties of the substituents on both of the aromatic rings showed no significant effect on the reaction efficiency. A series of substrates with electron-rich or electron-poor substituents worked well to give the desired products in good yields (**15a-15d**, 86–90% yield; **15i-15l**, 81–89% yield). In addition, 2-thiophenyl and 2-naphthyl-substituted 1,6-dihydropyridazines reacted well to give the corresponding pyridazine products **15d** and **15e** in 90% and 94% yield, respectively. Remarkably, the 1,6-dihydropyridazines bearing alkyl groups such as isopropyl, *tert*-butyl and cyclohexyl substituents, were well tolerated to deliver the desired products **15f-15h** in high yields (76–87%).

It has recently been documented that the pyridazine derivatives, such as diazinium salts bearing a dihydroxyacetophenone core, showed promising biological activities against a variety of microorganisms (germs and fungi)[56]. Thus, we further attempted to transform a range of representative pyridazines **15** into the corresponding diazinium salts **17** and preliminarily explored their potential structure–activity relationships (Fig. 9b). By refluxing a mixture of pyridazines **15** and 2-chloro-3′, 4′-dihydroxyacetophenone **16** in acetone for 12 h, a series of diazinium salts **17a-17e** were easily obtained in 63–85% yield after a simple filtration.

Over the past decades, the incidence of invasive fungal infections and the associated morbidity and mortality rates have risen remarkably due to the over-use of broad-spectrum antibiotics, serious medical interventions and immune deficiency disorders, such as AIDS[57,58]. Despite recent additions to the antifungal drug family, the limitations of the current antifungal drugs involve narrow activity spectra, detrimental drug–drug interactions and antifungal resistance, necessitating the development of new antifungal agents or leads. With diazinium salts **17a-17e** in hand, we evaluated the *in vitro* antifungal activities of these compounds against eight human pathogenic fungi, compared with commercially available fluconazole. In contrast to the antibacterial activities reported for related diazinium salts[56], our results demonstrated that some of these compounds showed promising activities against four common clinical pathogenic fungi (*Candida albicans*, *C. parapsilosis*, *C. neoformans* and *C. glabrata*; see Supplementary Tables 3 and 4 for details). These results also confirmed that the substitution patterns and electronic properties of the substituents at both of the phenyl rings are critical to their *in vitro* antifungal activities. Gratifyingly, the $MIC_{80}$ values of most of the compounds (**17b-17e**) against *C. parapsilosis*, *C. neoformans* and *C. glabrata* (0.5–4 μg ml$^{-1}$) were comparable to those of fluconazole, which should be valuable for our future biological studies.

## Discussion

We have developed a novel ODET/HAT strategy, which we used to directly convert the N–H bond of β,γ-unsaturated hydrazones to the N-centred radical, and developed an efficient catalytic N-radical cascade reaction. This mild protocol represents the first, to our knowledge, broadly applicable synthesis of 1,6-dihydropyradazines with good regioselectivity and yield, achieved by the merge of visible-light photocatalysis and TEMPO mediation. The 1,6-dihydropyridazines could also be conveniently transformed into biologically important diazinium salts bearing dihydroxyacetophenone core, which showed promising antifungal *in vitro* activities against various fungal pathogens. Control experiments and DFT calculations have been performed to help explain the mechanism. Owing to the wide occurrence of various N–H bonds, we believe that this strategy may find wide use for generation of other various N-centred radicals and new reaction developments with these reactive species[59].

## Methods

**Materials.** Unless otherwise noted, materials were purchased from commercial suppliers and used without further purification. All the solvents were treated according to general methods. Flash column chromatography was performed using 200–300 mesh silica gel. The manipulations for photocatalytic N-radical cascade reactions were carried out with standard Schlenk techniques under Ar by visible-light irradiation. See Supplementary Methods for experimental details.

**General methods.** $^1$H NMR spectra were recorded on 400 or 600 MHz spectrophotometers. Chemical shifts are reported in delta (δ) units in parts per million (p.p.m.) relative to the singlet (0 p.p.m.) for tetramethylsilane. Data are reported as follows: chemical shift, multiplicity (s = singlet, d = doublet, t = triplet, dd = doublet of doublets, m = multiplet), coupling constants (Hz) and integration. $^{13}$C NMR spectra were recorded on 100 or 150 MHz with complete proton-decoupling spectrophotometers ($CDCl_3$: 77.0 p.p.m. or DMSO-d$^6$: 39.5 p.p.m.). $^{19}$F NMR spectra were recorded on 376 MHz with complete proton-decoupling spectrophotometers. Mass spectra were measured on MS spectrometer (EI) or liquid chromatography-mass spectrometry (LC/MS), or electrospray ionization mass spectrometry (ESI-MS). High-resolution mass spectrometry was recorded on Bruker ultrafleXtreme matrix-assisted laser desorption/ionization–time-of-flight (TOF)/TOF mass spectrometer. $^1$H NMR, $^{13}$C NMR and $^{19}$F NMR spectra are supplied for all compounds: see Supplementary Figs 1–77.

**General procedure for catalytic nitrogen radical cascade reaction of hydrazones.** In a flame-dried Schlenk tube under Ar, **1c** (117.0 mg, 0.3 mmol), Ru(bpy)$_3$Cl$_2$.6H$_2$O

(0.006 mmol), TEMPO (46.9 mg, 0.3 mmol) and $K_2CO_3$ (61.2 mg, 0.45 mmol) were dissolved in $CHCl_3$ (6.0 ml). Then, the resulting mixture was degassed via 'freeze-pump-thaw' procedure (three times). After that, the solution was stirred at a distance of ~5 cm from a 3-W blue light-emitting diodes (450–460 nm) at room temperature ~5 h until the reaction was completed as monitored by thin-layer chromatography analysis. The crude product was purified by flash chromatography on silica gel (petroleum ether/ethyl acetate 20:1 ~ 10:1) directly to give the desired product **2c** in 84% yield as a white solid. Full experimental details and characterization of new compounds can be found in the Supplementary Methods.

## References

1. Chatgilialoglu, C. & Studer, A. *Encyclopedia of Radicals in Chemistry, Biology and Materials* (John Wiley & Sons, 2012).
2. Zard, S. Z. *Radical Reactions in Organic Synthesis* (Oxford Univ., 2003).
3. Sebren, L. J., Devery, III J. J. & Stephenson, C. R. Catalytic radical domino reactions in organic synthesis. *ACS Catal.* **4**, 703–716 (2014).
4. Wille, U. Radical cascades initiated by intermolecular radical addition to alkynes and related triple bond systems. *Chem. Rev.* **113**, 813–853 (2013).
5. Chen, J.-R., Yu, X.-Y. & Xiao, W.-J. Tandem radical cyclization of N-arylacrylamides: an emerging platform for the construction of 3,3-disubstituted oxindoles. *Synthesis* **47**, 604–629 (2015).
6. Zard, S. Z. Recent progress in the generation and use of nitrogen-centred radicals. *Chem. Soc. Rev.* **37**, 1603–1618 (2008).
7. Minozzi, M., Nanni, D. & Spagnolo, P. From azides to nitrogen-centered radicals: applications of azide radical chemistry to organic synthesis. *Chem. Eur. J* **15**, 7830–7840 (2009).
8. Nicolaou, K. C. *et al.* Iodine (V) reagents in organic synthesis. Part 3. New routes to heterocyclic compounds via o-iodoxybenzoic acid-mediated cyclizations: generality, scope, and mechanism. *J. Am. Chem. Soc.* **124**, 2233–2244 (2002).
9. Wang, Y.-F., Chen, H., Zhu, X. & Chiba, S. Copper-catalyzed aerobic aliphatic C-H oxygenation directed by an amidine moiety. *J. Am. Chem. Soc.* **134**, 11980–11983 (2012).
10. Zhou, L.-L. *et al.* Transition-metal-assisted radical/Radical cross-coupling: a new strategy to the oxidative $C(sp^3)$-H/N-H cross-coupling. *Org. Lett.* **16**, 3404–3407 (2014).
11. Li, Z.-D., Song, L.-Y. & Li, C.-Z. Silver-catalyzed radical aminofluorination of unactivated alkenes in aqueous media. *J. Am. Chem. Soc.* **135**, 4640–4643 (2013).
12. Liwosz, T. W. & Chemler, S. R. Copper-catalyzed oxidative amination and allylic amination of alkenes. *Chem. Eur. J* **19**, 12771–12777 (2013).
13. Dai, C., Narayanam, J. M. R. & Stephenson, C. R. J. Visible-light-mediated conversion of alcohols to halides. *Nat. Chem.* **3**, 140–145 (2011).
14. Xuan, J. & Xiao, W.-J. Visible-light photoredox catalysis. *Angew. Chem. Int. Ed.* **51**, 6828–6838 (2012).
15. Prier, C. K., Rankic, D. A. & Macmillan, D. W. Visible light photoredox catalysis with transition metal complexes: applications in organic synthesis. *Chem. Rev.* **113**, 5322–5363 (2013).
16. Schultz, D. M. & Yoon, T. P. Solar synthesis: prospects in visible light photocatalysis. *Science* **343**, 985–994 (2014).
17. Cecere, G., Konig, C. M., Alleva, J. L. & MacMillan, D. W. Enantioselective direct alpha-amination of aldehydes via a photoredox mechanism: a strategy for asymmetric amine fragment coupling. *J. Am. Chem. Soc.* **135**, 11521–11524 (2013).
18. Allen, L. J., Cabrera, P. J., Lee, M. & Sanford, M. S. N-Acyloxyphthalimides as nitrogen radical precursors in the visible light photocatalyzed room temperature C-H amination of arenes and heteroarenes. *J. Am. Chem. Soc.* **136**, 5607–5610 (2014).
19. Qin, Q.-X. & Yu, S.-Y. Visible-light-promoted redox neutral C-H amidation of heteroarenes with hydroxylamine derivatives. *Org. Lett.* **16**, 3504–3507 (2014).
20. Kim, H., Kim, T., Lee, D. G., Roh, S. W. & Lee, C. Nitrogen-centered radical-mediated C-H imidation of arenes and heteroarenes via visible light induced photocatalysis. *Chem. Commun.* **50**, 9273–9276 (2014).
21. Greulich, T. W., Daniliuc, C. G. & Studer, A. N-Aminopyridinium salts as precursors for N-centered radicals-direct amidation of arenes and heteroarenes. *Org. Lett.* **17**, 254–257 (2015).
22. Miyazawa, K., Koike, T. & Akita, M. Regiospecific intermolecular aminohydroxylation of olefins by photoredox catalysis. *Chem. Eur. J* **21**, 11677–11680 (2015).
23. Jiang, H. *et al.* Visible-light-promoted iminyl-radical rormation from acyl oximes: A unified approach to pyridines, quinolines, and phenanthridines. *Angew. Chem. Int. Ed.* **54**, 4055–4059 (2015).
24. Choi, G. J. & Knowles, R. R. Catalytic alkene carboaminations enabled by oxidative proton-coupled electron transfer. *J. Am. Chem. Soc.* **137**, 9226–9229 (2015).
25. Zou, Y.-Q. *et al.* Visible-light-induced oxidation/[3 + 2] cycloaddition/oxidative aromatization sequence: a photocatalytic strategy to construct pyrrolo[2,1-*a*] isoquinolines. *Angew. Chem. Int. Ed.* **50**, 7171–7175 (2011).
26. Xuan, J. *et al.* Visible-light-induced formal [3 + 2] cycloaddition for pyrrole synthesis under metal-free conditions. *Angew. Chem. Int. Ed.* **53**, 5653–5656 (2014).
27. Xuan, J. *et al.* Redox-neutral α-allylation of amines by combining palladium catalysis and visible-light photoredox catalysis. *Angew. Chem. Int. Ed.* **54**, 1625–1628 (2015).
28. Hu, X.-Q. *et al.* Photocatalytic generation of N-centered hydrazonyl radicals: a strategy for hydroamination of β,γ-unsaturated hydrazones. *Angew. Chem. Int. Ed.* **53**, 12163–12167 (2014).
29. Duan, X.-Y. *et al.* Hydrazone radical promoted vicinal difunctionalization of alkenes and trifunctionalization of allyls: synthesis of pyrazolines and tetrahydropyridazines. *J. Org. Chem.* **78**, 10692–10704 (2013).
30. Duan, X.-Y. *et al.* Transition from π radicals to σ radicals: substituent-yuned cyclization of hydrazonyl radicals. *Angew. Chem. Int. Ed.* **53**, 3158–3162 (2014).
31. Zhu, X. & Chiba, S. TEMPO-mediated allylic C-H amination with hydrazones. *Org. Biomol. Chem.* **12**, 4567–4570 (2014).
32. Maison, W. & Küchenthal, C.-H. Synthesis of cyclic hydrazino α-carboxylic acids. *Synthesis* **2010**, 719–740 (2010).
33. Potikha, L., Turelik, A. & Kovtunenko, V. Synthesis and properties of z-1, 3-bis-(aryl)-4-bromo-2-buten-1-ones. *Chem. Heterocycl. Compd* **45**, 1184–1189 (2009).
34. Gómez, A. M., Company, M. D., Uriel, C., Valverde, S. & López, J. C. 6-*Endo* versus 5-*exo* radical cyclization: streamlined syntheses of carbahexopyranoses and derivatives by 6-*endo-trig* radical cyclization. *Tetrahedron Lett.* **48**, 1645–1649 (2007).
35. Gómez, A. M., Uriel, C., Company, M. D. & López, J. C. Synthetic strategies directed towards 5a-carbahexopyranoses and derivatives based on 6-endo-trig radical cyclizations. *Eur. J. Org. Chem.* **2011**, 7116–7132 (2011).
36. Cuthbertson, J. D. & MacMillan, D. W. C. The direct arylation of allylic $sp^3$ C-H bonds via organic and photoredox catalysis. *Nature* **519**, 74–77 (2015).
37. Jin, J. & MacMillan, D. W. C. Alcohols as alkylating agents in heteroarene C-H functionalization. *Nature* **525**, 87–90 (2015).
38. Miller, D. C., Choi, G. J., Orbe, H. S. & Knowles, R. R. Catalytic olefin hydroamidation enabled by proton-coupled electron transfer. *J. Am. Chem. Soc.* **137**, 13492–13495 (2015).
39. Tarantino, K. T., Miller, D. C., Callon, T. A. & Knowles, R. R. Bond-weakening catalysis: conjugate aminations enabled by the soft homolysis of strong N-H bonds. *J. Am. Chem. Soc.* **137**, 6440–6443 (2015).
40. Zhou, Q.-Q. *et al.* Decarboxylative alkynylation and carbonylative alkynylation of carboxylic acids enabled by visible-light photoredox catalysis. *Angew. Chem. Int. Ed.* **54**, 11196–11199 (2015).
41. Jeffrey, J. L., Terrett, J. A. & MacMillan, D. W. C. O-H hydrogen bonding promotes H-atom transfer from C-H bonds for C-alkylation of alcohols. *Science* **349**, 1532–1536 (2015).
42. Mayer, J. M. Understanding hydrogen atom transfer: from bond strengths to marcus theory. *Acc. Chem. Res.* **44**, 36–46 (2011).
43. The reference for Harvey's programHarvey, J. N., Aschi, M., Schwarz, H. & Koch, W. The singlet and triplet states of phenyl cation. A hybrid approach for locating minimum energy crossing points between non-interacting potential energy surfaces. *Theor. Chem. Acc.* **99**, 95–99 (1998).
44. Tebben, L. & Studer, A. Nitroxides: applications in synthesis and in polymer chemistry. *Angew. Chem. Int. Ed.* **50**, 5034–5068 (2011).
45. Bagryanskaya, E. G. & Marque, S. R. A. Scavenging of organic C-centered radicals by nitroxides. *Chem. Rev.* **114**, 5011–5056 (2014).
46. Zhu, X., Wang, Y.-F., Ren, W., Zhang, F.-L. & Chiba, S. TEMPO-Mediated aliphatic C-H oxidation with oximes and hydrazones. *Org. Lett.* **15**, 3214–3217 (2013).
47. Li, Y. & Studer, A. Transition-metal-free trifluoromethylaminoxylation of alkenes. *Angew. Chem. Int. Ed.* **51**, 8221–8224 (2012).
48. Hartmann, M., Li, Y. & Studer, A. Transition-metal-free oxyarylation of alkenes with aryl diazonium salts and TEMPONa. *J. Am. Chem. Soc.* **134**, 16516–16519 (2012).
49. Maity, S. *et al.* Efficient and stereoselective nitration of mono- and disubstituted olefins with $AgNO_2$ and TEMPO. *J. Am. Chem. Soc.* **135**, 3355–3358 (2013).
50. Recupero, F. *et al.* Enhanced nucleophilic character of the 1-adamantyl radical in chlorine atom abstraction and in addition to electron-poor alkenes and to protonated heteroaromatic bases. Absolute rate constants and relationship with the Gif reaction. *J. Chem. Soc. Perkin Trans* **2**, 2399–2406 (1997).
51. Bertran, J., Gallardo, I., Moreno, M. & Saveant, J. M. Dissociative electron transfer. Ab initio study of the carbon-halogen bond reductive cleavage in methyl and perfluoromethyl halides. Role of the solvent. *J. Am. Chem. Soc.* **114**, 9576–9583 (1992).
52. Zhang, W., Yang, L., Wu, L.-M., Liu, Y.-C. & Liu, Z.-L. Photoinduced electron transfer retropinacol reaction of 4-(N,N-dimethylamino)phenyl pinacols in chloroform. *J. Chem. Soc. Perkin Trans* **2**, 1189–1194 (1998).
53. Costentin, C., Robert, M. & Savéant, J.-M. Successive removal of chloride ions from organic polychloride pollutants. mechanisms of reductive electrochemical elimination in aliphatic gem-polychlorides, α,β-polychloroalkenes, and α,β-polychloroalkanes in mildly protic medium. *J. Am. Chem. Soc.* **125**, 10729–10739 (2003).

54. Studer, A. & Curran, D. P. The electron is a catalyst. *Nat. Chem.* **6,** 765–773 (2014).
55. Seregin, I. V. & Gevorgyan, V. Direct transition metal-catalyzed functionalization of heteroaromatic compounds. *Chem. Soc. Rev.* **36,** 1173–1193 (2007).
56. Balan, A. M. *et al.* Diazinium salts with dihydroxyacetophenone skeleton: syntheses and antimicrobial activity. *Eur. J. Med. Chem.* **44,** 2275–2279 (2009).
57. Ostrosky-Zeichner, L., Casadevall, A., Galgiani, J. N., Odds, F. C. & Rex, J. H. An insight into the antifungal pipeline: selected new molecules and beyond. *Nat. Rev. Drug Discov.* **9,** 719–727 (2010).
58. Brown, G. D., Denning, D. W. & Levitz, S. M. Tackling human fungal infections. *Science* **336,** 647–647 (2012).
59. Chen, J.-R., Hu, X.-Q., Lu, L.-Q. & Xiao, W.-J. Visible light photoredox-controlled reactions of N-radicals and radical ions. *Chem. Soc. Rev.* http://dx.doi.org/10.1039/C5CS00655D (2016).

## Acknowledgements

We are grateful to the National Science Foundation of China (NO. 21272087, 21472058, 21472057 and 21232003), and the Youth Chen-Guang Project of Wuhan (No. 2015070404010180). This work was also financially supported by the self-determined research funds of CCNU from the colleges' basic research and operation of MOE (No. CCNU15A02009). X.Q. and Y.L. are grateful to the National Science Foundation of China (NO. 21372266 and 51302327) for financial support. We also thank the anonymous referees for helpful suggestions.

## Author contributions

X.-Q.H., J.-R.C., Q.-Q.Z. and Q.W. are responsible for the plan and implementation of the experimental work. X.Q. and Y.L. are responsible for the calculation study. J.-R.C. and W.-J.X. designed and guided this project and co-wrote the manuscript. All authors discussed the results and commented on the manuscript.

## Additional information

**Competing financial interests:** The authors declare no competing financial interests.

# 23

# High-flexibility combinatorial peptide synthesis with laser-based transfer of monomers in solid matrix material

Felix F. Loeffler[1,*], Tobias C. Foertsch[1,*], Roman Popov[1], Daniela S. Mattes[1,2], Martin Schlageter[3,4], Martyna Sedlmayr[1], Barbara Ridder[1,2], Florian-Xuan Dang[5], Clemens von Bojničić-Kninski[1], Laura K. Weber[1], Andrea Fischer[1], Juliane Greifenstein[1,2], Valentina Bykovskaya[1], Ivan Buliev[6], F. Ralf Bischoff[7], Lothar Hahn[1], Michael A.R. Meier[2], Stefan Bräse[2,8], Annie K. Powell[3,4], Teodor Silviu Balaban[5], Frank Breitling[1] & Alexander Nesterov-Mueller[1]

Laser writing is used to structure surfaces in many different ways in materials and life sciences. However, combinatorial patterning applications are still limited. Here we present a method for cost-efficient combinatorial synthesis of very-high-density peptide arrays with natural and synthetic monomers. A laser automatically transfers nanometre-thin solid material spots from different donor slides to an acceptor. Each donor bears a thin polymer film, embedding one type of monomer. Coupling occurs in a separate heating step, where the matrix becomes viscous and building blocks diffuse and couple to the acceptor surface. Furthermore, we can consecutively deposit two material layers of activation reagents and amino acids. Subsequent heat-induced mixing facilitates an *in situ* activation and coupling of the monomers. This allows us to incorporate building blocks with click chemistry compatibility or a large variety of commercially available non-activated, for example, posttranslationally modified building blocks into the array's peptides with >17,000 spots per $cm^2$.

[1] Karlsruhe Institute of Technology, Institute of Microstructure Technology (IMT), Hermann-von-Helmholtz-Platz 1, 76344 Eggenstein-Leopoldshafen, Germany. [2] Karlsruhe Institute of Technology, Institute of Organic Chemistry (IOC), Fritz-Haber-Weg 6, 76131 Karlsruhe, Germany. [3] Karlsruhe Institute of Technology, Institute for Inorganic Chemistry, Engesserstrasse 15, 76131 Karlsruhe, Germany. [4] Karlsruhe Institute of Technology, Institute of Nanotechnology (INT), Hermann-von-Helmholtz-Platz 1, 76344 Eggenstein-Leopoldshafen, Germany. [5] Aix-Marseille University, CNRS UMR 7313, Institut des Sciences Moléculaires de Marseille, Chirosciences, 13397 Marseille cedex 20, France. [6] Technical University of Varna, 1 Studentska, 9010 Varna, Bulgaria. [7] German Cancer Research Center, INF 580, 69120 Heidelberg, Germany. [8] Karlsruhe Institute of Technology, Institute of Toxicology and Genetics (ITG), Hermann-von-Helmholtz-Platz 1, 76344 Eggenstein-Leopoldshafen, Germany. * These authors contributed equally to this work. Correspondence and requests for materials should be addressed to F.F.L. (email: felix.loeffler@kit.edu), F.B. (email: frank.breitling@kit.edu) or to A.N.-M. (email: alexander.nesterov-mueller@kit.edu).

Laser direct-write approaches allow for versatile two- and three-dimensional structuring of a given workpiece in many different ways[1-3]. Laser-induced forward transfer (LIFT), a variant of the direct-write processes, uses laser irradiation to transfer liquid or solid material from a donor surface to defined areas of an acceptor surface. Thereby, a workpiece can be microstructured with different material patterns by simply employing several donor surfaces consecutively. The LIFT method and its variants are applied in the production of, for example, Samsung's organic light-emitting diode display[4] and other electronic materials and devices[5-7], nanoparticle[8] or hydrophobic/hydrophilic surface pattern generation[9], as well as several biological patterning applications[10], such as hydrogels[11], biomolecules[12-14] and even cells[15,16]. Very small feature sizes of down to 3 μm were reported[17]. This versatility sparked the idea to combine LIFT with combinatorial chemistry, where a number of building blocks are used to synthesize a vast number of different oligomeric molecules.

Most of today's array-based combinatorial peptide chemistry methods rely on Merrifield's solid-phase synthesis[18]: the SPOT synthesis[19], the light-directed lithographic synthesis[20-24], the electro-chemical synthesis[25] and the particle-based synthesis[26-29]. In the SPOT method, peptides are elongated in parallel on an array by spotting 20 different dissolved and chemically activated amino-acid building blocks at discrete locations. The light-directed lithographic synthesis uses light patterns to remove photolabile protecting groups from the growing peptides or photogenerated acids cleaving the acid-labile protecting groups at selected areas with a digital micromirror device. Yet, these methods have severe drawbacks: liquid solvents in the SPOT synthesis tend to evaporate or spread on the surface, which limits the spot density of arrays to some 25 peptides per cm$^2$. Lithographic methods offer a much higher density, but they allow for only one type of amino-acid building block to be coupled sequentially to the array, which yields lower synthesis quality and necessitates, for example, 200 coupling reactions to synthesize arrays of 10mer peptides using 20 different amino acids ($10 \times 20$).

The particle-based variant of the Merrifield synthesis method uses, for example, a laser printer or a microelectronic chip to structure a surface with 20 different types of solid polymer particles, each embedding a different type of amino-acid building block. This approach overcomes most of the mentioned drawbacks: structuring is achieved by electrical fields to precisely deposit patterns of the 20 different amino acid particle types, before the coupling reaction is induced for the whole pattern at once by heating. At 90 °C, the solid matrix material becomes viscous and the spots are transformed into gel-like droplets that serve as spatially confined reaction vessels, without evaporation or spreading on the surface. This enables the defined diffusion and coupling of the pre-activated monomers to free amino groups on the surface. However, some inherent drawbacks still remain: (i) the spot density is still not satisfying; (ii) the number of different amino-acid building blocks is limited due to process restrictions and high material consumption; and (iii) due to the rather large particle size and the slow diffusion within the melted matrix material, it is almost impossible to employ the large variety of commercially available non-activated amino-acid building blocks by in situ activation.

Our goal was to overcome the aforementioned problems with a combinatorial LIFT method (cLIFT), which allows us to synthesize affordable, very high-density ($>17,000$ spots per cm$^2$) and high-quality peptide arrays. Furthermore, its high flexibility enables us to employ a large variety of commercially available non-activated amino-acid building blocks and, thereby, synthesize peptide arrays with, for example, many different

posttranslationally modified peptides. We developed an automated machine setup, including a two-dimensional laser scanning system, which can currently accommodate over a hundred different donor surfaces. The laser rapidly and accurately transfers very small-sized and very thin material spots next to each other or on top of each other to defined locations on an acceptor surface. Each material spot can comprise a different amino-acid building block or the chemicals that are needed for an in situ activation of a non-activated amino-acid building block.

## Results

**Principle.** The principle of the cLIFT method is shown in Fig. 1. The method requires an acceptor slide and different donor slides, which are the sources of the different amino-acid building blocks (Fig. 1a). Donor slides are composed of a standard microscope glass slide, which is covered by a light-absorbing self-adhesive polyimide foil with a thickness of ~95 μm and a transfer material layer. The latter is generated by first dissolving a commercially available styrene-acrylic copolymer resin matrix and OPfp-activated amino acid or biotin building blocks in dichloromethane (DCM), followed by spin-coating this mixture on top of the polyimide foil. The solid matrix efficiently shields chemically activated amino-acid building blocks from decay[27]. The low-cost and easy-to-handle self-adhesive polyimide is attached to the glass slide's surface with a laminating machine.

As acceptor slides, we use commercially available amino-terminated poly (ethylene glycol) methacrylate (PEGMA)/methyl methacrylate 10/90 slides. For more details, please refer to the Methods section.

We automated the entire patterning procedure (see Fig. 2 and Supplementary Movie 1), schematically described in Fig. 1a–d. A robotic slide loader automatically handles the donor and acceptor slides, and positions the transfer layer of the donor slide directly on top of the N-terminated acceptor substrate as described in Fig. 1b. Next, selected spots of matrix material with a first type of embedded amino acid monomers are transferred to the acceptor slide by short laser pulses (Fig. 1c,d). Pulses are generated by an acousto-optic modulator, which rapidly switches the laser radiation of a 1-W continuous-wave 532 nm laser, whereas a commercially available laser scanning system directs the laser focus position (for details, see 'cLIFT machine setup' in Methods). We investigated the material transfer in a wide range of laser pulse durations from micro to milliseconds (standard irradiation time per spot: 2–7 ms).

During the short laser transfer process, time and temperature do not suffice to initiate the coupling reaction of activated monomers to the substrate (data not shown). This is achieved by heating the patterned acceptor slide in an oven for 60 min to 90 °C (Fig. 1e). The heating step initiates the diffusion of the monomers within the matrix material, which allows them to couple to free amino groups on the acceptor slide (for details on the solid phase chemistry, see Supplementary Fig. 1). Subsequent processing steps are identical to standard solid-phase peptide synthesis procedures: first, we remove excess monomers and resin (Fig. 1f), block unreacted amino groups on the surface ('capping'), remove the N-terminal 9-fluorenylmethoxycarbonyl (Fmoc) protecting groups and finally dry the acceptor slides (Fig. 1g). When repeated several times, we obtain an array of combinatorially synthesized peptides (Fig. 1h). We have also automated the chemical washing steps in a wet chemistry machine setup.

**Laser transfer.** Remarkably, and similar to laser ablation, laser-induced material transfer occurs through air over a distance of up to 60 μm (see Supplementary Methods and Supplementary Figs 2 and 3). A laser pulse is absorbed in the polyimide foil, where the

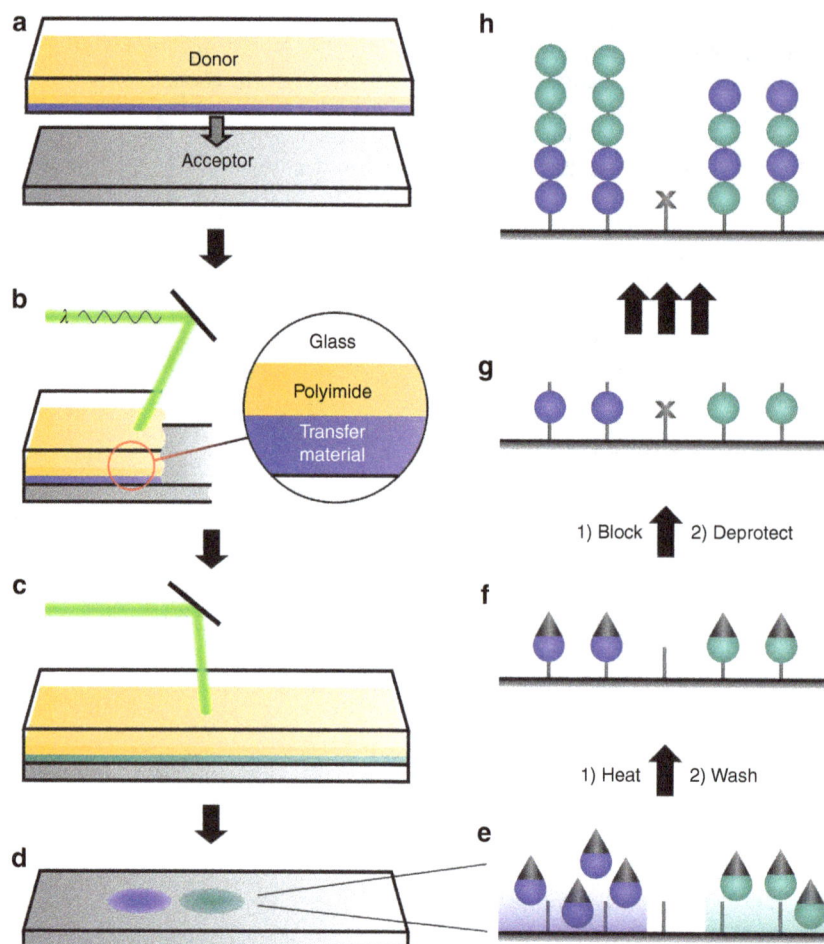

**Figure 1 | Principle of the combinatorial laser-induced forward transfer synthesis.** Donor slides bearing different monomer building blocks, embedded in a resin (different colours in **b** and **c**) are positioned on top of an acceptor slide (**a**). A laser scanning system transfers minute amounts of material to the acceptor slide. Repeating these steps with different donor slides results in a pattern of different amino acid types (**d**). The coupling reaction of monomers (**e**) is initiated by heating the surface. Next, uncoupled amino-acid building blocks are removed (**f**), uncoupled amino groups are blocked and then the protecting groups are removed (**g**). Repeating the cycle generates an array of combinatorially synthesized peptides (**h**).

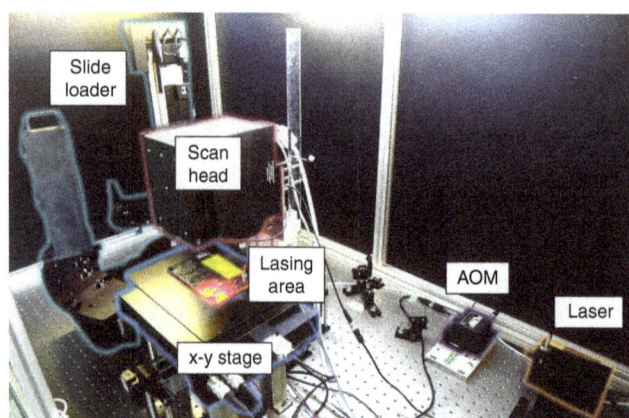

**Figure 2 | Setup of the combinatorial laser-induced forward transfer machine.** The laser is modulated by an acousto-optic modulator (AOM) and guided to a scan head system. The laser transfer is conducted on an *x-y* microscope stage, the lasing area is highlighted in yellow. Donor and acceptor slides are automatically handled and placed by the robotic slide loader.

energy is converted into heat, which causes the polyimide to quickly expand. Nevertheless, strong plastic deformations and blisters within the polyimide material are only observed when

using very high-energy laser pulses (see Supplementary Fig. 4). In parallel, the heat diffuses also into the less rigid transfer layer, melting the material layer and evaporating residual solvent in the transfer layer, which causes the ejection and transfer of material. When analysing the transfer layer in those regions that 'donated' material with atomic force microscopy, tiny impulse craters, and molten and re-solidified material can be observed (see Supplementary Figs 4 and 5). We also analysed the topography of the transferred material on the acceptor slide with a phase-shift interferometer and we found that the height of the transferred material spot on the acceptor slide is in the order of several nanometres (Fig. 3a). Analysing the coupled and stained amino acid monomers on the acceptor surface with a fluorescence scanner (Fig. 3b), we found that the spot dimensions approximately correspond to the crater size on the donor slide. Obviously, the material on the acceptor slide stems from the crater region that is visible on the donor slide.

Furthermore, by tuning the laser transfer parameters, we are able to adjust the amount of deposited material (see Fig. 3c). Interestingly, the amount of deposited material on the acceptor surface, which is approximately in the range between 1 and 50 nm in terms of layer thickness (10–500 pg), correlates linearly with the laser energy, ranging from 450 to 900 μJ (red marks in Fig. 3c). Above 900 μJ, the amount of deposited material reaches a plateau, until the laser starts to

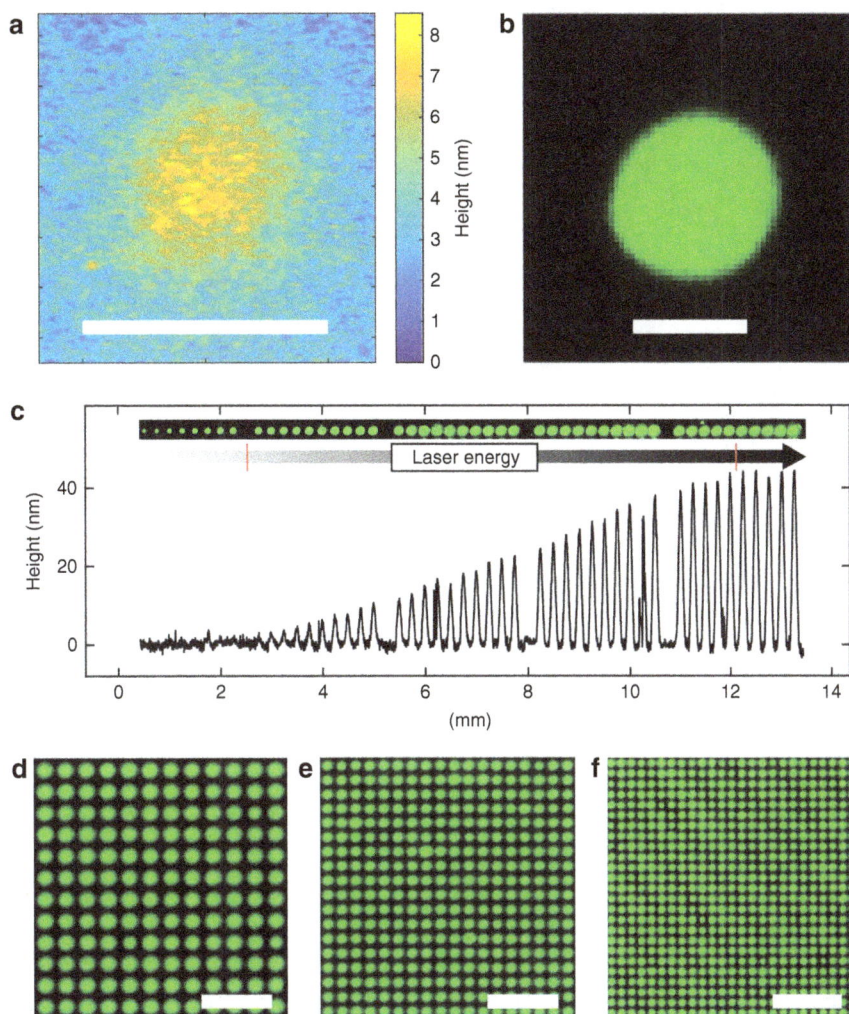

**Figure 3 | Analysis of transferred spots and spot resolution. (a)** Topography of the spot with activated leucine monomers in the polymer matrix on the acceptor slide before the coupling and washing steps, measured with phase-shift interferometry. The material spot is ~8 nm thick at its centre (scale bar, 100 μm). **(b)** The fluorescence image of the spot after coupling of the monomers and staining with a rhodamine dye. The amount of the transferred material is ~0.1 ng (scale bar, 100 μm). **(c)** Topography of transferred spot material in dependence of the laser energy. The height of the transferred spot material, containing an activated leucine building block, was measured with phase-shift interferometry, laser energy linearly increases from left to right in steps of ~15 μJ. The range of the linear correlation of deposited material (1–50 nm) and deposited laser energy (450–900 μJ) is marked by red lines. The corresponding fluorescence staining pattern of the coupled leucine was obtained with a rhodamine dye. **(d–f)** Fluorescence images of biotin patterns (scale bar, 500 μm), stained with labelled streptavidin, achieving different pitches: **(d)** 150 μm (4,444 spots per cm²), **(e)** 100 μm (10,000 spots per cm²), **(f)** 75 μm (17,777 spots per cm²).

burn the polyimide layer. Below 400 μJ, almost no material is transferred, although a weak fluorescence staining is visible down to about 300 μJ.

Another feature of our approach is the possibility to reuse the donor slides up to 20 times without loss in transfer quality, making the cLIFT process highly efficient (see Supplementary Methods and Supplementary Figs 6 and 7).

To assess the feasibility of various spot pitches, we transferred OPfp-activated biotin to a functionalized glass substrate with cLIFT using different lasing parameters. Owing to the heat diffusion within the 95-μm-thick self-adhesive polyimide layer, which limits the lowest possible pitch to ~100 μm, we replaced the very thick polyimide layer by a 5-μm thin layer of spin-coated and cured polyimide (Durimide 7520, Fujifilm). In this way, we achieved pitches from 150 to 75 μm (Fig. 3d–f). In principle, LIFT technology allows for very high-density patterns with resolutions of up to a few micrometres[17]. Details on the synthesis and staining experiments shown in Fig. 3 are described in 'cLIFT technique parameters' and 'General procedure' in Methods.

We used these experimental findings to derive an analytical heat diffusion model to define the conditions where laser-induced material transfer takes place (see Supplementary Methods and Supplementary Fig. 8). In a simplified model, we describe the heat diffusion in the light-absorbing layer. Laser pulse-induced heating competes with heat diffusion and conditions are tipped towards transfer conditions either by higher laser powers or by longer pulse durations. These transfer conditions are described by the formula:

$$P\tau^{0.5} \geq 4\pi\sqrt{kTD \cdot \rho cT\pi\sigma^2 D} \qquad (1)$$

Here, $P$ is the total laser power absorbed by the donor slide, $\tau$ is the pulse duration, $\sigma$ is the laser focus radius, $D$ is the thickness of the layers on the donor substrate, $T$ is the characteristic temperature of the transfer material layer where the transfer can take place, $k$ is the heat transfer coefficient, $\rho$ is the mass density and $c$ is the specific heat capacity.

**Synthesis.** First, we used the cLIFT machine (Fig. 2) to synthesize patterns of 3-mer, 6-mer and 9-mer peptides, with an alternating sequence of Ala and Gly, and a terminal biotin (Fig. 4a). These arrayed peptides were stained with fluorescently labelled streptavidin. We did not observe any significant decrease in fluorescence intensity, when we compared the staining intensity of the 3-mer, 6-mer and 9-mer peptides (for statistical analysis of spot fluorescence signals, see Supplementary Methods and Supplementary Fig. 9).

Next, we synthesized haemagglutinin (HA) (Tyr-Pro-Tyr-Asp-Val-Pro-Asp-Tyr-Ala) and Flag epitopes (Tyr-Asp-Tyr-Lys-Asp-Asp-Asp-Asp-Lys), as well as 62 permutation variants of these peptides, by exchanging two amino acids in each of the six differing layers of the synthesized peptides ($2^6 = 64$ variants). The selected peptides were synthesized in a pitch of 150 μm with measured spots sizes between 100 and 120 μm. Side-chain protecting groups were cleaved from the resulting peptides with a trifluoroacetic acid solution (for details, see 'General procedure' and 'Typical duration of synthesis for one layer' in Methods). Subsequently, the array was incubated with fluorescently labelled specific antibodies (Fig. 4b–d). Strong signals and negligible background indicate a good quality of the peptide spots, synthesized with cLIFT (for statistical analysis of spot fluorescence signals, see Supplementary Methods, Supplementary Fig. 10 and Supplementary Tables 1–4). Details on chemical analytics can be found in Supplementary Methods and Supplementary Figs 11–13. In our previous work, we did not observe any significant decay of activated monomers due to short laser irradiation[29] with comparable laser energies.

Click chemistry is of increasing interest in the synthesis of peptides and drug conjugates[30,31]. To demonstrate the chemical flexibility of cLIFT, we also patterned and coupled a synthetic amino acid on a large scale with an alkyne side group, a Fmoc-protected propargyl-glycine-OPfp to the amino-activated acceptor substrate (Fig. 5a). Then, we used a copper-catalysed click reaction to label this pattern with a styrylpyridinium fluorophore[32], functionalized with an azide group (Fig. 5b–d). We observed a bright fluorescence pattern (Fig. 5e) and a very low background signal. Thus, we can presume that our laser transfer does not harm the alkyne function of the activated Pra monomer (details in Supplementary Methods). Furthermore, X-ray photoelectron spectroscopy (XPS) analysis of the surface showed no residual copper catalyst on the synthesis slide.

Finally, we wanted to assess whether our cLIFT method can directly employ the plethora of commercially available Fmoc-protected, but non-activated, amino-acid building blocks. For the surface coupling reaction, building blocks have to be *in situ* activated. Therefore, we deposited material spots with non-activated amino acids on top of previously deposited material spots that contain suitable activation agents (Fig. 6a–c). First, we deposited material spots that contained a mixture of the activation reagent N,N'-diisopropylcarbodiimide (DIC) and the racemization suppressor hydroxybenzotriazole (HOBt), embedded in the solid matrix material (Fig. 6d,e). By increasing the lasing duration, we generated a gradient of increasing amounts of transferred activation reagents (blue arrow). Next, we positioned a second layer of material spots on top of the activation reagent spots, containing a non-activated glycine with an N-terminal protecting group (Fmoc-Gly-OH). Again, we generated a gradient of transferred materials by increasing the lasing time (gradient pattern from left to right, red arrow, perpendicular to the first layer; Fig. 6d). As controls, activation reagents only (Fig. 6e) and non-activated Fmoc-Gly-OH only (Fig. 6f) were used. An OPfp-ester-activated glycine (Fmoc-Gly-OPfp) was deposited as a positive control (Fig. 6g). After heat-induced coupling, we performed the washing and capping steps to block free amino groups on the surface and removed the Fmoc-protecting groups from the surface-bound amino acids. Then, we coupled a rhodamine N-hydroxysuccinimide ester dye to the free $NH_2$ groups on the array. We observed a strong fluorescent signal in those spots where non-activated amino acids were positioned on top of the activation reagents (Fig. 6d), but no or very weak signals for the two negative controls with only DIC/HOBt (Fig. 6e) and only Fmoc-Gly-OH (Fig. 6f). These observations indicate that non-activated amino-acid building blocks are efficiently converted to active esters, when diffusing through a very thin layer of melted activation reagents. As expected, the positive control with activated Fmoc-Gly-OPfp

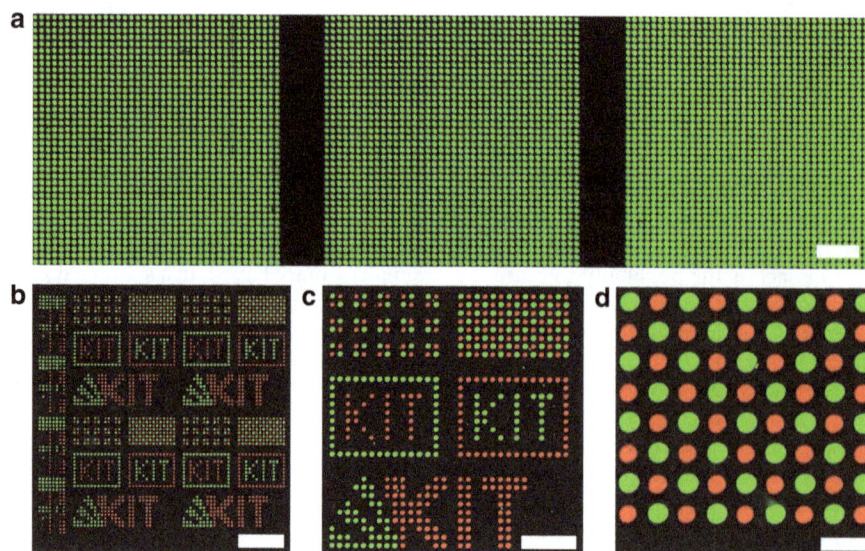

**Figure 4 | Combinatorial synthesis of peptide arrays with cLIFT.** (**a**) Three spot patterns of a 3-, 6- and 9-mer peptide (left, centre and right) with a biotin momoner as a terminal group (left Ala-Gly-biotin, centre Ala-Gly-Ala-Gly-Ala-biotin and right Ala-Gly-Ala-Gly-Ala-Gly-Ala-Gly-biotin); scale bar, 1 mm. No significant difference in staining intensity can be observed. (**b–d**) Array containing 64 different peptides with 4,444 peptide spots per cm² using cLIFT and different donor slides that bear the different amino-acid building blocks; scale bars, (**b**) 2 mm, (**c**) 1 mm, (**d**) 250 μm. Flag- and HA peptides (**c, d**) and 62 variants (**b**, left column) were stained with specific anti-Flag and anti-HA antibodies.

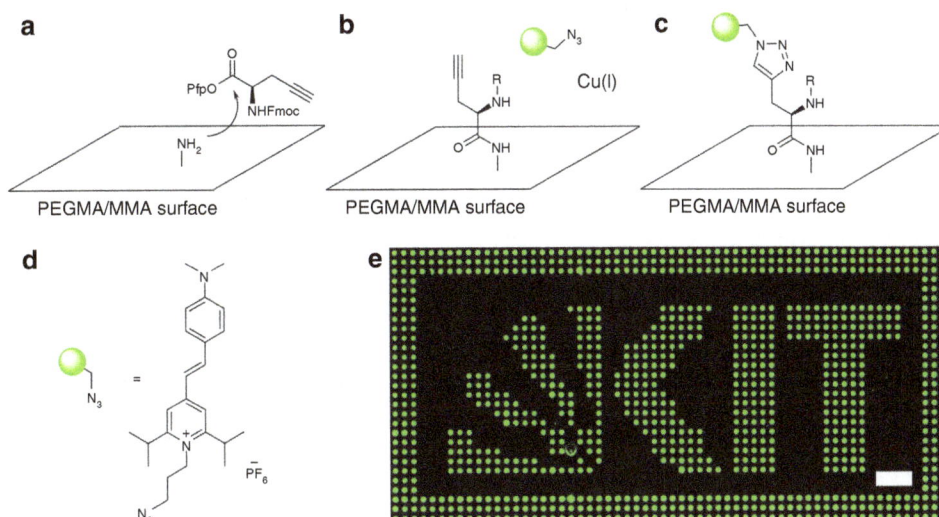

**Figure 5 | Schematic of the click chemistry reaction using an Fmoc protected propargyl-glycine-OPfp (Fmoc-Pra-OPfp).** (**a**) The synthetic amino acid (Pra, propargyl-glycine) was patterned and coupled with cLIFT. (**b-d**) Subsequently, the styrylpyridinium fluorophore was coupled to the Pra in a copper catalysed click reaction. (**e**) The fluorescent image of the patterned Pra coupled with the styrylpyridinium fluorophore via click chemistry; spot pitch 250 μm (scale bar, 1 mm).

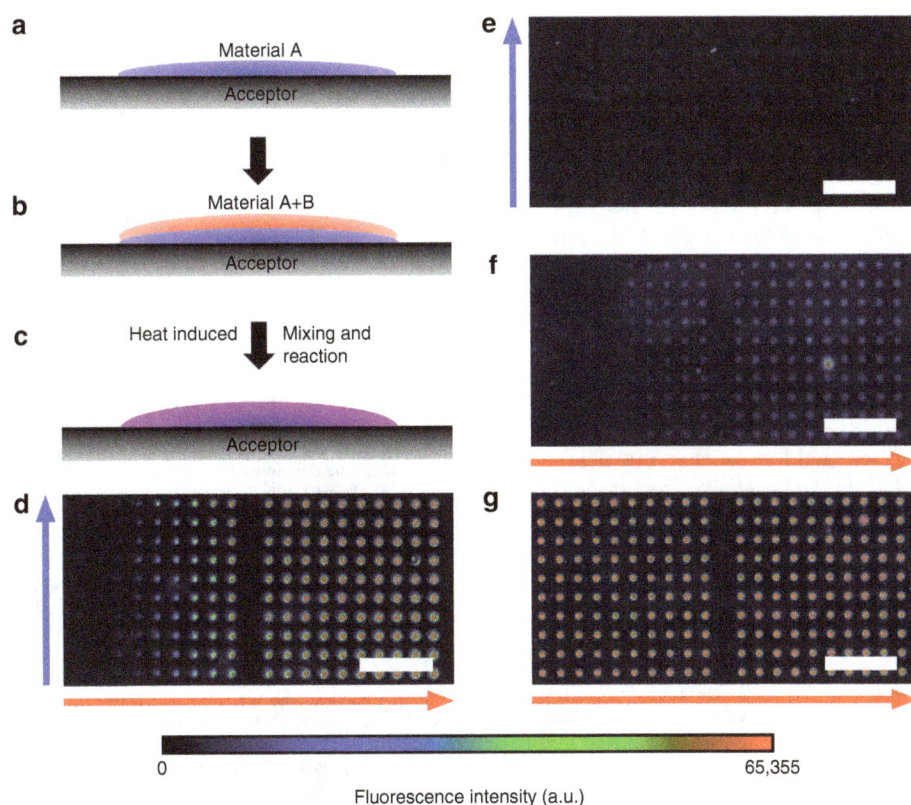

**Figure 6 | Two-step cLIFT gradient patterning experiment for solid-phase combinatorial synthesis.** (**a**) Deposition of first material. (**b**) Deposition of a second material on top of the previous pattern. (**c**) Heat-induced melting of material induces mixing and initiates the reaction. (**d-g**) Experimental results of an Fmoc-Gly-OH, reacting with DIC and HOBt activation reagents (250 μm pitch); scale bar, 1 mm. Free amino groups were stained with a rhodamine N-hydroxysuccinimide (NHS) ester dye, illustrated in rainbow colour scale. (**d**) Two layer reaction of a layer of DIC and HOBt (concentration increases in direction of blue arrow) with a second layer of Fmoc-Gly-OH (concentration increases in direction of red arrow); (**e**) one layer of DIC and HOBt as a negative control; (**f**) one layer of Fmoc-Gly-OH as a negative control; (**g**) one layer of Fmoc-Gly-OPfp as a positive control.

(Fig. 6g) also gave strong fluorescent signals. We presume that the weak fluorescent signals, which are only observed in the negative control of Fmoc-Gly-OH, are either due to nonspecific intercalation of the Fmoc-Gly-OH building block into the PEGMA/methyl methacrylate surface or result from a rare peptide bond formation

of the non-activated glycine with the amino groups on the surface, induced by the high coupling temperature (for experimental details, see 'Preparation of the donor and acceptor slides' and 'Activation reaction' in Methods). Thus, exploiting the cLIFT method, it is possible to separately pattern and precisely mix two or more types of reagents on the surface. By adjusting laser transfer (Fig. 3c) and spin-coating parameters (that is, initial concentration of reagents), it is also easily possible to predetermine the amounts and, thereby, the concentrations of the monomers and activation agents.

## Discussion

Life sciences strive to find out which molecules bind to each other, for example, which antigens are targeted by an antibody or which posttranslational modifications are crucial in specific cell signalling, for example, which acetyl-lysine-peptides are bound by a bromodomain[33]. One straightforward method to find out is the use of high-density peptide arrays, but all currently available methods to synthesize them have severe drawbacks: lithographic methods yield only short, low-quality peptides, due to the large number of coupling cycles. The SPOT synthesis yields high-quality, but low-density peptide arrays, due to solvents that tend to evaporate and spread on the surface. Particle-based methods yield medium-density peptide arrays that are limited to only 24 different building blocks, due to the limitation of the number of particle types and toner cartridges. However, simply by exchanging one donor foil for a different one, our cLIFT method can employ a theoretically unlimited number of different amino-acid building blocks. Moreover, the automated cLIFT machine does not require any expensive mechanical alignment, but rather relies on an inexpensive camera system, which reliably calibrates the laser beam with reference markers on the array surface (see Supplementary Figs 14 and 15). Thereby, we can currently synthesize arrays with a density of $>17,000$ spots per $cm^2$, which is certainly not the limit in terms of achievable array densities. It has been reported that LIFT-based structuring is possible with feature sizes of down to $3 \mu m^{17}$. Yet, another major advantage of our cLIFT method is its frugal consumption of expensive amino-acid building blocks: a few milligrams suffice to produce a donor slide, whereas the amount of transferred material is in the nanogram range. In particular, donor slides with expensive building blocks can be reused at least up to 20 times.

Although the exact mechanism of our laser-induced material transfer is still elusive, we could show that the size and the thickness of transferred spots can be adjusted by tuning the laser parameters. We exploited this feature by positioning two different and very thin material spots on top of each other, which consistently mixed on melting. As cLIFT structuring is performed with solid materials, both the *in situ* activation and the peptide elongation reaction are inhibited ('frozen'), until the heating step initiates the reaction by melting the material spots. This feature allows for structuring with many different materials that are positioned next to or on top of each other. After these structuring steps, heat-induced melting makes it possible for the amino acid building blocks to diffuse, the activation reagents to activate the amino acids and finally for the coupling of the activated amino acids to the growing peptides spot-by-spot on the array. As it is possible to control the amount of deposited materials, any combination of different reagents for peptide synthesis should be feasible. Thus, our method profits from an ever increasing number of commercially available Fmoc-protected amino-acid building blocks to synthesize, for example, peptides with different types of posttranslational modifications. Furthermore, we were able to include a synthetic amino-acid building block, showing that advanced postsynthesis functionalization (that is, click chemistry) is compatible with our synthesis approach. Currently, we have advanced our cLIFT method to nearly complete automation: we integrated a slide loader to automatically exchange donor slides, whereas

wet-chemistry processing is done in a fully automated ultra-sound-supported reactor.

## Methods

**Preparation of the donor and acceptor slides.** Donor slide preparation: microscope glass substrates were covered by self-adhesive polyimide foil (Kapton, DuPont, USA; cmc Klebetechnik GmbH, Frankenthal/Pfalz, Germany; thickness of polyimide layer $\sim 50 \mu m$, thickness of glue layer $\sim 45 \mu m$). The transfer material layer, similar to other solid material-based synthesis methods[26,27,29], was spin coated (80 r.p.s. for 45 s; solution was already applied during acceleration to 80 r.p.s.) on top of the polyimide foil. For this purpose, 10% w/w activated monomers, for example, pentafluorophenyl (OPfp)-activated amino acids with an N-terminal Fmoc-protecting group, and 90% w/w of the inert matrix polymer (SLEC PLT 7552, Sekisui Chemical GmbH, Düsseldorf/Germany) are dissolved in DCM: 15 mg of amino acid and 135 mg of resin are dissolved in 1 ml of DCM, or 8 mg of biotin and 68 mg of resin in 1 ml of DCM, respectively, due to the lower solubility of biotin in DCM. In the case of the coupling reagent donor slide, we dissolved 3.75 mg of HOBt in 12 µl of dry $N,N$-dimethylformamide (DMF) and then added 988 µl of DCM, 139 mg of matrix polymer and 9.35 µl (7,25 mg) of DIC. The donor slide with the non-activated glycine amino acid (Fmoc-Gly-OH) was prepared by dissolving 15 mg of Fmoc-Gly-OH in 25 µl of DMF and then adding 975 µl of DCM and 135 mg of matrix polymer.

Acceptor slides: The PEGMA slides were acquired from PEPperPRINT GmbH, Germany. The top surface of the acceptor slide was marked by laser ablation with a high-power laser[34]. The marks were used for determining the position of the acceptor slides with respect to the laser scanning system.

**cLIFT machine setup.** A robotic slide loader (PL200, Prior Scientific, UK) automatically handles and places the donor and acceptor slides. We use an acousto-optic modulator (1002AF1, Polytec GmbH, Germany) to switch the laser (FSDL-532-1000T, 1 W, Frankfurt Laser Company), a laser scanning system (hurrySCAN 10, Scanlab AG, Germany), an $x$–$y$ microscope stage (SCANplus $100 \times 100$, Maerzhaeuser, Germany) and a camera (DCC1645C, Thorlabs Inc., Newton, NJ, USA) with a microscope lens (PLN 4XCY, Olympus GmbH, Hamburg, Germany).

**cLIFT technique parameters.** For a pitch of 150 µm, we used the full power of the 1 W laser and a pulse duration of 5 ms. The scan head was set to a jump speed of 100 µs, a laser-off delay of 310 µs, a jump delay of 200 µs and a laser-on delay of 300 µs.

**General procedure.** After the patterning of one layer with different monomers, the coupling reaction is initiated by heating the acceptor slide in an oven to 90 °C for 60 min under argon atmosphere. Next, the acceptor slide was washed with acetone three times for 2 min (once in an ultrasonic bath). Then, the acceptor slide was dried in a jet of air. For the HA- and Flag-peptide synthesis, the patterning and coupling steps were repeated to increase the coupling yield.

To block the remaining free $NH_2$ groups on the acceptor slide, it was washed in a mixture of acetic anhydride ($Ac_2O$, 10%), $N,N$-diisopropylethylamine (DIPEA, 20%) and DMF (70%) for 30 min. Then, the slide was washed twice for 5 min with DMF.

The deprotection of $NH_2$ groups of the terminal amino acids was performed by washing the slide in a solution of piperidine (20%) and DMF (80%) for 20 min. Afterwards, the acceptor slide was washed with DMF for 5 min twice, then with acetone for 2 min twice and finally dried in a jet of air.

After the final coupling cycle, the side-chain protecting groups are cleaved from the amino acids by washing the slide three times for 30 min in a mixture of 51% v/v trifluoroacetic acid, 3% v/v triisobutylsilane (ultrapure, Sigma, USA), 44% v/v DCM and 2% v/v $H_2O$. Next, the slide was washed twice with DCM for 5 min and then with 5% DIPEA in DMF for 5 min. Finally, the slide was washed twice for 5 min each in DMF, subsequently in methanol twice for 5 min and then the substrate was dried in an air flow.

**Typical duration of synthesis for one layer.** The structuring process for 10,000 spots with 20 different donor slides on one acceptor slide currently requires $<15$ min: the transfer of one spot currently requires $<10$ ms (currently 3–10 ms, actual lasing time is 1–5 ms), which sums up to $<100$ s for 10,000 spots. The exchange of one donor slide for another currently requires about 40 s, which sums up to 800 s for 20 donor slides. Compared with this, the actual lasing time is very short. Therefore, we can easily increase the number of spots to 100,000 with only modest increase in process duration ($<30$ min). This is certainly not the end of optimization: it should be possible to decrease the donor slide handling time and the laser transfer time as well, by at least 50%. For the repetitive coupling yield and mass spectrometry experiment, lasing duration was 23 min for $\sim 215,000$ spots with 100 µm pitch and 5 ms lasing time per spot.

We only need one coupling step for all 20 amino acids in one layer, owing to the separation of transfer and chemical coupling reaction. Thus, the time required for one layer includes (1) the sequential physical patterning of the 20 different types of amino acid spots (<30 min for 100,000 spots) and (2) one coupling step (60 min at 90 °C). Next, the slide is shortly washed in acetone (<15 min) and the patterning and coupling step is repeated once before the transient Fmoc deprotection to increase the coupling efficiency (that is, double coupling). Thus, elongating all peptides on the acceptor surface by one layer with a pattern of all 20 amino acids, including coupling (twice), wet chemistry capping and Fmoc deprotection requires <5 h.

As structuring, which is rather time efficient, is separate from chemical coupling reaction, which is much more time intensive, it is easily possible to massively increase the throughput by conducting the coupling reaction and chemical washing steps in parallel with many acceptor slides at once. This is probably the most important factor, making our process more time efficient than other approaches. However, this is hard to quantify, as it depends on the actual cLIFT structuring time per array. We can currently process five acceptor slides in parallel in a chemical washing chamber.

**Activation reaction.** After the patterning of the different materials, the reaction was initiated by heating the acceptor slide in an oven to 90 °C for 90 min under argon atmosphere. To block (that is, acetylate) the remaining free $NH_2$ groups on the acceptor slide, it was incubated in a mixture of acetic anhydride (10%) + DIPEA (20%) + DMF (70%) for 30 min. Next, the slide was washed with DMF for 5 min twice. Deprotection of the $NH_2$ groups of the amino acids was achieved by washing the slide in a solution of piperidine (20%) and DMF (80%) for 20 min. Afterwards, the acceptor slide was washed twice with DMF for 5 min, twice with acetone for 2 min and then dried in a jet of air. The slide was then incubated in PBS with 0.05% Tween 20 (PBS-T) for 15 min. To stain the free amino groups, the slide was incubated with a rhodamine N-hydroxysuccinimide ester (5/6-carboxy-tetramethyl-rhodamine succinimidyl ester), diluted 1:10,000 in PBS-T and afterwards washed three times in PBS-T.

**Staining protocol with proteins.** For fluorescent staining, we use monoclonal mouse anti-HA antibodies (by Dr G. Moldenhauer, German Cancer Research Center) conjugated with a Cy5 fluorescent dye and monoclonal mouse anti-Flag M2 antibodies (Sigma) conjugated with a Cy3 dye. First, the slide is incubated in PBS-T for 15 min. Next, the surface is blocked in Rockland infrared blocking buffer (MB-070, Rockland Immunochemicals, USA) for 30 min and then washed with PBS-T for 1 min. Staining was performed for 1 h with a mixture of 10% Rockland blocking buffer (500 µl) in PBS-T (4,500 µl), adding 1:1,000 anti-HA and anti-Flag antibodies (5 µl each). Finally, the slide was washed with PBS-T three times for 3 min, briefly rinsed with distilled water and dried in a jet of air.

For the staining of the biotinylated peptides, Alexa Fluor 550-labelled streptavidin was used. The following steps were performed: the slide was washed with PBS-T for 15 min and then blocked with Rockland blocking buffer for 30 min. Afterwards, it was washed with PBS-T for 1 min and stained for 1 h in a mixture of 10% Rockland blocking buffer in PBS-T, and Alexa Fluor 550-labelled streptavidin, diluted 1:5,000, was added. Finally, the slide was washed with PBS-T three times for 3 min and washed with distilled water for 2 min.

**Image acquisition.** Fluorescent image acquisition was performed with two different fluorescent scanners: (1) a Molecular Devices (USA) Genepix 4000B fluorescent scanner at the wavelengths 532 and 635 nm with a laser power of 100%, a resolution of 5 µm and a photo multiplier gain of 470; (2) an Innopsys (France) InnoScan 1100 AL at the wavelengths 532 and 635 nm with a low laser power, the resolution set to 5 µm and a photo multiplier gain of 2. Surface topography was measured with a phase-shift interferometer Contour GT (Bruker, USA).

# References

1. Chrisey, D. B. Materials processing - the power of direct writing. *Science* **289**, 879–881 (2000).
2. Arnold, C. B. & Pique, A. Laser direct-write processing. *MRS Bull.* **32**, 9–11 (2007).
3. Hribar, K. C., Soman, P., Warner, J., Chung, P. & Chen, S. C. Light-assisted direct-write of 3D functional biomaterials. *Lab Chip* **14**, 268–275 (2014).
4. Lee, J. Y. & Lee, S. T. Laser-induced thermal imaging of polymer light-emitting materials on poly(3,4-ethylenedioxythiophene): silane hole-transport layer. *Adv. Mater.* **16**, 51–54 (2004).
5. Arnold, C. B., Wartena, R. C., Swider-Lyons, K. E. & Pique, A. Direct-write planar microultracapacitors by laser engineering. *J. Electrochem. Soc.* **150**, A571–A575 (2003).
6. Arnold, C. B., Kim, H. & Pique, A. Laser direct write of planar alkaline microbatteries. *Appl. Phys. A Mater.* **79**, 417–420 (2004).
7. Visser, C. W. *et al.* Toward 3D printing of pure metals by laser-induced forward transfer. *Adv. Mater.* **27**, 4087–4092 (2015).
8. Zywietz, U., Evlyukhin, A. B., Reinhardt, C. & Chichkov, B. N. Laser printing of silicon nanoparticles with resonant optical electric and magnetic responses. *Nat. Commun.* **5**, 3402 (2014).
9. Ellinas, K., Chatzipetrou, M., Zergioti, I., Tserepi, A. & Gogolides, E. Polymeric surfaces sustaining ultrahigh impact pressures of aqueous high-and low-surface-tension mixtures, tested with laser-induced forward transfer of drops. *Adv. Mater.* **27**, 2231–2235 (2015).
10. Chrisey, D. B. *et al.* Laser deposition of polymer and biomaterial films. *Chem. Rev.* **103**, 553–576 (2003).
11. Malda, J. *et al.* 25th Anniversary article: engineering hydrogels for biofabrication. *Adv. Mater.* **25**, 5011–5028 (2013).
12. Serra, P., Colina, M., Fernandez-Pradas, J. M., Sevilla, L. & Morenza, J. L. Preparation of functional DNA microarrays through laser-induced forward transfer. *Appl. Phys. Lett.* **85**, 1639–1641 (2004).
13. Serra, P. *et al.* Laser-induced forward transfer: a direct-writing technique for biosensors preparation. *J. Laser Micro Nanoen.* **1**, 236–242 (2006).
14. Fraunhofer Institute for Laser Technology, LIFTSYS prototype machine for transferring biomaterials–Press Release, Available at http://www.ilt.fraunhofer.de/content/dam/ilt/en/documents/Publication-and-Press/press_release/pr2013/PR_LIFTSYS_User-friendly_Machine_for_EPFL.pdf (2015).
15. Murphy, S. V. & Atala, A. 3D bioprinting of tissues and organs. *Nat. Biotechnol.* **32**, 773–785 (2014).
16. Guillotin, B. *et al.* Laser assisted bioprinting of engineered tissue with high cell density and microscale organization. *Biomaterials* **31**, 7250–7256 (2010).
17. Willis, D. A. & Grosu, V. Microdroplet deposition by laser-induced forward transfer. *Appl. Phys. Lett.* **86**, 244103 (2005).
18. Merrifield, R. B. Solid phase peptide synthesis. 1. Synthesis of a tetrapeptide. *J. Am. Chem. Soc.* **85**, 2149–2154 (1963).
19. Frank, R. Spot-synthesis - an easy technique for the positionally addressable, parallel chemical synthesis on a membrane support. *Tetrahedron* **48**, 9217–9232 (1992).
20. Fodor, S. P. A. *et al.* Light-directed, spatially addressable parallel chemical synthesis. *Science* **251**, 767–773 (1991).
21. Pellois, J. P. *et al.* Individually addressable parallel peptide synthesis on microchips. *Nat. Biotechnol.* **20**, 922–926 (2002).
22. Price, J. V. *et al.* On silico peptide microarrays for high-resolution mapping of antibody epitopes and diverse protein-protein interactions. *Nat. Med.* **18**, 1434–1440 (2012).
23. Buus, S. *et al.* High-resolution mapping of linear antibody epitopes using ultrahigh-density peptide microarrays. *Mol. Cell. Proteomics* **11**, 1790–1800 (2012).
24. Legutki, J. B. *et al.* Scalable high-density peptide arrays for comprehensive health monitoring. *Nat. Commun.* **5**, 4785 (2014).
25. Maurer, K., McShea, A., Strathmann, M. & Dill, K. The removal of the t-BOC group by electrochemically generated acid and use of an addressable electrode array for peptide synthesis. *J. Comb. Chem.* **7**, 637–640 (2005).
26. Beyer, M. *et al.* Combinatorial synthesis of peptide arrays onto a microchip. *Science* **318**, 1888 (2007).
27. Stadler, V. *et al.* Combinatorial synthesis of peptide arrays with a laser printer. *Angew. Chem. Int. Ed.* **47**, 7132–7135 (2008).
28. Loeffler, F. *et al.* Biomolecule arrays using functional combinatorial particle patterning on microchips. *Adv. Funct. Mater.* **22**, 2503–2508 (2012).
29. Maerkle, F. *et al.* High-density peptide arrays with combinatorial laser fusing. *Adv. Mater.* **26**, 3730–3734 (2014).
30. Tookmanian, E. M., Fenlon, E. E. & Brewer, S. H. Synthesis and protein incorporation of azido-modified unnatural amino acids. *RSC Adv.* **5**, 1274–1281 (2015).
31. Zimmerman, E. S. *et al.* Production of site-specific antibody-drug conjugates using optimized non-natural amino acids in a cell-free expression system. *Bioconjugate Chem.* **25**, 351–361 (2014).
32. Rudat, B. *et al.* Novel pyridinium dyes that enable investigations of peptoids at the single-molecule level. *J. Phys. Chem. B* **114**, 13473–13480 (2010).
33. Filippakopoulos, P. & Knapp, S. Targeting bromodomains: epigenetic readers of lysine acetylation. *Nat. Rev. Drug Discov.* **13**, 337–356 (2014).
34. Maerkle, F. *PhD thesis (German), Laserbasierte Verfahren zur herstellung hochdichter Peptidarrays* (Karlsruhe Institute of Technology, 2014).

# Acknowledgements

We thank Miriam Kaczynski, Richard Thelen, Alexandra Moritz and Heike Fornasier for technical assistance, and Dr Martina Schnoelzer for mass spectrometry analyses. This work was supported by funds from the ERC (grant number 277863), the HRJRG (grant number 316), the Carl-Zeiss-Stiftung (personal stipend of F.F.L.), the EU FP7 (grant number 256672) and the BMBF (grant number 031A170A and 03EK3030A). We also acknowledge the support by the Deutsche Forschungsgemeinschaft (DFG) and the Open Access Publishing Fund of the Karlsruhe Institute of Technology.

## Author contributions

F.F.L. and T.C.F. have contributed equally to this work and the order was chosen at random. F.F.L. conceived the method and conducted initial experiments. T.C.F. developed the automated machine and further advanced the method. F.F.L. and T.C.F. designed the experiments and analysed the results. R.P. supported the development and conducted the peptide synthesis experiments. M. Schlageter conducted and A.K.P. supervised the click chemistry experiment, F.-X.D. and T.S.B. provided the styrylpyridinium dye. M. Sedlmayr devised and constructed the automated wet chemistry machine and I.B. supported its mechatronic development. Under the supervision of S.B. and M.A.R.M., D.S.M., J.G. and B.R. conducted and supported chemical syntheses and analytics. C.v.B.-K. developed the laser-based generation of positioning markers. L.H., L.K.W., A.F. and V.B. supported the development of the LIFT method. F.R.B. supported the antibody stainings and fluorescence analyses. A.N.-M., F.B. and F.F.L. supervised the project. F.F.L., F.B., A.N.-M., T.C.F., D.S.M., B.R. and M. Schlageter wrote the manuscript.

## Additional information

**Competing financial interests:** F.B. is shareholder of PEPperPRINT GmbH. A.N.-M., F.F.L., C.v.B.-K. and F.B. are named on pending patent applications relating to molecule array synthesis (application number PCT/EP2013/001141, PCT/EP2014/001046, and US Patent Application 20160082406). All other authors declare no competing financial interests.

# 24

# An unforeseen polymorph of coronene by the application of magnetic fields during crystal growth

Jason Potticary[1], Lui R. Terry[1], Christopher Bell[2], Alexandros N. Papanikolopoulos[3], Peter C.M. Christianen[3], Hans Engelkamp[3], Andrew M. Collins[4], Claudio Fontanesi[5,6], Gabriele Kociok-Köhn[7], Simon Crampin[5], Enrico Da Como[5] & Simon R. Hall[1]

The continued development of novel drugs, proteins, and advanced materials strongly rely on our ability to self-assemble molecules in solids with the most suitable structure (polymorph) in order to exhibit desired functionalities. The search for new polymorphs remains a scientific challenge, that is at the core of crystal engineering and there has been a lack of effective solutions to this problem. Here we show that by crystallizing the polyaromatic hydrocarbon coronene in the presence of a magnetic field, a polymorph is formed in a β-herringbone structure instead of the ubiquitous γ-herringbone structure, with a decrease of 35° in the herringbone nearest neighbour angle. The β-herringbone polymorph is stable, preserves its structure under ambient conditions and as a result of the altered molecular packing of the crystals, exhibits significant changes to the optical and mechanical properties of the crystal.

[1] Complex Functional Materials Group, School of Chemistry, University of Bristol, Bristol BS8 1TS, UK. [2] School of Physics, HH Wills Physics Laboratory, Tyndall Avenue, Bristol BS8 1TL, UK. [3] High Field Magnet Laboratory (HFML-EMFL), Radboud University, Toernooiveld 7, 6525 ED Nijmegen, The Netherlands. [4] Bristol Centre for Functional Nanomaterials, HH Wills Physics Laboratory, Tyndall Avenue, Bristol BS8 1TL, UK. [5] Department of Physics, University of Bath, Claverton Down, Bath BA2 7AY, UK. [6] Dipartimento di Ingegneria Enzo Ferrari, Universita' di Modena e Reggio Emilia, Via Vivarelli 10, 41125 Modena, Italy. [7] Department of Chemistry, University of Bath, Claverton Down, Bath BA2 7AY, UK. Correspondence and requests for materials should be addressed to S.R.H. (email: simon.hall@bristol.ac.uk).

The ability to discover new phenomena and properties in materials depends on our ability to synthesize new structures. The discovery of new structures, however, should rely on variables that transcend the typical thermodynamic parameters of temperature and pressure. In the field of functional molecular materials, polymorphism, the presence of different crystal structures of the same molecular system, can be an opportunity to discover novel phenomena[1] and tune properties[2]. For example, charge carrier mobility in organic semiconductors can be increased by crystallizing molecules under pressure, resulting in shorter intermolecular distances favourable for transport[3]. One successful tool at the scientist's disposal to control crystal structure has been to perform crystallization in the presence of electric fields. This is a well-known process and is used extensively to prepare nonlinear optical materials through electric poling. Crystal growth in the presence of an external magnetic field is much less explored, although it has been proven to be efficacious in the melt texturing of alloys[4] or in cases where controlled convection of the crystallizing solution is required, for example, in the growth of high-quality protein crystals[5], chiral aggregates[6] or in the alignment of liquid crystal and block copolymer arrays[7,8]. A magnetic field has even been demonstrated to be able to separate polymorphs of crystals post-synthetically[9] and in one case to preferentially nucleate a monoclinic form of terpyridine[10]. Although it is known that magnetic forces can have an effect on solidification and subsequent physical behaviour in crystals, their use to create previously unknown polymorphs in single crystals has never been reported.

Polyaromatic hydrocarbons (PAHs) are commonly researched molecules due to their rigid planar structure, high stabilities and characteristic optical and electronic behaviour[10,11]. As molecular solids, PAHs crystallize in four basic structure types according to well-defined geometric and energetic considerations[12]. These are the herringbone structure, the gamma-herringbone ($\gamma$-) structure, the sandwich-herringbone (SHB) structure and the beta-herringbone ($\beta$-) structure (Supplementary Fig. 1). Comprehensive studies of PAHs have shown that the adoption of one of the four structure types depends ultimately on the relative strength of nonbonded $C \cdots C$ and $C \cdots H$ interactions[12–14]. Polymorphism in PAHs has been demonstrated previously by growing thin films on substrates, where variations in processing conditions direct the formation of different polymorphs[15,16], although in single crystals at ambient pressure, polymorphism in PAHs is rare with perylene and pyrene being two notable cases exhibiting both herringbone and SHB polymorphs[17]. In terms of applications, single crystals of PAHs are to be desired, however, as they have distinct physical advantages over thin-film versions of the same material. Most importantly, for conductive applications such as transistors, or optical applications in solar cells, single crystals typically have higher carrier mobilities than their thin-film analogues[18,19], which have to be defect free to achieve mobilities of the same order of magnitude.

Here we show that by crystallizing the PAH coronene in the presence of a magnetic field, it can be made to form as a new $\beta$-herringbone polymorph instead of the ubiquitous $\gamma$-herringbone form, with a change as large as 35° to the herringbone nearest neighbour angle. The $\beta$ polymorph is stable and can preserve its structure in ambient conditions and zero magnetic field. Dispersion corrected density functional theory (DFT-D) calculations indicate that the new form is energetically favoured at low temperature. Furthermore, we demonstrate how the new supramolecular structure generates remarkable changes of the electronic, optical and mechanical properties in the crystal.

## Results

**Crystal growth of a new polymorph.** Coronene is a PAH composed of six aromatic rings arranged in a planar discoidal geometry (Supplementary Fig. 2). The high molecular symmetry ($D_{6h}$) and 24 electron $\pi$-system has made coronene an ideal model system for the study of graphene, due to it being large enough to display exotic electronic behaviour, but not so large that contortion becomes a complicating factor[20]. Centimetre-long crystals of coronene (typically $\sim 0.75$ cm) were grown from a supersaturated solution of the molecules in toluene cooled slowly ($0.04$ K min$^{-1}$) from 328 K to 298 K over a period of 12 h (Fig. 1a, yellow crystal labelled '$\gamma$'). The very slow cooling rate was chosen to eliminate changes in solution shearing which have been shown to influence nucleation and growth with concomitant changes to the resultant crystal structure[21]. Single crystal X-ray diffraction (XRD) of these crystals indicated that the structure is the conventional $\gamma$-polymorph; $a = 10.02$ Å, $b = 4.67$ Å, $c = 15.60$ Å, $\beta = 106.7°$, $Z = 2$, space group P2$_1$/n, with an inter-planar distance ($d_\pi$) of 3.43 Å, which is consistent with parallel $\pi$-stacking[12] (Fig. 2a,b,e,g).

In the presence of an external magnetic field of 1 Tesla, however, significantly longer (typically $\sim 2.5$ cm) coronene crystals were grown from the same supersaturated solution and exhibit a different colour to normal $\gamma$-coronene crystals (Fig. 1a orange crystal labelled '$\beta$'). Single crystal XRD indicates that the unit cell of these crystals are consistent with a new $\beta$-coronene structure; $a = 10.39$ Å, $b = 3.84$ Å, $c = 17.23$ Å, $\beta = 96.24°$, $Z = 2$, space group P2$_1$/n and an inter-planar distance ($d_\pi$) of 3.48 Å (Fig. 2c,d,f,h). As a previously unreported polymorph of coronene, the crystallographic data has been lodged with the Cambridge Crystallographic Data Centre (deposition number CCDC 1409823).

By comparing the two crystal forms in Fig. 2c, it can be seen that the short axis, $b$, is substantially decreased in length for the $\beta$-coronene crystallized in the magnetic field when compared with the $\gamma$-coronene. This new polymorph also has a significantly smaller nearest neighbour herringbone angle of 49.71° compared with 95.86° in $\gamma$-coronene. The identification of the new crystal polymorph as a $\beta$-structure can be made by reference to a plot of the inter-planar angle of nearest neighbour molecules versus the unit cell short axis[12] (Fig. 3). From this, it can be seen that the new polymorph sits squarely with other PAHs identified as

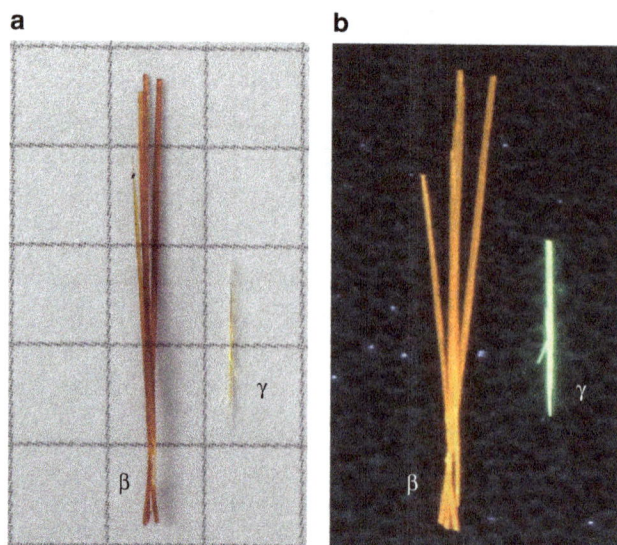

**Figure 1 | Optical images of the β- and γ- polymorphs of coronene. (a)** in daylight and (**b**) under ultraviolet ($\lambda = 365$ nm) illumination to show fluorescence. The squares on the grid in **a** are $0.5 \times 0.5$ cm$^2$.

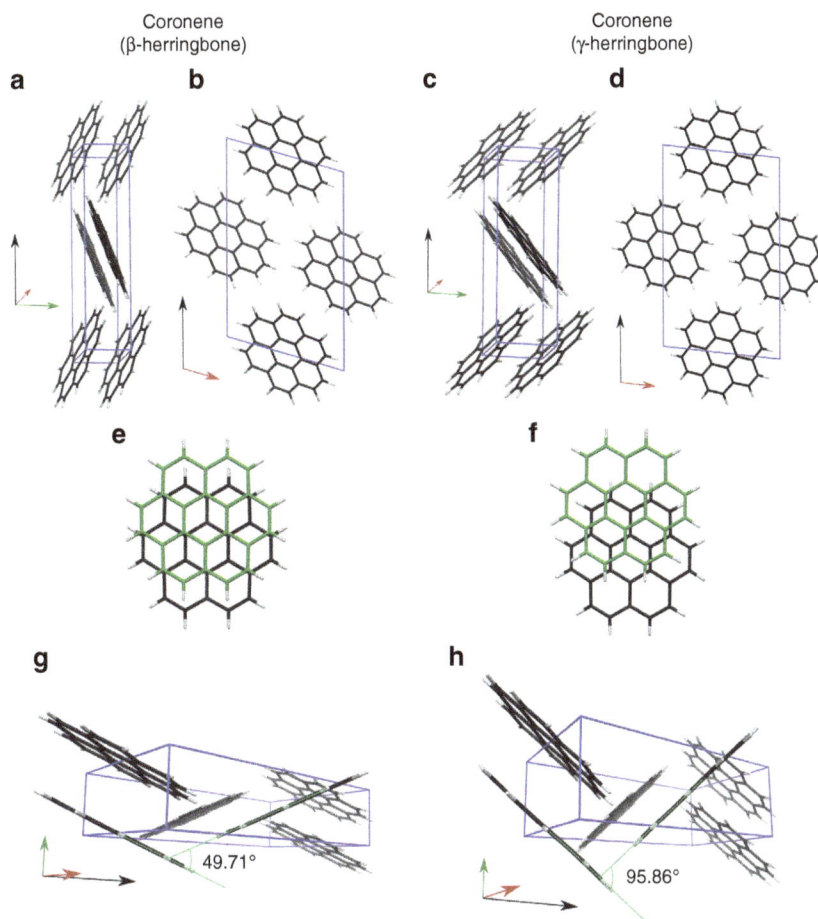

**Figure 2 | Representation of the β- and γ- polymorphs of coronene.** Differing perspectives of both unit cells (blue boxes) viewed slightly offset from along the *a*-axis (**a** and **c**) and along the *b*-axis (**b** and **d**). The relative shift of the molecules along the stacks are shown for β- (**e**) and γ- (**f**). (**g**) and (**h**) show an orientation of the unit cell clearly demonstrating the difference in nearest neighbour angle between the two polymorphs. Red green and black arrows indicate the direction of the *a-*, *b-* and *c*-axis respectively.

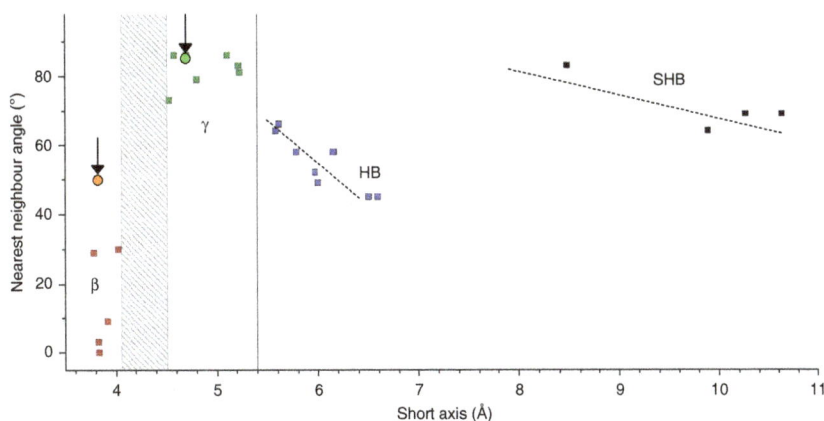

**Figure 3 | Grouping of PAHs into structure types.** Plotted on the graph are PAHs that adopt the herringbone structure (blue squares), the gamma-herringbone (γ-) structure (green squares), the SHB structure (black squares) and the beta-herringbone (β-) structure (red squares), according to the crystallographic short axis and nearest neighbour herringbone angle. The positions of both γ- and β-coronene polymorphs are indicated by circles and marked with arrows. Adapted from Desiraju and Gavezzotti[12], in which the names of the crystals corresponding to all of the marked squares can be found. Reproduced with permission of the International Union of Crystallography, http://journals.iucr.org/.

having the β-herringbone structure. Under an applied magnetic field of 1 T, the new β-coronene polymorph grows exclusively, is reproducible (>10 times to-date), stable under ambient conditions and also forms in the non-aromatic solvent hexane.

To determine the mechanism of crystal growth, we followed the inception of crystallization of coronene at 1 T and 0 T via *in situ* UV–vis spectrometry. At 358 nm, there are no peaks due to molecules of coronene in solution, nor in the solid state (as crystals). However, a peak will be observed due to the formation of nucleation clusters, which we determined via a single-point energy calculation using time-dependent self-consistent field density functional theory, using basis set

6-31G(D). Supplementary Figure 3 shows that there is a stark difference between the growth of coronene crystals under 1 T of applied field and at 0 T. At 1 T, there is suppression of nucleation which results in the crystallization of the least stable polymorph, that is, beta-coronene. This is entirely consistent with previously published work on the formation of meta-stable polymorphs and is epitomized in Ostwald's rules of stages[22]. Our experimental data therefore demonstrates that the magnetic field is achieving the effect of polymorphic control through suppression of nucleation.

As to the question of why an applied magnetic field of 1 T is having this effect in coronene and not in other molecules at this field strength, it is likely to be intimately linked to the magnetic susceptibility of the molecule. Based on inductively coupled plasma atomic absorption spectroscopy, we note that the level of magnetically active impurities such as cobalt, iron and nickel are at the parts per billion level ($Co = 0.138 \pm 0.98$ ppb; $Fe = 33.81 \pm 0.56$ ppb; and $Ni = 13.46 \pm 0.65$ ppb), and so are unlikely to play any role in polymorph selection. Instead we note that coronene has a very high diamagnetic susceptibility ($\chi$) of $-243 \times 10^{-6}$ emu mol$^{-1}$ (refs 23,24). Crystal growth performed at weaker field strengths of 0.2, 0.5 and 0.8 T resulted in crystals of the $\gamma$ – polymorph (Supplementary Fig. 4), suggesting that 1 T is close to a threshold for energetic selection between the two forms. In the case of other PAHs such as pyrene, which do not have such a strong diamagnetic

susceptibility, our attempts to control polymorphism at 1 T were unsuccessful. It would appear that for molecules with lower diamagnetic susceptibilities, higher fields would be required. Indeed, work reported in a thesis in 2013 has shown that a simple organic dye molecule with a lower diamagnetic susceptibility than coronene, namely an isoxazolone derivative, can be crystallized in a different known polymorph at fields $>2.5$ T (ref. 25).

**Optical behaviour.** Figure 1b shows optical images of the two different polymorphs taken under UV irradiation ($\lambda = 365$ nm). The remarkably different colours of the fluorescence of $\gamma$-coronene and $\beta$-coronene crystals is further confirmation that the molecular packing in the crystal has been transformed, leading to altered electronic behaviour. To quantify this, we measured the absorption spectrum of two single crystals of the two polymorphs by shining unpolarized light perpendicular to the $a$–$b$ plane of the crystals, that is, light propagation parallel to [001]. As shown in Fig. 4a (green plot), the $\gamma$-coronene single crystal is characterized by a first absorption resonance at 468 nm assigned to the free exciton in coronene, in agreement with previous studies[26]. The $\beta$-coronene spectrum is by stark contrast almost featureless, with an absorption onset at 780 nm and a maximum at $\sim 500$ nm (Fig. 4a, orange plot). This is a remarkable change in the optical properties of this material. These data can be plotted as the extinction molar coefficient

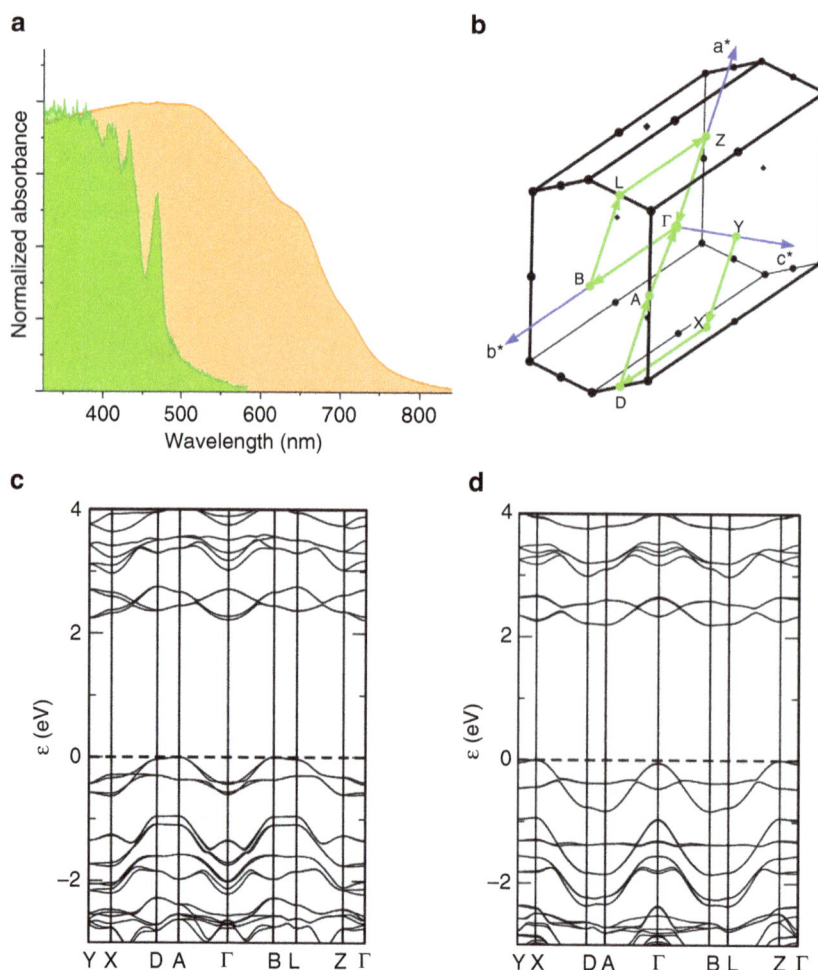

**Figure 4 | Electronic structure of γ- and β-coronene.** (**a**) Absorption spectra of γ-coronene (green) and β-coronene (orange) single crystals. Unpolarized light was irradiated perpendicular to the *a*–*b* plane at room temperature; (**b**) Brillouin zone with reciprocal lattice vectors and high symmetry points; (**c**) and (**d**) band dispersion along high symmetry points in γ- and β-coronene, respectively.

versus wavelength, to decouple any effects due to differences in the thickness of the crystals of the two polymorphs (50 microns and 76 microns for $\gamma$-coronene and $\beta$-coronene, respectively; Supplementary Fig. 5). From these spectra, it can be seen clearly that $\beta$-coronene is a more strongly absorbing polymorph than $\gamma$-coronene. Organic materials tend to be sensitive to a particular range of wavelengths, which can be seen in the features of the $\gamma$-coronene absorption spectrum and the lack of absorption of wavelengths >583 nm. In $\beta$-coronene, however, the crystal strongly absorbs over a wide band of radiation, from the UV into the near-IR (320–847 nm).

DFT-D calculations correctly describe the existence of the new stable polymorph and indicate that no appreciable difference in the indirect bandgap (Fig. 4b–d) exists between the two polymorphs, suggesting that the large shift in the light absorption onset is instead related to a change in the fundamental photoexcitations in the two structures[27,28]. In $\beta$-coronene it is difficult to identify sharp resonances from Frenkel excitons as observed for the $\gamma$-polymorph. Instead, the structureless absorption band of the $\beta$ – polymorph originates from charge transfer excitons that are favoured by the larger overlap between the molecular $\pi$ orbitals. Moreover, the transition dipole moment of a charge transfer exciton is likely to be oriented parallel to the $a$–$b$ plane, which would ensure optimal coupling with the electromagnetic radiation in our experiment.

**Structural stability.** The new $\beta$-polymorph structure is notable for the large change in the herringbone angle. We can gain insights into the relative stability of the $\beta$-coronene structure when compared with the $\gamma$-polymorph, by consideration of these changes in the $CH \cdots \pi$ hydrogen bonding motif. It is known that the stronger the hydrogen bond, the stronger the trend for linearity[29,30] depending on the strength of the proton donor. From XRD, we can see that the $CH \cdots \pi$ angle in $\gamma$-coronene of 95.86° suggests a strong, almost linear $CH \cdots \pi$ hydrogen bond, whereas in $\beta$-coronene, this angle becomes 49.71° (viz. Fig. 2). This change in angle would suggest a weakening of the hydrogen bonding in $\beta$-coronene, which is concomitant with the change in estimated $CH \cdots \pi$ hydrogen bond length[31] that increases from 2.5 Å in the $\gamma$-polymorph to 3.0 Å in the $\beta$-polymorph. This weakening of the $CH \cdots \pi$ hydrogen bonding should therefore manifest itself as a diminution of the physicomechanical properties of $\beta$-coronene. To confirm this, we have determined the melting point ($T_M$) and elastic modulus ($E$—the modal value measured on the ($\bar{1}011$ and $10\bar{1}1$ crystal faces) of both polymorphs and find that in $\gamma$-coronene, $T_M = 436.14 \pm 0.01$ °C and $E = 227$ GPa, whereas for $\beta$-coronene $T_M = 435.48 \pm 0.01$ °C and $E = 92$ GPa (Supplementary Figs 6 and 7). These data suggest weaker intermolecular forces in $\beta$-coronene at room temperature.

DFT-D calculations show that $\beta$-coronene is actually the more stable of the two polymorphs at 0 K, by 3.8 kJ mol$^{-1}$ based on lattice energy differences (Supplementary Table 1). Confirmation that this is the case comes from powder XRD of polycrystalline $\gamma$-coronene recorded through a temperature cycle from room temperature to 12 K. On cooling through a temperature of 150 K, three new peaks emerge which cannot be indexed to the $\gamma$ – polymorph (JCPDS card number 12-1611; Supplementary Fig. 8). These emergent peaks correspond to the (002), (101) and (112; $2\theta = 10.55°$, $10.67°$ and $27.72°$ respectively) reflections of the new polymorph, $\beta$-coronene. Thus $\beta$- and $\gamma$-coronene are enantiotropic polymorphs, with a critical temperature between 100 K and 150 K (Fig. 5). In general, in an enantiotropic pair of polymorphs, the more intrinsically stable member will have the greater specific enthalpy of fusion. As the differential scanning calorimetry (DSC) experiment to determine melting point

(Supplementary Fig. 6) is conducted at far higher temperatures than 150 K, it is the $\gamma$-coronene polymorph which will be the more stable at these temperatures and therefore be expected to have the greater specific enthalpy of fusion. From Supplementary Fig. 5, we can calculate the values for the heat of fusion for each of the polymorphs as 20.78 kJ mol$^{-1}$ and 25.26 kJ mol$^{-1}$ for $\beta$- and $\gamma$-coronene, respectively, confirming that these are enantiotropic polymorphs.

In summary, through judicious application of a magnetic field, we have demonstrated that a well-studied PAH can be grown as a new polymorph and confirmed that a range of physical properties have been significantly altered. The optical properties of the $\beta$-coronene crystal have been transformed to such an extent that it absorbs panchromatically from the UV into the near-IR. This new single crystal may therefore be of great interest in organic optoelectronics and photovoltaics.

## Methods

**Coronene.** Coronene crystals (Sigma Aldrich, purity 97%) were recrystallized from toluene and purity assayed by nuclear magnetic resonance ($C^{13}$ and $H^+$). Trace metals were tested for using inductively coupled plasma—atomic emission spectroscopy (Agilent 710 ICP-AES) after digestion in 3% nitric acid (Sigma Aldrich, $\geq$99.999%, trace metal basis). Structural analysis was conducted by optical microscopy and single crystal XRD. Analysis of the single crystal data of non-magnetic field grown crystals suggests favourable growth along the $b$-axis, which form the characteristic needle shaped crystals.

**Crystal growth under 1 T of magnetic field.** A supersaturated solution of coronene (2.5 mg ml$^{-1}$) in toluene was prepared and stored in an oven at 93 °C. The solution was then passed through a 0.22-μm PTFE filter directly into a 5-mm quartz cuvette with a stopper. Once sealed, the cuvette was placed in the magnetic cavity (Supplementary Fig. 9) and the whole system maintained at 93 °C for 4 h post cuvette insertion. The oven was then programmed to cool to 83, 73 and 63 °C then finally 50 °C, with a 24-h dwell at each temperature, before cooling to room temperature. Crystals of $\beta$-coronene are the sole polymorph to grow under the magnetic field and to-date the experiment has been repeated >10 times.

**Single crystal X-ray crystallography.** Intensity data for all coronene structures were collected at different temperatures on an Agilent SuperNova-E Dual diffractometer equipped with an Oxford Cryosystem, using CuKα radiation ($\lambda = 1.5418$ Å). Data were processed using the CrysAlisPro software (CrysAlisPro, Agilent Technologies, Version 1.171.37.35 (release 13-08-2014 CrysAlis171 .NET; compiled on 13 Aug 2014). For all structures a symmetry-related (multi-scan) absorption correction was applied. Crystal parameters are provided in Supplementary Table 2. See also Supplementary Data 1 and 2. Structure solution, followed by full-matrix least squares refinement was performed using the WINGX-v2014.1 suite of programs throughout.

**Physical characterization.** Optical absorption spectra were recorded with a ultraviolet-visible–near infrared spectrometer Agilent Cary 5000 measuring in transmission configuration. The spectrometer can measure absorbance up an optical density of 10. Single crystals of $\beta$- or $\gamma$-coronene were suspended in the sample beam path. Unpolarized light was shone perpendicular to the $a$–$b$ plane of the unit cells. High-quality crystals with flat surfaces were carefully selected to avoid light scattering effects. The beam size was narrowed to half the lateral size of crystals. All absorption spectra were recorded at room temperature in air. Elastic moduli were determined using an Asylum Research MFP-3D Infinity AFM operating in AMFM viscoelastic imaging mode, using an AC160TS-R3 silicon tip (9 $\pm$ 2 nm radius). Freshly cleaved mica was used as a calibration standard (measured at 178 GPa). Melting point determination was done using a TA-Instruments Q100 DSC with temperature ramp of 1 °C min$^{-1}$ between 35 and 460 °C with 2.2 mg of coronene hermetically sealed under a nitrogen atmosphere in an aluminium pan.

**Computational calculations.** Computational calculations have been performed with density functional codes CASTEP (ref. 32) and VASP (ref. 33), using the Perdew–Burke–Enzerhof exchange correlation functional with semi-empirical dispersion corrections (DFT-D) to account for van der Waals interactions. CASTEP calculations (version 7.03) used in-built ultra-soft pseudopotentials for C and H atoms, a plane wave cutoff of 600 eV, and Monkhorst-Pack[34] k-point samplings of $3 \times 4 \times 2$. Cell parameters and atomic coordinates were fully relaxed using the Broyden–Fletcher–Goldfarb–Shanno (BFGS) method, halting when residual forces fell below 1 meV Å$^{-1}$. Changing the cutoff energy from 450 to 600 eV caused structural parameters to change by <0.001 nm or <0.01°. Increasing k-point sampling to $6 \times 8 \times 4$ changed energies by <1 meV. Robustness

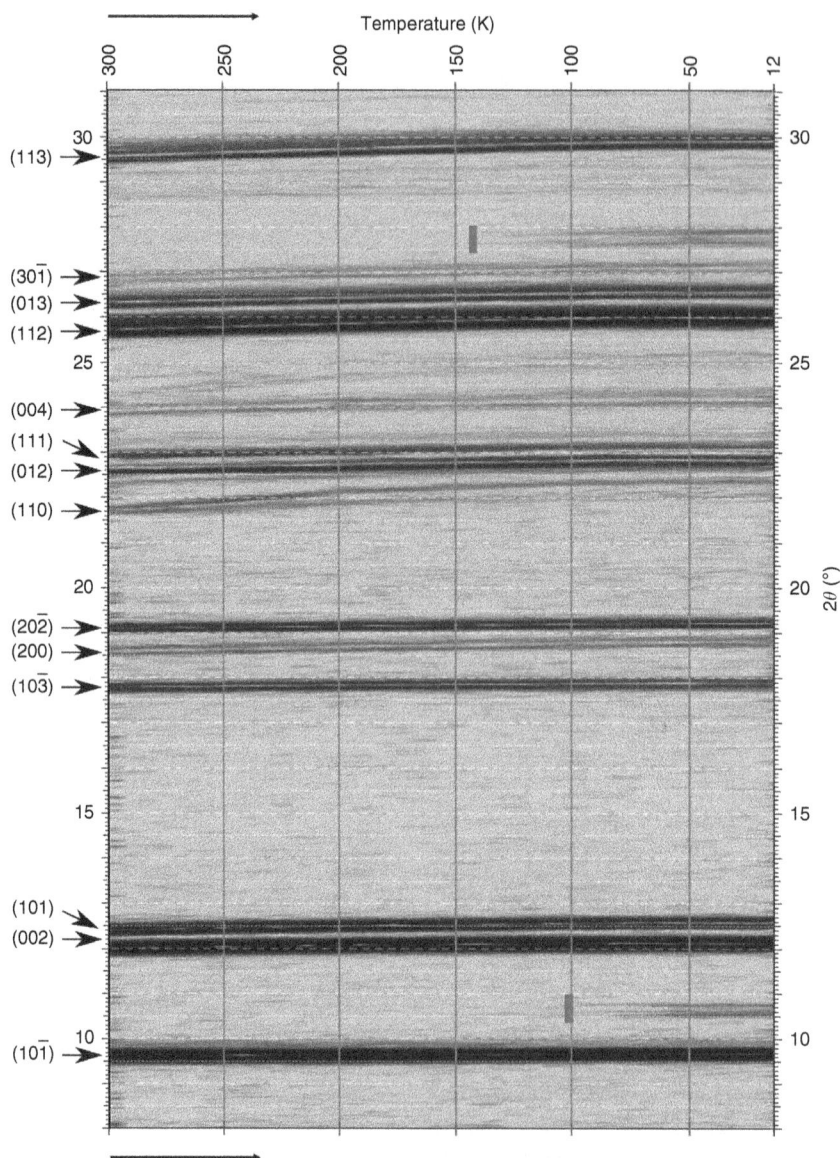

**Figure 5 | 2D map of XRD peak shifts as a function of temperature in γ-coronene.** Temperature is decreased from 300 to 12 K (left to right). Blue markers indicate the emergence of the new peaks due to β-coronene formation. The blue line near 10° 2θ encompasses two emergent peaks.

of results to choice of semi-empirical dispersion correction was assessed through the use[35] of using the Grimme scheme[36] with both default vdW radii ($R_H = 1.001$, $R_C = 1.452$) and experimental ($R_H = 1.090$, $R_C = 1.750$), as previously employed by Fedorov et al.[37] for γ-coronene. VASP calculations (version 5.3.3) use PAW potentials[38] with 500 eV energy cutoff, $3 \times 4 \times 2$ k-point sampling and the Tkatchenko-Scheffler[39,40] dispersion correction. Geometry optimization was performed starting from experimental cell parameters and atom coordinates, aboth with and without symmetry constraint. Stability of optimized geometries was verified by re-optimizing after randomly displacing atoms by 0.05 Å in $x$, $y$ and $z$.

### References

1. Ganin, A. Y. et al. Polymorphism control of superconductivity and magnetism in Cs₃C₆₀ close to the Mott transition. *Nature* **466,** 221–225 (2010).
2. Bernstein, J. *Polymorphism in Molecular Crystals* (Oxford University Press, 2002).
3. Giri, G. et al. Tuning charge transport in solution-sheared organic semiconductors using lattice strain. *Nature* **480,** 504–508 (2011).
4. Beaugnon, E. et al. Material processing in high static magnetic field. A review of an experimental study on levitation, phase separation, convection and texturation. *J. Phys. I France* **3,** 399–421 (1993).
5. Nakamura, A. et al. Improvement in quality of protein crystals grown in a high magnetic field gradient. *Cryst. Growth Des.* **12,** 1141–1150 (2012).
6. Micali, N. et al. Selection of supramolecular chirality by application of rotational and magnetic forces. *Nat. Chem.* **4,** 201–207 (2012).
7. Boamfa, M. I., Lazarenko, S. V., Vermolen, E. C. M., Kirilyuk, A. & Rasing, T. Magnetic field alignment of liquid crystals for fast display applications. *Adv. Mater.* **17,** 610–614 (2005).
8. Gopinadhan, M. et al. Thermally switchable aligned nanopores by magnetic-field directed self-assembly of block copolymers. *Adv. Mater.* **26,** 5148–5154 (2014).
9. Atkinson, M. B. J. et al. Using magnetic levitation to separate mixtures of crystal polymorphs. *Angew. Chem. Int. Ed.* **52,** 10208–10211 (2013).
10. Honjo, S., Yokota, M., Doki, N. & Shimizu, K. Magnetic field influence on the crystal structure of 2,2':6',2"-terpyridine. *Kagaku Kogaku Ronbun.* **34,** 383–387 (2008).
11. Rieger, R. & Müllen, K. Forever young: polycyclic aromatic hydrocarbons as model cases for structural and optical studies. *J. Phys. Org. Chem.* **23,** 315–325 (2010).
12. Desiraju, G. R. & Gavezzotti, A. Crystal structures of polynuclear aromatic hydrocarbons. Classification, rationalization and prediction from molecular structure. *Acta Crystallogr. Sect. B Struct. Sci.* **45,** 473–482 (1989).
13. Loots, L. & Barbour, L. J. A simple and robust method for the identification of π–π packing motifs of aromatic compounds. *CrystEngComm* **14,** 300–304 (2014).

14. Schatschneider, B., Phelps, J. & Jezowski, S. A new parameter for classification of polycyclic aromatic hydrocarbon crystalline motifs: a Hirshfeld surface investigation. *CrystEngComm* **11,** 7216–7223 (2011).
15. Mattheus, C. C. *et al.* Identification of polymorphs of pentacene. *Synth. Metals* **138,** 475–481 (2003).
16. Hiszpanski, A. M. *et al.* Tuning polymorphism and orientation in organic semiconductor thin films via post-deposition processing. *J. Am. Chem. Soc.* **136,** 15749–15756 (2014).
17. Botoshansky, M., Herbstein, R. H. & Kapon, M. Towards a complete description of a polymorphic crystal: the example of perylene redetermination of the structures of the (Z = 2 and 4) polymorphs. *Helv. Chim. Acta* **86,** 1113–1128 (2003).
18. Hu, W.-S., Weng, S.-Z., Tao, Y.-T., Liu, H.-J. & Lee, H.-Y. Oriented growth of rubrene thin films on aligned pentacene buffer layer and its anisotropic thin-film transistor characteristics. *Org. Electron.* **9,** 385–395 (2008).
19. Hamaguchi, A. *et al.* Single-crystal-like organic thin-film transistors fabricated from dinaphtho[2,3- b :2′,3′- f ]thieno[3,2- b ]thiophene (DNTT) precursor-polystyrene blends. *Adv. Mater.* **27,** 6606–6611 (2015).
20. Zhao, Y. & Truhlar, D. A prototype for graphene material simulation: structures and interaction potentials of coronene dimers. *J. Phys. Chem. C* **112,** 4061–4067 (2008).
21. Giri, G. *et al.* Effect of solution shearing method on packing and disorder of organic semiconductor polymers. *Chem. Mater.* **27,** 2350–2359 (2015).
22. Ostwald, W. Studien über die bildung und umwandlung fester körper. *Z. Phys. Chem.* **22,** 289–330 (1897).
23. Steiner, E., Fowler, P. W. & Jenneskens, L. W. Counter-rotating ring currents in coronene and corannulene. *Angew. Chem. Int. Ed.* **40,** 362–366 (2001).
24. Rogers, F. M. T. The magnetic anisotropy of coronene, naphthazarin, and other crystals. *J. Am. Chem. Soc.* **69,** 1506–1508 (1947).
25. Aret, E. Growth of Organic Dye Crystals, PhD Thesis, University of Radboud, Nijmegen. http://repository.ubn.ru.nl/handle/2066/112754 (2013).
26. Yamamoto, T. *et al.* Exciton-phonon coupling and pressure-induced structural phase changes in coronene crystals. *Chem. Phys.* **184,** 247–254 (1994).
27. Hummer, K., Puschnig, P. & Ambrosch-Draxl, C. Lowest optical excitations in molecular crystals: bound excitons versus free electron-hole pairs in anthracene. *Phys. Rev. Lett.* **92,** 147402 (2004).
28. Cudazzo, P., Sottile, F., Rubio, A. & Gatti, M. Exciton dispersion in molecular solids. *J. Phys.-Condes. Matter* **27,** 113204 (2015).
29. Nishio, M., Umezawa, Y., Suezawa, H. & Tsuboyama, S. in *The Importance of Pi-Interactions in Crystal Engineering: Frontiers in Crystal Engineering* 2nd edn (eds Tiekink, E. R. T. & Zukerman-Schpector, J.) 1-39 (Wiley (2012).
30. Steiner, T. & Desiraju, G. R. Distinction between the weak hydrogen bond and the van der Waals interaction. *Chem. Comm.* **8,** 891–892 (1998).
31. Allen, F. H. A systematic pairwise comparison of geometric parameters obtained by X-ray and neutron diffraction. *Acta Crystallogr. B* **42,** 515–522 (1986).
32. Clark, S. *et al.* First principles methods using CASTEP. *Z. Kristallogr.* **220,** 567 (2005).
33. Kresse, G. & Furthmüller, J. Efficient iterative schemes for ab initio total-energy calculations using a plane-wave basis set. *Phys. Rev. B* **54,** 11169 (1996).
34. Monkhorst, H. J. & Pack, J. D. Special points for Brillouin-zone integrations. *Phys. Rev. B* **13,** 5188 (1976).
35. McNellis, E. R., Meyer, J. & Reuter, K. Azobenzene at coinage metal surfaces: role of dispersive van der Waals interactions. *Phys. Rev. B* **80,** 205414 (2009).
36. Grimme, S. Semi-empirical GGA-type density functional constructed with a long-range dispersion correction. *J. Comput. Chem.* **27,** 1787 (2006).
37. Fedorov, I., Zhuravlev, Y. & Berveno, V. Properties of crystalline coronene: dispersion forces leading to a larger van der Waals radius for carbon. *Phys. Status Solidi B* **249,** 1438–1444 (2012).
38. Blöchl, P. E. Projector augmented-wave method. *Phys. Rev. B.* **50,** 17953 (1994).
39. Tkatchenko, A. & Scheffler, M. Accurate molecular Van der waals interactions from ground-state electron density and free-atom reference data. *Phys. Rev. Lett.* **102,** 073005 (2009).
40. Bučko, T., Lebègue, S., Hafner, J. & Ángyán, J. G. Tkatchenko-Scheffler van der Waals correction method with and without self-consistent screening applied to solids. *Phys. Rev. B.* **87,** 064110 (2013).

## Acknowledgements

S.R.H. and J.P. acknowledge the Engineering and Physical Sciences Research Council (EPSRC), UK (grant EP/G036780/1) and the Bristol Centre for Functional Nanomaterials for project funding. The work at Nijmegen was performed as part of the research programme of the 'Stichting voor Fundamenteel Onderzoek der Materie (FOM)', which is financially supported by the 'Nederlandse Organisatie voor Wetenschappelijk Onderzoek (NWO)'. In addition, S.R.H. and J.P. thank Stasja Stanišić for discussions on low-temperature X-ray crystallography and Spartacus Mills for estimations of magnitude. E.D.C. and S.R.H. are grateful to the GW4 for funding through an accelerator grant and L.R.T. is supported in part by a grant from the USAF European Office of Aerospace Research and Development (FA8655-12-1-2078). All authors thank Prof. John Evans and Liana Vella-Zarb of the University of Durham for low-temperature powder XRD, and Dr Chung Choi and Dr George Whittell of the University of Bristol for ICP-AES and DSC, respectively.

## Author contributions

S.R.H. initiated and supervised the project. J.P. performed the crystallization experiments at Bristol and characterized the samples optically and with C.B. undertook magnetic characterizations. J.P., A.N.P., P.C.M.C and H.E. performed the crystallization experiments at Nijmegen and undertook ultraviolet–visible characterizations. L.R.T. performed DSC experiments, E.D.C. and C.F. absorbance spectra experiments, G.K.K. single crystal X-ray characterization. S.C and C.F. undertook DFT-D calculations. A.M.C. performed elastic moduli experiments. All authors contributed to the discussion of the results, analysis of the structures and to manuscript preparation.

## Additional information

**Competing financial interests:** The authors declare no competing financial interests.

# 25

# Integrated catalysis opens new arylation pathways via regiodivergent enzymatic C–H activation

Jonathan Latham[1,2], Jean-Marc Henry[1,2], Humera H. Sharif[1,2], Binuraj R.K. Menon[1,2], Sarah A. Shepherd[1,2], Michael F. Greaney[1] & Jason Micklefield[1,2]

Despite major recent advances in C–H activation, discrimination between two similar, unactivated C–H positions is beyond the scope of current chemocatalytic methods. Here we demonstrate that integration of regioselective halogenase enzymes with Pd-catalysed cross-coupling chemistry, in one-pot reactions, successfully addresses this problem for the indole heterocycle. The resultant 'chemobio-transformation' delivers a range of functionally diverse arylated products that are impossible to access using separate enzymatic or chemocatalytic C–H activation, under mild, aqueous conditions. This use of different biocatalysts to select different C–H positions contrasts with the prevailing substrate-control approach to the area, and presents opportunities for new pathways in C–H activation chemistry. The issues of enzyme and transition metal compatibility are overcome through membrane compartmentalization, with the optimized process requiring no intermediate work-up or purification steps.

[1] School of Chemistry, The University of Manchester, Oxford Road, Manchester M13 9PL, UK. [2] Manchester Institute of Biotechnology, The University of Manchester, 131 Princess Street, Manchester M1 7DN, UK. Correspondence and requests for materials should be addressed to M.F.G. (email: michael.greaney@manchester.ac.uk) or to J.M. (email: jason.micklefield@manchester.ac.uk).

Selective activation of C–H positions has transformed synthesis in recent years. Using transition metal catalysis, a plethora of direct, fundamental bond formations are now achievable under mild, catalytic conditions, changing the way chemists make molecules[1]. The field is currently defined by substrate control, with a relatively small number of C–H motifs undergoing predictable and reliable metalation by a transition metal catalyst; for example, electron-rich heteroarenes, electron-poor fluoroarenes, activation ortho (and more recently, meta), to a directing group via cyclometalation all being dominant examples (Fig. 1a)[2–9]. Effective catalyst control, where choice of catalyst can discriminate between neighbouring C–H bonds in the absence of strong substrate-directing effects, is very rare and remains an outstanding challenge in the field[10–13]. Progress in this area will profoundly enhance the scope of C–H activation, as the majority of C–H bonds do not fit the current substrate-control criteria for chemocatalytic activation chemistry.

We were interested in addressing this problem using a combination of bio- and chemo-catalysis. The complex coordination sphere of an enzyme active site enables exquisite selectivities that cannot be achieved by simple metal coordination complexes. Correspondingly, the manifold power of noble metal catalysis in C–C bond formation presents synthesis pathways that are not possible through biocatalysis. Their integration offers a complementarity that could create new C–H activation strategies, whereby C–H positions beyond the scope of chemocatalysis alone can be targeted for C–C, C–N, C–Hal and other fundamental bond-forming steps. In addition, the merging of bio- and chemo-catalysis into single process, one-pot operations creates a step-change in efficiency through enhanced space-time yields, better energy consumption and elimination of auxiliary chemicals[14,15].

Despite the potential advantages of combining biocatalysts with chemocatalysts, their successful integration has been largely restricted to organic solvents[15–17], which are incompatible with many enzymes and cofactors—therefore limiting the number of biocatalysts that can be utilized. In contrast, integration of chemocatalysts with enzymes in their preferred aqueous media is relatively rare[14]. Indeed, the inherent differences in operating conditions of enzymes and chemocatalysts present serious challenges, often requiring compartmentalization or immobilization of at least one of the catalysts[18–23].

Our plans for a catalyst-controlled, integrated C–H activation system involve the merging of flavin-dependent halogenase (Fl-Hal) enzymes with palladium-catalysed cross-coupling. The pivotal role played by aryl halide functionality in metal-catalysed transformations suggested to us that a regioselective, biohalogenation would have great versatility, with the incipient Ar-X group being adaptable to a multitude of Pd-catalysed chemistries. We chose to integrate Fl-Hals and Suzuki–Miyaura coupling (SMC) with boronic acids in the first instance, given the critical role of SMC in producing biaryl products for pharmaceutical, agrochemical and materials science applications.

The most widely studied Fl-Hals are tryptophan halogenases, which are responsible for chlorinating different positions of the indole ring of tryptophan en route to various bacterial natural

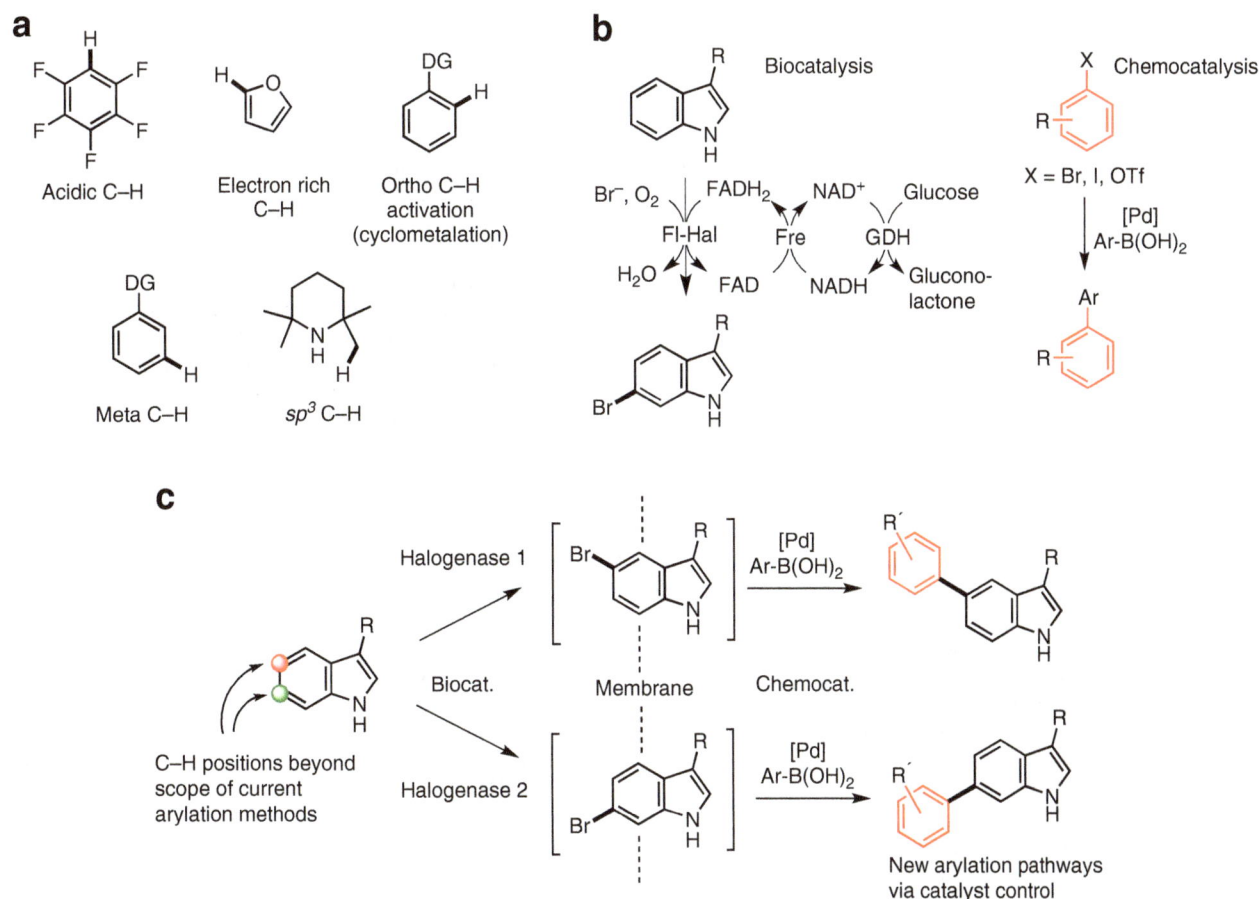

**Figure 1 | Previous work on C–H activation and proposed scheme of catalyst control using biohalogenation and Pd-catalysed cross-couplings.**
(**a**) Previous work on C–H activation utilizing substrate control of regioselectivity. (**b**) Catalytic cycle of halogenase biocatalysis using a Fl-Hal, along with flavin reductase (Fre) and glucose dehydrogenase (GDH) enzymes for cofactor recycling. (**c**) Proposed scheme of regioselective C–H activation using catalyst control by integration of a Fl-Hal and palladium-catalysed SMC. Biocat., biocatalysis; chemocat., chemocatalysis.

products[24–29]. Several fungal Fl-Hals have also been reported that chlorinate phenolic natural products[30–32]. Together this suite of enzymes have the potential to regioselectively halogenate a range of aromatic scaffolds. With appropriate cofactor-recycling systems (Fig. 1b) Fl-Hals can afford halogenation reactions using benign inorganic halide (typically $MgCl_2$ or NaBr), oxygen (from air), and glucose or isopropanol as the only stoichiometric reagents[33,34]. Studies on broadening the substrate scope of the tryptophan halogenases RebH, PrnA and PyrH[33–36], as well as improving their productivity and scaleability[37], have meant that Fl-Hals are becoming increasingly viable as biocatalysts for aromatic halogenation. In addition, Fl-Hals have been used to halogenate natural products and synthetic substrates, which have been isolated or extracted and then derivatized in separate cross-coupling reactions under standard synthetic operating procedures[38–40]. Biohalogenation is also attractive, given that traditional aromatic halogenation chemistry utilizes deleterious reagents, lacks regiocontrol and can afford mixtures, including halogenated by-products that can be toxic and persistent in the environment.

We anticipated that the unique regiocontrol inherent to Fl-Hal enzymatic halogenation could set-up an overall catalyst-controlled C–H arylation system, if a working SMC could be successfully integrated under aqueous conditions. The indole heterocycle has been central to the development of contemporary C–H activation chemistry, driving the discovery of new chemocatalytic methods for functionalizing the C2 and C3 positions (the 'innate' positions for functionalization with an electrophile)[41–46]. Direct C–C bond formation at the arene $C_6$–H and $C_7$–H positions, however, is not currently possible using chemocatalysis—something we hoped to achieve with the integrated approach.

## Results

### Biocatalytic halogenation.
To effect one-pot Fl-Hal-SMC reactions (Fig. 1c), we required enzymes and substrates that afford high yields of structurally diverse brominated products. First, we exploited our previous findings[33] that the tryptophan 5-halogenase PyrH can accept anthranilamide (1) to develop conditions for effective enzymatic synthesis of 5-bromoanthranilamide (2) (Fig. 2). Given the exquisite natural regio-complementarity of the flavin-dependent tryptophan halogenases we then set out to identify potential indole substrates for biocatalyst-controlled bromination and subsequent arylation. We found that tryptophol (3) could be efficiently brominated by RebH (a tryptophan 7-halogenase) to afford 7-bromotryptophol (4), after optimizing the conditions (Supplementary Fig. 1) that had been previously reported for chlorination of this substrate[34]. Initial attempts to brominate the 5-position of tryptophol (3) using PyrH in place of RebH gave very low conversion. The larger substrate 3-indolepropionic acid (5) however, was transformed by PyrH to give 5-bromo-3-indolepropionate (6) as a single regioisomer in good yield. In addition, the 6-bromo-3-indolepropionate (7) could be obtained in good regioisomeric excess and yield by using the tryptophan 6-halogenase SttH[27]. Collectively, these three halogenase enzymes established a clean, regiodivergent synthesis of $C_5$, $C_6$ and $C_7$ bromoindoles.

Finally, to extend the substrate diversity further we over-produced the radicicol halogenase RadH enzyme from the fungus *Chaetomium chiversii*[30]. RadH is similar in sequence to Rdc2—a related fungal halogenase that had been shown previously to chlorinate 6-hydroxy-isoquinoline (8)[31]. Accordingly conditions were developed whereby RadH could be used to prepare 5-bromo-6-hydroxyisoquinoline (9) in good yields.

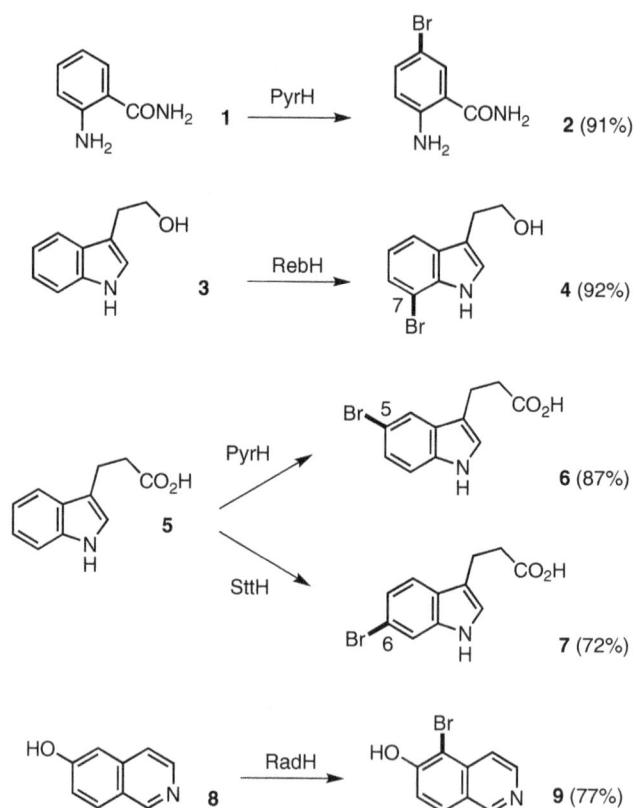

**Figure 2 | The regioselective and regio-divergent bromination of a range of aromatic scaffolds.** Regioselective halogenation of anthranilamide (1), tryptophol (3), 3-indole propionate (5) and 6-hydroxy isoquinoline (8) using a range of Fl-Hals to access different regiochemistries.

### Cross-coupling optimization.
We next sought conditions whereby the SMC could be integrated with the biocatalysts, using the coupling of 5-bromoanthranilamide (2) with phenyl boronic acid as a test reaction (Fig. 3). On the basis of previous studies of SMCs in aqueous media with an open atmosphere[47,48], a number of pyrimidine and guanidine ligands were selected for screening (L1–L4). Of these, L2 afforded highest conversion to 5-phenylanthranilamide (10) at 2.0 mM substrate (2) concentration, 2.5 mol% Pd loading, in potassium phosphate buffer in the absence of any cofactors or enzymes. In addition, a number of water-soluble phosphine ligands were also screened (L5–L8). Of these, tppts (L6) and sulfonated SPhos (L5) gave full conversion to the desired product (10). These two ligands were therefore carried forward in a base and palladium source screen (Supplementary Fig. 2). The reaction was found to be tolerant of many bases, including DIPEA, whilst the lack of any base afforded no product. In addition, conducting the reaction using L5 and L6 under air as opposed to inert atmosphere afforded no conversion, whilst the same reaction using L2 under open air afforded good conversion.

The tolerance of the selected Pd catalysts towards the biocatalytic halogenation conditions was then probed. As $NAD^+$/NADH and FAD/$FADH_2$ are rich in heteroatoms and redox active they could possibly interfere with Pd-catalysis either through redox chemistry or coordination to metal species. In addition, it has been previously reported that soft donor atoms (for example, sulfur) on the surface of proteins can coordinate to transition metals potentially inhibiting their catalytic activity. $Pd^{II}$ is also known to react with glucose under certain conditions (utilized to recycle NADH). The test reaction (Fig. 3) was therefore run in the presence of each of these components to

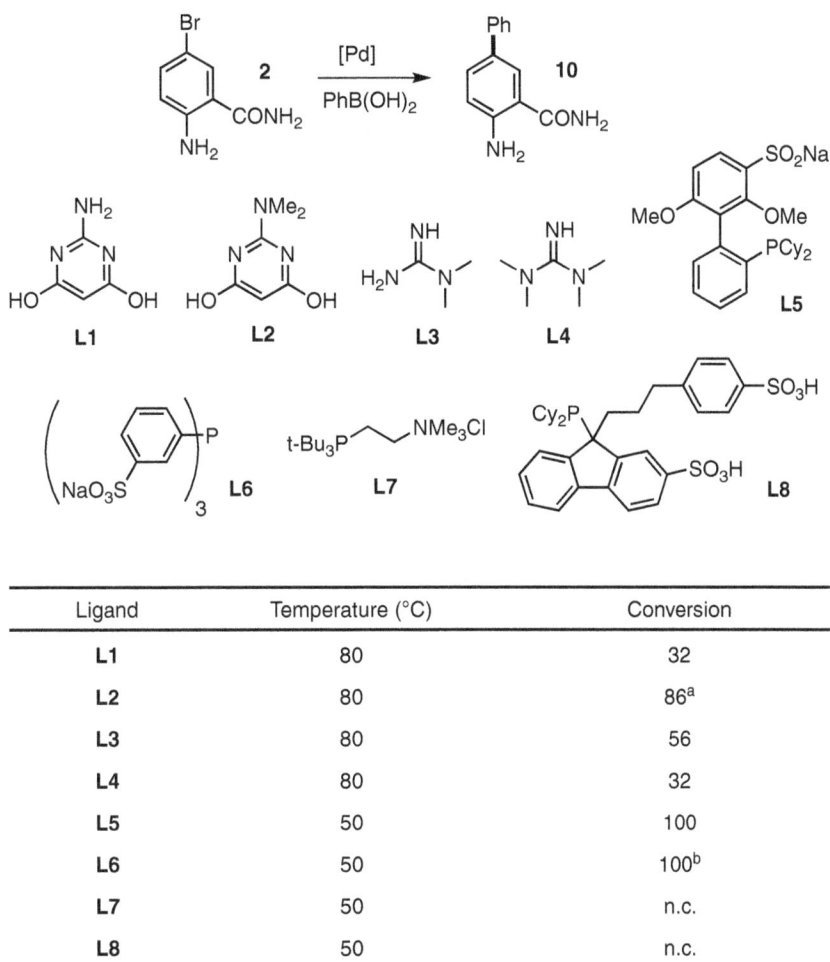

**Figure 3 | Ligand screening for the SMC of 2 with phenyl boronic acid.** $Pd(OAc)_2$ was used as the palladium source in each case in a 2:1 ligand:Pd ratio. Conversions determined by analytical high-performance liquid chromatography (HPLC) or liquid chromatography mass spectrometry (LC-MS). n.c., no conversion. [a]81% isolated yield; [b]80% isolated yield. Further optimization results are available in the Supplementary Material (Supplementary Fig. 2).

| Ligand | Temperature (°C) | Conversion |
|--------|------------------|------------|
| L1 | 80 | 32 |
| L2 | 80 | 86[a] |
| L3 | 80 | 56 |
| L4 | 80 | 32 |
| L5 | 50 | 100 |
| L6 | 50 | 100[b] |
| L7 | 50 | n.c. |
| L8 | 50 | n.c. |

determine their effect on SMC conversion. This revealed a somewhat deleterious effect from the presence of the cofactors and a major reduction in cross-coupling yields in the presence of proteins (Supplementary Fig. 3). Increasing the loading of palladium catalyst was generally advantageous in terms of conversion in the presence of NADH, FAD and glucose, but did not enhance cross-coupling when either the reductase or dehydrogenase proteins were present (Supplementary Fig. 4).

**Halogenase SMC integration.** Given that the presence of proteins caused the greatest reduction in cross-coupling yields, we reasoned that the enzymes would need to be compartmentalized or removed before cross-coupling, at least in the first instance. Pleasingly, we found that passing the PyrH biotransformation of anthranilamide (1) through a 10-kDa molecular-weight cut-off (MWCO) cellulose membrane to remove protein before cross-coupling was effective in this regard, affording very good conversion to arylated product (10). Following further optimization around this process (Supplementary Fig. 5), we could achieve an overall isolated yield of 79% of arylated product (10) using tppts ligand L6, $Pd(OAc)_2$ (50 mol%), and an excess of phenyl boronic acid (Fig. 4). Moreover, we showed that the enzymes that had been removed using the membrane filter could be recycled affording intermediate bromide (2) in 62% and 42% in the second and third cycles, respectively. Subsequent arylation of the combined biotransformation filtrates afforded arylated product (10)

in 54% overall yield; a ca. twofold increase in the amount of material that would be obtained from the same amount of enzyme in a single arylation reaction. After determining the scope of this method with respect to boronic acids (Supplementary Fig. 6) and finding the SMC compatible with the conditions required for RadH-catalysed bromination of 8 (Supplementary Fig. 7), we then extended this membrane filtration method (method A) to the selective 5-arylation of 6-hydroxyisoquinoline (8), using RadH to produce 15–17 in good yields.

With a workable integrated arylation process in hand, we looked at ways to improve our method, principally with a view to increasing scale and reducing the high Pd content in the SMC reaction. The requirement for high catalyst loadings appeared to arise from a combination of dilute reaction volumes and inhibitory effect of cofactors present in the reaction medium. Indeed, other attempts to conduct palladium-catalysed chemistry in biological media and in the presence of proteins have required very high loading or suffered from poor yields[47,48]. Enhancing enzyme efficiency would mitigate these two issues, by affording higher concentrations of aryl bromide for cross-coupling. The group of Sewald have recently reported improvements in RebH productivity through the use of crosslinked enzyme aggregates (CLEAs)[37]. The individual subunits of RebH are crosslinked with each other, as well as the partner reductase and dehydrogenase proteins, via surface lysines using glutaraldehyde forming a RebH-reductase-dehydrogenase CLEA that is more efficient and tractable to handle[37]. Accordingly, we successfully prepared these

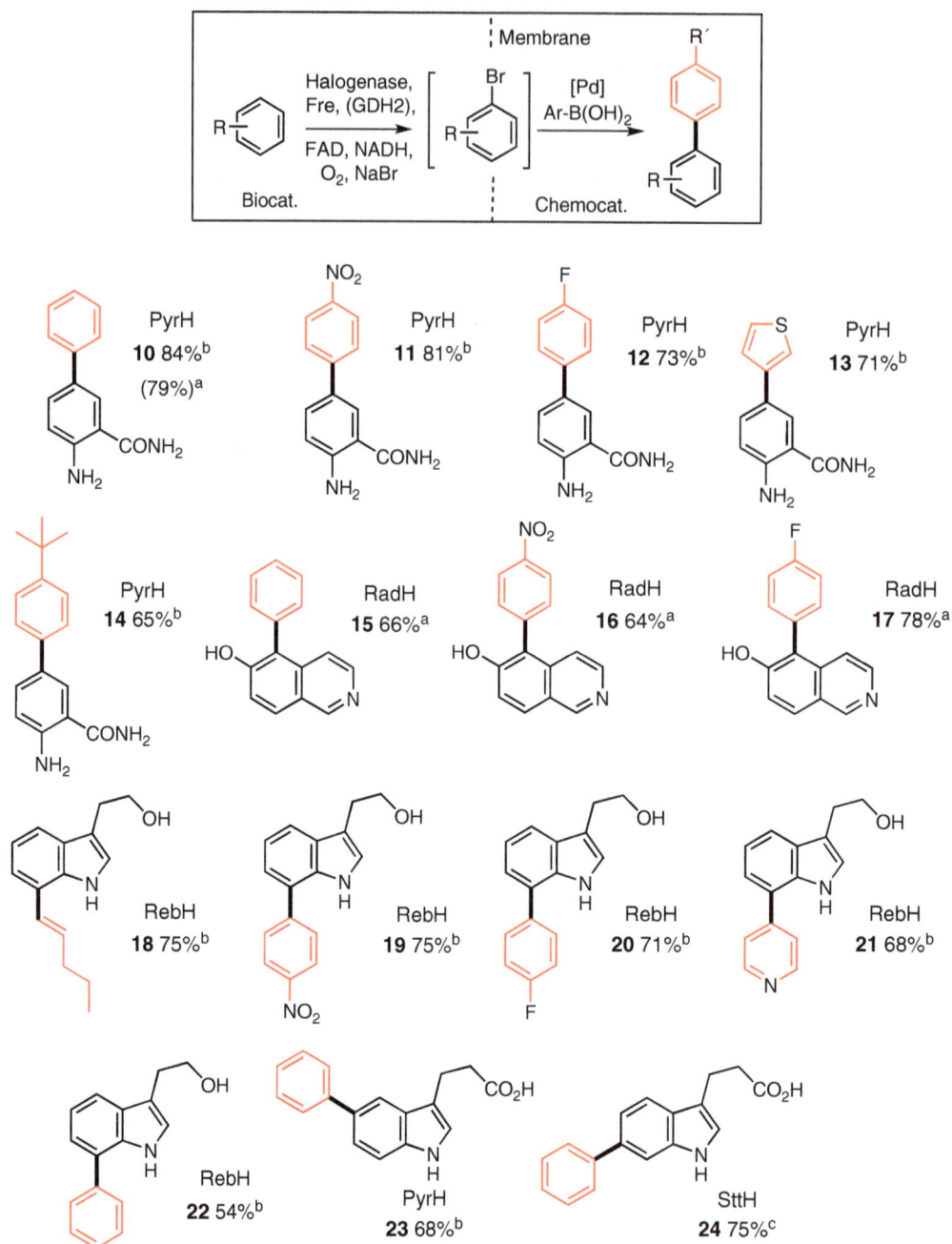

**Figure 4 | Isolated yield of the regioselective and regiodivergent arylation.** [a]Obtained using purified enzymes and a MWCO membrane (method A). [b]Obtained using a CLEA of the appropriate Fl-Hal and 10 mol% Pd-catalyst loading (method B). [c]Obtained from three successive cycles of halogenation using the same Fl-Hal. Biocat., biocatalysis; chemocat., chemocatalysis.

materials as active heterogeneous biocatalysts containing either PyrH, SttH or RebH. Although active CLEAs of the phenolic halogenase RadH were also obtained, they did not prove to be significantly more efficient than purified enzyme.

Initial screening of the PyrH-CLEA with anthranilamide (1) demonstrated improved efficiency, affording up to 3.0 mM aryl bromide. This enabled one-pot halogenase-SMC reactions to proceed to reasonable conversion (ca. 50%) without the need to remove or compartmentalize the CLEA from the Pd catalyst and other reagents, presumably as many heteroatoms of the protein had been effectively protected during the crosslinking process (Supplementary Fig. 8). We also found that the arylation of tryptophol (3) could be carried out in the same one-pot manner, with the RebH-CLEA and the Pd catalyst both present throughout (Supplementary Fig. 8). However, highest conversions

were obtained when the CLEA was removed, by centrifugation or filtration, before cross-coupling of the supernatant from biocatalytic halogenation (Supplementary Figs 8 and 9). The scope of this method (method B) was found to be broad with respect to boronic acids (Supplementary Fig. 10), allowing the efficient arylation of anthranilamide (1) with both electron-rich, electon-poor and heterocyclic boronic acids using PyrH-CLEA with reduced palladium loading (10 mol%), affording the 5-arylated compounds in very good yield (10–14, Fig. 4). Tryptophol (3) was also efficiently coupled using electron-rich, electron-poor, heterocyclic and alkenyl boronic acids to afford 7-substituted products 18 to 22 in good yield with RebH-CLEA at lower Pd loading. 3-Indole propionate (5) was likewise effectively arylated to biaryl products 23 and 24, with PyrH- and SttH-CLEAs, respectively. Along with the tryptophol examples

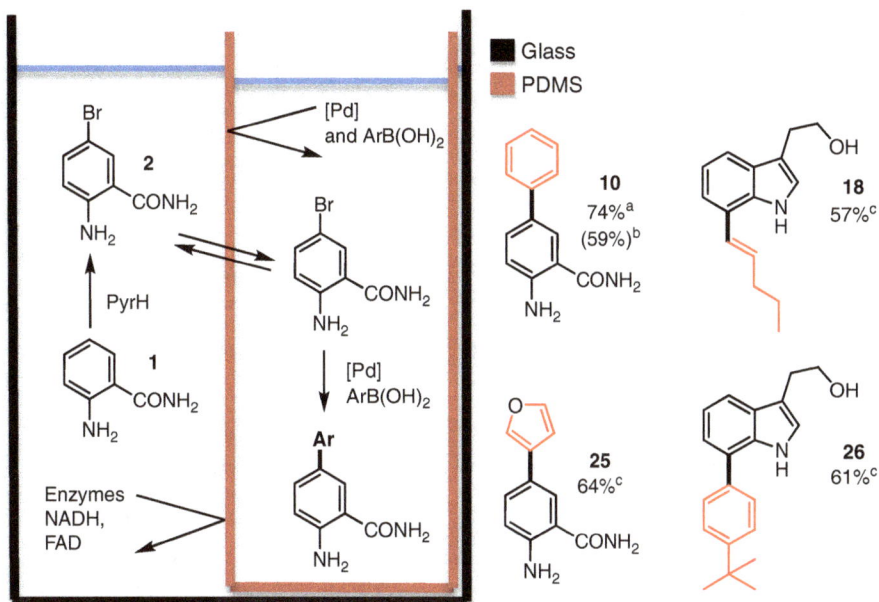

**Figure 5 | PDMS compartmentalization of the regioselective halogenation cross-coupling cascade.** [a]Obtained using purified enzyme and 10 mol% of **L2**.Pd(OAc)$_2$. [b]Obtained using halogenase CLEA and 2 mol% of **L2**.Pd(OAc)$_2$. [c]Obtained using halogenase CLEA and 10 mol% of **L2**.Pd(OAc)$_2$.

(**18–22**), this demonstrates the efficient and highly selective manipulation of each of the 5-, 6- and 7- C–H positions of the indole nucleus, under catalyst control, which is unattainable using existing synthetic methodology. In addition, the heterogeneous nature of the CLEA biocatalyst allowed it to be recycled over repeated cycles of bromination, with little loss in efficiency (Supplementary Fig. 11). In the case of the 6-arylation of 3-indolepropionic acid (**5**), this allowed an overall yield of 75% of **24** over three cycles with the SttH-CLEA. The increased stability and ease of preparation of the heterogeneous biocatalyst allowed this methodology to be scaled up to a 1.0-mmol scale to afford reasonable yields of **10** ($>100$ mg, 52%).

**Polydimethylsiloxane compartmentalization.** To further enhance the scope and efficiency of the halogenase-SMC process, we looked for alternative ways of separating the proteins from the palladium catalyst. While membrane filtration worked well, it did introduce an additional operation in the middle of the process and necessitated a deoxygenation step before SMC. To better streamline the process and develop a more efficient one-pot procedure, we examined polydimethylsiloxane (PDMS) thimbles as a compartmentalization strategy. This approach had been reported for a number of reactions that require the separation of incompatible reagents[19,49], including the combination of a Pd-catalyzed oxidation with a biocatalytic reduction[19].

Typically, one of the reactions is assembled inside a small thimble made of PDMS, which is then placed inside the larger bulk reaction vessel. As PDMS is a hydrophobic polymer, non-polar compounds (that is, substrates and intermediates in the cascade) should dissolve sufficiently well to pass through the PDMS walls, whilst charged reagents (Pd catalysts, enzymes and cofactors) would be unable to diffuse through PDMS and might therefore be compartmentalized. Initial tests showed that 5-bromoanthranilamide (**2**) would flux through PDMS well while glucose, NADH, FAD and the **L2**:Pd(OAc)$_2$ complex were all unable to effectively penetrate PDMS (Supplementary Fig. 12). SMC conditions using **L2**:Pd(OAc)$_2$ were therefore chosen to test the PDMS compartmentalization approach, as this catalyst functions well under open air and because alternative phosphine ligands have been shown to promote flux through PDMS.

Accordingly conditions were optimized, which allowed the PyrH-halogenation of anthranilamide (**2**) using pure protein to be carried out with the cross-coupling reagents within the PDMS thimble (Fig. 5), resulting in a 'one-pot' reaction to give 5-phenylanthranilamide (**10**) in 74% yield. The immediate advantages of this approach (method C) are that protein removal and deoxygenation of the reaction buffer are not required. Moreover a lower catalyst loading (10 mol%) is achievable under this regime, compared with the removal of pure protein using filtration (50 mol%), due to the compartmentalization of cofactors and enzymes from the Pd catalyst. By using a PyrH-CLEA to afford higher concentration of aryl bromide in conjunction with PDMS compartmentalization (method D), the Pd loading was further reduced to 2 mol% to afford **10** in good yield (59%). This combination also allowed the coupling of deactivated, electron-rich, boronic acids such as 3-furyl boronic acid to give **25** in good yield (64%), a reaction that was unsuccessful using a cellulose membrane (MWCO filter). Finally, this method was also extended to include coupling of the 7-position of tryptophol (**3**), using RebH-CLEA, to give 7-pent-1-enyl tryptophol (**18**) and 4'-tertbutyl-7-phenyl tryptophol (**26**) in good yields.

**Discussion**

In summary, we have combined four regioselective flavin-dependent halogenase enzymes with a palladium-catalysed Suzuki–Miyaura cross-coupling to afford the regio-controlled arylation of a number of aromatic scaffolds (benzamides, isoquinolines and indoles). In each case, the regioselective C–H arylation transformations were previously inaccessible using stand-alone chemocatalysis or biocatalysis methods. Specifically, we report the selective C–H arylation of the 5-, 6- and 7- positions of the indole nucleus under catalyst control.

To overcome issues of catalyst incompatibility, we initially developed a filtration method using a 10-kDa MWCO filter (method A). This allowed efficient biocatalyst recycling for a number of cycles, affording approximately double the overall productivity that the same batch would in a single cycle. Biocatalyst efficiency was further improved by using CLEAs, which afforded higher aryl bromide concentration—enabling lower palladium loading to be employed (method B). We also developed a PDMS thimble system to compartmentalize the chemo- and bio-catalysts. This method afforded similar yields of regioselectively arylated product as the MWCO filter system, but without the need for protein removal or solution deoxygenation before the cross-coupling chemistry (method C). Finally, the implementation of CLEAs with PDMS compartmentalization led to a further reduction in the required Pd loading down to 2 mol% (method D). Combined with directed evolution and other research aimed at improving the productivity and scalability of halogenase enzymes[37,50], the integrated method described here sets out a pathway to new C–H activation transformations that cannot be achieved through stand-alone biocatalysis or chemocatalysis. Research in this area is ongoing in our laboratories.

## Methods

**Method A: regioselective arylation 6-hydroxyisoquinoline.** 6-hydroxyisoquinoline (**8**) (0.5 mM), NaBr (10 mM), FAD (1 µM), Fre (4 µM), RadH (25 µM) and NADH (2.5 mM) were added to 10 mM potassium phosphate buffer (pH 7.2) containing 1% v/v EtOH (40 ml) and incubated with shaking overnight (30 °C) before filtration through a 10-kDa MWCO filter (Vivaspin 20). CsF (4.8 eq) and boronic acid (30 eq) were then added to the filtrate. Following freeze-thaw degassing and backfill with N$_2$, tppts (0.5 mM) and Na$_2$PdCl$_4$ (0.25 mM) were added as deoxygenated stock solutions in water. The resultant solution was heated to 80 °C with stirring for 24 h. After cooling, pH was adjusted to 6 using HCl, and the reaction partitioned into CH$_2$Cl$_2$ before the removal of solvent *in vacuo* and purification.

**Method B: regioselective arylation using CLEAs.** Substrate (3.0 mM), FAD (10 µM), NaBr (30 mM), NADH (100 µM) were dissolved in 15 mM sodium phosphate buffer with 5% v/v isopropanol (30 ml, pH 7.4). Halogenase CLEA from 1.5 l culture (prepared as described on page S46 of the Supplementary Methods) was then resuspended into the reaction buffer and the resultant suspension incubated at room temperature with orbital shaking overnight. After removal of CLEA by centrifugation (10,000 r.p.m., 4 °C, 30 min), K$_3$PO$_4$/K$_2$CO$_3$ (10 eq) and boronic acid (5 eq) were added to the supernatant before degassing of the solution with a stream of nitrogen. Tppts (20 mol%) and Na$_2$PdCl$_4$ (10 mol%) were then added and the solution was stirred at 80 °C overnight. On cooling, the reaction was partitioned into EtOAc/CH$_2$Cl$_2$ and concentrated before purification.

**Method C: regioselective arylation using PDMS thimbles.** To a solution containing anthranilamide (2.0 mM), NaBr (100 mM), FAD (1 µM), glucose (20 mM), PyrH (20 µM), Fre (2 µM) and GDH2 (12 µM) in 10 mM potassium phosphate buffer (30 ml, pH 7.2) was added NADH (100 µM). A PDMS thimble containing **L2**.Pd(OAc)$_2$ (10 mol%), CsF (10 eq) and PhB(OH)$_2$ (5 eq) in water was then placed inside the Erlenmeyer flask containing the biotransformation. After incubation at room temperature overnight, the reaction was heated to 80 °C for 8 h. After cooling, the inner and outer chambers were combined and basified with 4 N NaOH. The thimble was then soaked in EtOAc to flux out any residual product, and the reaction mixture extracted into EtOAc before concentration and purification.

**Method D: regioselective arylation using CLEAs and PDMS.** Halogenase CLEA from 3 l of culture (prepared as described on page S46 of the Supplementary Methods) was suspended into 30 ml of reaction buffer containing anthranilamide (**1**) or tryptophol (**3**) (5.0 mM), NADH (100 µM), FAD (10 µM) and NaBr (30 mM) in 15 mM sodium phosphate buffer with 5% v/v isopropanol (30 ml, pH 7.4). To this suspension was added a PDMS thimble containing **L2**.Pd(OAc)$_2$ (10 mol%), CsF (10 eq) and ArB(OH)$_2$ (5 eq). After incubation with shaking at room temperature overnight, the reaction was heated to 80 °C for a further 24 h. Reactions were then worked up as previously described. To facilitate the recycling of the biocatalysts, the CLEA can be removed using centrifugation (10,000 r.p.m., 30 min., 4 °C) after overnight incubation at room temperature before addition of the PDMS thimble containing the SMC components.

## References

1. Yu, J.-Q. & Shi, Z. *C-H Activation* Vol. 292 (Springer, 2010).
2. Clot, E. *et al.* C–F and C–H bond activation of fluorobenzenes and fluoropyridines at transition metal centers: how fluorine tips the scales. *Acc. Chem. Res.* **44**, 333–348 (2011).
3. Rossi, R., Bellina, F., Lessi, M. & Manzini, C. Cross-coupling of heteroarenes by C–H Functionalization: recent progress towards direct arylation and
heteroarylation reactions involving heteroarenes containing one heteroatom. *Adv. Synth. Catal.* **356**, 17–117 (2014).
4. Engle, K. M., Mei, T. S., Wasa, M. & Yu, J. Q. Weak coordination as a powerful means for developing broadly useful C–H functionalization reactions. *Acc. Chem. Res.* **45**, 788–802 (2012).
5. Kuhl, N., Hopkinson, M. N., Wencel-Delord, J. & Glorius, F. Beyond directing groups: transition-metal-catalyzed C–H activation of simple arenes. *Angew. Chem. Int. Ed.* **51**, 10236–10254 (2012).
6. Leow, D., Li, G., Mei, T. S. & Yu, J. Q. Activation of remote meta-C–H bonds assisted by an end-on template. *Nature* **486**, 518–522 (2012).
7. Wang, X. C. *et al.* Ligand-enabled meta-C–H activation using a transient mediator. *Nature* **519**, 334–338 (2015).
8. Saidi, O. *et al.* Ruthenium-catalyzed meta sulfonation of 2-phenylpyridines. *J. Am. Chem. Soc.* **133**, 19298–19301 (2011).
9. McNally, A., Haffemayer, B., Collins, B. S. L. & Gaunt, M. J. Palladium-catalysed C–H activation of aliphatic amines to give strained nitrogen heterocycles. *Nature* **512**, 338–338 (2014).
10. Ueda, K., Yanagisawa, S., Yamaguchi, J. & Itami, K. A general catalyst for the beta-selective C–H Bond arylation of thiophenes with iodoarenes. *Angew. Chem. Int. Ed.* **49**, 8946–8949 (2010).
11. Hickman, A. J. & Sanford, M. S. Catalyst control of site selectivity in the Pd-II/IV-catalyzed direct arylation of naphthalene. *ACS Catal.* **1**, 170–174 (2011).
12. Neufeldt, S. R. & Sanford, M. S. Controlling site selectivity in palladium-catalyzed C–H bond functionalization. *Acc. Chem. Res.* **45**, 936–946 (2012).
13. Gormisky, P. E. & White, M. C. Catalyst-controlled aliphatic C–H oxidations with a predictive model for site-selectivity. *J. Am. Chem. Soc.* **135**, 14052–14055 (2013).
14. Gröger, H. & Hummel, W. Combining the 'two worlds' of chemocatalysis and biocatalysis towards multi-step one-pot processes in aqueous media. *Curr. Opin. Chem. Bio.* **19**, 171–179 (2014).
15. Lohr, T. L. & Marks, T. J. Orthogonal tandem catalysis. *Nat. Chem.* **7**, 477–482 (2015).
16. Allen, J. V. & Williams, J. M. J. Dynamic kinetic resolution with enzyme and palladium combinations. *Tetrahedron Lett.* **37**, 1859–1862 (1996).
17. Larsson, A. L. E., Persson, B. A. & Bäckvall, J.-E. Enzymatic resolution of alcohols coupled with ruthenium-catalyzed racemization of the substrate alcohol. *Angew. Chem. Int. Ed.* **36**, 1211–1212 (1997).
18. Denard, C. A. *et al.* Cooperative tandem catalysis by an organometallic complex and a metalloenzyme. *Angew. Chem. Int. Ed.* **53**, 465–469 (2014).
19. Sato, H., Hummel, W. & Gröger, H. Cooperative catalysis of noncompatible catalysts through compartmentalization: Wacker oxidation and enzymatic reduction in a one-pot process in aqueous media. *Angew. Chem. Int. Ed.* **54**, 4488–4492 (2015).
20. Makkee, M., Kieboom, A. P. G., Van Bekkum, H. & Roels, J. A. Combined action of enzyme and metal catalyst, applied to the preparation of D-mannitol. *J. Chem. Soc. Chem. Commun.* **19**, 930–931 (1980).
21. Köhler, V. *et al.* Synthetic cascades are enabled by combining biocatalysts with artificial metalloenzymes. *Nat. Chem.* **5**, 93–99 (2013).
22. Wang, Z. J., Clary, K. N., Bergman, R. G., Raymond, K. N. & Toste, F. D. A supramolecular approach to combining enzymatic and transition metal catalysis. *Nat. Chem.* **5**, 100–103 (2013).
23. Tenbrink, K., Seßler, M., Schatz, J. & Gröger, H. Combination of olefin metathesis and enzymatic ester hydrolysis in aqueous media in a one-pot synthesis. *Adv. Synth. Catal.* **353**, 2363–2367 (2011).
24. Keller, S. *et al.* Purification and partial characterization of tryptophan 7-halogenase (PrnA) from *Pseudomonas fluorescens*. *Angew. Chem. Int. Ed.* **39**, 2300–2302 (2000).
25. Heemstra, J. R. & Walsh, C. T. Tandem action of the O$_2$- and FADH2-dependent halogenases KtzQ and KtzR produce 6,7-dichlorotryptophan for kutzneride assembly. *J. Am. Chem. Soc.* **130**, 14024–14025 (2008).
26. Yeh, E., Garneau, S. & Walsh, C. T. Robust in vitro activity of RebF and RebH, a two-component reductase/halogenase, generating 7-chlorotryptophan during rebeccamycin biosynthesis. *Proc. Natl Acad. Sci. USA* **102**, 3960–3965 (2005).
27. Zeng, J. & Zhan, J. Characterization of a tryptophan 6-halogenase from *Streptomyces toxytricini*. *Biotechnol. Lett.* **33**, 1607–1613 (2011).
28. Zehner, S. *et al.* A regioselective tryptophan 5-halogenase is involved in pyrroindomycin biosynthesis in *Streptomyces rugosporus* LL-42D005. *Chem. Biol.* **12**, 445–452 (2005).
29. Wagner, C., El Omari, M. & König, G. M. Biohalogenation: nature's way to synthesize halogenated metabolites. *J. Nat. Prod.* **72**, 540–553 (2009).
30. Wang, S. *et al.* Functional characterization of the biosynthesis of radicicol, an Hsp90 inhibitor resorcylic acid lactone from *Chaetomium chiversii*. *Chem. Biol.* **15**, 1328–1338 (2008).
31. Zeng, J., Lytle, A. K., Gage, D., Johnson, S. J. & Zhan, J. Specific chlorination of isoquinolines by a fungal flavin-dependent halogenase. *Bioorg. Med. Chem. Lett.* **23**, 1001–1003 (2013).

32. Zhou, H., Qiao, K., Gao, Z., Vederas, J. C. & Tang, Y. Insights into radicicol biosynthesis via heterologous synthesis of intermediates and analogs. *J. Biol. Chem.* **285,** 41412–41421 (2010).

33. Shepherd, S. A. *et al.* Extending the biocatalytic scope of regiocomplementary flavin-dependent halogenase enzymes. *Chem. Sci.* **6,** 3454–3460 (2015).

34. Payne, J. T., Andorfer, M. C. & Lewis, J. C. Regioselective arene halogenation using the FAD-dependent halogenase RebH. *Angew. Chem. Int. Ed.* **52,** 5271–5274 (2013).

35. Hölzer, M., Burd, W., Reißig, H.-U. & Pée, K.-H. V. Substrate specificity and regioselectivity of tryptophan 7-halogenase from *Pseudomonas fluorescens* BL915. *Adv. Synth. Catal.* **343,** 591–595 (2001).

36. Frese, M., Guzowska, P. H., Voß, H. & Sewald, N. Regioselective enzymatic halogenation of substituted tryptophan derivatives using the FAD-dependent halogenase RebH. *ChemCatChem.* **6,** 1270–1276 (2014).

37. Frese, M. & Sewald, N. Enzymatic halogenation of tryptophan on a gram scale. *Angew. Chem. Int. Ed.* **54,** 298–301 (2015).

38. Glenn, W. S., Nims, E. & O'Connor, S. E. Reengineering a tryptophan halogenase to preferentially chlorinate a direct alkaloid precursor. *J. Am. Chem. Soc.* **133,** 19346–19349 (2011).

39. Roy, A. D., Grüschow, S., Cairns, N. & Goss, R. J. M. Gene expression enabling synthetic diversification of natural products: chemogenetic generation of pacidamycin analogs. *J. Am. Chem. Soc.* **132,** 12243–12245 (2010).

40. Durak, L. J., Payne, J. T. & Lewis, J. C. Late-stage diversification of biologically active molecules via chemoenzymatic C–H activation. *ACS Catal.* **6,** 1451–1454 (2016).

41. Sandtorv, A. H. Transition metal-catalyzed C–H activation of indoles. *Adv. Synth. Catal.* **357,** 2403–2435 (2015).

42. Itahara, T., Ikeda, M. & Sakakibara, T. Alkenylation of 1-acylindoles with olefins bearing electron-withdrawing substituents and palladium acetate. *J. Chem. Soc. Perkin. Trans.* **1,** 1361–1363 (1983).

43. Grimster, N. P., Gauntlett, C., Godfrey, C. R. A. & Gaunt, M. J. Palladium-catalyzed intermolecular alkenylation of indoles by solvent-controlled regioselective C–H functionalization. *Angew. Chem. Int. Ed.* **44,** 3125–3129 (2005).

44. Stuart, D. R. & Fagnou, K. The catalytic cross-coupling of unactivated arenes. *Science* **316,** 1172–1175 (2007).

45. Kalyani, D., Dick, A. R., Anani, W. Q. & Sanford, M. S. A simple catalytic method for the regioselective halogenation of arenes. *Org. Lett.* **8,** 2523–2526 (2006).

46. Lebrasseur, N. & Larrosa, I. Room temperature and phosphine free palladium catalyzed direct C-2 arylation of indoles. *J. Am. Chem. Soc.* **130,** 2926–2927 (2008).

47. Chalker, J. M., Wood, C. S. C. & Davis, B. G. A convenient catalyst for aqueous and protein Suzuki-Miyaura cross-coupling. *J. Am. Chem. Soc.* **131,** 16346–16347 (2009).

48. Gao, Z., Gouverneur, V. & Davis, B. G. Enhanced aqueous Suzuki-Miyaura coupling allows site-specific polypeptide 18F-labeling. *J. Am. Chem. Soc.* **135,** 13612–13615 (2013).

49. Runge, M. B., Mwangi, M. T., Miller, A. L., Perring, M. & Bowden, N. B. Cascade reactions using LiAlH$_4$ and grignard reagents in the presence of water. *Angew. Chem. Int. Ed.* **47,** 935–939 (2008).

50. Poor, C. B., Andorfer, M. C. & Lewis, J. C. Improving the stability and catalyst lifetime of the halogenase RebH by directed evolution. *ChemBioChem.* **15,** 1286–1289 (2014).

## Acknowledgements

We acknowledge BBSRC (grant BB/K00199X/1), EPSRC, CoEBio3 and GlaxoSmithKline for support. We also thank Dr Anna-Winona Struck for helpful conversations.

## Author contributions

J.M. and M.F.G. conceived and supervised the project; J.L., J.-M.H. and H.H.S. performed integrated arylation reactions; J.L. optimized the integrated process whilst J.L. and H.H.S. screened conditions for the SMC; B.R.K.M. and S.A.S. developed expression and initial reaction conditions for RadH and SttH, respectively; J.L. and S.A.S. prepared functional CLEAS; J.L. prepared and carried out integrated reactions with PDMS thimbles; J.L., M.F.G. and J.M. wrote the manuscript.

## Additional information

**Competing financial interests:** The authors declare no competing financial interests.

# 26

# Spontaneous formation and base pairing of plausible prebiotic nucleotides in water

Brian J. Cafferty[1,2], David M. Fialho[1,2], Jaheda Khanam[1,2], Ramanarayanan Krishnamurthy[2,3] & Nicholas V. Hud[1,2]

The RNA World hypothesis presupposes that abiotic reactions originally produced nucleotides, the monomers of RNA and universal constituents of metabolism. However, compatible prebiotic reactions for the synthesis of complementary (that is, base pairing) nucleotides and mechanisms for their mutual selection within a complex chemical environment have not been reported. Here we show that two plausible prebiotic heterocycles, melamine and barbituric acid, form glycosidic linkages with ribose and ribose-5-phosphate in water to produce nucleosides and nucleotides in good yields. Even without purification, these nucleotides base pair in aqueous solution to create linear supramolecular assemblies containing thousands of ordered nucleotides. Nucleotide anomerization and supramolecular assemblies favour the biologically relevant β-anomer form of these ribonucleotides, revealing abiotic mechanisms by which nucleotide structure and configuration could have been originally favoured. These findings indicate that nucleotide formation and selection may have been robust processes on the prebiotic Earth, if other nucleobases preceded those of extant life.

[1] School of Chemistry and Biochemistry, Georgia Institute of Technology, 901 Atlantic Drive, Atlanta, Georgia 30332, USA. [2] NSF-NASA Center for Chemical Evolution, Atlanta, Georgia 30332, USA. [3] Department of Chemistry, The Scripps Research Institute, La Jolla, California 92037, USA. Correspondence and requests for materials should be addressed to N.V.H. (email: hud@chemistry.gatech.edu).

T he current role of mononucleotides and RNA polymers in numerous cellular functions gave rise to the long-standing hypothesis that these molecules were involved early in the emergence of life: the RNA World hypothesis[1]. Supporting this hypothesis, model prebiotic reactions and analyses of carbonaceous meteorites provide evidence that the canonical nucleobases of RNA (adenine, guanine, cytosine, uracil) were likely present on the prebiotic Earth[2-4]. In addition, progress has been made towards finding abiotic routes to ribose and related sugars from simple molecules (for example, formaldehyde, glyoxylate), as well as mechanisms for ribose phosphorylation and ribose selection from complex sugar mixtures[5-8]. Nevertheless, despite decades of effort, the chemical origin of nucleosides and nucleotides (that is, nucleobases glycosylated with ribose and phosphorylated ribose) remains an unsolved problem[9,10]. In the 1970s, Orgel and co-workers showed that adenosine (the nucleoside of adenine) can be formed in 1–5% yield if a solution of adenine and ribose is dried and heated[11,12], but no comparable reactions have been demonstrated for the other three canonical nucleosides. Frustrated by what became known as *The Nucleoside Problem*, Orgel proposed an alternative pathway for prebiotic pyrimidine nucleoside and nucleotide synthesis that bypassed glycosidic bond formation, a pathway in which the cytosine nucleobase is built stepwise on a sugar scaffold[13]. This approach was furthered by Sutherland and co-workers, who developed a synthetic route to cytidine and uridine nucleotides starting from glycolaldehyde and glyceraldehyde[13,14]. However, the necessity to temporally separate specific reagents and reaction steps caused some to question the relevance of this synthesis to the origin of RNA[15,16]. Sutherland and co-workers have since proposed a spatially separated geochemical scenario that includes an ordered delivery of reagents on the prebiotic Earth that would coincide with the sequential steps of their pyrimidine nucleotide synthesis, a scenario that is initiated by meteorite impacts[17].

The persistent challenge of finding a simple, robust and plausible prebiotic route to the canonical nucleosides—juxtaposed with the exquisite functionality of RNA—have caused many researches to consider RNA a product of chemical and/or biological evolution[18,19]. Inspired by the possibility that RNA evolved from a *proto*-RNA with alternative nucleobases that more easily formed nucleosides, Miller and co-workers demonstrated that urazole (a triazole analog of uracil) is efficiently glycosylated by ribose in water[20]. Subsequent demonstrations of nucleoside formation with different plausible prebiotic heterocycles[12,21,22] suggest that other nucleosides may have been common on the prebiotic Earth. While encouraging, a model prebiotic reaction has yet to be reported that produces two extant nucleosides that form a Watson–Crick base pair or two noncanonical nucleosides that form a similar base pair—a property used by extant life for information transfer and, arguably, essential for the emergence of RNA-based life.

As part of our search for the possible components of proto-RNA, we evaluated a series of heterocycles for the ability to be spontaneously glycosylated by ribose and to selectively assemble in water[23,24]. In this work, we report the glycosylation reactions and nucleotide assembly/mutual selection of two candidate proto-nucleobases, barbituric acid (BA) and melamine. These heterocycles are well suited to function as the recognition units of a primitive informational system, as BA and melamine: (i) are produced in the same model prebiotic reaction[25]; (ii) can form Watson–Crick-like base pairs with each other[26]; (iii) have H-bond donor and acceptor groups that are complementary with uracil (for melamine) and adenine (for BA), making these heterocycles 'forward compatible' for base pairing with two extant nucleobases[9]; (iv) possess chemical properties that indicate

favourable glycosylation by ribose, specifically at the C5 position of BA and the exocyclic amines of melamine. Additional reasons have been previously put forth for why BA and melamine should be considered potential ancestral nucleobases of RNA[27,28].

Here, we show that glycosylation of melamine and BA by ribose-5-phosphate (R5P) occurs spontaneously in water to produce nucleotides in yields of up to 55% and 82%, respectively. When combined, the nucleotides form supramolecular assemblies with Watson–Crick-like base pairs, even within the crude reaction mixtures. These assemblies are shown to preferentially incorporate and increase the fraction of the β-anomer of the melamine nucleotide over the α-anomer. These findings demonstrate prebiotically plausible mechanisms for the selection of nucleotides in both nucleobase and sugar structure.

## Results

**Spontaneous formation of nucleosides and nucleotides.** Given previous reports that phosphorylated sugars can be produced in model prebiotic reactions[7], we were motivated to explore the potential for BA and melamine to be glycosylated by R5P, which could represent a model prebiotic route to nucleotides that can form base pairs that are similar to those formed by the canonical nucleobases (Fig. 1a). Nucleotides spontaneously form when BA is mixed in water with one equivalent of R5P at 20 °C (without the need for drying). The reaction is surprisingly efficient with BA + R5P conjugates exceeding 80% in the unpurified (crude) reaction mixture after 24 h and with significant yields for reactions performed over the range of pH 3–11 (Fig. 1b and Supplementary Fig. 1). Reactions between melamine and R5P are also productive, forming melamine + R5P conjugates at yields ranging from 33 to 55% when the reaction was carried out for 24 h at 65 °C (Fig. 1c). Glycosylation of melamine with R5P was observed from pH 3 to 9 and at 20 °C (Supplementary Figs 2 and 3). Glycosylation of BA and melamine by (unphosphorylated) ribose was also found to occur spontaneously in water, producing four nucleoside isomers for each reaction (Supplementary Figs 4 and 5). The robustness (for example, wide pH and temperature range) and good yields for both nucleoside and nucleotide formation with BA and melamine are noteworthy among model prebiotic reactions, especially considering that none of the four canonical nucleobases form nucleosides in detectable yields when heated with R5P in water (Supplementary Fig. 6). On the contrary, the canonical nucleosides and nucleotides of RNA are thermodynamically disfavoured (but kinetically stable) in water[29].

To characterize the nucleotides formed from the BA reactions with R5P, the products formed in water after 24 h at 20 °C were isolated by column chromatography for further analysis. Two-dimensional (2D) NMR spectroscopy confirmed that the BA + R5P conjugates are C-nucleotides, with a C–C glycosidic bond between ribose and BA (5-ribofuranosyl-C-barbiturate-5′-monophosphate, C-BMP; Fig. 2b and Supplementary Fig. 7). One-dimensional (1D) rotating frame NOE (ROE) spectroscopy confirmed that the β-anomer is preferentially formed in a 67:33 ratio over the α-anomer (Fig. 2c). Similar to our previous report that reactions between 2,4,6-triaminopyrimidine and ribose yield the C-linked β-ribofuranoside as the major product[10], BA reactions with R5P again demonstrate that the biologically significant β-anomeric sugar form can be preferentially selected through nucleosidation with an alternative nucleobase. Intermediate products suggest that the BA nucleotide-formation reaction proceeds through a Knoevenagel condensation (Supplementary Fig. 8a). We note that β-C-BMP shares a close structural relationship with the C-nucleotide pseudouridine, the most common posttranscriptional modification of RNA in biology[30].

We next isolated and characterized the melamine + R5P conjugates formed from a reaction between melamine and R5P

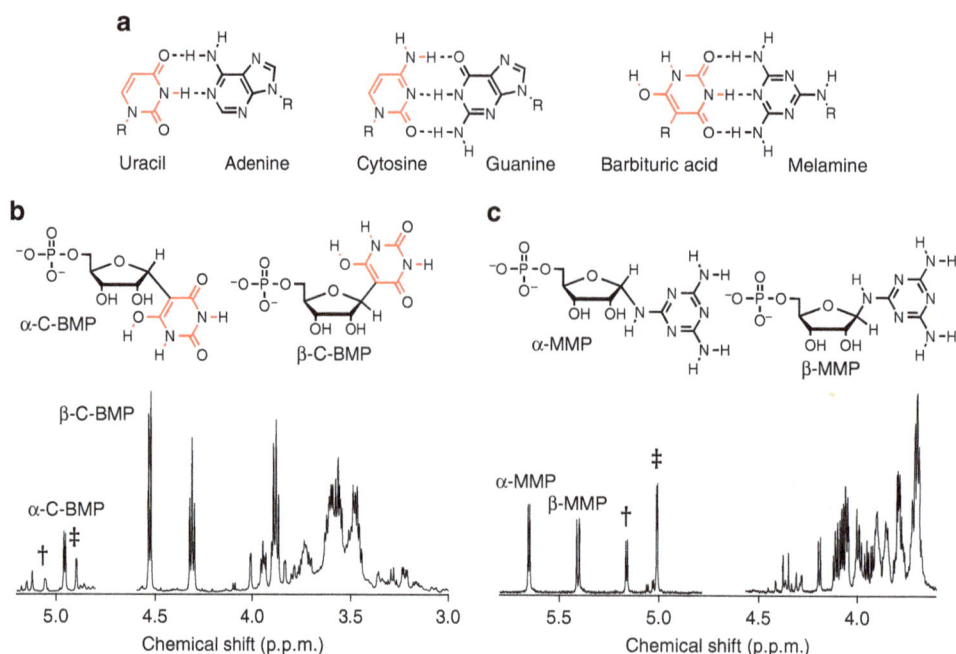

**Figure 1 | Spontaneous formation of nucleotides by barbituric acid (BA) and melamine in water.** (a) Chemical structures of the four canonical nucleobases of RNA shown in their Watson–Crick base pairs, and BA with melamine in an analogous Watson–Crick-like base pair. The R group on all nucleobases, which is H for the free nucleobases, indicates the position of ribose attachment through a glycosidic bond on the canonical bases and for melamine and BA in the current work. (b) Chemical structures of the two C-nucleotide anomers of BA-ribosyl-monophosphate (C-BMP) and $^1$H NMR spectrum of a BA + R5P crude reaction mixture revealing the formation of α-C-BMP and β-C-BMP. (c) Chemical structures of the two anomers of melamine-ribosyl-monophosphate (MMP) and the $^1$H NMR spectrum of a melamine + R5P crude reaction mixture revealing the formation of α-MMP and β-MMP. The anomeric proton resonances for each nucleotide are labelled, and those for R5P are marked with † indicating α-R5P, and ‡ indicating β-R5P. Relative integrated intensities of the nucleotide anomeric resonances show that, for these two reactions, the total C-BMP yield was 82%, and the total MMP yield was 55%. The HOD peaks have been removed from the NMR spectra for clarity. See the Methods for reaction details.

that was performed at 65 °C over 24 h. 2D NMR analysis of the melamine + R5P conjugates confirmed that glycosylation of melamine occurs at an exocyclic amine (Fig. 3b and Supplementary Fig. 9). Glycosylation proceeds through a reversible Schiff base intermediate (Supplementary Fig. 8b), which is partially stabilized by ring closure, to form two N-nucleotides (N-ribofuranosyl-melamine-5′-monophosphate (MMP)). 1D ROE analysis confirms that the α- and β-anomers of MMP equilibrate to approximately equal amounts in aqueous solution (Fig. 3c). The stability of the glycosidic bond of exocyclic N-linked triaminotriazine nucleosides in water had previously been shown to be quite poor, with total hydrolysis occurring on the order of minutes[27]. The hydrolytic stability of a 5 mM solution of MMP was evaluated to determine the rate of hydrolysis at pH 5 and 5 °C. Under these conditions, the hydrolysis rate was determined to be $3 \times 10^{-6}$ min$^{-1}$, with a half-life of 6 months (Supplementary Fig. 10). Remarkably, the dissociation constant between melamine and R5P was found to be 3.7 mM. For comparison, the work of Miller and co-workers[20] revealed that the $K_d$ for the ribonucleoside of urazole is 700 mM at 5 °C, whereas the $K_d$ for the ribonucleoside of uracil is estimated to be around 700 M. Thus, the equilibrium of nucleoside formation between melamine and ribose could be several orders of magnitude more favourable than that of the extant nucleobases.

**MMP and C-BMP form supramolecular assemblies.** We next tested the ability of MMP and C-BMP to exhibit base pairing in aqueous solution. Base pairing is *not* exhibited by the canonical mononucleotides[31], but such a property would have been

advantageous for the prebiotic mutual selection and co-localization of base-pairing nucleotides, particularly if proto-biopolymers emerged from a complex 'prebiotic soup'[32,33]. The crude products of melamine and BA reactions with R5P were mixed and the mixtures were analysed by circular dichroism (CD) spectroscopy as an initial test of whether melamine and BA nucleotides can form chiral assemblies even without purification. To maximize assembly of the two heterocycles, solution pH was adjusted to between pH 4 and 5. This pH range was chosen due to $pK_a$ considerations of the parent heterocycles (BA $pK_a = 4$; melamine $pK_a = 5$), which would be expected to minimize the difference in relative ionization between the bases[34]. This mixture of nucleotides, side products and unreacted starting materials from the two reactions (50 mM each in total BA and melamine) exhibits a substantial CD signal, whereas separately the crude products of each reaction exhibit either no CD signal (the melamine-R5P reaction) or a much lower CD signal and with a different wavelength profile (the BA-R5P reaction; Fig. 4a). The loss of the CD signal upon heating the mixture of the crude reactions above 30 °C, and return of signal upon cooling to 5 °C (Supplementary Fig. 11a) indicates the reversible formation of non-covalent assemblies.

The assemblies formed upon combining the crude reaction mixtures were imaged by atomic force microscopy (AFM), which revealed linear supramolecular polymers with diameters of ca 2 nm (Fig. 4b). These structures are fully consistent with the presence of stacked hydrogen-bonded hexads with paired BA and melamine bases (Fig. 4f)[23], assemblies that have been observed previously with analogous molecules that also form hexads[10,35]. The length of these supramolecular polymers (typically >1 μm) indicates that tens of thousands of heterocycles are paired within

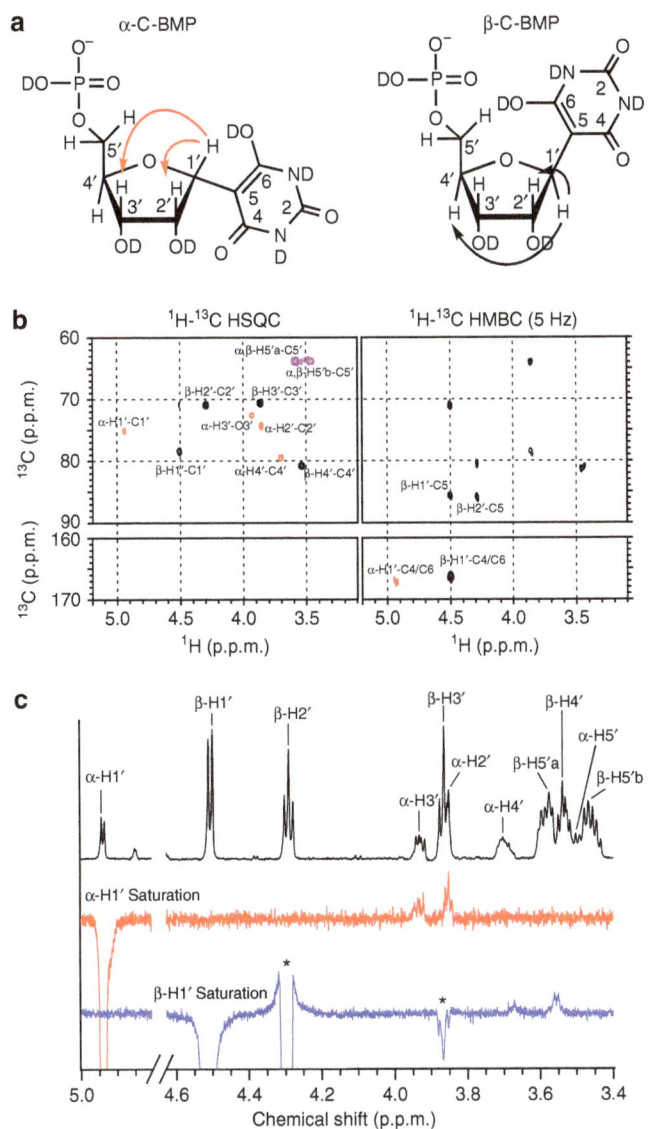

**Figure 2 | NMR characterization of C-BMP nucleotides. (a)** Chemical structure of α-C-BMP and β-C-BMP with arrows indicating through-space proton–proton magnetization transfer (ROE) as shown in **c**. **(b)** Heteronuclear single-quantum correlation (HSQC) and heteronuclear multiple bond correlation (HMBC) spectra showing $^1$H-$^{13}$C couplings for α-C-BMP and β-C-BMP. $^1$H-$^{13}$C correlation observed between H1' and C5 of β-C-BMP and H1'-C4/C6 of α-C-BMP and β-C-BMP in the HMBC as well as the absence of H5-C5 correlations in the HSQC support the C-nucleoside assignment. α-Anomer cross peaks are shown in red, β-anomer cross peaks in blue and overlapping cross peaks in purple. $^1$H Correlation spectroscopy (COSY) spectrum of a mixture of C-BMP nucleotides is provided in the Supplementary Information. **(c)** $^1$H NMR and 1D ROE spectra of a solution containing a 1:2 ratio of α-C-BMP to β-C-BMP. (Top) $^1$H NMR spectrum with resonance assignments as indicated in **a**. (Middle) Irradiation of the H1' of α-C-BMP results in through space magnetization transfer to the H2' and H3' of α-C-BMP. (Bottom) Irradiation of the H1' of β-C-BMP results in through space magnetization transfer to the H4' of β-C-BMP. * Indicates TOCSY transfer from β-H1' to β-H2' and β-H3'.

**Figure 3 | NMR characterization of MMP nucleotides. (a)** Chemical structure of α-MMP and β-MMP with arrows indicating through-space proton–proton magnetization transfer as shown in **c**. **(b)** HSQC and HMBC spectra showing $^1$H-$^{13}$C couplings for α-MMP and β-MMP. $^1$H-$^{13}$C correlation observed in the HMBC (H1'-C5) of α-MMP and β-MMP show coupling between ribose and melamine. α-Anomer cross peaks are shown in red, β-anomer cross peaks in blue and overlapping cross peaks in purple. $^1$H COSY spectrum of a mixture of MMP nucleotides is provided in the Supplementary Information. **(c)** $^1$H NMR and 1D ROE spectra of a solution containing an approximately equal concentration of α-MMP to β-MMP. (Top) $^1$H NMR spectrum with resonance assignments as indicated in **a**. (Middle) Irradiation of the H1' of α-MMP results in through space magnetization transfer to the H3' of α-MMP. (Bottom) Irradiation of the H1' of β-MMP results in through space magnetization transfer to the H2' and H4' of β-MMP. * Indicates TOCSY transfer from β-H1' to β-H3'.

a single assembly. When solutions containing purified MMP and C-BMP were combined and analysed by AFM, supramolecular polymers with a diameter of 2 nm were also observed (Fig. 4c and Supplementary Fig. 11b). As can be seen in Fig. 4b,c, the assemblies formed from purified MMP and BMP are noticeably shorter than those formed upon combining the crude reaction

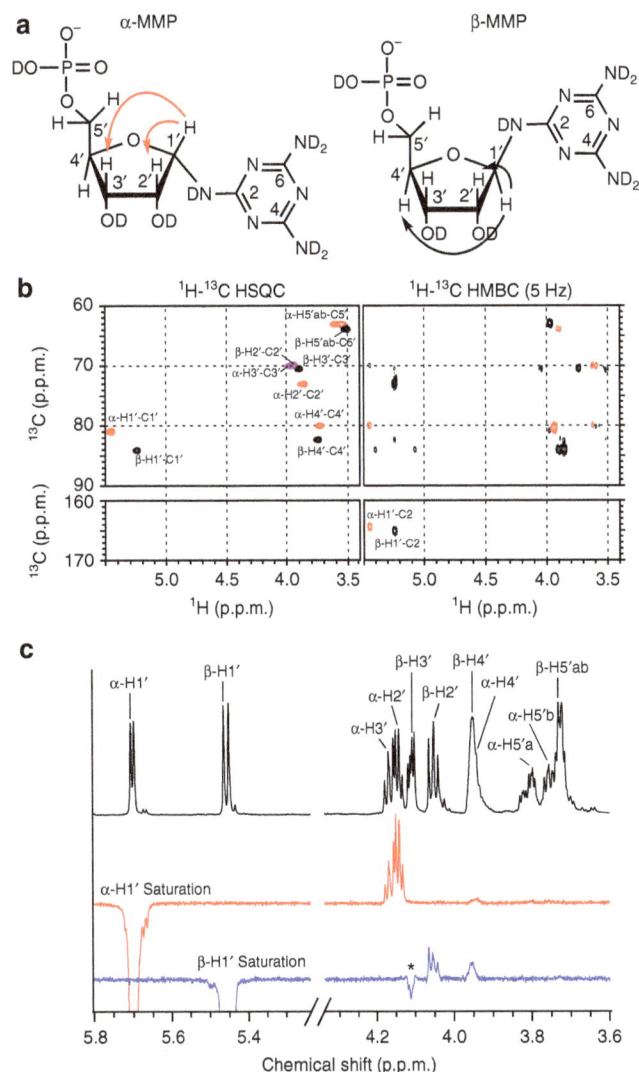

mixtures. This observation is consistent with length being limited by greater peripheral charge[36], an effect that is expected to be greater for supramolecular polymers formed from the purified nucleotides than it is for assemblies containing both nucleotides and the parent heterocycles (that is, those present in the crude reaction mixtures).

C-BMP and MMP nucleotides also form water-soluble supramolecular assemblies when mixed with free melamine and free BA, respectively. These C-BMP-melamine and MMP-BA assemblies were visualized by AFM (Fig. 4d,e and Supplementary Fig. 11c,d), revealing 2 nm fibres that are again indicative of

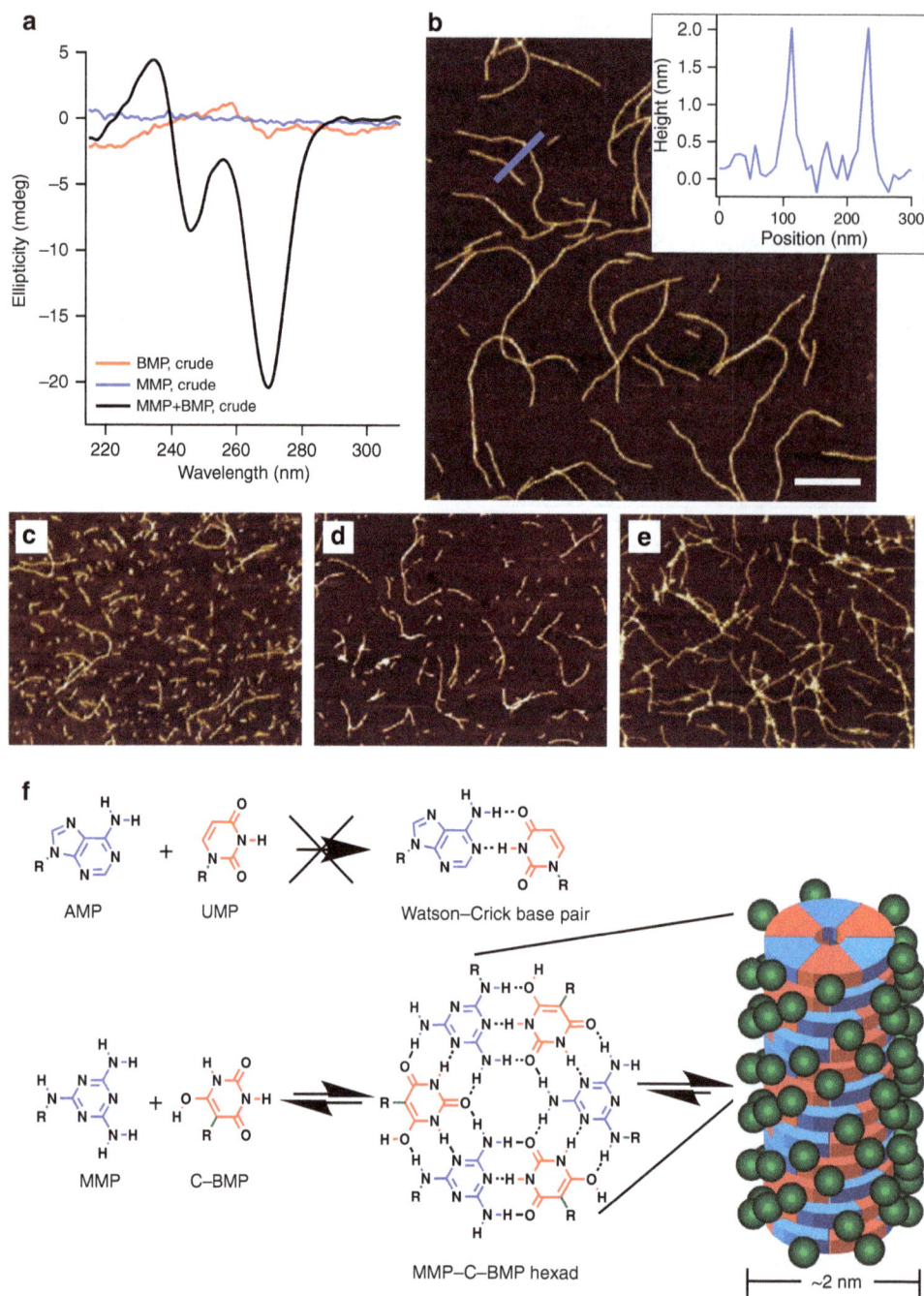

**Figure 4 | Nucleotides assemble into supramolecular polymers.** (**a**) CD spectra of melamine-R5P and BA-R5P crude reaction mixtures, separate and combined at 5 °C. Black curve is mixture of melamine-R5P and BA-R5P crude reaction mixtures; red curve is BA-R5P crude reaction mixture; blue curve is melamine-R5P crude reaction mixture. Ultraviolet spectra associated with CD spectra are provided in Supplementary Fig. 11. (**b**–**e**) AFM topographic images of (**b**) mixture of melamine-R5P and BA-R5P crude reaction mixtures, (**c**) mixture of purified C-BMP and MMP, (**d**) purified C-BMP mixed with melamine, (**e**) purified MMP mixed with BA. Inset in **b** shows height measurements of blue line in image. Scale bar in **b** is 300 nm, and all AFM images are shown at the same magnification. (**f**) Chemical structures of MMP and C-BMP and their association into 2 nm wide stacked hexads. The green R and green spheres indicate R5P. The similarity of AMP to MMP and UMP to C-BMP, and the inability of these canonical nucleosides to assemble in aqueous solution is also illustrated. All solutions contained 1 M NaCl.

stacked hexad assemblies. Although free BA and free melamine form insoluble precipitates when mixed in aqueous solution (Supplementary Fig. 12), the steric bulk and charge provided by conjugation with R5P on one nucleobase favours the formation of the water soluble, linear assemblies of stacked hexads[35].

**Supramolecular assemblies preferentially incorporate β-MMP.** [1]H-NMR spectroscopy was used to further characterize C-BMP-melamine and MMP-BA assemblies (Supplementary

Figs 13 and 14). Nucleotides incorporated into these supramolecular assemblies exhibit extreme [1]H NMR line broadening (to baseline), which render them invisible to solution state NMR spectroscopy. In contrast, free nucleotides that exist in equilibrium with the assemblies exhibit virtually no change in [1]H resonance chemical shift or line width. Quantitative analysis of [1]H resonance intensity can be used to determine the fraction of free nucleotides, and subtraction of these values from the known total concentration of nucleotides in a sample reveals the concentration of assembled nucleotides[35]. Analysis of $^1$H spectra of solutions containing MMP + BA, 50 mM in nucleotide and heterocycle, showed temperature-dependent assembly of nucleotides from 5 to 40 °C (Fig. 5a,b and Supplementary Fig. 14). Unexpectedly, the preferential incorporation for β-MMP over α-MMP was observed at all temperatures where assemblies are present. In particular, a twofold preference for β-MMP incorporation is observed at 5 °C, and only incorporation of β-MMP at 20 °C (Fig. 5b).

As the β-anomer is selectively incorporated into the assemblies over the α-anomer, and anomerization of MMP occurs in the solution (see the Methods for details), we next explored if supramolecular assembly will affect the anomeric ratio of MMP. As noted above, at equilibrium, a solution of MMP contains a mixture of α- and β-anomers (45% α and 55% β), however, when 50 mM MMP was incubated with 50 mM BA at 5 °C, we observe the conversion of α-MMP to β-MMP, a conversion that reached a maximum of 63% β-MMP after 8 days (Fig. 5c). This conversion was not observed when BA was omitted or when MMP was below the minimal concentration required for MMP-BA assembly. Therefore, we have found that the assemblies formed with BA preferentially select/stabilize the β-anomer of MMP, and as a result, with anomerization, the assemblies enrich β-MMP (Fig. 5d). These observations indicate that solutions containing MMP and BA assemblies will change overtime in both molecular (that is, enrichment of β-MMP) and supramolecular composition (that is, more MMP and BA assembled).

## Discussion

The data presented here demonstrate the efficient single-step syntheses of complementary nucleosides and nucleotides, starting with the plausible proto-nucleobases melamine and BA and ribose or R5P. Although R5P forms an exocyclic N-glycosidic bond with melamine (MMP) and R5P forms a C-nucleotide with BA (C-BMP), both of these nucleotides favour their β anomers. Specifically, for the free nucleotides in solution, the β and α anomers of C-BMP exist at equilibrium in a 67:33 ratio, and for MMP the β-anomer is favoured in a 55:44 ratio over the α-anomer. Perhaps coincidentally, extant life uses the β-anomeric form of ribonucleotides, and our observations indicate that this form may have been enriched on the early Earth for β-ribonucleotides for which glycosidic bond anomerization is under equilibrium control. This nucleotide structural preference is apparently not limited to the nucleotides of melamine and BA, as previous studies have also revealed that nucleosides formed by drying and heating ribose with two other pyrimidines, 2-pyrimidinone[21,37] and 2,4,6-triaminopyrimidine[23], also favour the β-anomer.

Unlike the mononucleotides of extant RNA, we observe that MMP and C-BMP will pair as monomers in aqueous solution, with each other and with their unmodified pairing partners (that is, MMP with C-BMP, MMP with BA, and C-BMP with melamine), producing, in all cases, supramolecular structures that indicate the highly efficient stacking of H-bonded hexads that are themselves composed of Watson–Crick-like base pairs.

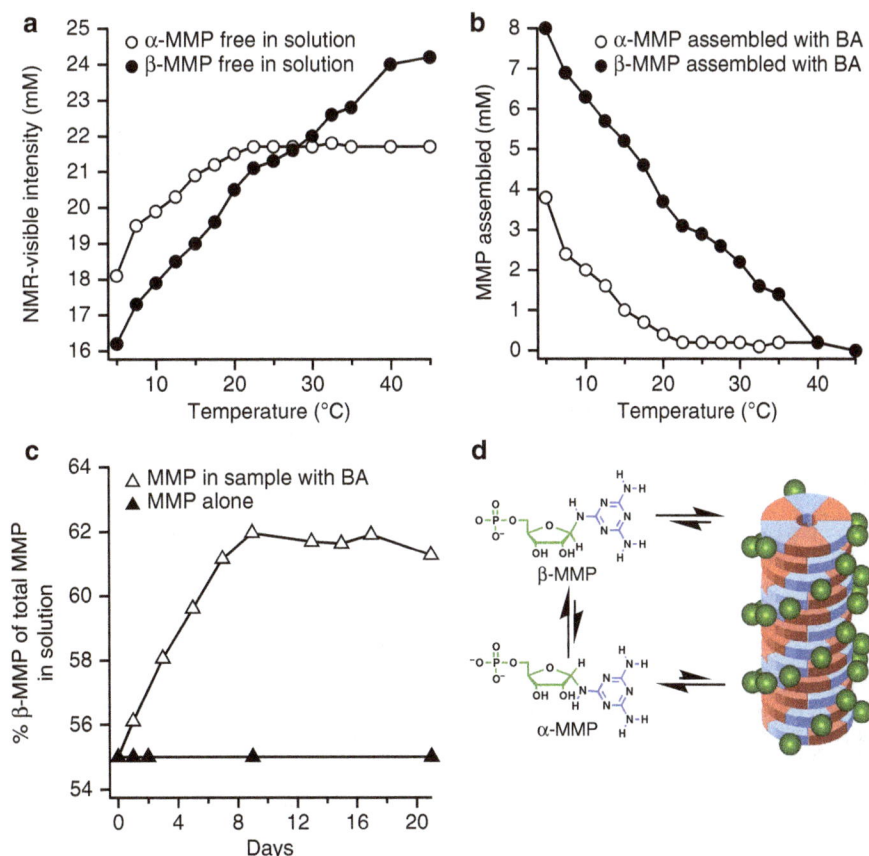

**Figure 5 | MMP assembly and anomerization in the presence of BA.** (a) Plot of NMR visible resonance intensity of unassembled α-MMP and β-MMP as a function of temperature in a solution containing 50 mM MMP and BA. (b) Plot showing fraction of both MMP anomers assembled at various temperatures (plot generated by subtracting data shown in a from measured total concentration of α-MMP and β-MMP in each sample). (c) Plot showing the change in anomeric ratio (by percent) of β-MMP as a function of time in solutions containing both MMP and BA, or MMP alone. Samples were maintained at 5 °C during the experiment and diluted just prior to analysis to disassemble MMP in order to enable quantification of the total MMP in solution by NMR. (d) Schematic showing preferential assembly (stacked hexads) and anomerization. All solutions contained 0.3 M NaCl.

This pairing and stacking is sufficiently robust to drive assembly in the presence of the side products and unreacted starting materials of the crude nucleotide reactions. Based on these observations, and because C-BMP and MMP structurally resemble two nucleotides found in life today (UMP and AMP, respectively) and have been reported to pair with extant, complementary nucleobases[38,39], it is tempting to speculate that these heterocycles could represent ancestral nucleotides of the contemporary genetic polymers. In particular, the ability for C-BMP and MMP to form *noncovalent* supramolecular assemblies could have facilitated the prebiotic localization, organization and subsequent linking of these (or similar) nucleotides into *covalent* polymers that were then capable of storing and transferring information (for example, by templating the formation of sequence-specific assemblies for the polymerization of additional monomers[40]).

The ability of C-BMP and MMP to form supramolecular assemblies might have also facilitated the emergence of early RNA-like polymers by selecting nucleotides with sugars (or earlier trifunctional linkers[19]) that were structurally compatible with the assemblies and their subsequent coupling into covalent polymers. In the present study, we have, for practical reasons, used D-ribose and D-R5P for our nucleoside and nucleotide reactions with melamine and BA, but L-ribose or L-R5P would exhibit equivalent reactivity with these two heterocycles. Nevertheless, it has been often postulated that a racemic mixture of nucleotides would have inhibited the prebiotic synthesis of RNA polymers[41], and so the question of how the present system might address this challenge deserves some discussion. Although we have not shown chiral nucleotide selection, in the current study we have demonstrated that the β-anomer of MMP is enriched in supramolecular assemblies over the α-anomer of MMP, and this selection leads to a detectable increase in the ratio of the β-anomer over the α-anomer of MMP in the entire solution (presumably due to anomerization and selective stabilization by the assembly). As a recent example of the ability of supramolecular polymers to promote local chiral resolution, Aida and co-workers demonstrated that racemic solutions of chiral macrocycles self-sort into homochiral supramolecular polymers[42]. It is therefore possible that supramolecular assemblies, formed by nucleotides with different sugars, including different anomers and enantiomers, could have been selectively enriched in individual supramolecular assemblies before polymerization. Current investigations of this possibility are actively being pursued in our laboratory.

## Methods

**Materials.** Melamine and BA were purchased from Acros Organic, D-R5P disodium salt and D-ribose were purchased from Sigma-Aldrich. All chemicals were used as received.

**Synthesis of C-BMP.** BA (2.5 mmol) and R5P (2.5 mmol) were dissolved in 5 ml of $H_2O$ and the pH was adjusted to 9 with NaOH (unless otherwise noted). The solution was stirred for 24 h at 20 °C, at which time a clear, pale yellow solution was present. This solution is referred to as the crude BA-R5P reaction mixture. To isolate C-BMP, the crude reaction mixture was loaded onto a gravity column containing QAE Sephadex A-25 anion exchange media and eluted with a gradient of $NH_4HCO_3$ buffer (pH 9.4) from 50 mM to 0.5 M. The fractions containing product were lyophilized, redissolved in water and pooled. When the reaction was performed at pH 9, yield of C-BMP was 82% (α-C-BMP 22%, β-C-BMP 60%); $^1H$ NMR of the crude reaction mixture is shown in Fig. 1b. HRMS (m/z): [M]$^-$ calculated for $C_9H_{12}N_2O_{10}P^-$, 339.0235; found, 339.0243. Ultraviolet-visible: (50 mM $NaH_2PO_4$, pH 7) $\lambda_{max} = 260$ nm; $\varepsilon_{260} = 23{,}000$ mol l$^{-1}$ cm$^{-1}$.

*α-C-BMP.* $^1H$ NMR (500 MHz, $D_2O$): δ 4.94 (d, J = 4.2 Hz, H1′), 3.85 (dd, J = 4.2, X Hz, H2′), 3.93 (dd, J = 4.1, 3.5 Hz, H3′), 3.70 (m, H4′), 3.59 (m, H5′a), 3.46 (m, H5′b); $^{13}C$ NMR (126 MHz, $D_2O$): δ 167.1 (C4/C6), 152.7 (C2), 85.5 (C5), 79.5 (d, J = 8.2 Hz, C4′), 75.1 (C1′), 74.3 (C2′), 72.7 (C3′), 63.6 (d, J = 4.4 Hz, C5′).

*β-C-BMP.* $^1H$ NMR (500 MHz, $D_2O$): δ 4.51 (d, J = 5.7 Hz, H1′), 4.29 (t, J = 5.7 Hz, H2′), 3.86 (t, J = 6.2 Hz, H3′), 3.54 (m, H4′), 3.58 (m, H5′a); 3.46

(m, H5′b); $^{13}C$ NMR (126 MHz, $D_2O$): δ 166.4 (C4/C6), 153.1 (C2), 85.7 (C5), 80.8 (d, J = 7.9 Hz, C4′), 78.6 (C1′), 70.9 (C2′), 70.5 (C3′), 63.8 (d, 4.6 Hz C5′).

**Synthesis of MMP.** Melamine (1 mmol) and R5P (1 mmol) were dissolved in 5 ml of $H_2O$ and the pH was adjusted with HCl to pH 5, unless otherwise noted. The solution was stirred at 65 °C for 24 h, at which point a clear, light brown solution was present. This solution is referred to as the melamine-R5P crude reaction mixture. To isolate MMP, the crude reaction mixture was loaded onto a gravity column containing SP Sephadex C-50 cation-exchange media and eluted with $NH_4OAc$ buffer, 50 mM, pH 4.3. The fractions containing product were lyophilized, redissolved in water and pooled. When the reaction was performed at pH 5, yield of MMP was 55% (α-MMP 26.4%, β-MMP 28.6%); $^1H$ NMR of the crude reaction mixture is shown in Fig. 1c. HRMS (m/z): [M]$^-$ calculated for $C_8H_{14}N_6O_7P^-$, 337.0667; found, 337.0659. Ultraviolet–visible: (50 mM $NaH_2PO_4$, pH 7) $\lambda_{max} = 235$ nm; $\varepsilon_{235} = 5{,}500$ mol l$^{-1}$ cm$^{-1}$.

*α-MMP.* $^1H$ NMR (500 MHz, $D_2O$): δ 5.46 (d, J = 4.1 Hz, H1′), 3.97 (dd, J = 4.7, 1.9 Hz, H3′), 3.94(dd, J = 4.7, 4.1 H2′), 3.74 (m, H4′), 3.64 (m, H5′a), 3.55 (m, H5′b); $^{13}C$ NMR (126 MHz, $D_2O$): δ 164.4 (C2), 164.2 (C4/C6), 81.0 (C1′), 79.5 (d, J = 8.2 Hz, C4′), 70.0 (C3′), 69.9 (C2′), 63.4 (d, J = 5.3, C5′).

*β-MMP.* $^1H$ NMR (500 MHz, $D_2O$): δ 5.25 (d, J = 6.6 Hz, H1′), 3.89 (dd, J = 5.4, 3.2 Hz, H3′), 3.85 (dd, J = 6.6, 5.4 Hz, H2′), 3.75 (m, H4′), 3.49 (m, H5′a,b); $^{13}C$ NMR (126 MHz, $D_2O$): δ 165.1 (C2), 164.9 (C4/C6), 84.2 (C1′), 82.3 (d, J = 8.0 Hz, C4′), 73.1 (C2′), 70.5 (C3′), 64.2 (d, 4.5 Hz C5′).

**General assembly protocol.** All solutions were 50 mM in each heterocycle. Final concentration of solutions containing crude reaction mixtures of either nucleotide was 50 mM in total heterocycle (that is, both unmodified and glycosylated). Upon combining melamine and BA nucleotides and/or parent heterocycles, the pH of solution was adjusted with HCl or NaOH to pH 4.5–5. This pH range was chosen as it would be expected to maximize assembly of the heterocycles due to $pK_a$ considerations ($pK_a$ of protonated melamine is 5 and the $pK_a$ of BA is 4)[34]. All solutions contained either 0.3 or 1.0 M NaCl as noted. Anomerization experiments were performed with MMP that was first incubated at 5 °C for 24 h to enable the anomeric ratio to equilibrate under the conditions tested. Variable temperature NMR experiments were performed on samples containing MMP and BA that were mixed and stored at 5 °C for 15 h before analysis to ensure that the association between MMP and BA had reached equilibrium. Stock solutions of nucleotides and crude reaction mixtures were stored at − 20 °C.

**Spectroscopic analysis of the supramolecular assemblies.** CD analysis was carried out on a Jasco J-810 CD spectrometer equipped with a six-cell Quantum Northwest peltier temperature controller. Strain-free 0.01 mm demountable cells from Starna were used for all CD analysis. $^1H$ NMR spectra were collected on a Bruker DRX-500 NMR and were the sum of 32 transients. All NMR samples were $D_2O$ exchanged and lyophilized before analysis in $D_2O$ with an internal standard of 3-(trimethylsilyl)-2,2′,3,3′-tetradeuteropropionic acid at 1.11 mM. Variable temperature NMR experiments were performed by gradually heating the solutions in order to disassemble supramolecular assemblies present, or slow cooling to enable reassembly, followed by short incubation at the desired temperature to allow for the assemblies to reach equilibrium. To analyse anomerization of MMP over time, a sample of MMP and BA (50 mM each) was stored at 5 °C, and at specified time points, samples were removed and diluted with $D_2O$ to 5 mM in MMP total (a concentration below the minimal assembly concentration) and analysed by NMR at 20 °C

**Atomic force microscopy.** Imaging was performed with a Nanoscope IIIa (Digital Instruments) in tapping mode, using Si tips (MicroMash, 16 N m$^{-1}$). A freshly cleaved mica substrate was pre-activated by incubation with a solution of 20 mM $MgCl_2$ that was then rinsed with water and dried under $N_2$ (g). A 3-μl sample (initially stored on ice) was deposited on the mica substrate and spread with $N_2$ (g) followed by the addition of 50 μl of cold water to remove sample that was not adsorbed to the mica surface and quickly dried again with $N_2$ (g).

**Analytical HPLC.** Analysis of nucleotide samples was performed by running a linear gradient of 90% acetonitrile/10% water to 50% acetonitrile/$H_2O$ for 7 min, followed by isocratic 90% aqueous acetonitrile for 6 min on a Waters XBridge Amide (150 × 2.5 × 3.5 μm$^3$) column. Analytical HPLC analysis of nucleoside samples was performed by running a linear gradient of 85% acetonitrile/15% water to 40% acetonitrile/60% water over 25 min on a Waters XBridge Amide (150 × 2.5 × 3.5 μm$^3$) column.

## References

1. Joyce, G. F. The antiquity of RNA-based evolution. *Nature* **418**, 214–221 (2002).
2. Callahan, M. P. *et al.* Carbonaceous meteorites contain a wide range of extraterrestrial nucleobases. *Proc. Natl Acad. Sci. USA* **108**, 13995–13998 (2011).

3. Barks, H. L. *et al.* Guanine, adenine, and hypoxanthine production in UV-irradiated formamide solutions: Relaxation of the requirements for prebiotic purine nucleobase formation. *ChemBioChem.* **11**, 1240–1243 (2010).

4. Hayatsu, R. Orgueil Meteorite: organic nitrogen contents. *Science* **146**, 1291–1293 (1964).

5. Ricardo, A., Carrigan, M. A., Olcott, A. N. & Benner, S. A. Borate minerals stabilize ribose. *Science* **303**, 196–196 (2004).

6. Springsteen, G. & Joyce, G. F. Selective derivatization and sequestration of ribose from a prebiotic mix. *J. Am. Chem. Soc.* **126**, 9578–9583 (2004).

7. Krishnamurthy, R., Guntha, S. & Eschenmoser, A. Regioselective alpha-phosphorylation of aldoses in aqueous solution. *Angew. Chem. Int. Ed. Engl.* **39**, 2281–2285 (2000).

8. Ponnamperuma, C. & Mack, R. Nucleotide synthesis under possible primitive Earth conditions. *Science* **148**, 1221–1223 (1965).

9. Orgel, L. E. Prebiotic chemistry and the origin of the RNA world. *Crit. Rev. Biochem. Mol. Biol.* **39**, 99–123 (2004).

10. Cafferty, B. J. & Hud, N. V. Abiotic synthesis of RNA in water: a common goal of prebiotic chemistry and bottom-up synthetic biology. *Curr. Opin. Chem. Biol.* **22**, 146–157 (2014).

11. Fuller, W. D., Sanchez, R. A. & Orgel, L. E. Studies in prebiotic synthesis. VI. Synthesis of purine nucleosides. *J. Mol. Biol.* **67**, 25–33 (1972).

12. Fuller, W. D., Sanchez, R. A. & Orgel, L. E. Studies in prebiotic synthesis: VII. Solid-state synthesis of purine nucleosides. *J. Mol. Evol.* **1**, 249–257 (1972).

13. Sanchez, R. A. & Orgel, L. E. Studies in prebiotic synthesis. V. Synthesis and photoanomerization of pyrimidine nucleosides. *J. Mol. Biol.* **47**, 531–543 (1970).

14. Powner, M. W., Gerland, B. & Sutherland, J. D. Synthesis of activated pyrimidine ribonucleotides in prebiotically plausible conditions. *Nature* **459**, 239–242 (2009).

15. Benner, S. A., Kim, H.-J. & Carrigan, M. A. Asphalt, water, and the prebiotic synthesis of ribose, ribonucleosides, and RNA. *Acc. Chem. Res.* **45**, 2025–2034 (2012).

16. Eschenmoser, A. Etiology of potentially primordial biomolecular structures: From vitamin B12 to the nucleic acids and an inquiry into the chemistry of life's origin: A retrospective. *Angew. Chem. Int. Ed. Engl.* **50**, 12412–12472 (2011).

17. Patel, B. H., Percivalle, C., Ritson, D. J., Duffy, C. D. & Sutherland, J. D. Common origins of RNA, protein and lipid precursors in a cyanosulfidic protometabolism. *Nat. Chem.* **7**, 301–307 (2015).

18. Joyce, G. F., Schwartz, A. W., Miller, S. L. & Orgel, L. E. The case for an ancestral genetic system involving simple analogs of the nucleotides. *Proc. Natl Acad. Sci. USA* **84**, 4398–4402 (1987).

19. Hud, N. V., Cafferty, B. J., Krishnamurthy, R. & Williams, L. D. The origin of RNA and 'My Grandfather's Axe'. *Chem. Biol.* **20**, 466–474 (2013).

20. Kolb, V. M., Dworkin, J. P. & Miller, S. L. Alternative bases in the RNA world: the prebiotic synthesis of urazole and its ribosides. *J. Mol. Evol.* **38**, 549–557 (1994).

21. Bean, H. D. *et al.* Formation of a *β*-pyrimidine nucleoside by a free pyrimidine base and ribose in a plausible prebiotic reaction. *J. Am. Chem. Soc.* **129**, 9556–9557 (2007).

22. Kim, H. J. & Benner, S. A. Prebiotic glycosylation of uracil with electron-donating substituents. *Astrobiology* **15**, 301–306 (2015).

23. Chen, M. C. *et al.* Spontaneous prebiotic formation of a *β*-ribofuranoside that self-assembles with a complementary heterocycle. *J. Am. Chem. Soc.* **136**, 5640–5646 (2014).

24. Cafferty, B. J. & Hud, N. V. Was a pyrimidine-pyrimidine base pair the ancestor of Watson-Crick base pairs? Insights from a systematic approach to the origin of RNA. *Israel J. Chem.* **55**, 891–905 (2015).

25. Menor-Salvan, C., Ruiz-Bermejo, D. M., Guzman, M. I., Osuna-Esteban, S. & Veintemillas-Verdaguer, S. Synthesis of pyrimidines and triazines in ice: Implications for the prebiotic chemistry of nucleobases. *Chem. Eur. J* **15**, 4411–4418 (2009).

26. Whitesides, G. M. *et al.* Noncovalent synthesis: using physical-organic chemistry to make aggregates. *Acc. Chem. Res.* **28**, 37–44 (1995).

27. Hysell, M., Siegel, J. S. & Tor, Y. Synthesis and stability of exocyclic triazine nucleosides. *Org. Biomol. Chem.* **3**, 2946–2952 (2005).

28. van Vliet, M. J., Visscher, J. & Schwartz, A. W. Hydrogen-bonding in the template-directed oligomerization of a pyrimidine nucleotide analog. *J. Mol. Evol.* **41**, 257–261 (1995).

29. Rios, A. C., Yu, H. T. & Tor, Y. Hydrolytic fitness of N-glycosyl bonds: comparing the deglycosylation kinetics of modified, alternative, and native nucleosides. *J. Phys. Org. Chem.* 173–180 (2014).

30. Hamma, T. & Ferre-D'Amare, A. R. Pseudouridine synthases. *Chem. Biol.* **13**, 1125–1135 (2006).

31. Ts'o, P. O. P. in *Basic Principles in Nucleic Acid Chemistry* Vol. 1 (ed. Ts'o, P. O. P.) 453–584 (Academic Press, 1974).

32. Engelhart, A. E. & Hud, N. V. Primitive genetic polymers. *Cold Spring Harb. Perspect. Biol.* **2** doi: 10.1101/cshperspect.a002196 (2010).

33. Krishnamurthy, R. On the emergence of RNA. *Israel J. Chem.* **55**, 837–850 (2015).

34. Cafferty, B. J., Avirah, R. R., Schuster, G. B. & Hud, N. V. Ultra-sensitive pH control of supramolecular polymers and hydrogels: pK(a) matching of biomimetic monomers. *Chem. Sci.* **5**, 4681–4686 (2014).

35. Cafferty, B. J. *et al.* Efficient self-assembly in water of long noncovalent polymers by nucleobase analogues. *J. Am. Chem. Soc.* **135**, 2447–2450 (2013).

36. Besenius, P. *et al.* Controlling the growth and shape of chiral supramolecular polymers in water. *Proc. Natl Acad. Sci. USA* **107**, 17888–17893 (2010).

37. Sheng, Y., Bean, H. D., Mamajanova, I., Hud, N. V. & Leszczynski, J. A comprehensive investigation of the energetics of pyrimidine nucleoside formation in a model prebiotic reaction. *J. Am. Chem. Soc.* **131**, 16088–16095 (2009).

38. Xia, X., Piao, X. & Bong, D. Bifacial peptide nucleic acid as an allosteric switch for aptamer and ribozyme function. *J. Am. Chem. Soc.* **136**, 7265–7268 (2014).

39. Voet, D. Crystal and molecular structure of intermolecular complex 9-ethyladenine-5,5-diethylbarbituric acid. *J. Am. Chem. Soc.* **94**, 8213–8222 (1972).

40. Hud, N. V. & Anet, F. A. L. Intercalation-mediated synthesis and replication: a new approach to the origin of life. *J. Theor. Biol.* **205**, 543–562 (2000).

41. Joyce, G. *et al.* Chiral selection in poly(C)-directed synthesis of oligo(G). *Nature* **310**, 602–604 (1984).

42. Sato, K., Itoh, Y. & Aida, T. Homochiral supramolecular polymerization of bowl-shaped chiral macrocycles in solution. *Chem. Sci.* **5**, 136–140 (2014).

## Acknowledgements

We thank J. Forsythe for accurate mass analysis, L.A. Bottomley for use of AFM, L.D. Williams and G.B. Schuster for discussions. This work was supported by the NSF and the NASA Astrobiology Program under the NSF Center for Chemical Evolution (CHE-1504217).

## Author contributions

B.J.C., R.K. and N.V.H. designed the research, D.M.F. and J.K. synthesized and purified MMP and BMP, B.J.C. and D.M.F. developed and characterized the nucleotide reactions, B.J.C. performed the NMR, CD and AFM analysis, all authors analysed and discussed the results, B.J.C., R.K. and N.V.H. wrote the paper.

## Additional information

**Competing financial interests:** The authors declare no competing financial interests.

# Permissions

All chapters in this book were first published in NC, by Nature Publishing Group; hereby published with permission under the Creative Commons Attribution License or equivalent. Every chapter published in this book has been scrutinized by our experts. Their significance has been extensively debated. The topics covered herein carry significant findings which will fuel the growth of the discipline. They may even be implemented as practical applications or may be referred to as a beginning point for another development.

The contributors of this book come from diverse backgrounds, making this book a truly international effort. This book will bring forth new frontiers with its revolutionizing research information and detailed analysis of the nascent developments around the world.

We would like to thank all the contributing authors for lending their expertise to make the book truly unique. They have played a crucial role in the development of this book. Without their invaluable contributions this book wouldn't have been possible. They have made vital efforts to compile up to date information on the varied aspects of this subject to make this book a valuable addition to the collection of many professionals and students.

This book was conceptualized with the vision of imparting up-to-date information and advanced data in this field. To ensure the same, a matchless editorial board was set up. Every individual on the board went through rigorous rounds of assessment to prove their worth. After which they invested a large part of their time researching and compiling the most relevant data for our readers.

The editorial board has been involved in producing this book since its inception. They have spent rigorous hours researching and exploring the diverse topics which have resulted in the successful publishing of this book. They have passed on their knowledge of decades through this book. To expedite this challenging task, the publisher supported the team at every step. A small team of assistant editors was also appointed to further simplify the editing procedure and attain best results for the readers.

Apart from the editorial board, the designing team has also invested a significant amount of their time in understanding the subject and creating the most relevant covers. They scrutinized every image to scout for the most suitable representation of the subject and create an appropriate cover for the book.

The publishing team has been an ardent support to the editorial, designing and production team. Their endless efforts to recruit the best for this project, has resulted in the accomplishment of this book. They are a veteran in the field of academics and their pool of knowledge is as vast as their experience in printing. Their expertise and guidance has proved useful at every step. Their uncompromising quality standards have made this book an exceptional effort. Their encouragement from time to time has been an inspiration for everyone.

The publisher and the editorial board hope that this book will prove to be a valuable piece of knowledge for researchers, students, practitioners and scholars across the globe.

# List of Contributors

**Xuesong Wu, Yan Zhao and Haibo Ge**
Department of Chemistry and Chemical Biology, Indiana University-Purdue University Indianapolis, 402 N. Blackford Street, Indianapolis, Indiana 46202, USA

**Ke Yang and Hao Sun**
Institute of Chemistry and BioMedical Sciences, Nanjing University, Nanjing 210093, P.R. China

**Guigen Li**
Institute of Chemistry and BioMedical Sciences, Nanjing University, Nanjing 210093, P.R. China
Department of Chemistry and Biochemistry, Texas Tech University, Lubbock, Texas 79409-1061, USA

**Xiaoyu Li, Yang Gao, Charlotte E. Boott and Ian Manners**
School of Chemistry, University of Bristol, Bristol BS8 1TS, UK

**Mitchell A. Winnik**
Department of Chemistry, University of Toronto, Toronto, Ontario M5S 3H6, Canada

**Raj Kumar Roy, Anna Meszynska, Chloé Laure, Claire Verchin and Jean-François Lutz**
Precision Macromolecular Chemistry, Institut Charles Sadron, UPR22-CNRS, BP84047, 23 rue du Loess, 67034 Strasbourg Cedex 2, France

**Laurence Charles**
Aix-Marseille Université – CNRS, UMR 7273, Institute of Radical Chemistry, 13397 Marseille Cedex 20, France

**David C. Leitch, Laure V. Kayser, Zhi-Yong Han, Ali R. Siamaki, Evan N. Keyzer, Ashley Gefen and Bruce A. Arndtsen**
Department of Chemistry, McGill University, 801 Sherbrooke Street West, Montreal, Quebec, Canada H3A 0K8

**Panpan Tian**
Hefei National Laboratory for Physical Sciences at the Microscale and Department of Chemistry, University of Science and Technology of China, Hefei 230026, China

**Chao Feng and Teck-Peng Loh**
Hefei National Laboratory for Physical Sciences at the Microscale and Department of Chemistry, University of Science and Technology of China, Hefei 230026, China
Division of Chemistry and Biological Chemistry, Nanyang Technological University, 50 Nanyang Avenue, Singapore 639798, Singapore

**Guangfan Zheng, Yan Li, Jingjie Han, Tao Xiong and Qian Zhang**
Department of Chemistry, Northeast Normal University, Changchun 130024, China

**Lorena Mendive-Tapia and Sara Preciado**
Institute for Research in Biomedicine, Barcelona Science Park, Baldiri Reixac 10-12, Barcelona 08028, Spain

**Can Zhao and Nick D. Read**
Manchester Fungal Infection Group, Institute of Inflammation and Repair, University of Manchester, CTF Building, Grafton St, Manchester M13 9NT, UK

**Ahsan R. Akram and Marc Vendrell**
MRC/UoE Centre for Inflammation Research, University of Edinburgh, 47 Little France Crescent, Edinburgh EH16 4TJ, UK

**Fernando Albericio**
Institute for Research in Biomedicine, Barcelona Science Park, Baldiri Reixac 10-12, Barcelona 08028, Spain
Department Organic Chemistry, University of Barcelona, Martí Franqués 1-11, Barcelona 08028, Spain
CIBER-BBN, Networking Centre for Bioengineering, Biomaterials and Nanomedicine, Baldiri Reixac 10-12, Barcelona 08028, Spain
School of Chemistry, University of KwaZulu-Natal, Durban 4001, South Africa

**Nicola Kielland**
Laboratory of Organic Chemistry, Faculty of Pharmacy, University of Barcelona, Barcelona Science Park, Baldiri Reixac 10-12, Barcelona 08028, Spain

**Martin Lee and Alan Serrels**
Edinburgh Cancer Research Centre, University of Edinburgh, Crewe South Road, Edinburgh EH4 2XR, UK

**Rodolfo Lavilla**
CIBER-BBN, Networking Centre for Bioengineering, Biomaterials and Nanomedicine, Baldiri Reixac 10-12, Barcelona 08028, Spain
Laboratory of Organic Chemistry, Faculty of Pharmacy, University of Barcelona, Barcelona Science Park, Baldiri Reixac 10-12, Barcelona 08028, Spain

**Mathieu Colomb-Delsuc, Elio Mattia, Jan W. Sadownik and Sijbren Otto**
Centre for Systems Chemistry, Stratingh Institute, University of Groningen, Nijenborgh 4, 9747 AG, Groningen, The Netherlands

**Diederik W.R. Balkenende, Christophe A. Monnier, Gina L. Fiore and Christoph Weder**
Adolphe Merkle Institute, University of Fribourg, Chemin des Verdiers 4, CH-1700 Fribourg, Switzerland

**Pratap S. Patil, Ting-Jen Rachel Cheng, Medel Manuel L. Zulueta, Shih-Ting Yang, Larry S. Lico and Shang-Cheng Hung**
Genomics Research Center, Academia Sinica, No. 128, Section 2, Academia Road, Taipei 115, Taiwan

**Daniel Görl, Xin Zhang, Vladimir Stepanenko and Frank Würthner**
Institut für Organische Chemie and Center for Nanosystems Chemistry, Universität Würzburg, Am Hubland, Würzburg 97074, Germany

**Masakazu Nambo**
Institute of Transformative Bio-Molecules (WPI-ITbM), Nagoya University, Chikusa, Nagoya 464-8602, Japan

**Cathleen M. Crudden**
Queen's University, Department of Chemistry, Chernoff Hall, Kingston, Ontario, Canada K7L 3N6
Institute of Transformative Bio-Molecules (WPI-ITbM), Nagoya University, Chikusa, Nagoya 464-8602, Japan.

**Christopher Ziebenhaus, Jason P.G. Rygus, Kazem Ghozati, Phillip J. Unsworth, Samantha Voth, Marieke Hutchinson, Veronique S. Laberge and Daisuke Imao**
Queen's University, Department of Chemistry, Chernoff Hall, Kingston, Ontario, Canada K7L 3N6

**Yuuki Maekawa**
Queen's University, Department of Chemistry, Chernoff Hall, Kingston, Ontario, Canada K7L 3N6
Department of Chemistry and Biomolecular Science, Faculty of Engineering, Gifu University, 1-1 Yanagido, Gifu 501-1193, Japan

**Zhi-Wei Jiao, Qing Zhang, Wen-Xing Liu, Shu-Yu Zhang, Shao-Hua Wang, Fu-Min Zhang and Sen Jiang**
State Key Laboratory of Applied Organic Chemistry & College of Chemistry and Chemical Engineering, Lanzhou University, Lanzhou 730000, P. R. China

**Yong-Qiang Tu**
State Key Laboratory of Applied Organic Chemistry & College of Chemistry and Chemical Engineering, Lanzhou University, Lanzhou 730000, P. R. China
Collaborative Innovation Center of Chemical Science and Engineering, Tianjin 300071, P. R. China

**James Allen Frank, Martin Sumser and Dirk Trauner**
Department of Chemistry and Center for Integrated Protein Science, Ludwig Maximilians University Munich, Butenandtstrasse 5–13, Munich 81377, Germany

**Mirko Moroni and Gary R. Lewin**
Molecular Physiology of Somatic Sensation, Max Delbrück Center for Molecular Medicine, Berlin 13125, Germany

**Rabih Moshourab**
Molecular Physiology of Somatic Sensation, Max Delbrück Center for Molecular Medicine, Berlin 13125, Germany
Department of Anesthesiology, Campus Charité Mitte und Virchow Klinikum, Charité Universitätsmedizin Berlin, Augustburgerplatz 1, Berlin 13353, Germany

**Sang Yeon Park, Ji-Woong Lee and Choong Eui Song**
Department of Chemistry, Sungkyunkwan University, 2066, Seobu-ro, Jangan-gu, Suwon-si, Gyeonggi-do 440-746, Korea

**Lorena Mendive-Tapia**
Institute for Research in Biomedicine, Barcelona Science Park, Baldiri Reixac 10-12, 08028 Barcelona, Spain
Department of Organic Chemistry, University of Barcelona, Martí i Franqués 1-11, 08028 Barcelona, Spain
CIBER-BBN, Networking Centre on Bioengineering, Biomaterials and Nanomedicine

**Sara Preciado**
Department of Organic Chemistry, University of Barcelona, Martí i Franqués 1-11, 08028 Barcelona, Spain

**Jesús García**
Institute for Research in Biomedicine, Barcelona Science Park, Baldiri Reixac 10-12, 08028 Barcelona, Spain

**Rosario Ramón and Nicola Kielland**
Barcelona Science Park, Baldiri Reixac 10-12, 08028 Barcelona, Spain

**Fernando Albericio**
Institute for Research in Biomedicine, Barcelona Science Park, Baldiri Reixac 10-12, 08028 Barcelona, Spain
Department of Organic Chemistry, University of Barcelona, Martí i Franqués 1-11, 08028 Barcelona, Spain
CIBER-BBN, Networking Centre on Bioengineering, Biomaterials and Nanomedicine
School of Chemistry, Yachay Tech, Yachay City of Knowledge, 100119 Urcuqui, Ecuador

**Rodolfo Lavilla**
Barcelona Science Park, Baldiri Reixac 10-12, 08028 Barcelona, Spain
Laboratory of Organic Chemistry, Faculty of Pharmacy, University of Barcelona, Avda. Joan XXII s.n., 08028 Barcelona, Spain

**Qineng Xia, Zongjia Chen, Yi Shao, Xueqing Gong, Haifeng Wang, Xiaohui Liu and Yanqin Wang**
Key Laboratory for Advanced Materials, Research Institute of Industrial Catalysis, East China University of Science and Technology, Shanghai 200237, China

**Stewart F. Parker**
ISIS Facility, STFC Rutherford Appleton Laboratory, Chilton, Oxfordshire OX11 0QX, UK

**Xue Han**
School of Chemistry, University of Nottingham, Nottingham, NG7 2RD, UK
School of Chemistry, University of Manchester, Manchester M13 9PL, UK

**Sihai Yang**
School of Chemistry, University of Manchester, Manchester M13 9PL, UK

**Erik B. Pinxterhuis, Massimo Giannerini, Valentín Hornillos and Ben L. Feringa**
Faculty of Mathematics and Natural Sciences, Stratingh Institute for Chemistry, University of Groningen, Nijenborgh 4, 9747AG Groningen, The Netherlands

**Xinjiang Cui, Annette-Enrica Surkus, Kathrin Junge, Christoph Topf, Jörg Radnik, Carsten Kreyenschulte and Matthias Beller**
Leibniz-Institute for Catalysis, University of Rostock, Albert Einstein Street, 29a, Rostock 18059, Germany

**Jun Yuan, Xin He and Liang Zhao**
Key Laboratory of Bioorganic Phosphorus Chemistry and Chemical Biology (Ministry of Education), Department of Chemistry, Tsinghua University, Beijing 100084, China

**Tingting Sun, Ke An and Jun Zhu**
State Key Laboratory of Physical Chemistry of Solid Surfaces and Collaborative Innovation Center of Chemistry for Energy Materials (iChEM), Fujian Provincial Key Laboratory of Theoretical and Computational Chemistry and Department of Chemistry, College of Chemistry and Chemical Engineering, Xiamen University, Xiamen 361005, China

**Yongjun Li, Zhiyu Jia, Huibiao Liu and Yuliang Li**
CAS Key Laboratory of Organic Solids, Beijing National Laboratory for Molecular Science (BNLMS), Institute of Chemistry, Chinese Academy of Sciences, Beijing 100190, China

**Shengqiang Xiao**
State Key Laboratory of Advanced Technology for Materials Synthesis and Processing, Wuhan University of Technology, Wuhan 430070, China

**Xiao-Qiang Hu, Jia-Rong Chen, Quan-Qing Zhao, Qiang Wei and Wen-Jing Xiao**
CCNU-uOttawa Joint Research Centre, Key Laboratory of Pesticide and Chemical Biology, Ministry of Education, College of Chemistry, Central China Normal University, 152 Luoyu Road, Wuhan 430079, China

**Xiaotian Qi and Yu Lan**
School of Chemistry and Chemical Engineering, Chongqing University, Chongqing 400030, China

**Felix F. Loeffler, Tobias C. Foertsch, Roman Popov, Martyna Sedlmayr Valentina Bykovskaya, Lothar Hahn, Frank Breitling, Alexander Nesterov-Mueller, Clemens von Bojničić-Kninski, Laura K. Weber and Andrea Fischer**
Karlsruhe Institute of Technology, Institute of Microstructure Technology (IMT), Hermann-von-Helmholtz-Platz 1, 76344 Eggenstein-Leopoldshafen, Germany

**Daniela S. Mattes, Barbara Ridder and Juliane Greifenstein**
Karlsruhe Institute of Technology, Institute of Microstructure Technology (IMT), Hermann-von-Helmholtz-Platz 1, 76344 Eggenstein-Leopoldshafen, Germany
Karlsruhe Institute of Technology, Institute of Organic Chemistry (IOC), Fritz-Haber-Weg 6, 76131 Karlsruhe, Germany

**Martin Schlageter and Annie K. Powell**
Karlsruhe Institute of Technology, Institute for Inorganic Chemistry, Engesserstrasse 15, 76131 Karlsruhe, Germany
Karlsruhe Institute of Technology, Institute of Nanotechnology (INT), Hermann-von-Helmholtz-Platz 1, 76344 Eggenstein-Leopoldshafen, Germany

**Florian-Xuan Dang and Teodor Silviu Balaban**
Aix-Marseille University, CNRS UMR 7313, Institut des Sciences Moléculaires de Marseille, Chirosciences, 13397 Marseille cedex 20, France

**Ivan Buliev**
Technical University of Varna, 1 Studentska, 9010 Varna, Bulgaria

**F. Ralf Bischoff**
German Cancer Research Center, INF 580, 69120 Heidelberg, Germany.

**Michael A.R. Meier**
Karlsruhe Institute of Technology, Institute of Organic Chemistry (IOC), Fritz-Haber-Weg 6, 76131 Karlsruhe, Germany

**Stefan Bräse**
Karlsruhe Institute of Technology, Institute of Organic Chemistry (IOC), Fritz-Haber-Weg 6, 76131 Karlsruhe, Germany
Karlsruhe Institute of Technology, Institute of Toxicology and Genetics (ITG), Hermann-von- Helmholtz-Platz 1, 76344 Eggenstein-Leopoldshafen, Germany

**Jason Potticary, Lui R. Terry, Simon R. Hall**
Complex Functional Materials Group, School of Chemistry, University of Bristol, Bristol BS8 1TS, UK

**Christopher Bell**
School of Physics, HH Wills Physics Laboratory, Tyndall Avenue, Bristol BS8 1TL, UK

**Alexandros N. Papanikolopoulos, Peter C.M. Christianen and Hans Engelkamp**
High Field Magnet Laboratory (HFML-EMFL), Radboud University, Toernooiveld 7, 6525 ED Nijmegen, The Netherlands

**Andrew M. Collins**
Bristol Centre for Functional Nanomaterials, HH Wills Physics Laboratory, Tyndall Avenue, Bristol BS8 1TL, UK

**Claudio Fontanesi**
Department of Physics, University of Bath, Claverton Down, Bath BA2 7AY, UK
Dipartimento di Ingegneria Enzo Ferrari, Universita' di Modena e Reggio Emilia, Via Vivarelli 10, 41125 Modena, Italy

**Gabriele Kociok-Köhn**
Department of Chemistry, University of Bath, Claverton Down, Bath BA2 7AY, UK

**Simon Crampin and Enrico Da Como**
Department of Physics, University of Bath, Claverton Down, Bath BA2 7AY, UK

**Jonathan Latham, Jean-Marc Henry, Humera H. Sharif, Binuraj R.K. Menon, Sarah A. Shepherd and Jason Micklefield**
School of Chemistry, The University of Manchester, Oxford Road, Manchester M13 9PL, UK
Manchester Institute of Biotechnology, The University of Manchester, 131 Princess Street, Manchester M1 7DN, UK

**Michael F. Greaney**
School of Chemistry, The University of Manchester, Oxford Road, Manchester M13 9PL, UK

**Brian J. Cafferty, David M. Fialho, Jaheda Khanam and Nicholas V. Hud**
School of Chemistry and Biochemistry, Georgia Institute of Technology, 901 Atlantic Drive, Atlanta, Georgia 30332, USA
NSF-NASA Center for Chemical Evolution, Atlanta, Georgia 30332, USA

**Ramanarayanan Krishnamurthy**
NSF-NASA Center for Chemical Evolution, Atlanta, Georgia 30332, USA Department of Chemistry, The Scripps Research Institute, La Jolla, California 92037, USA Xuesong Wu, Yan Zhao and Haibo Ge
Department of Chemistry and Chemical Biology, Indiana University-Purdue University Indianapolis, 402 N. Blackford Street, Indianapolis, Indiana 46202, USA

**Ke Yang and Hao Sun**
Institute of Chemistry and BioMedical Sciences, Nanjing University, Nanjing 210093, P.R. China

**Guigen Li**
Institute of Chemistry and BioMedical Sciences, Nanjing University, Nanjing 210093, P.R. China
Department of Chemistry and Biochemistry, Texas Tech University, Lubbock, Texas 79409-1061, USA

**Xiaoyu Li, Yang Gao, Charlotte E. Boott and Ian Manners**
School of Chemistry, University of Bristol, Bristol BS8 1TS, UK

**Mitchell A. Winnik**
Department of Chemistry, University of Toronto, Toronto, Ontario M5S 3H6, Canada

**Raj Kumar Roy, Anna Meszynska, Chloé Laure, Claire Verchin and Jean-François Lutz**
Precision Macromolecular Chemistry, Institut Charles Sadron, UPR22-CNRS, BP84047, 23 rue du Loess, 67034 Strasbourg Cedex 2, France

**Laurence Charles**
Aix-Marseille Université – CNRS, UMR 7273, Institute of Radical Chemistry, 13397 Marseille Cedex 20, France

**David C. Leitch, Laure V. Kayser, Zhi-Yong Han, Ali R. Siamaki, Evan N. Keyzer, Ashley Gefen and Bruce A. Arndtsen**
Department of Chemistry, McGill University, 801 Sherbrooke Street West, Montreal, Quebec, Canada H3A 0K8

**Panpan Tian**
Hefei National Laboratory for Physical Sciences at the Microscale and Department of Chemistry, University of Science and Technology of China, Hefei 230026, China

**Chao Feng and Teck-Peng Loh**
Hefei National Laboratory for Physical Sciences at the Microscale and Department of Chemistry, University of Science and Technology of China, Hefei 230026, China
Division of Chemistry and Biological Chemistry, Nanyang Technological University, 50 Nanyang Avenue, Singapore 639798, Singapore

**Guangfan Zheng, Yan Li, Jingjie Han, Tao Xiong and Qian Zhang**
Department of Chemistry, Northeast Normal University, Changchun 130024, China

**Lorena Mendive-Tapia and Sara Preciado**
Institute for Research in Biomedicine, Barcelona Science Park, Baldiri Reixac 10-12, Barcelona 08028, Spain

**Can Zhao and Nick D. Read**
Manchester Fungal Infection Group, Institute of Inflammation and Repair, University of Manchester, CTF Building, Grafton St, Manchester M13 9NT, UK

**Ahsan R. Akram and Marc Vendrell**
MRC/UoE Centre for Inflammation Research, University of Edinburgh, 47 Little France Crescent, Edinburgh EH16 4TJ, UK

**Fernando Albericio**
Institute for Research in Biomedicine, Barcelona Science Park, Baldiri Reixac 10-12, Barcelona 08028, Spain
Department Organic Chemistry, University of Barcelona, Martíi Franqués 1-11, Barcelona 08028, Spain
CIBER-BBN, Networking Centre for Bioengineering, Biomaterials and Nanomedicine, Baldiri Reixac 10-12, Barcelona 08028, Spain
School of Chemistry, University of KwaZulu-Natal, Durban 4001, South Africa

**Nicola Kielland**
Laboratory of Organic Chemistry, Faculty of Pharmacy, University of Barcelona, Barcelona Science Park, Baldiri Reixac 10-12, Barcelona 08028, Spain

**Martin Lee and Alan Serrels**
Edinburgh Cancer Research Centre, University of Edinburgh, Crewe South Road, Edinburgh EH4 2XR, UK

**Rodolfo Lavilla**
CIBER-BBN, Networking Centre for Bioengineering, Biomaterials and Nanomedicine, Baldiri Reixac 10-12, Barcelona 08028, Spain
Laboratory of Organic Chemistry, Faculty of Pharmacy, University of Barcelona, Barcelona Science Park, Baldiri Reixac 10-12, Barcelona 08028, Spain

**Mathieu Colomb-Delsuc, Elio Mattia, Jan W. Sadownik and Sijbren Otto**
Centre for Systems Chemistry, Stratingh Institute, University of Groningen, Nijenborgh 4, 9747 AG, Groningen, The Netherlands

**Diederik W.R. Balkenende, Christophe A. Monnier, Gina L. Fiore and Christoph Weder**
Adolphe Merkle Institute, University of Fribourg, Chemin des Verdiers 4, CH-1700 Fribourg, Switzerland

**Pratap S. Patil, Ting-Jen Rachel Cheng, Medel Manuel L. Zulueta, Shih-Ting Yang, Larry S. Lico and Shang-Cheng Hung**
Genomics Research Center, Academia Sinica, No. 128, Section 2, Academia Road, Taipei 115, Taiwan

**Daniel Görl, Xin Zhang, Vladimir Stepanenko and Frank Würthner**
Institut für Organische Chemie and Center for Nanosystems Chemistry, Universität Würzburg, Am Hubland, Würzburg 97074, Germany

**Masakazu Nambo**
Institute of Transformative Bio-Molecules (WPI-ITbM), Nagoya University, Chikusa, Nagoya 464-8602, Japan

**Cathleen M. Crudden**
Queen's University, Department of Chemistry, Chernoff Hall, Kingston, Ontario, Canada K7L 3N6
Institute of Transformative Bio-Molecules (WPI-ITbM), Nagoya University, Chikusa, Nagoya 464-8602, Japan.

**Christopher Ziebenhaus, Jason P.G. Rygus, Kazem Ghozati, Phillip J. Unsworth, Samantha Voth, Marieke Hutchinson, Veronique S. Laberge and Daisuke Imao**
Queen's University, Department of Chemistry, Chernoff Hall, Kingston, Ontario, Canada K7L 3N6

**Yuuki Maekawa**
Queen's University, Department of Chemistry, Chernoff Hall, Kingston, Ontario, Canada K7L 3N6
Department of Chemistry and Biomolecular Science, Faculty of Engineering, Gifu University, 1-1 Yanagido, Gifu 501-1193, Japan

**Zhi-Wei Jiao, Qing Zhang, Wen-Xing Liu, Shu-Yu Zhang, Shao-Hua Wang, Fu-Min Zhang and Sen Jiang**
State Key Laboratory of Applied Organic Chemistry & College of Chemistry and Chemical Engineering, Lanzhou University, Lanzhou 730000, P. R. China

**Yong-Qiang Tu**
State Key Laboratory of Applied Organic Chemistry & College of Chemistry and Chemical Engineering, Lanzhou University, Lanzhou 730000, P. R. China
Collaborative Innovation Center of Chemical Science and Engineering, Tianjin 300071, P. R. China

**James Allen Frank, Martin Sumser and Dirk Trauner**
Department of Chemistry and Center for Integrated Protein Science, Ludwig Maximilians University Munich, Butenandtstrasse 5-13, Munich 81377, Germany

**Mirko Moroni and Gary R. Lewin**
Molecular Physiology of Somatic Sensation, Max Delbrück Center for Molecular Medicine, Berlin 13125, Germany

**Rabih Moshourab**
Molecular Physiology of Somatic Sensation, Max Delbrück Center for Molecular Medicine, Berlin 13125, Germany
Department of Anesthesiology, Campus Charité Mitte und Virchow Klinikum, Charité Universitätsmedizin Berlin, Augustburgerplatz 1, Berlin 13353, Germany

**Sang Yeon Park, Ji-Woong Lee and Choong Eui Song**
Department of Chemistry, Sungkyunkwan University, 2066, Seobu-ro, Jangan-gu, Suwon-si, Gyeonggi-do 440-746, Korea

**Lorena Mendive-Tapia**
Institute for Research in Biomedicine, Barcelona Science Park, Baldiri Reixac 10-12, 08028 Barcelona, Spain
Department of Organic Chemistry, University of Barcelona, Martí i Franqués 1-11, 08028 Barcelona, Spain
CIBER-BBN, Networking Centre on Bioengineering, Biomaterials and Nanomedicine

**Sara Preciado**
Department of Organic Chemistry, University of Barcelona, Martí i Franqués 1-11, 08028 Barcelona, Spain

**Jesús García**
Institute for Research in Biomedicine, Barcelona Science Park, Baldiri Reixac 10-12, 08028 Barcelona, Spain

**Rosario Ramón and Nicola Kielland**
Barcelona Science Park, Baldiri Reixac 10-12, 08028 Barcelona, Spain

**Fernando Albericio**
Institute for Research in Biomedicine, Barcelona Science Park, Baldiri Reixac 10-12, 08028 Barcelona, Spain
Department of Organic Chemistry, University of Barcelona, Martí i Franqués 1-11, 08028 Barcelona, Spain
CIBER-BBN, Networking Centre on Bioengineering, Biomaterials and Nanomedicine
School of Chemistry, Yachay Tech, Yachay City of Knowledge, 100119 Urcuqui, Ecuador

**Rodolfo Lavilla**
Barcelona Science Park, Baldiri Reixac 10-12, 08028 Barcelona, Spain
Laboratory of Organic Chemistry, Faculty of Pharmacy, University of Barcelona, Avda. Joan XXII s.n., 08028 Barcelona, Spain

Qineng Xia, Zongjia Chen, Yi Shao, Xueqing Gong, Haifeng Wang, Xiaohui Liu and Yanqin Wang
Key Laboratory for Advanced Materials, Research Institute of Industrial Catalysis, East China University of Science and Technology, Shanghai 200237, China

Stewart F. Parker
ISIS Facility, STFC Rutherford Appleton Laboratory, Chilton, Oxfordshire OX11 0QX, UK

Xue Han
School of Chemistry, University of Nottingham, Nottingham, NG7 2RD, UK
School of Chemistry, University of Manchester, Manchester M13 9PL, UK

Sihai Yang
School of Chemistry, University of Manchester, Manchester M13 9PL, UK

Erik B. Pinxterhuis, Massimo Giannerini, Valentín Hornillos and Ben L. Feringa
Faculty of Mathematics and Natural Sciences, Stratingh Institute for Chemistry, University of Groningen, Nijenborgh 4, 9747AG Groningen, The Netherlands

Xinjiang Cui, Annette-Enrica Surkus, Kathrin Junge, Christoph Topf, Jörg Radnik, Carsten Kreyenschulte and Matthias Beller
Leibniz-Institute for Catalysis, University of Rostock, Albert Einstein Street, 29a, Rostock 18059, Germany

Jun Yuan, Xin He and Liang Zhao
Key Laboratory of Bioorganic Phosphorus Chemistry and Chemical Biology (Ministry of Education), Department of Chemistry, Tsinghua University, Beijing 100084, China

Tingting Sun, Ke An and Jun Zhu
State Key Laboratory of Physical Chemistry of Solid Surfaces and Collaborative Innovation Center of Chemistry for Energy Materials
(iChEM), Fujian Provincial Key Laboratory of Theoretical and Computational Chemistry and Department of Chemistry, College of Chemistry and Chemical Engineering, Xiamen University, Xiamen 361005, China

Yongjun Li, Zhiyu Jia, Huibiao Liu and Yuliang Li
CAS Key Laboratory of Organic Solids, Beijing National Laboratory for Molecular Science (BNLMS), Institute of Chemistry, Chinese Academy of Sciences, Beijing 100190, China

Shengqiang Xiao
State Key Laboratory of Advanced Technology for Materials Synthesis and Processing,Wuhan University of Technology, Wuhan 430070, China

Xiao-Qiang Hu, Jia-Rong Chen, Quan-Qing Zhao, Qiang Wei and Wen-Jing Xiao
CCNU-uOttawa Joint Research Centre, Key Laboratory of Pesticide and Chemical Biology, Ministry of Education, College of Chemistry, Central China Normal University, 152 Luoyu Road, Wuhan 430079, China

Xiaotian Qi and Yu Lan
School of Chemistry and Chemical Engineering, Chongqing University, Chongqing 400030, China

Felix F. Loeffler, Tobias C. Foertsch, Roman Popov, Martyna Sedlmayr Valentina Bykovskaya, Lothar Hahn, Frank Breitling, Alexander Nesterov-Mueller, Clemens von Bojničić-Kninski, Laura K. Weber and Andrea Fischer
Karlsruhe Institute of Technology, Institute of Microstructure Technology (IMT), Hermann-von-Helmholtz-Platz 1, 76344 Eggenstein-Leopoldshafen, Germany

Daniela S. Mattes, Barbara Ridder and Juliane Greifenstein
Karlsruhe Institute of Technology, Institute of Microstructure Technology (IMT), Hermann-von-Helmholtz-Platz 1, 76344 Eggenstein-Leopoldshafen, Germany
Karlsruhe Institute of Technology, Institute of Organic Chemistry (IOC), Fritz-Haber-Weg 6, 76131 Karlsruhe, Germany

Martin Schlageter and Annie K. Powell
Karlsruhe Institute of Technology, Institute for Inorganic Chemistry, Engesserstrasse 15, 76131 Karlsruhe, Germany
Karlsruhe Institute of Technology, Institute of Nanotechnology (INT), Hermann-von-Helmholtz-Platz 1, 76344 Eggenstein-Leopoldshafen, Germany

Florian-Xuan Dang and Teodor Silviu Balaban
Aix-Marseille University, CNRS UMR 7313, Institut des Sciences Moléculaires de Marseille, Chirosciences, 13397 Marseille cedex 20, France

Ivan Buliev
Technical University of Varna, 1 Studentska, 9010 Varna, Bulgaria

**F. Ralf Bischoff**
German Cancer Research Center, INF 580, 69120 Heidelberg, Germany.

**Michael A.R. Meier**
Karlsruhe Institute of Technology, Institute of Organic Chemistry (IOC), Fritz-Haber-Weg 6, 76131 Karlsruhe, Germany

**Stefan Bräse**
Karlsruhe Institute of Technology, Institute of Organic Chemistry (IOC), Fritz-Haber-Weg 6, 76131 Karlsruhe, Germany
Karlsruhe Institute of Technology, Institute of Toxicology and Genetics (ITG), Hermann-von-Helmholtz-Platz 1, 76344 Eggenstein-Leopoldshafen, Germany

**Jason Potticary, Lui R. Terry, Simon R. Hall**
Complex Functional Materials Group, School of Chemistry, University of Bristol, Bristol BS8 1TS, UK

**Christopher Bell**
School of Physics, HH Wills Physics Laboratory, Tyndall Avenue, Bristol BS8 1TL, UK

**Alexandros N. Papanikolopoulos, Peter C.M. Christianen and Hans Engelkamp**
High Field Magnet Laboratory (HFML-EMFL), Radboud University, Toernooiveld 7, 6525 ED Nijmegen, The Netherlands

**Andrew M. Collins**
Bristol Centre for Functional Nanomaterials, HH Wills Physics Laboratory, Tyndall Avenue, Bristol BS8 1TL, UK

**Claudio Fontanesi**
Department of Physics, University of Bath, Claverton Down, Bath BA2 7AY, UK
Dipartimento di Ingegneria Enzo Ferrari, Universita' di Modena e Reggio Emilia, Via Vivarelli 10, 41125 Modena, Italy

**Gabriele Kociok-Köhn**
Department of Chemistry, University of Bath, Claverton Down, Bath BA2 7AY, UK

**Simon Crampin and Enrico Da Como**
Department of Physics, University of Bath, Claverton Down, Bath BA2 7AY, UK

**Jonathan Latham, Jean-Marc Henry, Humera H. Sharif, Binuraj R.K. Menon, Sarah A. Shepherd and Jason Micklefield**
School of Chemistry, The University of Manchester, Oxford Road, Manchester M13 9PL, UK
Manchester Institute of Biotechnology, The University of Manchester, 131 Princess Street, Manchester M1 7DN, UK

**Michael F. Greaney**
School of Chemistry, The University of Manchester, Oxford Road, Manchester M13 9PL, UK

**Brian J. Cafferty, David M. Fialho, Jaheda Khanam and Nicholas V. Hud**
School of Chemistry and Biochemistry, Georgia Institute of Technology, 901 Atlantic Drive, Atlanta, Georgia 30332, USA
NSF-NASA Center for Chemical Evolution, Atlanta, Georgia 30332, USA

**Ramanarayanan Krishnamurthy**
NSF-NASA Center for Chemical Evolution, Atlanta, Georgia 30332, USA Department of Chemistry, The Scripps Research Institute, La Jolla, California 92037, USA

# Index

www.ingramcontent.com/pod-product-compliance
Lightning Source LLC
Chambersburg PA
CBHW080527200326
41458CB00012B/4360

* 9 7 8 1 6 8 2 8 5 3 7 4 0 *